T0336747

Claims Reserving in General Insurance

A comprehensive and accessible reference source that documents the theoretical and practical aspects of all the key deterministic and stochastic reserving methods that have been developed for use in general insurance. Worked examples and mathematical details are included, along with many of the broader topics associated with reserving in practice. The key features of reserving in a range of different contexts in the UK and elsewhere are also covered.

The book contains material that will appeal to anyone with an interest in claims reserving. It can be used as a learning resource for actuarial students who are studying the relevant parts of their professional bodies' examinations, as well as by others who are new to the subject. More experienced insurance and other professionals can use the book to refresh or expand their knowledge in any of the wide range of reserving topics covered in the book.

DAVID HINDLEY is an actuary who has spent over 30 years specialising in general insurance, during which time he has gained extensive practical experience of claims reserving in the UK and elsewhere. He is currently a Non-Executive Director with a number of general insurance companies. He has been a member of the Council of the Institute of Actuaries and Chair of its General Insurance Board. He is currently a member of the Actuarial Council, part of the UK Financial Reporting Council. He received an Outstanding Contribution award at the UK Actuarial Profession's 2012 General Insurance conference, in recognition of his contribution to General Insurance research.

INTERNATIONAL SERIES ON ACTUARIAL SCIENCE

The *International Series on Actuarial Science*, published by Cambridge University Press in conjunction with the Institute and Faculty of Actuaries, contains textbooks for students taking courses in or related to actuarial science, as well as more advanced works designed for continuing professional development or for describing and synthesizing research. The series is a vehicle for publishing books that reflect changes and developments in the curriculum, that encourage the introduction of courses on actuarial science in universities, and that show how actuarial science can be used in all areas where there is long-term financial risk.

A complete list of books in the series can be found at www.cambridge.org/statistics. Recent titles include the following:

Claims Reserving in General Insurance
David Hindley

Financial Enterprise Risk Management (2nd Edition)
Paul Sweeting

Insurance Risk and Ruin (2nd Edition)
David C.M. Dickson

Computation and Modelling in Insurance and Finance
Erik Bølviken

Predictive Modeling Applications in Actuarial Science, Volume 1: Predictive Modeling Techniques
Edited by Edward W. Frees, Richard A. Derrig & Glenn Meyers

Actuarial Mathematics for Life Contingent Risks (2nd Edition)
David C.M. Dickson, Mary R. Hardy & Howard R. Waters

Solutions Manual for Actuarial Mathematics for Life Contingent Risks (2nd Edition)
David C.M. Dickson, Mary R. Hardy & Howard R. Waters

Risk Modelling in General Insurance
Roger J. Gray & Susan M. Pitts

CLAIMS RESERVING IN GENERAL INSURANCE

DAVID HINDLEY
FIA

CAMBRIDGE
UNIVERSITY PRESS

CAMBRIDGE
UNIVERSITY PRESS

University Printing House, Cambridge CB2 8BS, United Kingdom

One Liberty Plaza, 20th Floor, New York, NY 10006, USA

477 Williamstown Road, Port Melbourne, VIC 3207, Australia

314-321, 3rd Floor, Plot 3, Splendor Forum, Jasola District Centre, New Delhi - 110025, India

79 Anson Road, #06-04/06, Singapore 079906

Cambridge University Press is part of the University of Cambridge.

It furthers the University's mission by disseminating knowledge in the pursuit of education, learning and research at the highest international levels of excellence.

www.cambridge.org
Information on this title: www.cambridge.org/9781107076938

First published 2018

A catalogue record for this publication is available from the British Library

ISBN 978-1-107-07693-8 Hardback

Contents

Preface

Legal disclaimer

This book is intended to provide information on the subject of claims reserving in general insurance, using descriptive text, worked examples, software code, data and mathematical formulations. The content is intended solely as a learning aid, and should not be used in any particular application without independent testing and verification by the person making the application. In addition, the author and publisher are not offering this book as actuarial, statistical or other professional services advice, and no reliance should be placed on it as such.

For these reasons, the author and publisher make no warranties, express or implied, that the descriptions, worked examples, software code, data, mathematical formulations, or other information in this volume are free of error, that they are consistent with industry standards, or that they will meet the requirements for any particular application. THE AUTHOR AND PUBLISHER EXPRESSLY DISCLAIM ANY IMPLIED WARRANTIES OF SATISFACTORY QUALITY MERCHANTABILITY AND OF FITNESS FOR ANY PARTICULAR PURPOSE, even if the author or publisher has been advised of a particular purpose, and even if a particular purpose is indicated in the book. The author and publisher also disclaim all liability for direct, indirect, incidental, or consequential damages that result from any use of the descriptions, worked examples, software code, data, mathematical formulations, or other information in this book.

Permission to use the third-party content in this volume was obtained for use in this volume only. Neither the publisher nor the author make any representation regarding whether re-use of third-party descriptive text, worked examples, software code, data, mathematical formulation, or other information in this volume might infringe others' intellectual property rights, including US and foreign patents. It is the reader's sole responsibility to ensure that he or she is not infringing any third-party intellectual property rights, even for use which is considered to be experimental in nature. By using any of the third-party descriptive text, worked examples, software code, data, mathematical formulations or other information in this volume, the reader has agreed to assume all liability for any damages arising from or relating to such use, regardless of whether such liability is based on intellectual property or any other cause of action, and regardless of whether the damages are direct, indirect, incidental, consequential, or any other type of damage. The authors and publisher disclaim any such liability for re-use of third-party content.

Nothing in the foregoing paragraphs is intended to limit or restrict liability that cannot be lawfully limited or restricted.

Author's note

Claims reserving in general insurance often involves a considerable degree of judgement, with many different approaches being possible in any particular situation. The author has endeavoured to include examples of a range of reserving methods and approaches, and to highlight some of the associated matters that apply when they are used in practice.

Various opinions, statements and views on certain aspects of reserving are given in many parts of the book. Some of these are based on those given in other sources, in which case the source is identified. In other cases, unless stated otherwise, the views, etc. are from the author's personal perspective only, and do not represent formal or recommended best practice, standards or guidance. In addition, they do not necessarily represent the views or practices of any companies or organisations with whom the author is, or has been, associated.

There will inevitably be situations in practice where different methods and approaches can be used to those described here, due, for example, to the particular circumstances that apply. Any views or opinions expressed in this book are not therefore intended to be absolute, and practitioners and others involved in any particular reserving exercise will need to decide upon the appropriate approach to use in each case.

The author would welcome feedback or comments on any aspect of this book.

Acknowledgements

I am indebted to the numerous colleagues and clients with whom I have worked during my career, and with whom I have enjoyed discussing how to approach the many and varied aspects of the general insurance reserving projects that we have worked on together during this time.

In drafting the individual sections on reserving in specific contexts in Chapter 7, I have benefited from the assistance of a number of individuals who have shared with me their specialist knowledge and expertise in the relevant areas. In particular, in relation to the two non-UK contexts included in that chapter, I would like to thank Bill Van Dyke and Dave Foley, who helped me with the US Workers' Compensation section, and Jeremy Wall and the team at his firm who drafted the initial version of the Asia Motor section. Several other people with experience and knowledge of the relevant contexts also reviewed drafts of sections in that chapter. These included Derek Adamson, Jahan Anzsar, Simon Black, Robert Brooks, Gauri Dandeker, Darren Michaels, Shreyas Shah, Klaas Sijbrandij, Duncan Spooner and Seema Thaper. I would like to thank them all for their very helpful comments on drafts of the relevant sections.

Other chapters or individual sections were reviewed by a number of different people. I would particularly like to thank Peter England for providing me with his detailed and helpful comments on a draft of the stochastic reserving methods chapter. I would also like to thank the following people for their valuable contributions in reviewing either that chapter or other chapters or sections: Philip Archer-Lock, Neil Bruce, Andrzej Czernuszewicz, Gauri Dandeker, Susan Dreksler, Wan Hsien Heah, Philip Hobbs, Alex Marcuson, Markus Gesmann, Darren Michaels, Derek Newton, Gae Robinson, Richard Rodriguez, Daniel Sevitt, Chris Short, Klaas Sijbrandij, Andrew Smith, Seema Thaper and Professor Richard Verrall.

I would like to thank the librarians at the Institute and Faculty of Actuaries, who have been very helpful in sourcing various journal articles and books that I have used when researching material for this book. I would also like to thank David Tranah and his colleagues at Cambridge University Press for their patience and assistance with various aspects of the production of this book.

I am very grateful to all the people mentioned above who have kindly given their time to contribute to this book. However, I am fully responsible for the content and for any omissions or inaccuracies.

Finally, I would like to thank Frank Guaschi, who started it all, and my wife, Sally, who put up with it all.

1

Introduction

The process of claims reserving is at the core of the financial management of general insurance organisations. It determines what is held on the balance sheet for claims that are not yet settled, affects the premiums that are charged and impacts on the capital that is held to support the solvency of the organisation. Thus, the selection of appropriate reserving methodologies and assumptions, and the application to practical situations, often with imperfect data, are of critical importance to the operations of an insurance organisation. This book has been designed for both those who are relatively new to the subject of reserving in general insurance, and for experienced practitioners who want a single reference source on the subject. It is also anticipated that it will be of interest to those wanting to develop new reserving techniques. In writing this book, I had four key objectives in mind:

1. To produce a comprehensive description of both the theory and practice of claims reserving in general insurance. Although there are already a large number of papers and books on this subject, I have always felt that creating a single reference work covering a wide range of reserving topics would be of benefit to those new to the subject and to experienced practitioners, partly to help avoid the reinvention of previously developed approaches that I have seen occur several times during my career (by others and myself). Although most reserving is now carried out in bespoke or proprietary reserving software, which will usually have associated user guides and manuals, this book is intended to supplement those sources by covering a variety of reserving topics and techniques, the latter of which may not all be implemented in the relevant software products.

2. To provide those who wish to advance the more theoretical aspects of reserving with a clear and thorough description of the many practical issues that frequently occur when applying reserving methods in the real world. My hope is that, by doing this, more effective new approaches can be developed in future which make appropriate allowance for these practical issues, which in turn should encourage their wider application. In so doing, this should improve the reliability of reserve estimates and assist with understanding the uncertainty surrounding those estimates. More accurate reserve estimates should lead to fairer insurance/reinsurance premiums and to more reliable financial statements, both of which must be in the public interest.

1

3. To provide reserving practitioners with an accessible reference source for some of the more theoretical aspects of claims reserving (e.g. stochastic reserving) so that more of them might consider investigating the greater use of such techniques in practice. Such practitioners quite rightly spend most of their time focusing on the many and varied difficulties that occur when applying reserving methodologies in practice, and so do not always have time to search for (let alone fully digest) a reference source of the more theoretical aspects of reserving.

4. To provide a reference source for actuarial students studying the general insurance reserving components of a number of the global professional actuarial bodies' examination syllabi. Although some parts of the material go beyond what is within the relevant syllabi, this will provide students with a better overall context for their studies, and will also enable those students who wish to develop their knowledge further to do so, without having to cross-reference significant volumes of other material.

Reserving in general insurance[1] is still a developing subject, particularly in respect of both the practical and theoretical detail of the methods. In many situations, there is no right or wrong approach, and practitioners often have their preferences as to how to apply methods in practice, with several bespoke modifications to the standard approaches described here being used to suit the particular circumstances at the time. Whilst some of these bespoke modifications are discussed here, there will be many more that exist in the real world. To reflect this, in a number of places in the book there are suggestions, rather than definitive statements, as to what approach might be used in specific circumstances. One feature of a reserving exercise that it is possible to be definitive about, however, is that it is necessary to apply judgement at each stage of the process.

The book is designed to be of relevance to all those with an interest in claims reserving in respect of general insurance risk carrying entities, including actuaries, statisticians, reserving technicians, senior management and board members, risk management specialists, underwriters, finance staff and claims managers. It covers both general material as well as some of the more mathematical elements of reserving. However, readers who are less interested in the mathematical aspects can simply ignore those sections, as the majority of the other sections do not rely on an understanding of these parts of the book.

After covering introductory background material in the remainder of this chapter, the next chapter covers the data requirements, followed by two key chapters that provide a detailed description of a wide range of both straightforward and more advanced reserving methods. The next two chapters cover applying these methods in practice and a selection of additional reserving topics, not covered in the other chapters. The final chapter describes the specific features of reserving in a range of different contexts, including selected classes of business or claim types in the UK, US and Asia.

---

[1] The word "general" in this context is synonymous with "Non-Life" and "Property & Casualty", and is intended to mean all classes of insurance other than life insurance. Insurance is deemed to include reinsurance business as well as insurance business. Further notes on terminology are given in Section 1.5.

Readers who are new to the subject who wish to gain an initial overview would be advised to read the first two chapters, followed by the first four sections of Chapter 3 and then Chapter 5, which covers applying reserving methods in practice. The remainder of Chapter 3 and the other three chapters can then be studied as required. A review of the headings of each chapter in the contents page should provide a self-explanatory overview of the structure of the book. This should enable readers who are more familiar with the subject to select chapters and sections based on their own specific requirements.

I have tried to make the book as self-contained as possible, so that the reader does not have to refer to other sources in order to gain a reasonable understanding of the material being covered, although there are some areas where only high-level descriptions are given. At numerous points in the text, I have included references to original sources on which the content is based and in some cases to suggested further reading, which can be used by readers wishing to explore the relevant topics in more detail. However, since the volume of published material on claims reserving is extensive, there are many sources that could be relevant to a particular reserving exercise or research study which may not be cited here. Furthermore, the theory and practice of claims reserving continues to evolve, with new methods and approaches being developed from time to time. The practitioner can use various ways to identify sources relevant to their particular requirements beyond those cited here, and to keep up to date with developments. One such source is the bibliography at Schmidt (2015).

1.1 Basic Concepts

Although this book assumes that the reader has a good basic knowledge of general insurance, a brief introduction to the concepts underlying reserving is given in this section, to provide some context for the remainder of the book. Readers who need a more comprehensive introduction to general insurance should be able to access suitable training materials through their local insurance or actuarial institutional bodies (e.g. in the UK, the Chartered Insurance Institute and the Institute and Faculty of Actuaries).

The fundamental process that occurs in an insurance entity is that it charges a premium in return for promising to pay claims if they are covered by the relevant policy. The premium is either collected at the beginning of the policy period, or via instalments over the policy period. However, claims can be paid either during the policy period, or in some cases up to several years after the end of the policy period. Although claims are usually covered only if they occur within the period of the policy,[2] there is typically a delay between the date of the claim and the date by which the insurer knows the full amount that it will pay on the claim. The delay can occur for three main reasons. First, there can be a delay between the occurrence date and the date on which the claim is advised to the insurer. Second, even if the claim is advised quickly after the occurrence date, it can take time to agree the value of the claim. Finally, some claims are payable as recurring payments, for which the duration may

---

[2] Other forms of coverage are also possible, such as so-called "claims-made" policies, which cover claims made or notified during the policy period.

be unknown, according to the terms of the relevant insurance policy. These reasons mean that, whenever an insurer is seeking to produce financial records or statements, it needs to determine the total claims that it expects to pay out, in respect of the relevant insurance policies that it has issued. This determination will inevitably involve a degree of estimation, since, by definition, it will not know the total value of claims at a particular date in respect of the policies that it has issued. The estimation process will give rise to an amount that needs to be "reserved" for the claims that are not yet paid – i.e. the future claims outgo – hence the term "reserving". When this amount is added to the amount that the insurer has already paid, it produces what is generally referred to as the "ultimate" claims – i.e. the ultimate amount that the insurer is estimated to pay in respect of a particular group of policies or claims.

For many years, the reserve (or "provision" or "technical provision" as it is sometimes called) was determined using relatively straightforward approaches, such as simply adding up all the amounts that had been estimated for each individual outstanding (i.e. not settled) claim on a case by case basis. In the UK, this changed in the early 1970s, and more numerical or statistical type approaches started to be used, based on various forms of aggregated claims data.[3] Nowadays in the UK and elsewhere, most reserving is done using these approaches and much of this book is devoted to a description of the different techniques that have been developed, and their application in practice. Consideration of the techniques can give the impression that reserving is a science, but in reality it involves an element of art as well, since in practice there is a significant degree of judgement involved in the application of the techniques.

The fact that the insurer will not know the total value of claims at a particular date in respect of the policies that it has issued can be summarised by simply stating that there is uncertainty in the future claims outgo. This uncertainty translates into uncertainty over the accuracy of the reserves, or in other words, how will the actual future claims outgo compare with the reserves established at a particular date? This characteristic of the process of reserving is also sometimes expressed in the statement that the only thing that is certain about an established reserve is that it will be wrong (i.e. that it will not exactly match the actual future outgo). A key aspect of reserving is to seek to understand how wrong it could be. Although that will not be known accurately until all relevant claims have been settled, an important part of many reserving exercises will involve estimating the uncertainty surrounding a single reserve estimate, and then communicating this to the relevant stakeholders.

There are several different types of reserve referred to by insurers. The terminology and associated definition can vary depending on the context and location of the insurer, as discussed further in Section 1.5. A brief description of some common types is given below. In some contexts the word "provision" is used instead of "reserve" for a number of these.

---

[3] The reasons for this are believed to include the recognition that regulators needed to be able to derive independent analytical techniques to better assess the solvency of insurers and also an acknowledgement that case reserving may not adequately capture the amounts required for claims that have occurred but have not yet been reported to the insurer. These and other factors are discussed in Ackman et al. (1982).

The descriptions given here are designed to convey the general meaning of the terms used for each type of reserve, when they are used in a reserving context. This meaning can be different to that which applies, for example, in an accounting context, as explained further in Section 1.5. Hence, the descriptions given here do not represent formal legal or accounting definitions that may apply in any particular jurisdiction.

Case Reserves: The amount in respect of claims that have been advised, but which have not yet settled. The Case Reserves are usually established either by assessment on a per-claim basis by the insurer's claims staff, or for some claims during the very early period after the claim has been advised, by an automated procedure.

Outstanding Claims Reserve: Either used as another term for Case Reserves, or, in the case of the accounts of insurers, can often relate to the sum of the Case Reserves, IBNR Reserve and IBNER Reserve. Outstanding Claims Reserve may be abbreviated to "OCR" or "OS".

Unearned Premium Reserve: The proportion of the written premiums, as at a particular date, that relate to future policy periods (i.e after the end of the latest reporting period covered by the relevant financial statements – usually the valuation date for the reserving exercise). It is often shortened to "UPR". It is typically viewed as an accounting item, rather than a reserve that is calculated by applying the types of reserving method described in this book.

Unexpired Risk Reserve ("URR"): The amount in excess of the UPR (after deduction of any Deferred Acquisition Costs, or DAC[4]) that needs to be held to cover claims and expenses that relate to future policy periods in respect of policies that the insurer has already written. It is sometimes also referred to as the "Additional Amount for Unexpired Risk", the "Additional Unexpired Risk Reserve" or the "Premium Deficiency Reserve". It is only needed if the UPR (less DAC) is deemed insufficient to cover claims and expenses in respect of the unexpired policies. As for UPR, it is typically viewed as an accounting item.

IBNR Reserve: Stands for Incurred But Not Reported Reserve. Usually relates to the amount in respect of claims that have occurred, but which, at a particular date, have not yet been advised or reported to the insurer. Can sometimes also be deemed to include the IBNER Reserve, as explained below. Depending on the context it can also sometimes include an estimate of the claims payable on the unexpired period of all relevant policies that are included in the data used for a particular reserving exercise. A related term that might be used is "Pure IBNR", which generally refers to the element of IBNR that relates only to unreported claims – i.e. it excludes any IBNER element related to reported claims.

IBNER Reserve: This stands for Incurred But Not Enough Reported Reserve. If the Case Reserves have been assessed as being too high or too low in aggregate (e.g. based on some form of statistical analysis), then an amount is estimated to adjust them so that the

---

[4] In simple terms, DAC are an asset which represent the proportion of the prepaid acquisition expenses (e.g. broker commissions), which are unearned at the valuation date.

aggregate amount is more appropriate. The IBNER Reserve can sometimes be included as part of the IBNR Reserve.

Equalisation Reserve: An amount established, usually by reference to a formula or procedure specified by a regulator, that is designed to smooth or "equalise" reserves over a certain period. Typically only used for certain types of low frequency/high severity events and only in certain territories.

ALAE Reserve: Stands for Allocated Loss Adjustment Expenses Reserve and relates to future expenses (as opposed to Indemnity amounts) associated with specific claims. It is often included as part of the Case Reserves, IBNR and IBNER, rather than being identified separately.

ULAE Reserve: Stands for Unallocated Loss Adjustment Reserve and relates to future expenses related to claims in general, as opposed to specific claims.

There can be other types of reserve, depending on the context, such as Reopened Claims Reserve and Reserve for Claims in Transit (i.e. Incurred and Reported, but not yet recorded on the insurer's systems).

The term **"Claims Reserve"** is also used in practice, as is the equivalent term **"Loss Reserve"**. This can mean any one or combination of the above types of reserve. One of the more common usages, however, would be that it represents the sum of Case Reserves, IBNR Reserve and IBNER Reserve; this is the definition used in this book, unless stated otherwise. In some instances, future (unpaid) premiums are deducted from the Claims Reserve, in which case the term "Claims Reserve Net of Future Premiums" or "Loss Reserve Net of Future Premiums" might be used.

Other terms may be used in different accounting, regulatory and solvency contexts in general insurance around the world. For example, the terms "Claims Provision" or "Claim Liabilities" might be used to mean one or more of the above types of reserve, but related only to claims that have already occurred at a particular valuation date. Similarly, "Premium Provision" or "Premium Liabilities" might be used to refer to the corresponding reserve in respect of the unexpired contractual obligations associated with relevant policies at that date (i.e. future claims on such policies at the valuation date).

The size of the IBNR and IBNER Reserve relative to the Case Reserves will vary depending on the circumstances. It will generally be higher for types of business where there is a long delay between the underlying incident occurring and the claim being reported to the insurer and/or settled by the insurer. Industry data at an aggregate level can provide some broad indication of the relationship. For example, A.M. Best (2016a) shows that for the US Insurance industry as a whole, IBNR was 111 % of Case Reserves as at 31 December 2015.[5] In other words, on this measure, IBNR is approximately equal to the Case Reserves. For specific insurers the relationship may be materially different from this, but in most cases the process of determining the Claims Reserve will rarely involve just using the total reserve for reported but not settled claims (i.e. the Case Reserves) and so a

---

[5] Using the ratio of IBNR for Loss and ALAE combined to the Case for Loss and ALAE combined for US Property/Casualty companies in aggregate, taken from Page 12 of the Quantitative Analysis Report in A.M. Best (2016a).

process will be needed to estimate the IBNR/IBNER component (which is usually, but not always, positive); that process is the focus of much of this book.

Most types of reserve can be either gross (i.e. before) or net (i.e. after) of outwards reinsurance, with the difference being the reinsurer's share of the reserve. The gross reserve will usually appear as a liability in the balance sheet of an insurer, whereas the reinsurer's share will usually appear as an asset – with the difference being the net reserve. There can also be a distinction between reserves that are booked in an insurer's financial accounts and those established on a specified basis. A common basis that is used is referred to as either an "Actuarial Best Estimate" ("ABE"), "Actuarial Central Estimate" ("ACE") or just "Best estimate", which represents the reserves established by the insurer's actuarial/reserving team on a best estimate basis. In this context, the term "best estimate" is intended to represent the mean (i.e. the average or the expected value) of the range of potential outcomes for future claims. Although this is the intended meaning used in this book, in certain markets or contexts, different definitions of best estimate or central estimate might be used. See Chapter 21 of Friedland (2013) for some alternative examples.

The best estimate will usually be above the median (i.e. the 50/50 value) and the mode (i.e. the most likely value) because the distribution of potential outcomes for the future claims tends to be positively skewed (sometimes referred to as "skewed to the right" or "long-tailed") and such distributions often (but not always) have this relationship between the mean, median and mode. In practice, depending on the reserving methods used, it may not be possible to demonstrate that the selected reserve represents the best estimate as defined in this way. However, the definition is still helpful, as it effectively expresses the intent of the practitioner when selecting the reserve estimate. The booked reserve can be referred to as the Booked Reserve, Carried Reserve or Management Best Estimate (or "MBE" Reserve). If the MBE exceeds the ABE then the difference might be referred to as a management loading or margin. The margin can be regarded as an additional amount to introduce a degree of prudence into the booked reserves, to allow for example for the inherent uncertainty involved in the reserve estimation process and/or to represent an explicit buffer against adverse developments in respect of identified and unidentified sources.

The context in which reserves are being established can influence certain aspects of the reserving process. A common context will be the determination of financial results of the insurer, for management, statutory or regulatory reporting purposes. Other important contexts include transactional (e.g. Sale or Purchase in a Merger/Acquisition), Pricing, Business Planning, Tax reporting, reinsurance purchasing and input to capital modelling. The context may also define or influence the "basis" that is used in the reserving exercise. For example, in some countries the regulations require reserves to be established at a certain level of confidence (e.g. 75th percentile in Australia) and in some regulatory contexts there can be specific requirements related to the margin that must be held in excess of a best estimate (e.g. in a European context, Solvency II requires a risk margin to be determined using a defined approach). Other aspects of the basis that can be influenced by the context include whether or not the reserves are discounted for the time value of money. In some contexts, the purpose of the reserving work is to provide a formal opinion on the reserves, usually for regulatory purposes. If this is the case, then the relevant regulations will usually define the basis of this opinion (e.g. whether the reserves are "reasonable" or whether they are at least as large as a best estimate) and this can influence the scope of the reserving work.

In most parts of this book, the subject of a reserving exercise is usually referred to as an insurer. This is intended to mean any relevant risk-bearing entity, which can include a reinsurer, Lloyd's syndicate, captive insurer, self-insured entity, insurance/reinsurance pool, mutual, Protection and Indemnity Club or government body.

1.2 Elements of a Reserving Exercise

Any reserving exercise involves far more than simply applying one or more technical methods to a dataset. The key elements involved are summarised in Table 1.1. This is set out in the approximate sequence of a typical exercise, although in practice there will inevitably be a degree of iteration between the different stages. In addition, the steps involved will depend on the purpose and scope of the exercise; this summary assumes the aim is to produce a single point estimate for the reserves.

A high-level summary of these elements is given below. The purpose of setting this out here is so that the additional detail in the subsequent chapters can be seen in the overall framework of a reserving exercise. In practice, the procedure will vary significantly between

Table 1.1 *Key elements of a reserving exercise*

Element	Key stages
Background	Understand context
	Understand the business
	Gather background information
	Discuss with key personnel
	Determine basis of exercise
	Planning
Data	Determine availability
	Determine suitability
	Determine reserving categories
	Process data
	Check data
Analysis	Select method(s)
	Initial data review
	Apply methods
	Review results – numerical
	Review results – graphical
	Actual vs Expected
	Select results
Reporting	Summary tables
	Graphs
	Presentation
	Reporting

exercises. Chapter 5 contains more detail on applying reserving methods in practice and Chapter 7 provides a summary of the particular features of reserving for a selected range of reserving class/country combinations.

Background

The first task will be to understand the context for why the reserving exercise is being carried out. For example, there could be a particular regulatory requirement that is being met, or there could be a potential transaction for which a view on the reserves of the target is required. This context will affect issues such as the timing and whether a best estimate or other basis is appropriate for the exercise.

Understanding the business that is subject to the reserving exercise is also critical at an early stage of the process. As far as is practical, this can involve matters such as the underlying process that gives rise to claims and the key drivers that influence the frequency and severity of claims, together with consideration of whether this has changed over time. This will help determine key issues such as whether standard reserving methods will be appropriate or not and the approximate length of the development tail. The extent to which historical data is appropriate for predicting future claims, and hence reserves, can also be considered at this stage. Gathering relevant background information, such as previous reserving reports, underwriter commentaries on the business and market-wide reports can also be helpful in understanding the business. There can rarely be a substitute, however, for meeting with relevant staff at the insurer, such as underwriting, claims and finance. This is the case even with a regular reserving exercise, as there can be underlying changes that need to be allowed for. Interaction between the relevant staff and the reserving practitioner at regular points will further enhance the overall reserving process.

Most reserving exercises of any scale will usually involve an element of planning before the detailed work is carried out, in order to determine timescales, review and sign-off stages etc. Where relevant, the impact of any applicable professional standards on the work will often be considered as part of this planning phase, so that any particular requirements (e.g. such as peer review perhaps) can be built into the work programme.

Chapter 5 contains additional detail on the background stage, including examples of approaches that can be used to gain an understanding of the business, and a discussion of other topics relevant to this stage, such as governance and compliance with professional and regulatory requirements.

Data

Consideration of the data will initially involve establishing the data that is available for the relevant insurer within the required timescales, as well as deciding upon the most appropriate type of data to collect, taking into account the nature of the business and the type of reserving method(s) that will be used. In practice, a compromise may be necessary, taking into account these sometimes opposing factors. This determination will include defining the categories by which the reserving exercise will be carried out. In practice, these categories will depend on the specific reserving exercise, but in general terms, examples of the categories which can be used to group the data include:

- Class of business: e.g. Property, Liability, Motor, Household etc.
- Size of claim: e.g. in size band groupings or split between Attritional, Large and Catastrophe.
- Type of claim: e.g. Property damage, Third Party Liability, Latent etc.
- Gross, Net or Reinsurance.
- Premiums, Claims and/or Commissions.

The above categories are not mutually exclusive – for example, some classes of business might be subdivided by claim type. The goal in determining these groupings will be to establish homogeneous categories, with sufficient volumes of data, to which appropriate methods can be applied.

Often there will be an obvious default type of data to collect, and hence corresponding reserving method(s) to use, based, for example, on the previous reserving exercise. However, it will usually be advisable to review this periodically to assess whether that default approach is in fact still the most appropriate.

Within any chosen grouping, a decision will need to be made on whether to collect claim amounts (including whether Paid, Outstanding and/or Incurred Claims) and/or claim numbers (including whether settled, reported, including nil claims etc.) and/or premiums and/or commissions.

For most reserving exercises, there will usually be some form of existing estimated ultimate claims (and premiums, if relevant) that would be collected at an early stage of the process. The source of these estimates will vary depending on the context, but will include the booked estimates derived from the previous valuation and estimates made by internal or external actuaries at different points in time.

As well as considering the data that is available from the specific insurer or entity that is the subject of the reserving exercise, consideration of available external or benchmark data at an early stage can also be beneficial. Such data can include industry data derived from regulatory returns, published reports and analyses from industry commentators/consultants/market bodies or data collated by consulting firms. This is discussed further in Section 6.9.1, which includes examples of possible sources for selected types of business.

Once the data that will be used has been determined it is processed and checks carried out to ensure that it is reasonable. Such checks include considering matters such as the explanation for any unusual movements (e.g. particularly large or small increases or decreases in claims over recent time periods), comparison to the data from previous exercises and reconciliation to other sources such as finance systems.

Chapter 2 discusses data in more detail.

Analysis

This is the element that many would regard as the core of a reserving process, and, although it should ideally involve the most amount of time, in practice other elements such as dealing with data issues can unfortunately sometimes take up disproportionate time.

If not already established at the data definition stage, determining the reserving method(s) to use will be one of the first key stages of this part of the process. Method selection will involve considering many factors such as the data available, the underlying claims process, whether a point estimate or assessment of variability is required and if so whether just a Standard Error or full distribution is required. Furthermore, consideration of whether the assumptions that are implicit within each potential reserving method (such as whether all cohorts have the same development pattern) are valid for each chosen reserving category, will also be relevant.

Some form of initial data review can be helpful at this stage, before the methods are applied to the data. A comparison between the actual and expected claims development since the previous reserving exercise may also form part of this initial review. In some cases, it can also include reviewing the claims development data graphically to gain an understanding of the development tail and volatility of the data. Such graphs can be used before or after the reserving methods have been applied to the data, as explained in Section 2.5.5.

The methods are then applied to the data. There is a wide variety of software available that make this part of the process relatively straightforward to carry out in practice, as discussed further in Section 1.3. This can be both an advantage, in that methodologically correct results can be obtained quickly, but also a disadvantage because it can result in less careful consideration of the appropriateness of particular methods, and hence in some cases can produce results that might initially appear reasonable, but on closer inspection may be based on invalid assumptions. Although the use of such software can make the reserving process appear somewhat mechanical, in reality there is a considerable degree of judgement involved in the application of any reserving method, regardless of the software used. The nature of this judgement varies according to the method being used, and will be apparent from the detailed description of the methods given in Chapters 3 and 4.

In most cases, more than one method will be applied to the data, or the same method will be applied to different subsets of the overall data (e.g. to both the paid and incurred claims data). In other cases, using some form of "roll-forward" of estimates established at the previous valuation date can represent one of the methods that is applied to the data.

The results from applying the methods will then be summarised, possibly in both numerical and graphical form. An initial selection will be made of the results, sometimes by blending or weighting the results from the different methods. This selection may be modified following review by other personnel involved in the reserving process, by application of methods to a different breakdown of the business or due to separate consideration of large/event claims or application of additional reserving methods. This stage of the process can involve a number of iterations until a final set of results are selected.

Chapter 5 covers various aspects of the analysis stage of a reserving process in more detail.

Reporting

Once the results have been finalised, there will normally be a need to produce some form of summary, presentation and/or formal report on the analysis. The form and content of this will depend on several factors such as the context for the reserving exercise, any applicable professional standards and regulatory requirements, and the nature of the audience for the reporting etc. If a presentation or formal report is required, it will typically include sections covering the scope and purpose of the work, the data and methods used, summary tables, graphical analyses of results, limitations/caveats attaching to the analysis, a discussion of uncertainty, a comparison with previous results, a commentary on the results and methods /assumptions used for at least each major reserving category and on the key judgements. A full written report would normally also include appendices with, for example, additional detail on the methodology and detailed numerical and graphical presentation of the results etc.

As well as formal reporting, there may also be other items of documentation related to the analysis, such as a more detailed description of the underlying methods, assumptions and models that have been used. Although this material would typically not be provided outside the team doing the reserving work, it can form a useful part of the record of the work that has been done. In some cases, professional standards may influence the items included in this documentation. Depending on the purpose and scope of the reserving exercise, the results that are summarised in this stage may represent an input to a further stage to determine the reserves to be used in the relevant context. For example, in a financial reporting context, the agreed reserving process may involve management making the final selection of reserves that will be booked, taking into account broader considerations.

Chapter 5 provides some additional material on this topic.

1.3 Software

Most reserving carried out in practice is done either using one of a range of proprietary reserving software packages or in bespoke spreadsheets or computer programmes developed for a particular insurer. In most cases where proprietary software is used, it will have been obtained on a licensed basis from specialist vendors.

Even with only a basic knowledge of reserving, it is possible to produce results relatively easily using specialist reserving software. However, whilst these results might appear perfectly reasonable, and may indeed be reasonable, the use of such software is far more likely to produce reasonable results if the user has a full understanding of the underlying methodology used in the software. This can usually be obtained through a combination of in-house and external training, user manuals/documentation and learning how to use the software by applying it in practice. This book describes a wide range of reserving methods, many of which will have been implemented in such software. However, the specific implementation will vary between different software products and may have important differences to the description of the methods in this book. This will mean that the material contained here will need to be supplemented by the user manuals and technical documentation for the relevant software packages if a full understanding of the software is required.

There are a number of specialist reserving software packages sold by actuarial and other firms. In addition, several practitioners have produced implementations of specific reserving methods, which are available online, often to accompany an article or paper in a journal.

Some of these implementations are referenced in the description of the relevant reserving methods in Chapters 3 and 4.

As well as specialist reserving software packages, numerous generic statistical software products are available, some of which are freely available online. In theory at least, the various methods used for general insurance claims reserving could be implemented in such software. One such product where a number of specific reserving packages have been written is the "R" software. Many of the worked examples in this book use this software, which is available by referring to R Core Team (2013). Appendix B contains the R code for most of these examples. This code is included for purely illustrative purposes; it has not been fully tested by the author, and is not designed to be suitable for use in practical reserving exercises. Its use in this book is not intended to endorse the "R" software or its use in claims reserving. Useful references for R in a reserving context include Gesmann et al. (2014), Chapter 16 of Charpentier (2014) and Chapter 10 of Kaas et al. (2009).

1.4 Mathematical Notation

Table 1.2 summarises the notation used in this book for key data items and reserving methods. The notation used only for specific aspects of certain methods is not included here; this is provided in the relevant sections for those methods. Since consistent mathematical notation has been used for all methods, this will inevitably mean that the mathematical description of some reserving methods will not be the same as the original reference source for the method. Readers who do not wish to study the mathematical content in subsequent chapters can ignore this section if they wish.

Table 1.2 *Mathematical notation*

Item	Description
i	Index for rows of triangle. Ranges from 0 to I
j	Index for columns of triangle. Ranges from 0 to J
$I + 1$	Number of rows in triangle
$J + 1$	Number of columns in triangle
	Unless otherwise stated $J = I$
$C_{i,j}$	Cumulative claims for cohort i at development point j
$X_{i,j}$	Incremental claims for cohort i at development point j
$C_{i,\text{ult}}$	Ultimate claims for cohort i
	$C_{i,\text{ult}} = C_{i,J}$ when no tail factor
D_I	Part of data triangle relating to "known" data
	$= \{C_{i,j} : i + j \leq I; 0 \leq j \leq J\}$
$C_{i,I-i}$	Cumulative claims at leading diagonal of triangle
	for cohort i, when $J = I$
R_i	Reserve for cohort i
	$= C_{i,J} - C_{i,I-i}$ when no tail factor
R	Reserve across all cohorts
$F_{i,j}$	Development factor from j to $j + 1$ for cohort i
$\hat{f}_j$	Link ratio estimator from j to $j + 1$
	where $0 \leq j \leq J - 1$ when no tail factor

1.5 A Note on Terminology

Reserving terminology can vary both within and between different regions. Within regions, for example, it can vary depending on whether the terms are being used in a purely reserving-focused context or in an accounting context. This can sometimes lead to confusion over the exact meaning of specific terms. The terms used in this book are based largely on those currently in use in the UK within a reserving, rather than accounting context.

Selected key terms used in this book, along with commonly used synonyms and examples of equivalent terms that may be in use in some other regions are shown in Table 1.3. The meaning of these terms is intended to be that which applies in a reserving context; this may be different to the meaning of the same term in an accounting context. For example, "claims outstanding" in a UK accounting context will usually include IBNR, whereas in a purely reserving context it will usually only relate to case reserves. Similarly, "incurred claims" in a UK accounting context usually refers to the paid claims and movements in outstanding claims (including IBNR) in the relevant period, whereas in a reserving context it usually just means cumulative paid claims plus case reserves. Throughout this book, the meaning of the relevant terms is that which generally applies in a UK reserving rather than in an accounting context, unless stated otherwise.

There will inevitably be other equivalent terms not listed in Table 1.3, as well as many more terms that are specific to particular reserving methods or to other aspects of individual reserving exercises. Different terms are also used for the same or similar types of insurance

Table 1.3 *Terminology*

UK term used here	Equivalent terms
Claim	Loss
Reserves	Technical Provisions or Provisions or Unpaid claim estimates
Reserving	Estimating unpaid claims or Unpaid claim estimate analysis
General Insurance	Non-Life or Property and Casualty
Incremental Development factor	Link ratio, Development, Age-to-Age or Report to Report factor, or Development ratio
Cumulative Development Factor	Age to Ultimate or Ultimate Development Factor
Average Development Factor	Estimator or Link ratio estimator – e.g Column Sum Average Estimator
Cohort	Experience period or Origin period – e.g. Accident Year
Unallocated Loss Adjustment Expenses	Internal Loss Adjustment Expenses
Booked reserve	Held reserve or Carried reserve
Case reserves	Outstanding claims reserve or Case outstanding or Claims outstanding or Unpaid case
Claim numbers	Claim counts
Incurred claims	Reported claims or Case Incurred or Incurred on reported

coverage across the world. The range of coverages and the different terms is very wide and further details are not given here, as from a reserving practitioner's perspective, the key issue is to understand the nature of the coverage that is subject to the reserving process.

A further point to note regarding terminology is that, in some contexts, there may be uncertainty over whether the recipients of the results of a reserving exercise have sufficient understanding of the various terms that may have been used in the relevant documents or presentation of results. Where this is the case, it will often be appropriate to include an explanation of the key terms, to avoid any misinterpretation of the results.

2

Data

2.1 Introduction

The main purpose of this chapter is to describe the types of data that are used for estimating claims reserves, using one or more of the methods described in Chapters 3 and 4. In this context, claims reserves are assumed to represent case reserves plus IBNER and IBNR, as defined in Section 1.1. The data that are needed for estimating categories of reserves other than claims reserves (e.g. reinsurance bad debt or unallocated loss adjustment expenses) and for certain specific types of claim (e.g. latent claims) or business are covered in Chapters 5 and 7.

The term "data" as used here is intended to mean both quantitative and qualitative information – in effect, any type of information that is used as part of the reserving process. However, since most reserving methods are applied to quantitative information, this chapter focuses on that type of data. Nevertheless, collation of qualitative information is an essential part of any reserving exercise, and so reference is also made to that type of data, with further details in general terms being given in Chapter 5 and in specific contexts in Chapter 7.

After first introducing the claims development and data triangle concepts, the remainder of this chapter gives a description of the different types of reserving data, followed by some further details related to data triangles, and then a discussion of the important topic of grouping and subdivision of reserving data. The chapter concludes with some examples of the common data quality issues encountered in practice. A more complete understanding of these issues and of other data-related practical points is best achieved when considered in the context of the application of the key reserving methods, which are described in Chapters 3 and 4. This is also partly why some of the detailed data topics related to specific methods are dealt with in these later chapters, rather than in this chapter, which focuses on more generic data topics. Hence, readers who are new to reserving may need to re-read this chapter after gaining at least an initial understanding of the key reserving methods.

2.2 The Claims Development Concept

One of the major types of data that is collected for reserving purposes is related to the claims experience of the insurer. This and other types of data are discussed in the next section, but it is first helpful to understand the concept of claims development, since much of the claims data that is used for reserving is collated in this format.

Building on the observation made in Chapter 1 that there can be delays between the incident date of a claim and the final settlement of that claim, the analysis of the pattern of these delays over time for a specified group or "cohort" of claims is the cornerstone of many claims reserving methods. There are several different specifications for the grouping of claims that can be used for this purpose, but a common example would be to use what is referred to as an "accident year". This simply means that claims are grouped together according to the calendar year (or other 12 month period) in which the underlying incident occurred. The use of the term "accident" is just for convenience; the underlying incident does not have to be an accident per se (e.g. as in a motor vehicle road accident), it can be whatever fortuitous event is covered by the relevant insurance policy, as long as a date (or period) can be assigned to the event. Other groupings are discussed in Section 2.5.2.

Staying with the accident year example, the development of claims is analysed by observing either a claim amount or claim number related figure for each accident year at certain delay periods after the start of the accident year. For example, the figure might be observed after one year's development (i.e as at the end of the accident year) and then at yearly intervals thereafter. There are numerous types of claim-related figures that can be measured in this way, as discussed in Section 2.4, but one example would be the claims paid by the insurer. So, for example, for accident year 2010, the data collated as at 31 December 2014 might be as shown in Table 2.1 in respect of a particular class of business and/or type of claim.

Hence, after one year's development of the accident year, the total amount of claims paid by the insurer, for all accidents that occurred in 2010, was 396,000. One year later, this amount had increased to 1,333,000 and after four years it had reached 2,986,000. One further year later, it had reached 3,692,000. This does not mean that 3,692,000 is necessarily the final amount that will be paid in respect of that accident year; it is simply what has been paid in total up to 31/12/14. When the data is organised in this way it is often referred to as being in a "cumulative" format – i.e. cumulative up to successive points in development time. It can also be organised in "incremental" format, as explained further in Section 2.5.3.

Table 2.1 *Example of claims development for Accident Year 2010*

Delay (years)	Implied date	Claims paid ('000)
1	31/12/2010	396
2	31/12/2011	1,333
3	31/12/2012	2,181
4	31/12/2013	2,986
5	31/12/2014	3,692

2.3 The Data Triangle Concept

Observing the development of single groups or cohorts of claims, such as those related to individual accident years, can help understand how those claims might develop towards the final or "ultimate" amount for each cohort, which is effectively the goal of a reserving exercise. However, organising the data in a way that enables the development of older cohorts to be used to estimate how more recent cohorts might develop in future is far more powerful. It is this simple idea that is behind the concept of the data triangle.

Continuing with the example given in Table 2.1, assume that data was also available as at 31/12/14 in the same format for accident year 2009, which might be as shown in Table 2.2.

Table 2.2 *Example of claims development for Accident Year 2009*

Delay (years)	Implied date	Claims paid ('000)
1	31/12/2009	443
2	31/12/2010	1,136
3	31/12/2011	2,128
4	31/12/2012	2,898
5	31/12/2013	3,403
6	31/12/2014	3,873

Notice that this accident year has one further development period than 2010; earlier accident years would have correspondingly more development periods and later accident years fewer. The data triangle concept is explained by first rearranging the data for 2009 and 2010 in row format as shown in Table 2.3. It should be obvious how the layout shown in Table 2.3 has been obtained from the original data shown in Tables 2.1 and 2.2. Then, if data for further subsequent accident years is added in the same way, the data might be as shown in Table 2.4. Table 2.4 can be seen to form a triangle-type shape, hence the term "data triangle". It should be noted that the terms "data triangle", "triangle" and "development triangle" may be used interchangeably by reserving practitioners, and that is also the case in this book. Although accident year is used in the simple example in Table 2.4, the term "cohort" is used in the subsequent sections to mean the grouping that is used for this aspect of the triangle. Further details relating to data triangles are given in Section 2.5, but this brief description is sufficient to enable the concept to be used in the subsequent section on data types.

Table 2.3 *Example data triangle: Accident years 2009 and 2010 only*

Accident yr	Development year					
	1	2	3	4	5	6
2009	443	1,136	2,128	2,898	3,403	3,873
2010	396	1,333	2,181	2,986	3,692	

Table 2.4 *Example data triangle: Accident years 2009 to 2014*

Accident yr	Development year					
	1	2	3	4	5	6
2009	443	1,136	2,128	2,898	3,403	3,873
2010	396	1,333	2,181	2,986	3,692	
2011	441	1,288	2,420	3,483		
2012	359	1,421	2,864			
2013	377	1,363				
2014	344					

2.4 Data Types

The data that will be used for a particular reserving exercise will depend on a wide range of factors including the classes of business being analysed, types of claim, scope of the exercise etc. Consideration of the underlying process that drives the claims experience should, at least in theory, govern the data that is collected. Although this will in many cases result in data triangles being the core data that is used, in some cases they will not be a suitable format (e.g. for some types of latent claims) and alternative types of data will need to be collected.

This section summarises some of the key data types in general terms that are used in reserving. Chapter 7 discusses more specific data requirements in a range of different reserving contexts. The data types have been divided into two groups – Key Quantitative Data and Other Data. Data can be either internal or external to the entity that is the subject of the reserving exercise, and so both data types are included here.

2.4.1 Key Quantitative Data

This can be divided into the following categories:

- Claim amount triangles (Paid, Case reserves and Incurred).
- Claim number triangles.
- Loss ratio (= Claims/Premiums) triangles.
- Premiums by cohort, including possibly also premium triangles.
- Exposure information by cohort (e.g. turnover, vehicle years).
- Commissions by cohort, including possibly also commission triangles.
- Individual large losses or catastrophes, both by cohort.
- Allocated Loss Adjustment expenses (ALAE). Note that these will often be included as part of the claim amounts, rather than being separately identified.
- Unallocated Loss Adjustment expenses (ULAE).
- Reinsurance programme details.
- Historical estimates of ultimate claims or loss ratios by reserving category and cohort.
- Estimates made at previous valuation dates of expected incremental claim movements (paid and/or incurred) by cohort.
- Other miscellaneous quantitative data.

Not all of these will be collected for every reserving exercise. For example, in some situations claim number data will either not be readily available or even if it is, will not be used. The subdivisions of the data that will be used for categories of data such as the claim amount and claim number data triangles will vary between exercises, but possible examples are as follows:

- Class of business (e.g. Motor, Liability, Property).
- Claim type within class of business (e.g for Motor – Injury, Physical damage, Windscreen etc.).
- Claim size.
- Territory/Region/Country.
- Individual contracts or groups of contracts.
- Distribution channel (e.g. broker, direct).
- Currency.
- Outwards reinsurance basis: Gross, Net, Reinsurance and possibly by type of outwards reinsurance.
- Legal entity.

The precise specification for each data category within each of these subdivisions will also vary between exercises, but some common examples for selected categories are given in Table 2.5.

The final entry in the list of Key Quantitative Data categories – Other miscellaneous quantitative data – will vary considerably depending on the context, with examples including:

- Historical premium rate change information.
- Historical claim inflation indices.
- Individual policy/contract details
- Details of commuted inwards and outwards contracts and their impact on relevant data.
- Output from specific reserving analyses on selected subsets of the business.
- As-if data triangles, with the development history adjusted so that it is more representative of recent cohorts where the reserve is most material (e.g. adjustment to current excess levels or removal of selected cancelled business).
- Benchmark data – e.g. ULRs, development factors, claims frequency, average cost.
- Schedules of individual claims.

2.4.2 Other Data

A wide variety of other data and information, both quantitative and qualitative, may be used in practice in reserving, depending on the context. Some common examples are given in Table 2.6. Many of the items would be required over, say, the last 2–3 years, where relevant. Section 5.2 covers the information gathered at the initial background stage of a reserving

Table 2.5 *Quantitative data types: Examples of data for selected categories*

Data category	Example of data
Claim amount triangles	Paid, Outstanding, Incurred
Claim number triangles	Number of claims reported or settled, with or without zero claims
Premiums	Earned, written, received/signed and estimated ultimate by cohort
Premium triangles	Received/signed
Commissions	Either netted off premiums, or shown separately in similar way as for premiums
Exposure information	By cohort, depending on class of business. Examples might be vehicle years (for Motor) and total sum assured for property business
Individual large losses	Current Paid and Incurred amount and possibly also historical development
Individual catastrophes	Current Paid and Incurred amount and possibly also historical development, and detailed exposure information.
ALAE	Either included with claim data, or shown separately in similar way as for claims
ULAE	Schedules by category of expense
Reinsurance programme	Schedules of programme for each cohort where relevant, showing contract details and possibly also latest utilisation data

exercise in more detail and Chapter 7 discusses more specific data requirements in a range of different reserving contexts.

2.5 Further Details Related to the Data Triangle Format

This section provides additional descriptive details related to the triangle format and introduces a mathematical notation for the triangle, which is used extensively in the mathematical description of reserving methods given in Chapters 3 and 4.

2.5.1 "Squaring" The Data Triangle and the "Tail"

As explained in Section 2.3, the basic idea behind the data triangle is to analyse the development over time of successive cohorts. This analysis is implemented through the use of one or more reserving methods, with the objective being to project or estimate the future claims and hence the reserves. This is sometimes summarised by the expression "squaring

Data

Table 2.6 *Examples of Other Data*

Category	Details
Key Background data	Classes of business written over what years; ownership changes; key staff; overall reserving process etc.
Reserving reports	Internal and external reserving reports produced by actuaries and other reserving practitioners
Board/Other committee papers	Extract from Board and other relevant committee papers relevant to reserving
Financial reports	Report and Accounts
Regulatory returns	Returns to insurance regulator and any relevant correspondence with the regulator(s)
Reserving governance documents	Reserving policy/philosophy documents, including case reserving philosophy
Class of business description	Summary of business written in each class, including any major changes (e.g. in terms and conditions) over time
Ongoing litigation	Summary of any material litigation (e.g. on certain claims) likely to affect reserving
Data Reconciliations	E.g. between data triangles and financial statements
Market environment	E.g. Legal and regulatory issues, underwriting cycles, general market commentary
Inception data information	Pattern of inception dates of contracts by cohort

the triangle", the logic for which can be seen by using the data in Table 2.4 and then highlighting the unknown future element in the lower right-hand corner using asterisks, as shown in Table 2.7. If these unknown future amounts are estimated by applying the reserving methods, then the squaring of the triangle will have been completed; it is no longer a triangle shape but more like an "array" or matrix of data.

If the cohort period and development period are different, or there are fewer cohorts than development periods (or vice versa) then the triangle is not being projected to a "square" as such, but the principle remains that the unknown future element in the bottom right-hand corner needs to be "filled in" – that is, estimated by applying the chosen reserving methods.

Another reason why the estimation of reserves using a data triangle will not simply involve just projecting the lower right-hand corner is if the claims development is expected to extend beyond the final column of the triangle. In other words, if one or more of the first few cohorts are still showing signs of upward (or downward in some cases) development in the last few columns, then it is possible that further development will occur beyond the last column. For the example in Table 2.7, the claims for the first accident year, 2009, increased

Table 2.7 *Example data triangle: Squaring the triangle*

Accident yr	Development year					
	1	2	3	4	5	6
2009	443	1,136	2,128	2,898	3,403	3,873
2010	396	1,333	2,181	2,986	3,692	*
2011	441	1,288	2,420	3,483	*	*
2012	359	1,421	2,864	*	*	*
2013	377	1,363	*	*	*	*
2014	344	*	*	*	*	*

by 14% (i.e. 3,873/3,403 − 1) between development year 5 and 6, so it is reasonable to at least consider whether there might be further development for this accident year beyond development year 6, and if so how much. If all accident years are expected to follow a similar development pattern, then this feature may apply to the years after 2009 as well, so that the completed triangle after applying the reserving method will be a rectangle, rather than a square, with the rectangle extending out to the right of the triangle as far as is deemed necessary. The concept of projecting beyond the domain of the observed triangle is commonly referred to as projecting, or estimating, the "tail" or "tail factor". There are various approaches to estimating the tail, which will depend in part on the specific reserving method that is being used. The details are therefore covered in the description of each method in Chapters 3 and 4, where relevant. For example, the subject of tail factors in the context of the commonly used development factor or Chain Ladder method is covered in Section 3.2.4.

2.5.2 Cohorts and Development Frequency

Data triangles can be constructed using a range of cohort and development frequency definitions. The cohort can be described by reference to a grouping definition and the length or period of the cohort. Common examples for the grouping definition are claim accident date, policy inception date and claim reporting date. Typical cohort periods are a year, quarter or month. They will often, but not always, relate to calendar periods, leading to commonly used groupings such as Accident Year and Policy Year (or sometimes "Underwriting Year"). Typical development frequencies are similarly annual, quarterly or monthly. Examples of the combinations of cohort periods and development frequencies include Annual/Annual, Annual/Quarterly, Quarterly/Quarterly and Monthly/Monthly.

The definition of the cohort and development frequency will depend on the particular circumstances and type of business being analysed. It might also depend on data availability, although in many cases gaining access to the underlying transactional claims and policy records (in electronic format, clearly) will give considerable flexibility to the reserving practitioner, and in some cases such access will be deemed necessary. For some individual reserving exercises it might be deemed appropriate to include consideration of the impact on the results of using different options for one or both of the cohort and development frequency definitions. For example, using a shorter development frequency period (e.g. quarterly) can sometimes reveal features of the development that may not be apparent from longer periods, such as seasonal patterns of development or a flattening of the claims development. In other situations, use of a shorter development frequency can make the application of some claims reserving methods unnecessarily cumbersome and it may be more difficult to identify the overall development pattern.

The cohort definition used will affect the pattern of development that might be expected. This can be seen by considering the example triangle shown in Table 2.7. If the cohort used had been a policy or underwriting year, rather than accident year, then this will mean that each row in the triangle will show the cumulative paid claims in respect of all policies that have an inception (or renewal) date in the relevant year. This will mean, for example, that a claim that occurred and was paid in full in 2010, on a policy that incepted in 2009, will be included in the increase in cumulative claims between the first and second year of development for the 2009 cohort. Hence, other things being equal, a policy inception date grouping will develop for longer than an accident date grouping.

The cohort definition will also affect how the results of applying reserving methods to the relevant data triangles should be interpreted. For example, applying such methods to data triangles that use an accident date grouping will produce a reserve that corresponds only to claims that have already occurred (but have not necessarily been reported) at the relevant valuation date. Such claims will also be included in a reserve derived from applying reserving methods to data triangles that use a policy inception date grouping (such as underwriting year), as well as those that are expected to occur during the unexpired period of the relevant policies at the relevant valuation date. In effect, some of the claims movement in the early development periods for cohorts with a policy date grouping may be due to the earning of exposure in those periods, rather than development on claims that occurred prior to the valuation date. For such groupings, the additional element of the reserve will be included in the calculated IBNR or reserve derived by applying the relevant reserving methods.

Some examples of different combinations of cohort and development frequency definitions in a number of specific contexts are given in Chapter 7.

2.5.3 Miscellaneous Practical Points Related to Data Triangles

Format for Triangles

The first point to note is that the arrangement of the triangle used in this section and in the remainder of the book, with the cohorts placed in the rows and the development placed in the columns, can equally well be swapped round. This is chosen based purely on personal preference. The data can also be shown in either incremental or cumulative format, with incremental representing the movement of the relevant item in the development period, and cumulative representing the total amount of the relevant item as at the end of the development period. Both have their uses, although the cumulative format is more commonly used in practice, partly because some of the frequently used methods use this format. Some data items, such as outstanding claims (i.e. case reserves), are more naturally displayed in a cumulative "as-at" format, whereas others, such as paid claims or number of claims settled, are displayed in either format. The organisation of the data in a triangle format can be construed as suggesting that different cohorts are expected to have similar development patterns, and indeed some key reserving methods make this assumption. However, whilst this may not be an unreasonable assumption in many circumstances, it may not be valid for all reserving categories in an individual reserving exercise.

Calendar Period Dimension

The array format referred to earlier in this section might suggest that the data triangle has two "dimensions", and many reserving methods are designed with this in mind. The first dimension – the cohort – represents the relevant grouping of claims (or other item), and the second – the development period – represents the development over time of each cohort. However, there is a third dimension that can sometimes be very important – this is the calendar period dimension which exists along the diagonals of the triangle. Working with the example data in Table 2.7, the first diagonal is just the single cell in the top left-hand corner of the triangle (i.e. 443) which is calendar year 2009; the second diagonal contains the values 396 (from the first development year of 2010 accident year) and 1,136 (from the second development year of 2009) – both of which are in calendar year 2010. The other diagonals then progress in a similar way. Different drivers affecting the triangle will impact one or more of the three dimensions. For example, other things being equal, a growth in business volumes will impact the volumes in each row and hence the cohort dimension, whereas a speeding up of development will impact the column and hence the development dimension. An inflationary impact or a change in case reserving practices, for example, that affects all cohorts at the same time will have an impact along the diagonals of the triangle – that is, the calendar period dimension. Different reserving methods seek to capture one or more of these dimensions in their underlying structure, or make assumptions about the impact of various drivers on each dimension of the triangle.

Understanding the Business Represented by the Triangles

Although the triangle is a very commonly used and convenient way of organising claims data, this does not remove the need to understand various aspects of the underlying business and data, before applying reserving methods. The aspects related to the underlying business include the nature of the business represented by the triangle (e.g. policy coverage, potential types of claim, territories, legislative environment, impact of inflation etc.) and whether there have been any changes in these aspects across cohorts. Understanding the business in the context of a reserving exercise is discussed further in Section 5.2.1. A simple visual inspection of the triangle data can also be beneficial and may reveal unusual features such as missing items, obvious outliers (high or low) and sudden changes in volume. Such inspection can be insightful if done on both an incremental and cumulative format, and by using graphs, as explained in Section 2.5.5.

Understanding How the Triangles Have Been Constructed

In many situations, the data will be made available to the reserving practitioner in triangle format. The extent of the involvement of the practitioner in the process that led to the construction of the data triangles can vary depending on the context. Where the involvement is minimal and/or where there is uncertainty over certain aspects of how the triangles have been constructed, it may be appropriate to make enquiries regarding the process. This can sometimes reveal unexpected aspects which impact upon the usefulness of the triangles for reserving purposes, or upon the interpretation of the results using the triangles.

In other situations, for example in new or developing insurance markets, the practitioner will need, or might choose to construct the data triangles using detailed transactional data, rather than simply being presented with the triangles. This has the advantage that the precise specification of how the triangles are constructed can be defined. In some cases, however, this can be time consuming, since it will require a detailed understanding of the relevant transactional databases to ensure they are interpreted correctly.

In practice, even if the underlying development appears to be relatively smooth and stable, a wide range of factors can distort the development observed in data triangles. The factors will vary depending on the context, but examples include the impact of multiple currencies, unusually large claims, incomplete historical development and known or unknown data errors. These and other data quality issues are discussed further in Section 2.7.

Subdivision of Business for Reserving Purposes

For virtually all reserving exercises that use data triangles, a decision will need to be made regarding how the overall insurance business that is being analysed is subdivided for reserving purposes. As referred to in Section 2.4.1, there are various different types of subdivisions used in practice, with the most common being class of business and claim type. For reserving methods that use data triangles, it is important that these subdivisions are selected based on similar underlying characteristics (i.e. that they are "homogeneous"), whilst at the same time ensuring that there is enough volume to give reasonable stability in the observed development. Homogeneity can relate to aspects such as type of business, the expected development pattern or "tail", and, where loss ratios are a relevant consideration, the overall profitability of the business. If some of the chosen subdivisions are known to contain varied underlying characteristics (i.e. they have an element of heterogeneity), which in some cases may be necessary to ensure a reasonable level of stability, then this can be less of a concern if the mix of the these characteristics does not change materially across the cohorts. In practice, there can often be elements of heterogeneity in individual data triangles. For example, one or more cohorts might be different in some way to the remainder (e.g.

expected shorter-tailed development). This will need to be dealt with by adjusting the standard application of the relevant reserving methods. Some of these practical issues are discussed in Chapter 5. In addition, allowing for changes in business mix is considered in Section 6.9.2.

2.5.4 Examples of Published Data Triangles

Most reserving work will be based on privately held data triangles that are not generally made available publicly. However, there are various sources which provide data triangles for individual insurance companies, including those published by the companies themselves and those collated from regulatory returns. In addition, in many countries, triangles are available on a whole market basis, usually by class of business, but sometimes only to subscribers or members of particular associations or market groups. Partly to demonstrate the use of the data triangle format in practice, and partly to give a real example of what is available publicly, this section provides some examples of published data triangles for individual companies and of market-wide data sources. The individual company data triangles included here are not intended for use in practice; they are merely included for illustrative purposes.

The specifics of what is available will vary by country and type of insurer, and hence the usage that can be made of published data sources will also vary. For example, under certain International Financial Reporting Standards[1] and UK Financial Reporting Standards[2] the "loss development tables" disclosure requires development of ultimate claims to be published. This provides a measure, albeit it a very high level, of the company's reserving performance, by showing how estimates of ultimate claims for individual cohorts have changed over time.

Some types of data might be used by the practitioner for benchmarking purposes or by investment analysts seeking to review a company's reserving strength or historical accuracy. Aggregate market data might also be used to analyse market trends, for example. For some contexts, the data will not be granular enough, and other private sources will need to be used. Using external data sources for benchmarking purposes is discussed further in Section 6.9.1.

Some examples of types of data that individual companies publish are given in Tables 2.8 to 2.11 inclusive, with details as follows:

- Tables 2.8 and 2.9 show an extract of the data triangles (of loss ratios) published annually by the international reinsurer, SCOR. This company publishes data triangles by class of business, along with an accompanying reference document that gives background details of the company's approach to reserving. The triangles as at 31 December 2013 are available in Excel format at SCOR (2014a) and the accompanying document at SCOR

---

[1] Specifically IFRS 4 para 39(c)(iii) that was in force for UK listed companies from 1 January 2005.
[2] Specifically FRS 103 para 4.8(b)(iii) that applies to relevant companies for accounting periods beginning on or after 1 January 2015.

Table 2.8 *Example of published data triangles: SCOR Worldwide Casualty Non-Proportional and Facultative – Incurred Loss Ratios (%)*

		Development year[a]														
Yr	Prm[b]	1	2	3	4	5	6	7	8	9	10	11	12	13	14	15
99	84	9	26	41	57	67	85	126	130	137	141	144	143	147	146	145
00	106	4	53	77	96	109	118	115	124	131	137	145	153	154	150	
01	152	15	47	59	67	75	88	110	116	121	128	131	135	137		
02	237	5	14	21	28	34	51	58	59	63	64	62	64			
03	246	3	10	16	22	28	35	44	47	50	53	63				
04	167	6	13	19	25	28	29	33	35	36	39					
05	111	4	11	19	23	31	30	31	45	47						
06	135	2	10	15	19	22	23	23	26							
07	144	4	16	22	32	35	37	41								
08	135	4	11	18	20	24	29									
09	136	4	12	22	26	31										
10	143	3	9	19	24											
11	169	3	10	16												
12	177	3	11													
13	184	4														

[a] The values in each development year are Incurred Loss Ratios (%)
[b] Premiums are ultimate values in €millions.

Table 2.9 *Example of published data triangles: SCOR Worldwide Casualty Non-Proportional and Facultative – Paid Loss Ratios (%)*

		Development year[a]														
Yr	Prm[b]	1	2	3	4	5	6	7	8	9	10	11	12	13	14	15
99	84	3	6	14	20	28	36	79	81	87	91	96	107	110	112	114
00	106	0	1	18	30	42	60	66	75	81	85	90	94	97	101	
01	152	0	9	22	32	39	49	65	74	80	90	96	107	111		
02	237	0	1	4	7	13	18	29	34	38	40	42	44			
03	246	0	1	3	6	9	14	19	28	30	33	36				
04	167	0	1	3	6	10	13	16	19	22	23					
05	111	0	1	3	9	19	20	20	26	29						
06	135	0	1	3	7	9	9	10	13							
07	144	0	2	6	9	19	23	26								
08	135	0	3	6	9	13	15									
09	136	0	1	2	7	15										
10	143	0	1	3	5											
11	169	0	1	4												
12	177	0	2													
13	184	0														

[a] The values in each development year are Paid Loss Ratios (%)
[b] Premiums are ultimate values in €millions.

(2014b). The triangles are divided into eight classes of business, and the data also includes the company's estimated ultimate loss ratio (or "ULR") by underwriting year as at both 31 December 2013 and 31 December 2012 (although note that these are not shown in Tables 2.8 and 2.9).

Table 2.10 *Example of published data triangles: Aviva plc (all GI classes) from Annual report and accounts 2015 – Gross paid claims (£m)*

	Accident year									
Dev yr	2006	2007	2008	2009	2010	2011	2012	2013	2014	2015
1	3,653	4,393	4,915	3,780	3,502	3,420	3,055	3,068	3,102	2,991
2	5,525	6,676	7,350	5,464	5,466	4,765	4,373	4,476	4,295	
3	5,971	7,191	7,828	6,102	5,875	5,150	4,812	4,916		
4	6,272	7,513	8,304	6,393	6,163	5,457	5,118			
5	6,531	7,836	8,607	6,672	6,405	5,712				
6	6,736	8,050	8,781	6,836	6,564					
7	6,936	8,144	8,906	6,958						
8	7,015	8,224	8,986							
9	7,062	8,257								
10	7,077									

Table 2.11 *Example of published data triangles: Aviva plc (all GI classes) from Annual report and accounts 2015 – Gross ultimate claims (£m)*

	Accident year									
Dev yr	2006	2007	2008	2009	2010	2011	2012	2013	2014	2015
1	7,533	8,530	9,508	7,364	6,911	6,428	6,201	6,122	5,896	5,851
2	7,318	8,468	9,322	7,297	7,006	6,330	6,028	6,039	5,833	
3	7,243	8,430	9,277	7,281	6,950	6,315	6,002	6,029		
4	7,130	8,438	9,272	7,215	6,914	6,292	5,952			
5	7,149	8,409	9,235	7,204	6,912	6,262				
6	7,167	8,446	9,252	7,239	6,906					
7	7,167	8,381	9,213	7,217						
8	7,176	8,381	9,207							
9	7,184	8,378								
10	7,207									

- Tables 2.10 and 2.11 show the Gross data triangles published by the UK insurer, Aviva, in its 2015 Annual report and accounts. The first shows the paid claims triangle and the second a triangle of the estimated ultimate claims at successive year-ends. The latter of these allows the change in the estimated ultimate claims over time to be monitored. The published triangles use the alternative format with accident years in columns, rather than rows. Net of reinsurance triangles are also published, and, like the Gross triangles, are for all classes of business combined. Hence, their main use is to be able to see how the estimated ultimate claims have changed over time on a whole company basis. This is a type of disclosure currently required under IFRS/UK FRS 103, as referred to above.

Some examples of the type of data that is available on a market-wide basis are as follows:

- For some countries, databases are available for purchase which contain regulatory return data for individual companies, including data triangles. For example, in the UK, A.M. Best make such data available via their "Statement File" product, from which data triangles for individual companies by class of business can be obtained, and then aggregated to provide market-wide data. Standard & Poor's make the same data available via their "SynThesis Non-Life" product. In the US, similar data is available.

- There are some publicly available sources of industry-wide data in certain countries. For example, in the US, Meyers and Shi (2011b) provides data triangles of six lines of business for all US property casualty insurers (individually) for accident years 1988 to 1997. This data comes from Schedule P Analysis of Losses and Loss Expenses in the National Association of Insurance Commissioners (NAIC) database. It was used in Meyers and Shi (2011a) to carry out some retrospective testing on Stochastic claims reserving methods, as referred to in Section 4.4.7.
- In the UK, data triangles by "Risk Code" are available for Lloyd's of London business for all syndicates combined. This data is made available to members of the Lloyd's Market Association, and is used by some practitioners for benchmarking purposes.
- In the US, the Reinsurance Association of America publishes the Historical Loss Development Study, which includes historical loss development patterns by accident year for companies writing casualty excess reinsurance for automobile liability, general liability, medical malpractice and workers compensation. It has been published since 1969 and the data triangles are available in Excel format to purchasers of the study.

2.5.5 Graphical Representation of Data Triangles

Even before applying any reserving methods to a data triangle, graphs can be used to provide insight into the underlying development process. In many contexts, this will just mean graphs of the relevant claims or loss ratio triangles, but in some contexts (e.g. London Market) reviewing graphs of the development of the premiums received or "signed" can also be useful. Reviewing graphs might, for example, help identify if there have been any significant changes in the paid and/or claims development patterns across the cohorts, which may need to be allowed for in subsequent stages of the reserving analysis.

Examples of the types of graph that can be used before applying reserving methods, and hence before estimated reserves or ultimate values have been derived, are shown in Figures 2.1 to 2.5. All these graphs are taken from the data for the SCOR Worldwide Property Fire class in SCOR (2014a), with details and brief observations as follows:

- Figure 2.1 shows cumulative Paid claims – with all years on the same graph. This shows a relatively short-tailed development pattern, with materially higher claims in 2010 and 2011 Underwriting years, reflecting their relatively higher catastrophe claims, as referred to in SCOR (2014b).
- Figure 2.2 shows incremental Paid claims. This type of graph can sometimes reveal features that are not as obvious in cumulative graphs. Here, there is a clear peak volume of claims in the second development year, with a sharp reduction thereafter.
- Figure 2.3 shows Outstanding claims. This shows a very similar shape to the incremental paid claims graph, reflecting the short-tail nature of the business.

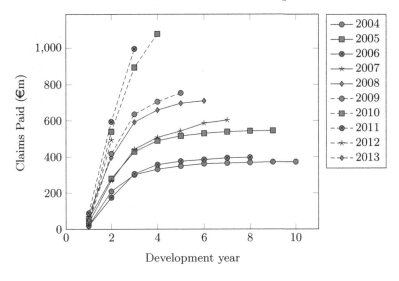

Figure 2.1 Example graph: Cumulative Claims Paid (all years)

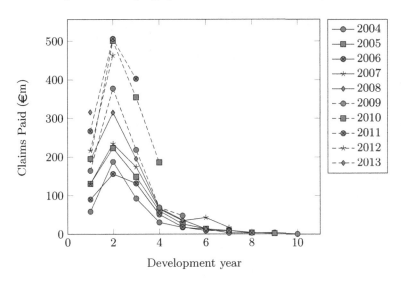

Figure 2.2 Example graph: Incremental Claims Paid (all years)

- Figure 2.4 shows a single year, with Paid, Incurred and Outstanding claims on the same graph. These can helpful in reviewing how paid and incurred claims develop over time for individual cohorts.
- Figure 2.5 shows cumulative Incurred Loss ratios – with all years on the same graph. These can be used to assess development patterns and variations in overall profitability. In this graph, 2005, 2010 and 2011 have the highest loss ratios, reflecting their relatively high exposure to natural catastrophe events, as referred to in SCOR (2014b).

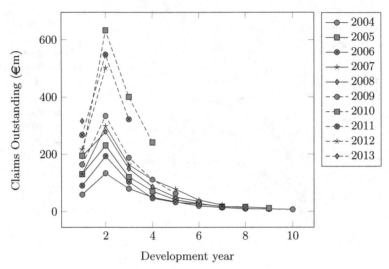

Figure 2.3 Example graph: Outstanding Claims (all years)

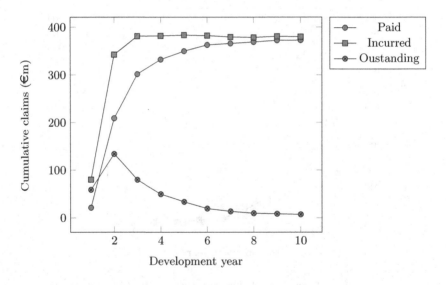

Figure 2.4 Example graph: Single Year – 2004

In practice, these types of graphs may be viewed in specialist reserving software, some of which provides useful additional flexibility, such as only viewing selected groups of cohorts, highlighting individual cohorts and magnifying certain parts of the graph to identify specific features more easily.

As well as reviewing the overall shape and length of the development pattern of the premium and claims (both paid and incurred), and to consider whether there are any

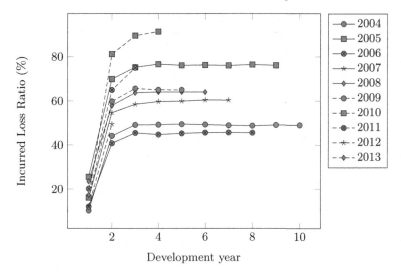

Figure 2.5 Example graph: Incurred Loss Ratios (all years)

obvious changes in this pattern across the cohorts, these graphs can also be used to identify unusual movements in the data in particular development periods (e.g. due to data errors), changes in profitability across cohorts and potential exposure to a range of features such as material catastrophe/large claims or latent claims. Whilst graphs that are created before applying reserving methods to the data can be very helpful as an exploratory data tool, graphs can also be constructed after such application, and provide a powerful diagnostic tool for reviewing the results. These graphs are discussed further in the context of the Chain Ladder method in Section 3.2.7 and in more general terms in Section 5.8.

2.5.6 Mathematical Notation for Triangle Format

The triangle format can be represented using a simple algebraic notation, as shown in Table 2.12, which shows a triangle with ten cohorts and ten points of development.

In this triangle, $C_{i,j}$ represents the item being shown in the triangle for cohort i and development period j. As explained in Section 2.5.2, the cohorts and development periods can be a range of different periods (e.g. years, quarters or months) and need not necessarily be the same.

The triangle can either show movements in particular development periods (i.e be incremental) or totals as at the end of each period (i.e. cumulative). The notation used here is $C_{i,j}$ for cumulative and $X_{i,j}$ for incremental, with i ranging from 0 to I and j from 0 to J so that the triangle has $I + 1$ rows and $J + 1$ columns. The rationale for starting the index value for development at zero is partly because it is consistent with that used in certain other sources that provide mathematical descriptions of reserving methods (e.g. Merz and Wüthrich, 2008c). In addition, when the length of the cohort and development frequency

Table 2.12 *Mathematical notation for data triangle: Example*

| Cohort i | \multicolumn{10}{c}{Development period j} |
	0	1	2	3	4	5	6	7	8	9
0	$C_{0,0}$	$C_{0,1}$	$C_{0,2}$	$C_{0,3}$	$C_{0,4}$	$C_{0,5}$	$C_{0,6}$	$C_{0,7}$	$C_{0,8}$	$C_{0,9}$
1	$C_{1,0}$	$C_{1,1}$	$C_{1,2}$	$C_{1,3}$	$C_{1,4}$	$C_{1,5}$	$C_{1,6}$	$C_{1,7}$	$C_{1,8}$	
2	$C_{2,0}$	$C_{2,1}$	$C_{2,2}$	$C_{2,3}$	$C_{2,4}$	$C_{2,5}$	$C_{2,6}$	$C_{2,7}$		
3	$C_{3,0}$	$C_{3,1}$	$C_{3,2}$	$C_{3,3}$	$C_{3,4}$	$C_{3,5}$	$C_{3,6}$			
4	$C_{4,0}$	$C_{4,1}$	$C_{4,2}$	$C_{4,3}$	$C_{4,4}$	$C_{4,5}$				
5	$C_{5,0}$	$C_{5,1}$	$C_{5,2}$	$C_{5,3}$	$C_{5,4}$					
6	$C_{6,0}$	$C_{6,1}$	$C_{6,2}$	$C_{6,3}$						
7	$C_{7,0}$	$C_{7,1}$	$C_{7,2}$							
8	$C_{8,0}$	$C_{8,1}$								
9	$C_{9,0}$									

are the same (e.g. years), this means that, in a claims paid triangle for example, the claims that are paid by the end of the cohort will have a delay to payment since the start of the cohort of zero (measured in whole time units of the period of the cohort).[3] When defined in this way, the triangle format means that values in the upper left part of the array exist only where $i + j \leq I$. The notation implies that

$$C_{i,j} = \sum_{k=0}^{j} X_{i,k}. \tag{2.1}$$

Throughout this book, when referring to data triangles, I is assumed to be equal to J and the unit of the development time is assumed to be the same as the period of the cohort (e.g. years, quarters, months), unless stated otherwise. In practice, this will not always be the case, but most of the method descriptions apply equally well to data triangles with different structures.

The following notation is used for the triangle of data:

$$D_I = \{C_{i,j} : i + j \leq I; 0 \leq j \leq J\}. \tag{2.2}$$

The leading diagonal in this type of triangle will be $C_{i,I-i}$, with $0 \leq i \leq I$.

Using this notation, the generic form of the 10×10 example triangle in Table 2.12 is as shown in Table 2.13.

Many different types of data can be shown in a triangle format. Typical items are Claim Amount Paid, Claim Amount Outstanding, Claim Amount Incurred, Claim Number Notified, Claim Number Settled, Average Claim Incurred, Incurred Loss Ratio and Premium Amount received. The word "Amount" is usually dropped in practice, with the default item therefore being an amount rather than a number or count. The $C_{i,j}$ and $X_{i,j}$ notation is used here for generic triangle items, rather than specifically for Paid or Incurred claim amounts for example. This is because the method descriptions are largely independent of the type of

---

[3] Note that, in practice, any convenient notation can be used. For example, for the purpose of introducing the first reserving method in the next chapter, the initial index value will be shown as 1, but all the detailed worked examples start the index value at 0 to be consistent with the mathematical formulation given here.

Table 2.13 *Mathematical notation for triangle data: General version*

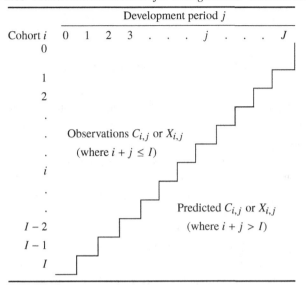

data that is contained within the triangles. Additional notation is used where it is necessary to denote specific items within data triangles.

When the data is organised in this format, one key component of the reserving process is the estimation of the ultimate claims for each cohort – that is, to estimate $C_{i,\text{ult}}$ for each value of i. Many of the methods explained in this book are designed to estimate this ultimate amount. Estimating the ultimate is equivalent to estimating the lower right-hand part of the triangle (or "squaring" the triangle), as shown in Table 2.7.

When the triangle represents claims paid data, and there is assumed to be no claims development after development point J (i.e. there is no "tail factor" required, so that $C_{i,\text{ult}} = C_{i,J}$), the claims reserve is defined as follows:

$$R_{i,J} = \sum_{k=I-i+1}^{J} X_{i,k} = C_{i,J} - C_{i,I-i}.$$

This is sometimes abbreviated to R_i. Note that unless otherwise stated, when the term "reserve" is used in this book, it is deemed to be the claims reserve (or loss reserve), rather than any other definition such as reserve net of future premiums, which can be used in some contexts.

The reserve across all cohorts combined is denoted by R, so that

$$R = \sum_{i=0}^{I} R_i.$$

Whilst some other papers and books on claims reserving use the same notation as that used in this book, others do not. Since a consistent notation is used to describe all the reserving methods here, the mathematical formulation will differ to these other sources in some cases.

2.6 Grouping and Subdivision of Data

The need to balance homogeneity with stability of data, when deciding on the subdivisions for data triangles was mentioned in Section 2.5.3. There are other considerations that are relevant, depending on the context. These include separate identification of large claims or catastrophe events, which can distort the overall development. Such identification, and subsequent separate analysis, is reasonably common in practice for business exposed to such claims or events. For example, in the survey at ASTIN Committee (2016), over 50% of participants indicated that they make a special treatment of large claims.

If the business is written in multiple currencies, then a subdivision by major currency (within class of business) can also be appropriate, particularly if the development pattern or profitability varies by currency. Even if the currencies exhibit similar patterns and profitability within classes of business, they might still be separated for reserving purposes, since there will often be a requirement to report results by currency.

In many reserving exercises involving several classes of business with many years' history, it will often be necessary to identify separately one or more selected individual contracts or groups of contracts. This will arise, if for example certain contracts are no longer written, and it is deemed appropriate to remove them from a particular data triangle, to avoid influencing the development patterns and/or loss ratio assumptions for the cohorts where the relevant contracts were not written.

Finally, a decision will need to be made whether to analyse the data on a net of reinsurance basis, or separately for gross and reinsurance.

The details of the subdivisions used will vary considerably depending on the context, with many more variations existing than those identified here. Consideration of the subdivisions will always be an important step in the early stages of any reserving exercise, regardless of the context.

Further discussion of how to allow for changes in business characteristics across cohorts is given in Section 6.9.2 and examples of the data subdivisions that might be used in a range of specific contexts are given in Chapter 7.

2.7 Data Quality Issues

Experienced reserving practitioners will be well aware of the wide range of data quality issues that arise in practice according to the circumstances of each reserving exercise. This section identifies some examples of the main themes that arise, rather than specifying all such issues, which would be an impossible task. It is worth noting at the outset that identification of these issues and development of approaches to overcome them can sometimes involve a considerable proportion of the overall effort of a reserving exercise. To avoid this becoming disproportionate, consideration of the potential materiality of each issue can be helpful, as measured against the overall level of reserves being estimated.

In general terms, the overall quality of the data used for reserving can be summarised as relating to three different aspects of the data – accuracy, appropriateness and completeness.[4] These can relate to data that is internal or external to the relevant entity. Accuracy refers to how reliable the data is, including for example how it reconciles with other data sources within the insurer. Appropriateness refers to the data being suitable for its intended purpose. Completeness refers to a broader range of features such as whether all relevant claims have been processed before collating the reserving data and to the types of data that are required to estimate reserves in the relevant context using suitable reserving methods. The extent to which the practitioner will need to establish the appropriateness, completeness and accuracy of the data will depend at least partly on the context and scope of the reserving exercise. This in turn will influence, for example, the degree to which reliance can be placed on data provided by others and the extent to which it is necessary to conduct reasonableness checks on the available data.

Ideally, data that is appropriate for the required purpose and which is of sufficient accuracy and completeness will be collated, taking into account the context of the particular exercise. However, it is difficult to define precise rules that can be used to determine whether this is the case, and in many reserving exercises there may be a degree of compromise necessary, as the data available is rarely perfectly accurate, appropriate and complete. Hence, the practitioner will need to use their experience and judgement when considering the available data. Completeness and appropriateness in particular will depend critically on the context of the exercise, and the practitioner will often have their own view of what a "complete" and "appropriate" set of data looks like, perhaps based in part on what has been done in the past in the relevant context or in other similar contexts.

Where suitable information can be referenced that documents the range of different types of reserving data that might be collated, both in a general sense (e.g. as per Section 2.4) and in the specific context (e.g. as per the examples in Chapter 7), then reviewing the set of data available for a particular exercise against those sources can be helpful in making a judgement regarding data completeness and appropriateness. Where relevant, documenting any issues related to data completeness, appropriateness and accuracy may help the users of the output from the reserving exercise understand the impact on the work carried out. In addition, maintaining an itemised list of all data deficiencies, with an agreed process and timetable for their resolution, may help ensure reserving data improves over time, which will in turn benefit the overall robustness of the reserving process.

A number of insurers will have formal data governance frameworks in place, which will cover reserving data as well as that related to a wide range of other processes within the company. Such frameworks can help improve the overall quality of reserving data, and ensure its consistency with data used for other business processes. They will include, for

---

[4] These three aspects are generally relevant when considering data quality in reserving work, but it is worth noting that they are also specifically referenced in the Solvency II regulations in respect of Technical Provisions, in Article 19 of European Parliament (2015).

example, the use of clearly defined data controls, with associated owners in the business, and dataflow diagrams, which itemise the data inputs and outputs at all stages of the reserving process.

The remainder of this section identifies some examples of issues that can arise in relation to the quality of the data, and in particular its accuracy.

Some data issues will be known at the outset of the exercise, whereas others only surface during the course of the work. The latter are usually more difficult to handle, and hence some level of data checking and reconciliation is advisable at the start. Where relevant, this can involve comparing historical data with that supplied previously to identify any changes in the data and understanding the cause, particularly where these are material. Reviewing the numerical data and graphs can help identify any obvious data anomalies, which can then be investigated. This early checking can in some situations require the data to be re-processed to correct any errors, and in others allowance may need to be made during the application of the reserving methods.

Depending on the context of the exercise, reconciliation of the data used for reserving with another source within the company (e.g. the financial statements) can be an essential check to ensure that the output from the reserving work can be used with confidence for its intended purpose. There can sometimes be known adjustments that enable the reserving data to reconcile with these other sources, but these may not be detailed until the reserving analysis is complete. Where this is the case, the practitioner may need to consider whether the adjustments should affect the results of the reserving exercise, or whether they can be directly added/subtracted to the relevant figures from the reserving exercise. For example, if an adjustment relates to corrections to the case reserves that have already been used in the reserving data, then it is likely that reserve estimates (including IBNR) produced by applying the relevant methods will also need adjusting, perhaps by correcting the case reserves and then re-applying the methods.

Missing data points can sometimes exist, but unless widespread, it is usually relatively straightforward to adjust the reserving methodology to avoid them distorting the results. A more common problem is where there have been changes in underlying processes that affect the claims (or premium) development. This type of issue does not represent a particular problem or error with the data, but it does mean that care will be needed when applying the reserving methods and interpreting the results. Examples of such processes include case reserving philosophy, speed of claims payment, backlog of processing etc. The changes can affect data triangles in one or more of the three dimensions of the triangle – cohort, development and calendar period. This is discussed further in Section 6.9.3, including the identification of such changes and adjustment of the data and/or methodology to allow for them. There can also be known misallocation of claims to reserving categories or cohorts, which can mean that results need to be adjusted accordingly.

Some datasets can have other unusual characteristics within individual triangles, such as variable cohort and development period definition or length, and artificial cut-off of claims development due to particular accounting conventions used for the underlying business. In addition, where data has been converted from one or more currencies to another currency, then the way in which this conversion has been carried out can sometimes cause distortions to the development pattern.

There are many variations of the specific issues identified in this section, and also many others that arise in practice. Further details of data issues that may arise in specific contexts are given in Chapter 7.

3

Deterministic Reserving Methods

3.1 Introduction

This chapter describes several of the reserving methods that can be used to derive reserve estimates. The methods are labelled as "deterministic", to distinguish them from stochastic methods, which are described in Chapter 4. The term deterministic in this context is intended to mean that the method will typically produce a single or "point" estimate of the reserves, without an assessment of the statistical variability around that estimate. Such methods can, however, be used to provide a range of results by varying the assumptions used within the method.

Although deterministic methods are different in nature to stochastic methods, they are still sometimes referred to as "statistical" methods, to distinguish them from case by case type approaches, in which the reserves are determined by merely adding up all the individual case reserves for each reported claim. The use of the term statistical in this way is not intended to mean that the methods are based on a formal statistical model.

In practice, deterministic methods are used in most cases where a single point estimate of reserves is required, notwithstanding the fact that stochastic methods can also be used to produce a point estimate. For both types of method, they will usually be applied within the context of an overall reserving process, as opposed to a confined theoretical context. This process is outlined in Chapter 1. Although this chapter focuses on explaining the methods themselves, it does also include some high-level comments relating to their practical application, with additional details being given in Chapters 5, 6 and 7.

The goal of applying the reserving methods described in this chapter is assumed to be to estimate the ultimate claims and hence the corresponding Claims Reserve (as defined in Chapter 1) – that is the total amount relating to Case Reserves, IBNR and IBNER. Different methods are usually needed to derive the other types of reserve described in that chapter, such as ULAE and UPR. Where relevant, reference is made to the methods that can be used for these other reserve types in Chapter 6.

These methods are assumed to be applied to data organised in a triangle format (which was introduced in Chapter 2), and to one triangle at a time, unless stated otherwise. If the triangle is a claims paid triangle, then the estimated ultimate less the claims paid to date will be the reserve. If the triangle is a claims incurred triangle, then the estimated ultimate less the incurred claims to date will be the IBNR. Other results will be produced for other types of triangle.

As mentioned in Section 2.5.2, the definition of the cohort in the data triangle to which a claims reserving method is applied will determine how the resulting ultimate claims and reserves should be interpreted (e.g. whether they relate only to claims that have already occurred or to both these claims and the future claims in respect of the relevant unexpired policies). However, this does not directly affect most of the aspects of the description or application of the methods included in this chapter; if it does have an impact, a reference is made at the appropriate point.

For each method, a description is given, and in many cases a mathematical formulation is also provided. The latter is not required in order to understand each method and can be ignored if necessary. One or more worked examples are also given for each method, along with other details of the method.

When applying reserving methods to real datasets, it is usual for more than one method to be used, both for individual data triangles, and between different data triangles. In some cases, the final selected results will be based on a combination of the results from a range of methods. The related subjects of method and results selection are not covered in this chapter, but rather are dealt with in Chapter 5.

The remainder of this chapter begins with an explanation of the Chain Ladder, Initial Expected Loss Ratio and Bornhuetter–Ferguson methods. These are three of the most commonly used deterministic methods in many countries (e.g. see ASTIN Committee, 2016, where these were reported as the top three methods used amongst survey respondents). The chapter continues with separate sections for other selected methods such as Frequency–Severity and Cape Cod, with a range of other deterministic methods being described briefly in a further section.

3.2 The Chain Ladder Method

3.2.1 Introduction

In most, if not all insurance markets, the Chain Ladder (or "CL") method is by far the most commonly used claims reserving technique in general insurance. Various other names are sometimes used to describe the method, including "Link Ratio", "Development Factor" method, or just "Development technique". It relies on organising the data into a triangular format, although aspects of the method can be applied to a single cohort of data.

In simple terms, the CL method involves using the observed development pattern over time of claims from earlier cohorts, and assuming that this pattern will also apply to later cohorts. In essence, the past development is used as a guide to the future development. For example, consider a particular claims paid data triangle where, on average, the older cohorts suggest that, after two years have elapsed since the start of a cohort, the total claims paid for a single cohort increase by a factor of 30% between that time point and their ultimate settled value. Then, for a cohort that has developed for two years, the ultimate claims are estimated by multiplying the paid claims at this time point by 130%. The Chain Ladder method provides an approach for estimating this factor, to "project" the claims for each cohort from their current position to their estimated ultimate position. Further details of the steps involved in applying the method in practice are given in the later subsections.

The CL method was originally developed as a simple numerical algorithm for estimating reserves, but was later formulated as a statistical model, as explained in Section 4.1.5. The origins of the method go back to at least the 1970s. For example, Taylor (1986) suggests that its name is attributable to Professor R. E. Beard, who worked as a consultant to the UK Department of Trade in the early 1970s. Also around this time, in Skurnick (1973b), a so-called "projection method" is described, which is very similar to the CL method, and which is attributed to an accounting paper from 1966 (Harnek, 1966). However, there is some evidence that the underlying idea of analysing the historical development of claims over time existed well before then. For example, Benedikt (1969) refers to "chain relatives" as being suggested in an economics book from 1941 (Davis, 1941). Even earlier than this, as mentioned in Hindley et al. (2000), the 1908 "Instructions for the guidance of auditors" at Lloyd's of London (see Lloyd's of London, 1908) describe a method of determining a value to be placed upon each year of account's liabilities which can be described as a type of simple average chain ladder method.

Regardless of the provenance of the CL method, it is certainly true that the concept of IBNR has been in existence for a long time – for example, a very early method for estimating IBNR was given in a 1934 paper – Tarbell (1934). Despite its widespread use over many years, the CL method does have a number of limitations, not least of which is the assumption that the past is a good guide to the future. Most practitioners will be very well aware of these limitations and will take them into account when applying the method in practice, and when interpreting the results. These limitations are discussed further in Section 3.2.5.

Given its popularity as a reserving method, it is perhaps not surprising that there has been considerable analysis of whether there is a stochastic equivalent to the deterministic CL method. A number of suggestions have been made, which are explained in Section 4.1.5.

3.2.2 Description of the Basic Chain Ladder Method

This section describes the CL method in its most basic form, using a worked example. Subsequent sections build on this to explain the various modifications that are used in practice.

The idea behind the CL method derives from the notion of completing the lower right-hand corner of the triangle (or "squaring the triangle"), as introduced in Section 2.5.1. Using the same example data triangle as used there, the concept of the "chain" can be seen in Table 3.1.

Each arrow, or chain, represents a factor that takes the claims development from one development year to the next. When these factors are multiplied together, they allow the

Table 3.1 *The concept behind the Chain Ladder method*

Year	Development year					
	1	2	3	4	5	6
2009	443	1,136	2,128	2,898	3,403	3,873
2010	396	1,333	2,181	2,986	3,692	
2011	441	1,288	2,420	3,483		
2012	359	1,421	2,864			
2013	377	1,363				
2014	344	–	–	–	–	–
	1→2	2→3	3→4	4→5	5→6	

lower right hand corner of the triangle to be completed, and hence the reserves to be estimated. The essence of the CL method is the arithmetic procedure for deriving these factors, which are known as development factors, "link ratios" or "age-to-age" factors. The term "link ratio" will be used in most cases in this book. An obvious starting point for derivation of these factors is to calculate the individual link ratios for each cohort individually. These are calculated for each cohort (i.e. accident year in this example) as the ratio of the cumulative value at a particular development point, divided by the cumulative value at the previous development point. They represent the proportionate change (i.e. increase or decrease) in the cumulative values over development time. They should eventually converge to a value of 1.0 for each cohort, once the relevant item has stabilised (e.g. if the data were claims paid, this would be when all the claims have been paid for that cohort).

The calculated link ratios for this example are shown in Table 3.2. In this table, the values in column 2, for example, show the link ratios from development point 1 to 2. For example, for accident year 2009, the 1→2 ratio is calculated as $1,136/443 = 2.56$. The other values are calculated in a similar way and should be self-explanatory.

Table 3.2 *CL method example: Individual link ratios*

Accident yr	Development year					
	1	2	3	4	5	6
2009	–	2.56	1.87	1.36	1.17	1.14
2010	–	3.37	1.64	1.37	1.24	
2011	–	2.92	1.88	1.44		
2012	–	3.95	2.02			
2013	–	3.62				
2014	–					

The CL method requires a single link ratio "estimator" to be chosen for each column, based for example, on some form of average of the individual link ratios. In this book, these are referred to as link ratio estimators, or sometimes just link ratios or estimators. When the term link ratio is used, it should be obvious from the context whether this refers to individual link ratios or to some form of link ratio estimator. Once derived, the estimators are then used to complete the lower right-hand corner of the triangle. There are several different types of estimator that have been suggested for use in practice. Perhaps the most obvious would be to take the simple average of the individual link ratios, but this suffers from the problem that it can be distorted by either very low or very high values, which might arise for cohorts where there is very little data. Hence, a more commonly used approach is to use a weighted average of the individual link ratios, with the weight being the value in the denominator of each ratio. This is known as the "Column Sum Average", or "CSA", since, as shown in the example calculation that follows, the weighted average is equivalent to summing the values in each column and dividing by the sum of the values in the preceding column (excluding the final value). Other types of estimator are described in Section 3.2.3.

Table 3.3 shows the sum of the relevant figures for the first two columns of the data triangle.

Table 3.3 *CL method using CSA: First two column sums*

Year	Development year					
	1	2	3	4	5	6
2009	443	1,136	2,128	2,898	3,403	3,873
2010	396	1,333	2,181	2,986	3,692	
2011	441	1,288	2,420	3,483		
2012	359	1,421	2,864			
2013	377	1,363				
2014	344	–	–	–	–	–

Sum = 2,016   Sum = 6,542

Hence, the first CSA link ratio estimator is

$$\text{CSA 1 to 2} = 6{,}542/2{,}016 = 3.24.$$

In effect, this is an estimate of the multiplicative factor by which the cumulative claims are expected to change between the first and second development year. So, for the 2014 accident year, this would suggest that the cumulative claims will be $1{,}116$ ($= 344 \times 3.24$)[1] by the end of the second development year. Hence, this allows us to begin to complete the lower right-hand corner of the triangle.

The relevant figures that are used in the next CSA link ratio estimator (i.e. for 2→3) are shown in Table 3.4, so that

$$\text{CSA 2 to 3} = 9{,}593/5{,}179 = 1.85.$$

[1] The development ratio is only shown here to 2 decimal places, whereas the full value has been used in the calculation, which explains any differences if the calculation is repeated using the quoted figure. Similar differences may arise for other calculations given in this section, such as the summations in Tables 3.3 and 3.4.

Table 3.4 *CL method using CSA: Second two column sums*

		Development year				
Year	1	2	3	4	5	6
2009	443	1,136	2,128	2,898	3,403	3,873
2010	396	1,333	2,181	2,986	3,692	
2011	441	1,288	2,420	3,483		
2012	359	1,421	2,864			
2013	377	1,363				
2014	344	–	–	–	–	–
		Sum = 5,179		Sum = 9,593		

The complete set of calculations is shown in the first four columns of Table 3.5. The Cumulative Ratio column is explained further below.

Table 3.5 *CL method: Calculation of CSA link ratio estimators*

	Column sums			Cumulative
Link ratio	First	Second	CSA	Ratio
1→2	2,016	6,542	3.24	11.48
2→3	5,179	9,593	1.85	3.54
3→4	6,729	9,367	1.39	1.91
4→5	5,884	7,094	1.21	1.37
5→6	3,403	3,873	1.14	1.14

At this stage of the process, all of the "chain" factors have been calculated, as shown in Table 3.6. This table also shows the projected values to the right of the stepped line. The values in each successive column are calculated by multiplying the relevant CSA link ratio estimator shown below the triangle by the value in the preceding column. For example,

Table 3.6 *CL method: Projected triangle*

			Development year			
Accident yr	1	2	3	4	5	6
2009	443	1,136	2,128	2,898	3,403	3,873
2010	396	1,333	2,181	2,986	3,692	4,202
2011	441	1,288	2,420	3,483	4,200	4,781
2012	359	1,421	2,864	3,987	4,808	5,473
2013	377	1,363	2,525	3,515	4,239	4,825
2014	344	1,116	2,068	2,878	3,471	3,951
		×3.24	×1.85	×1.39	×1.21	×1.14

to derive the estimated value at the end of the third development year for accident year 2014, the value of 1,116 is multiplied by the 2→3 link ratio estimator of 1.85, to get 2,068. This process continues until each accident year has been projected to its estimated value at the end of the sixth development year. In this simple example, since the development is assumed to be complete after 6 years (i.e. the "tail factor", as explained in Section 2.5.1, is 1), this will produce the estimated ultimate value for each accident year.

It should be clear from this description that the ultimate value for each cohort can also be derived by first multiplying together all the link ratio estimators beyond the latest point for that cohort (i.e. the leading diagonal) in the triangle. So, for example, to project to ultimate the value at development point 2, the "chain" is linked from development point 2 to development point 6 by multiplying together the link ratio estimators of 1.85, 1.39, 1.21 and 1.14, which gives a value of 3.54. The implication of this is that for any particular cohort, the link ratio estimators prior to the latest point for that cohort have no direct impact on the results produced by the CL method.

These are the cumulative ratios shown in Table 3.5. To use these directly to derive the ultimate values for a particular accident year, it is simply a matter of determining the "age" of the accident year (i.e. how many years it has developed so far) and then using the relevant cumulative ratio that projects from that development point to ultimate. Hence, accident year 2013 is at development year 2, and hence the cumulative ratio from 2 to ultimate is used, which is 3.54. The calculations are set out in Table 3.7.

Table 3.7　*CL method: Cumulative development pattern and % developed*

Accident yr	Age of acc yr	Factor to Ult	% Developed	Estimated Ult
2014	1	11.48	9%	3,951
2013	2	3.54	28%	4,825
2012	3	1.91	52%	5,473
2011	4	1.37	73%	4,781
2010	5	1.14	88%	4,202
2009	6	1.00	100%	3,873

Hence, it is not necessary to complete each future cell in the lower right-hand corner of the triangle; the cumulative "factors to ultimate" can be derived first and then used directly to calculate the ultimate values for each accident year. However, if the incremental cash flows in each future period are required, then deriving the full lower right-hand triangle will be necessary.

Some practitioners use the inverse of the factor to ultimate to describe the development pattern, and this is usually referred to as the "Percentage developed", or just "% developed". The % developed values for this example are shown in the relevant column of Table 3.7. So, for example, the 52% figure is simply 1/1.91. This can then be used to derive the ultimate for 2012 accident year, which is at development point 3, by dividing the claims to date by 52%, that is 2,864/52% = 5,473.[2] The % developed concept has a convenient

---

[2] The difference to the calculated figure using the quoted numbers in the text is due to the latter being rounded.

and intuitive interpretation, as it represents how close the cohort is estimated to be to its ultimate value, or in other words how mature the relevant cohort is, on its development path to ultimate. A low value represents a "young" cohort where there is likely to be relatively more uncertainty regarding the ultimate value compared to a more mature "older" cohort that has a higher % developed. Although the % developed will increase from 0% at the start of a cohort to 100% at ultimate, it may not be necessarily stay below 100% through all development points. This is because the actual claims can reduce in some development periods in certain circumstances; e.g. for claims incurred, this can occur if the case reserves turn out to be higher than the final settled values, and for claims paid it may occur due to salvage and subrogation. As for the factor to ultimate calculation, the age of the cohort is first determined and then the relevant % developed is determined at that development point, which is then divided into the cumulative claims value for the cohort to derive the ultimate.

The final step in the CL method is to derive the estimated reserve. In the example used here, as the data triangle represents claims paid, this is just the estimated ultimate value less the claims paid to date. The calculation is shown in Table 3.8. The ordering of the accident years in this table has been changed from that used in Table 3.7, as this is more logical for reserving summaries.

Table 3.8 *CL method example: Calculation of reserve*

Accident year	Claims paid	Estimated ultimate	Claims reserve
2009	3,873	3,873	–
2010	3,692	4,202	511
2011	3,483	4,781	1,298
2012	2,864	5,473	2,609
2013	1,363	4,825	3,462
2014	344	3,951	3,607
Total	15,620	27,105	11,485

3.2.3 Link Ratio Estimators

A range of alternatives to the Column Sum Average link ratio estimators have been suggested for use in practice. Some examples of these are given in Table 3.9. The mathematical formulation for two of these is given in Section 3.2.6. Of the estimators listed in Table 3.9, those most commonly used in practice (at least in the UK currently) are probably the CSA and CSA Recent (e.g. CSA4), although some practitioners might prefer to use other estimators in particular reserving exercises.

Variations on the estimators contained in Table 3.9 are also used. For example, rather than deriving the ratio from data for, say, the last four cohorts, as in "CSA4", a different group of four cohorts from amongst all the relevant cohorts might be selected, perhaps by excluding data for individual cohorts that are deemed to be outliers. In fact, a review of the individual development ratios for the whole triangle, and exclusion of the data underlying

Table 3.9 *Chain Ladder method: Types of link ratio estimator*

Type	Abbreviation	Description
Arithmetic Average	AM	Average of individual link ratios for all cohorts
Weighted Arithmetic Average	WAM	As AM, except user-defined weights are used
Arithmetic Average Recent	AMn	As AM except only over the most recent "n" cohorts
Weighted Arithmetic Recent	WAMn	As WAM except only over the most recent "n" cohorts
Column Sum Average	CSA	As AM, but each ratio is weighted by its denominator
Column Sum Average Recent	CSAn	As CSA, except only over the most recent "n" cohorts
Arithmetic Average Excluding	AM(Ex)	As AM, but excluding minimum and maximum values
Minimum	MIN	Minimum individual link ratio
Maximum	MAX	Maximum individual link ratio

any possible outliers, is often a useful first step when using the CL method. In addition, a different estimator could be used for different columns, although that is not thought to be done frequently in practice. Sometimes, however, one or more of the calculated estimators is replaced with a user-defined ratio, based on judgement. Graphs can be useful in deciding on the groups of cohorts to use for each estimator, as well as to identify potential parts of the development for individual cohorts that might be excluded when deriving the estimators.

Depending on the circumstances, the impact of any such exclusions or user-defined selections on the final results may, however, need to be considered before the results are finalised.

Estimators can either be "incremental", meaning that they represent factors that project the claims development from one column to the next in the triangle, or "cumulative", which represent factors that do so from one column to ultimate. These may also be referred to respectively as individual development factors and development factors to ultimate.

3.2.4 Tail Factor Estimation

The description of the CL method in this section has so far assumed that no projection beyond the triangle is required, and hence the tail factor will be 1. In many situations, this projection will be necessary if the claims are still developing for one or more of the oldest cohorts. Whilst for most applications of the CL method, the cumulative claims development will be increasing, for some it can be decreasing. In either case, a judgement will need to be made regarding whether that development is complete, or if a tail factor other than 1 is appropriate. This judgement will include consideration of the actual claims development itself (for which graphs can be useful, as discussed further in Section 3.2.7), any relevant benchmark data for business similar to that in the particular triangle being considered, as well as the practitioner's knowledge and experience. Where it is deemed appropriate to estimate a tail factor, various approaches can be used, which have been grouped here under three headings – Curve fitting, Benchmarking and Other methods. These are described further below. Unless stated otherwise, the methods can in theory be applied to any type of data triangle, such as claims or premiums, amounts or numbers, and paid or incurred claims.

Curve Fitting

This approach is included in most commercially available reserving software packages. It involves fitting one or more mathematical curves to the data, in some form, typically the incremental link ratio estimators. A range of mathematical curves can be used, including Exponential Decay, Inverse Power and Weibull.

As well as providing an estimate of the tail factor, curve fitting can also be a useful way of smoothing the observed estimators derived in the CL method. This can be particularly helpful where the data is relatively sparse, as can often be the case for the later development periods in the triangle, because the link ratios depend upon a relatively small number of cohorts and so can sometimes be quite volatile. Hence, even where a tail factor is not thought to be necessary, curve fitting can still be used as a smoothing mechanism. In some cases, this can change the results materially compared with using just the "raw" estimators.

Selecting one or more curves to use and fitting them to the data involves a degree of judgement, which will influence the resulting tail factors (and possibly the estimators), and the hence reserves. This judgement can include, for example, deciding the part of the overall development to which the curve is fitted – since fitting to just the later link ratio estimators, for example, can sometimes produce a materially different tail factor to fitting over all development periods. The mathematical details of some examples of the types of curves used are given in the relevant paragraphs of Section 3.2.6, which presents a mathematical formulation of the CL method. Section 3.2.7 discusses using graphs in the context of tail factors, which can be a useful way of assessing the reasonableness of fitted curves.

Curve fitting can also be used when the development data does not have regular evaluation points, or where it is available at different points for different cohorts. For example, the cumulative incurred claims might be available for a number of cohorts at, say, 18 months, 24 months, 30 months and 48 months development perhaps. In such circumstances, the development factors derived from such data are difficult to use directly for other cohorts if they are at different development points. In such cases, some form of interpolation is necessary, either using approximate methods, or through curve fitting. This is discussed further in Section 3.2.6.

In situations where the claims development for the older cohorts is still developing very steeply upwards or downwards (which can easily be seen by examining either the link ratios themselves or the graphs of the incremental or cumulative claims development), then curve fitting is unlikely to be a suitable method for determining a tail factor. This is because the lack of a slowing down of development will usually mean that curve fitting will produce a very wide range of results. In contrast, where the development does appear to have stabilised, curve fitting will typically project continuation of that stability, which in many cases will be reasonable. However, if there is not sufficiently complete historical data, the observed stabilisation can be misleading and the resulting tail factors can be inappropriate. Therefore, whenever curve fitting is at least partly relied upon to determine tail factors, comparison of the results with those from other approaches described here may be advisable, and in some cases placing a greater reliance on benchmarking, for example, may be appropriate.

Benchmarking

Where the practitioner has access to suitable benchmarks regarding the development of the type of business that is the subject of the reserving exercise, then they can be a useful source from which benchmark tail factors can be derived. As discussed in Section 6.9.1, which covers the subject of benchmarking in general terms, the appropriateness of any selected benchmark tail factors is an important consideration. In the context of tail factor selection, it can be useful to compare the observed link ratios from the data triangle with the corresponding benchmark values at the same development points to assess whether they appear consistent. This can be done both numerically and graphically. Small differences are common, but where these are more significant, it might be possible to apply some form of scaling factor to the benchmark pattern, which will therefore impact the resulting tail factor. The need for such scaling might suggest that the benchmark pattern is not appropriate, however, and other approaches should perhaps be explored before finalising the selected tail factor. Sources of suitable benchmarks will of course depend on the context. Examples of external data that could be used for this purpose are given for some of the reserving categories described in Chapter 7.

Other Approaches

Various other approaches to deriving a tail factor have been suggested by practitioners and authors of relevant papers. In some circumstances, one or more of these might be the only option, particularly where the data is sparse and when limited or no suitable benchmark data is available. These include:

- Using a group of approaches referred to in Herman et al. (2013) and Boor (2006) as "Algebraic" methods, which rely on both the paid and incurred data triangles, without any form of curve fitting or benchmarking. One approach derives a tail factor specifically for the claims paid data triangle by using the incurred data triangle. The latter can either be assumed to be fully developed, or a tail factor estimated using a suitable method. In both cases, the ratio of the ultimate incurred claims to the paid claims to date for the oldest cohort is used as the tail factor for the paid claims data triangle. This then results in the ultimate claims that are derived from the claims paid data triangle for the oldest cohort being equal to the ultimate derived from the incurred claims data. This approach is arguably no different from the comparison that is often made in practice between ultimate claims derived separately from paid claims and incurred claims. A further algebraic type approach is described in Sherman (1984) and referred to in Herman et al. (2013) as the "Sherman–Boor" method, which can be used, for example, when the incurred claims show a consistent downward development. This involves using the ratio of incremental paid claims to incremental outstanding claims in each cell of the triangle to derive an estimate of the total paid claims for the oldest cohort, and hence the tail factor. Further details of these algebraic and some other methods are given in Sherman (1984), Herman et al. (2013) and Boor (2006).

- Using a group of relatively simple methods that are based around the idea of using the latest link ratio as the tail factor. These are referred to in Boor (2006) and Herman et al. (2013) as "Bondy"-type methods, based on those developed by Martin Bondy in the 1960s. Variations involve using a chosen function of the latest link ratio, such as a multiple or power. This and other generalisations of the Bondy method are described in Herman et al. (2013).

- Using experience and judgement, perhaps partly based on the use of graphs to assess how the cumulative claims development is likely to develop beyond the latest development point, and hence what tail factor is appropriate. Some practitioners prefer to select the % developed that is appropriate for the oldest cohort, rather than the tail factor. Using judgement and experience to select the % developed or tail factor is sometimes the only approach that is feasible where the data is sparse and where there are no suitable benchmarks.

In many situations, the selected tail factor will be based on a combination of the results of using one or more of the methods described in this section. If it is clear that a tail factor is required, but application of one or more of these methods produces results that have a very wide range or are otherwise judged to be unreliable, then this can be an indication that a CL method is not suitable for that particular reserving category. In such circumstances, the use of other methods such as IELR, which are not dependent on the claims development, might be preferable. In many cases, however, it will be possible to produce a reasonable tail factor, which is certainly preferable to using a CL method and ignoring the need for a tail factor, which could result in over- or under-estimation of the reserves.

The process of fitting curves to the link ratio estimators and the selection of tail factors is explained further in the worked example in Section 3.2.8.

3.2.5 Limitations and Assumptions of the CL Method

In its purest form, the CL method is subject to a number of limitations and assumptions that affect whether it is likely to produce reasonable reserve estimates for a particular dataset. Limitations and assumptions are considered together here, as they are very closely related, in that the appropriateness of the method will depend on the validity of the assumptions, which itself becomes a limitation. The most significant limitations and assumptions are identified below, together with some examples of the common ways in which these can be overcome in practice. They are in no particular order of priority, as this will vary between different applications of the method to individual datasets. A number of these limitations also apply to other reserving methods.

- **Cohorts are assumed to have the same development pattern:** This is an inherent feature of the method itself. It can relate to the overall pattern or to specific movements in individual development periods of one or more cohorts. It is quite common for this assumption to be at least partially invalid for particular datasets, but there are many ways in which this is dealt with in practice. These include removing individual cohorts that have different patterns to the others (e.g. as may be identified by viewing the development patterns using graphs) and applying a separate reserving method to those cohorts. Specific movements which are not expected to be repeated for later cohorts may sometimes be excluded from the derivation of the estimators.

- **The results are highly geared to the observed data:** Because the method is multiplicative in nature, the estimated ultimate values are highly dependent on the magnitude of the data in the leading diagonal of the triangle (e.g. the latest Claims Paid or Incurred). This can be a particular problem for the more recent cohorts of longer-tailed classes of business, for example, where the development factor to ultimate can be very high. The resulting estimated ultimate claims can then either be unreasonably very high or very low, or even zero if the to-date figure is zero. This limitation is generally overcome by using an alternative reserving method for the more recent cohorts, such as the Initial Expected Loss Ratio method or the Bornhuetter–Ferguson method, which are described in Sections 3.3 and 3.4 respectively. In some cases, it can also be overcome by excluding distorting features such as amounts related to large claims (and then considering these separately) or by applying a cumulative development factor to the claims at development points prior to the latest point.

- **Calendar period trends are not modelled explicitly:** Because the CL method does not have any explicit identification of the calendar period dimension within the calculations, it therefore assumes that any trends (e.g. due to inflation) in that dimension are implicitly captured appropriately. This may not be the case if these trends are subject to individual calendar period "shocks" or if the overall trend has a particular pattern. For example, if a change in case reserving practice has affected one or more calendar periods in particular, and is not expected to be repeated in future, then a simple application of the CL method could extrapolate these trends in an inappropriate way. This limitation can be overcome, for example, by either adjusting the data to remove the calendar period trend before applying the CL method, and then adding an assumed future trend back in to the subsequent results, or by careful selection of individual development factors to avoid invalid extrapolation of calendar period trends. Where the calendar period trend is inflation related, then the inflation-adjusted CL method can be used, as described in Section 3.2.9. As an alternative, methods which allow explicitly for calender period trends can be used. Some stochastic models have such a feature, including the PTF model described in Section 4.7.8, Wright's model described in Section 4.7.12 and a type of Generalised Linear Model suggested in Björkwall et al. (2011), which is outlined in Section 4.7.4. A further discussion of the impact of calendar period trends on the CL method is given in Brydon and Verrall (2009), which challenges the commonly held view that the CL method effectively uses an average value for past calendar year trend (e.g. due to inflation) as the constant trend in future. In many practical contexts, it can be difficult to establish what calendar period effects exist and, if they do, what impact they are having on the data. This will therefore often contribute to the overall uncertainty surrounding a selected reserve estimate.

- **Results can be unreasonable if the development is likely to continue beyond the data triangle:** Where the data for the oldest cohort is clearly still showing increasing or decreasing development at the latest point, then unless a tail factor is selected, the CL method will, other things being equal, correspondingly under- or over-state the estimated ultimate values. Where the development for the oldest cohort has only recently stabilised, the estimators for the later development periods will be based only on a small number of observations, which can therefore be quite volatile. Both of these limitations can be at least partly overcome by fitting a curve to the data, as described in Section 3.2.4, which also describes other methods for selecting a tail factor.

- **Results may be unreliable if the data is too volatile or sparse:** In many reserving exercises, there will be one or more categories where the data is simply too volatile or sparse for the CL method to produce meaningful results, and other methods will need to be used which are not dependent on the development data. Sometimes, the sparsity can be overcome by adding data together for reserving categories, but the risk of breaching the homogeneity requirements discussed in Section 2.5.3 needs careful consideration. In other situations, it can be possible to apply the CL method to individual or small groups of cohorts within a volatile data triangle, and to use curve fitting to smooth the development factors.

- **Application to Claims Paid and Incurred data can produce divergent results:** The CL method is designed for application to a single data triangle. In some instances, applying it to each of the Claims Paid and Claims Incurred data for a particular reserving category can produce materially different results for individual cohorts and/or in aggregate across all cohorts. This is often overcome by using judgement and experience to select results for each cohort based on using one or other of Claims Paid or Claims Incurred, or a weighted average between the two. However, the underlying cause of the divergent results can merit investigation, since it might reveal some important information regarding, for example, the change in speed of claims payment and/or in case reserving practice.[3] It is also possible to use an alternative method which explicitly projects Claims Paid and Incurred within a single framework. For example, the Munich Chain Ladder approach described in Section 3.8.1 has this feature.

---

[3] See Section 6.9.3 for a discussion of how to allow for these changes.

A more mathematical consideration of some of the key assumptions underlying the CL method is given in Section 4.2.7, which addresses this topic in the context of the Mack method. This is a stochastic claims reserving method that is very closely related to the CL method. That section also considers some specific tests that can be used to assess the validity of these key assumptions.

The assessment of whether it is appropriate to use the CL method for any particular data triangle effectively involves considering the relevant limitations and assumptions, and determining which of them are of concern, and if so whether they can be overcome through modifications to the standard CL procedure.

3.2.6 Mathematical Formulation of the Chain Ladder Method

The CL method can be easily understood without any reference to a mathematical formulation, which is perhaps one its key strengths. However, a mathematical formulation of the method, in its most basic form, does help to understand what lies behind the approach. The formulation is also a useful building block for some of the stochastic methods covered in Chapter 4, although the description here is not intended to be a stochastic representation of the CL method, which is covered in detail in Section 4.1.5. Rather, this section provides a mathematical notation for the deterministic CL method, which is then used in the remainder of the book.

Using the notation introduced in Section 2.5.6, where $C_{i,j}$ represents the cumulative claims for cohort period i as at development period j, then the basic chain-ladder assumption is that there are factors

$$\{f_j : j = 0, 1, \ldots, J - 1\},$$

so that, for all $0 \le i \le I$ and all $1 \le j \le J$,

$$E(C_{i,j}|C_{i,0}, C_{i,1}, \ldots, C_{i,j-1}) = f_{j-1}C_{i,j-1}. \tag{3.1}$$

Hence, the factors, f_j, are used to estimate the value between successive development points, or to "link" the claims from development point $j - 1$ to j, thus the term "link-ratios" or "age-to-age" factors.

As an example, in the 10×10 triangle as per Table 2.12, where the rows and columns each range over 0 to 9 inclusive, there will be f_j values over the range $j = 0, 1, \ldots, 8$, which will be used to estimate successive values of $C_{i,j}$ for all values of i between 0 and 9 and j from 1 to 9 (i.e. for the second column onwards).

If the triangle is "complete" after $J + 1$ development periods, i.e. no tail factor is required, then the above holds, but if a tail factor is required then an additional value for f_J needs to be estimated. Tail factors are discussed in general terms in Section 3.2.4, and some of the mathematical curves that can be used are given in the curve fitting paragraphs later in this section.

The factors, f_j, need to be estimated in some way using the data triangles. The arithmetic procedures for this estimation that are described in Section 3.2.2 are used for this purpose, with the standard notation, $\hat{f}_j$, being used to represent the estimators. For example, the Column–Sum–Average link ratio estimator (or CSA, as defined in Section 3.2.3), used in the classic mathematical formulation of the CL method, is

$$\hat{f}_j = \frac{\sum_{i=0}^{I-j-1} C_{i,j+1}}{\sum_{i=0}^{I-j-1} C_{i,j}}, \qquad \text{where } 0 \le j \le J - 1. \tag{3.2}$$

The individual link ratios are defined as

$$F_{i,j} = \frac{C_{i,j+1}}{C_{i,j}}. \tag{3.3}$$

Then (3.2) can easily be rewritten as

$$\hat{f}_j = \frac{\sum_{i=0}^{I-j-1} C_{i,j} F_{i,j}}{\sum_{i=0}^{I-j-1} C_{i,j}}. \tag{3.4}$$

In other words, the CSA estimator is simply the weighted average of the individual link ratios, with the weight being the amount in the first column of the ratio. The arithmetic average estimator, as per Section 3.2.3, is

$$\hat{f}_j = \frac{1}{I-j} \sum_{i=0}^{I-j-1} F_{i,j}, \qquad \text{where } 0 \le j \le J - 1. \tag{3.5}$$

If required, other formulae can be defined for the other estimators given in Table 3.9, so that, for example, the "recent" versions of the above estimators just involve summing over the relevant cohorts.

If the triangle is deemed to be "complete", so that it is not necessary to estimate a tail factor beyond the triangle, then, assuming $J = I$, the ultimate claims based on the CL method for cohort i are estimated using

$$\hat{C}_{i,J} = C_{i,I-i} \hat{f}_{I-i} \cdots \hat{f}_{J-1}. \tag{3.6}$$

As an example, in the 10×10 triangle again (so that $I = J = 9$), the estimated ultimate claims for, say, cohort $i = 4$, represented by $\hat{C}_{4,9}$, are derived by multiplying the actual claims in the leading diagonal, represented by $C_{4,5}$, by the values for $\hat{f}_5$, $\hat{f}_6$, $\hat{f}_7$ and $\hat{f}_8$.

Similarly, if the triangle consists of claims paid, the reserve for cohort i, at development point J, which is denoted by $R_{i,J}$, is estimated by

$$\hat{R}_{i,J} = C_{i,I-i} \left(\hat{f}_{I-i} \cdots \hat{f}_{J-1} - 1 \right). \tag{3.7}$$

It is helpful to denote the product of the individual development factors as follows:

$$\hat{\lambda}_j = \hat{f}_j \cdots \hat{f}_{J-1}. \tag{3.8}$$

Thus, $\hat{\lambda}_j$ represents the factor by which the known claims at development point j are multiplied to estimate the ultimate claims. If there is a tail factor, denoted by $\hat{\lambda}_J$, which allows for development between J and ultimate, then this is just included as an additional term in the multiplication in (3.8). The inverse of $\hat{\lambda}_j$ represents the proportion developed at time j that is implied by the Chain Ladder estimation process – i.e. the "% developed" referred to in the description of the CL method in Section 3.2.2. This is denoted by $\hat{\alpha}_j$, so that $\hat{\alpha}_j = 1/\hat{\lambda}_j$.

Using this notation, to derive the ultimate claims for cohort i at the latest time point $I - i$, the known claims at that time point are divided by the estimated % developed, that is, by $\hat{\alpha}_j$, so that

$$\hat{C}_{i,J} = \frac{C_{i,I-i}}{\hat{\alpha}_{I-i}}. \tag{3.9}$$

It is sometimes also useful to define the incremental % developed, which are denoted here by $\hat{\gamma}_j$. This represents, for example, the estimated proportion paid or incurred (or whatever else, according to the triangle being analysed) in development period j – i.e. the cash flow if using claims paid. Hence, $\hat{\gamma}_j = \hat{\alpha}_j - \hat{\alpha}_{j-1}$. If the cumulative development factors, $\hat{\lambda}_j$ are being used, then the incremental % developed in period j can alternatively be calculated as

$$\hat{\gamma}_j = \frac{1}{\hat{\lambda}_j} - \frac{1}{\hat{\lambda}_{j-1}}.$$

The Chain Ladder Method as a Regression Model

The CL method can be thought of as using linear regression on each of the successive pairs of columns. The form of the linear regression has zero intercept and is written as

$$C_{i,j} = b_j C_{i,j-1} + \epsilon_{i,j}, \tag{3.10}$$

where b_j is the slope and $\epsilon_{i,j}$ is the error term with mean zero and variance σ^2.

Using the usual regression notation, $C_{i,j-1}$ is the "x" variable and $C_{i,j}$ is the "y" variable. If the parameter b is fitted to the data using least squares methodology, then the standard regression formula for b of $\sum x_i y_i / \sum x_i^2$ becomes

$$b_j = \frac{\sum\limits_{i=0}^{I-j-1} C_{i,j-1} C_{i,j}}{\sum\limits_{i=0}^{I-j-1} C_{i,j-1}^2}. \tag{3.11}$$

This can be rewritten as

$$b_j = \frac{\sum\limits_{i=0}^{I-j-1} C_{i,j-1}^2 \frac{C_{i,j}}{C_{i,j-1}}}{\sum\limits_{i=0}^{I-j-1} C_{i,j-1}^2} = \frac{\sum\limits_{i=0}^{I-j-1} C_{i,j-1}^2 F_{i,j-1}}{\sum\limits_{i=0}^{I-j-1} C_{i,j-1}^2}. \tag{3.12}$$

In other words, b_j becomes the weighted average of the individual development ratios, $F_{i,j-1}$, with a weight equal to the squared values in the first column (i.e. $C_{i,j-1}^2$).

If the b parameters are alternatively derived using a weighted least squares approach, with weights w_i equal to $1/C_{i,j-1}$, then the formula for b_j becomes

$$b_j = \frac{\sum_{i=0}^{I-j-1} C_{i,j}}{\sum_{i=0}^{I-j-1} C_{i,j-1}}.$$
(3.13)

This is the same as the CSA Chain Ladder formula defined in (3.2), which therefore shows that this version of the CL method is equivalent to using linear regression with weights equal to $1/C_{i,j-1}$. If instead, the least squares regression is carried out using a weight of the inverse of the square of the first column (i.e. $1/(C_{i,j-1}^2)$), then this produces the same value for $\hat{f}_j$ as the simple average of the link-ratios, as defined in Section 3.5. This equivalence is discussed further in Mack (1997) and can be seen in practice for the Taylor and Ashe (1983) dataset by referring to the R software code in Appendix B.2.2. As a simple illustration of the concept, Figure 3.1 shows a graphical representation of the data for a particular dataset.[4]

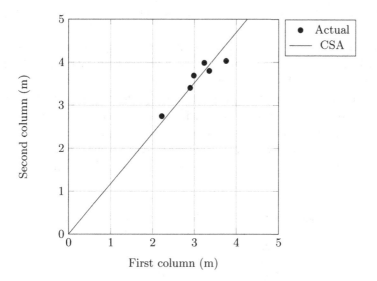

Figure 3.1 Chain Ladder as regression

Each dot on the graph, labelled "Actual", represents a pair of consecutive cumulative claim amounts for a particular cohort, with its x-coordinate being the value in the first column and its y-coordinate being that in the second column. The straight line shows the Chain Ladder represented as a linear regression through the data, with the calculated CSA estimator being the slope of the line (1.17 in this case). This type of graph provides a visual

---

[4] This shows the data for Columns 4 (x-axis) and Column 5 (y-axis) for the worked example given in Section 3.2.8, with the figures shown in millions.

tool for reviewing the degree of variability of the data around the selected CSA estimator, as well as for identifying outliers.

One advantage of expressing the CL method in terms of a statistical framework such as linear regression, is that this can be used, for example, to estimate the variability of the link ratio estimators. This idea is developed further in the very well-known stochastic reserving method – The Mack method, which is explained in Section 4.2. The fact that the CL method can be expressed in a statistical framework also means that the validity of the method for a particular data triangle can, in theory, be tested using standard statistical diagnostic tests and residual plots. Whether this is done in practice for a particular deterministic reserving exercise is a matter for individual judgement. In many cases, consideration of the limitations and assumptions of the CL method, as set out in Section 3.2.5, will also be important.

Representing the CL method in a regression framework is discussed, for example, in Murphy (1994), Barnett and Zehnwirth (2000) and Mack (1993b).

Curve Fitting

As explained in Section 3.2.4, curve fitting can be used as one of the methods for estimating tail factors and also for smoothing purposes. Curve fitting can be applied to the incremental or cumulative link ratio estimators, or, for individual cohorts, to the data itself. This section focuses mainly on fitting to the incremental estimators. A range of functional forms can be used depending on the development patterns observed for each data triangle. This section provides some examples, but others may be more suitable in certain circumstances.

First, define $g(t)$ as a curve that will be fitted through the incremental estimators, $\hat{f}_j$. An index of t is used for the curve to indicate that it is a continuous function, but it will be fitted to the discrete values of the estimators. Table 3.10 shows two common examples of this curve that can be used when the claims development curve is showing a generally positive upward development in the latter stages of development (i.e. where the link ratio estimators are greater than 1).

Table 3.10 *CL method: Examples of fitted curve types*

Type	Formula
Exponential Decay	$g(t) = 1 + a/b^t$
Inverse Power	$g(t) = 1 + a/t^b$

The functional form of the curves given in Table 3.10 follows that used in Sherman (1984) and Herman et al. (2013). Other equivalent functional forms may be used in some contexts. For example, the exponential decay curve may be expressed as $g(t) = 1 + A\exp(-Bt)$, so that $A = a$ and $B = \ln(b)$.

The Inverse Power curve (or "Loglogistic" or "Polynomial Decay" as it is sometimes called) in particular can also be expressed with a transformation of the time variable, by using $t + c$ instead of t. Several other curves can be used, including a Weibull curve and curves involving higher order terms.

The Exponential curve assumes that the estimators decay at a constant rate, whereas for the Inverse Power curve, the rate of decay decreases over time. As a simple example, assume that $a = 1$ and $b = 3$, in which case the curves and their associated rates of decay are as shown in Table 3.11.

Table 3.11 *Example of rates of decay for Exponential and Inverse Power curves*

	Exponential		Inverse Power	
Time (t)	Dev't factor	Decay rate	Dev't factor	Decay rate
1	1.333		2.000	
2	1.111	67%	1.125	88%
3	1.037	67%	1.037	70%
4	1.012	67%	1.016	58%
5	1.004	67%	1.008	49%
6	1.001	67%	1.005	42%

The rate of decay expresses how fast the incremental link ratio changes over time, so only the "projection" component is included in the calculation. For example, in the above table, the rate of decay for the Inverse Power curve between time 2 and 3 (shown at time 3 in the table) is calculated as

$$\left[1 - \frac{1.037 - 1}{1.125 - 1} \right] = 70\%.$$

In other words, for the Inverse Power curve, the projection part of the incremental development ratio decreases by 70% between development points 2 and 3, and then by a reduced percentage in successive periods. However, for the Exponential curve, the decay rate is constant at 67%.

These curves can be fitted, for example, by taking logarithms and then using linear regression methods. The simple transformations required for this approach are shown in Table 3.12. a' and b' are just the fitted linear regression values. This is straightforward to implement (for example in a spreadsheet), but, as pointed out in Lowe and Mohrman (1984), is not always appropriate, and so other fitting methods can also be used such as weighted or unweighted least squares. Modifications will be needed when there is negative development in the latter stages of the triangle (i.e. when the link ratio estimators are less than one), and so the tail factor is likely to be less than 1.

Table 3.12 *CL method: Linear regression formulae for curve fitting*

Type	Log	y	x	Fitted estimator
Exponential Decay	$\ln(g(t) - 1)$ $= \ln(a) - t \ln(b)$	$\ln(\hat{f}_j) - 1$	j	$1 + \exp(a' + b' \times j)$
Inverse Power	$\ln(g(t) - 1)$ $= \ln(a) - b \ln(t)$	$\ln(\hat{f}_j) - 1$	$\ln(j)$	$1 + \exp(a') \times j^{b'}$

A number of the proprietary reserving software packages enable these and other curves to be fitted to the data for both increasing and decreasing claims development.[5] Some of them provide a range of different fitting algorithms and statistical goodness of fit measures, as well as graphs of the actual and fitted values. The "R" software (R Core Team, 2013) also has many curve fitting procedures that can be used, and this is the software used in the worked example in Section 3.2.8.

When fitting any type of curve to a particular group of link ratio estimators, the results can vary considerably depending on which estimators are included in the fit. If the main purpose of fitting the curve is to estimate a tail factor (rather than to smooth the estimators in certain parts of the development), then fitting the curve only to the estimators towards the end of the data triangle may be appropriate. In doing so, it will be important to use enough data points, though, to avoid over-fitting to the data.

Once a curve has been fitted to the estimators, as well as being used to estimate the tail factor, it can also be used to provide a smoothed pattern of link ratio estimators at chosen parts of the development curve. For example, if one particular link ratio estimator is judged to be an outlier, then this could possibly be replaced by the appropriate value from the fitted curve, rather than let it unduly influence the results. Since the curve will be continuous, it can be also be applied to cohorts where the development data is evaluated at different development points to the cohorts that have been used to derive the link ratio estimators. This can be an alternative to interpolating between the observed estimators.

It is also possible to fit curves through the claims development data itself to derive a tail factor, rather than through the link ratio estimators. An example of such an approach is the "Craighead" curve, which is an exponential-type curve that, for example, was used in the 1980's for certain types of London Market business. Further details are given in Craighead (1979) and Benjamin and Eagles (1997). Curves with other functional forms can also be used in practice, depending on the observed claims development pattern.

When fitting a curve to either the data or the link ratio estimators, if this is being used mainly to smooth the pattern, then the goodness of fit (visually and/or statistically) in the relevant part of the development will be the main selection criteria. Whilst this will also be important when the curve is being used to estimate the tail factor, the value of the tail factor itself will obviously also be important. For each curve fitted to the link ratio estimators, the tail factor will be derived by taking the product of the fitted values for development periods beyond the data triangle. The period taken for the values to converge to 1 will influence the resulting tail factor, and in some cases it may be necessary to only use fitted values up to a selected development point. Applying other approaches, such as benchmarking and graphical examination of the claims development using alternative tail factors, can help with the judgement that is made in this part of the curve fitting process.

Curve fitting is discussed further in Herman et al. (2013), Sherman (1984) and Boor (2006).

---

[5] However, the functional form of the similarly named, but equivalent curve types might be different in these packages to that shown in Table 3.10.

Table 3.13 *CL method: Example calculation of scaled development for use in graphs – Claims Paid*

Accident yr	1	2	3	4	5	6
2009	11.4%	29.3%	54.9%	74.8%	87.8%	100.0%
2010	9.4%	31.7%	51.9%	71.0%	87.8%	
2011	9.2%	27.0%	50.6%	72.9%		
2012	6.6%	26.0%	52.3%			
2013	7.8%	28.3%				
2014	8.7%					

3.2.7 Using Graphs as part of the Chain Ladder Method

As for all reserving methods, the use of judgement is a vital part of every stage of the application of the CL method, and whilst this can be achieved through close scrutiny of the numerical schedules, graphs can also be used to supplement this. The types of graph that can be used are similar to those described in Section 2.5.5, where it was explained that graphs can be helpful even before any reserving methods are applied to the data. Once the CL method has been applied to the data to produce estimated ultimates, both the individual cohort graphs and the combined cohort graphs can be used, but with the development being shown against the estimated ultimate values. If the example given in Section 3.2.2 is used, with the CSA ultimates shown in Table 3.8 being selected, then a number of different graphs can be used to scrutinise the results. Some of these involve scaling the claims development (paid or incurred) for each cohort to its selected ultimate. An example of the straightforward calculation required for this is given in Table 3.13, which shows the cumulative claims paid development, scaled to ultimate. Similar calculations can be done to produce incremental scaled graphs, as referred to in Section 2.5.5.

As an illustration of the calculation, consider the 2012 accident year, which has an estimated ultimate, taken from Table 3.8, of 5,473. At development year 1, the Claims Paid (taken from Table 3.6) were 359, giving a scaled proportion developed of 6.6%, as shown in Table 3.13. Similarly, at development year 2, the Claims Paid were 3,010, so the figure becomes 26%. All other cohorts are scaled by their selected ultimates. Any other graph shown here that depicts the development as a percentage of ultimate will use an equivalent calculation, which allows all cohorts to be placed on the same graph and the development patterns compared. This type of graph is referred to subsequently as "Scaled Paid" or "Scaled Incurred" etc. depending on the type of data being depicted. Such labelling may or may not also refer to whether the data being scaled is incremental or cumulative.

The graph of the cumulative data in Table 3.13 is shown in Figure 3.2.

Although it is difficult to distinguish between the lines representing the different accident years on this graph, this suggests that the development pattern is very consistent between the accident years (and in any case in practice such graphs would often be viewed using software that enables individual years to be identified). It also shows that the Claims Paid for the oldest year, 2009, are still developing quite strongly in the sixth development year, implying

Table 3.14 *CL method: Example Claims Incurred data*

Accident yr	1	2	3	4	5	6
2009	764	2,317	2,918	3,541	3,874	3,912
2010	695	2,451	3,284	3,879	4,155	
2011	905	2,797	3,540	4,435		
2012	995	3,199	4,185			
2013	908	3,412				
2014	1,013					

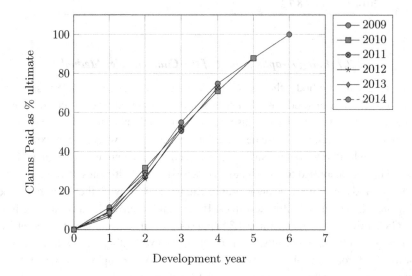

Figure 3.2 Example scaled graph: Cumulative Claims Paid (all years)

that a tail factor is necessary, otherwise the ultimate claims for this and all subsequent years derived using the CL method will be understated.

Table 3.14 shows the Claims Incurred data for this dataset.[6]

If this data is also scaled to the same ultimates as for the Paid graph, then the graph in Figure 3.3 is produced.

As for the Paid graph, this shows a reasonably consistent pattern for all accident years. It is shorter-tailed than the Claims Paid graph, which is to be expected. However, this graph also shows that the development for both the 2013 and 2014 accident years is positioned above the other years. This results from the fact that the estimated ultimate claims that have been used to scale this graph were based on a Claims Paid data triangle and in addition that the ratio between the Paid and Incurred claims for the 2013 and 2014 accident years is lower than the other accident years at their respective latest development points. This can be seen in Table 3.15, which shows this ratio for each cohort at each development point,

---

[6] Although the Claims Paid data for this example is based on "real" data, taken from Taylor and Ashe (1983) (but with amended cohort labels), the Claims Incurred data has been created solely for this book to illustrate how graphs can be used as part of the reserving process.

Table 3.15 *CL method: Paid to Incurred ratio for Example*

Accident yr	1	2	3	4	5	6
2009	58.0%	49.0%	72.9%	81.8%	87.8%	99.0%
2010	57.0%	54.4%	66.4%	77.0%	88.8%	
2011	48.7%	46.1%	68.4%	78.5%		
2012	36.1%	44.4%	68.5%			
2013	41.5%	40.0%				
2014	34.0%					

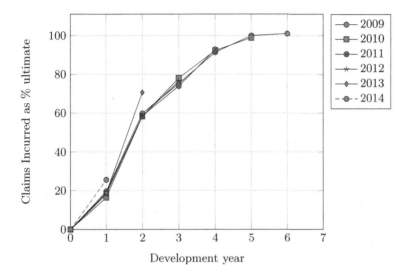

Figure 3.3 Example scaled graph: Cumulative Claims Incurred (all years)

using the cumulative data. This shows that, at development point 1, the average ratio for cohorts prior to 2014 is 48.3%, whereas for 2014 it is only 34%. Similarly, at development point 2, the average ratio for cohorts prior to 2013 is 48.5%, compared to 40% for 2013.

If the CL method were instead applied to the Incurred Claims and the resulting estimated ultimate claims were used to produce a similar scaled graph, then the 2013 and 2014 accident years in this example would not be expected to sit above the other accident years, and the corresponding estimated ultimate claims for these years would be higher than those produced using the Claims Paid data (other things being equal). This does not necessarily mean that the estimated ultimate claims derived using the Claims Incurred data are more appropriate than those using the Claims Paid data, but it suggests that a judgement needs to be made on what weight should be assigned to the two sets of ultimates when selecting the final results. The topic of using these graphs to scrutinise the results in this way, as part of a results selection process, is discussed further in Section 5.8.

The next two graphs in this section are shown in Figures 3.4 and 3.5. These are examples of single cohort graphs, which show the Claims Paid, Incurred and Outstanding development

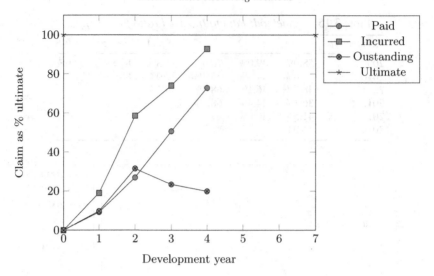

Figure 3.4 CL method – Example scaled graph: Single Year – 2011

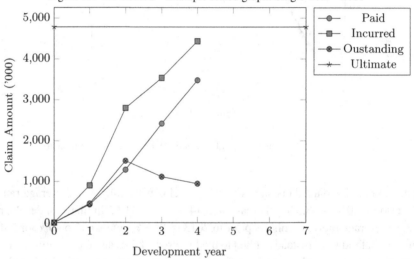

Figure 3.5 CL method – Example unscaled graph: Single Year – 2011

all on the same graph. The only difference between them is that the first graph has the development data scaled to the selected ultimate claims for that cohort, which is shown at the 100% point, whereas the second graph shows the unscaled data, and hence has the ultimate shown at its absolute level. The unscaled graph can also be shown in Loss Ratio format, if preferred. These graphs can be used to review the development pattern of individual cohorts, and to assess the reasonableness of a particular estimated ultimate claim value against that development. It is a matter of personal preference whether the scaled,

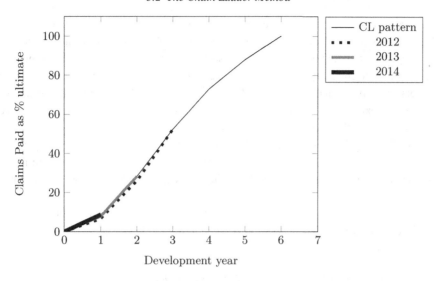

Figure 3.6 Example scaled graph: Cumulative Claims Paid with CL pattern

unscaled or loss ratio graphs are used for individual cohorts, as the development pattern is obviously the same. The use of unscaled graphs where more than one cohort is displayed can also be beneficial, although the ultimate values are typically not also shown on the same graph.

The final graph that is shown here is in Figure 3.6. This is the same as the Scaled Paid graph in Figure 3.2, except that the CL pattern has been added to the graph. This is the % developed pattern implied by the CL method. Only 2012 to 2014 cohorts are shown in this example, merely so that it is easier to see the fitted pattern. The end point of the line for each cohort will, by definition, lie on the CL pattern curve. These graphs can be helpful in comparing the development across the different cohorts with each other and with the CL pattern, and also for visualising the expected future development implied by the CL method. This example also shows a slightly different format of graph to the others included in this section, merely for illustration purposes; this will obviously be a matter of personal preference. In practice they will usually be viewed in colour in reserving software.

Further discussion of the use of graphs as part of the reserving process is included in Section 5.8.

3.2.8 Worked Example of the Chain Ladder Method

The data triangle used to explain the CL method in the earlier parts of this section was intentionally relatively small, for ease of presentation. This section summarises the results of applying the method to an expanded version of the same dataset. The data is taken from Taylor and Ashe (1983) and will be used in the worked examples for a number of the other methods in this chapter and for many of the stochastic methods in Chapter 4, to facilitate comparison of the results between methods.[7] The claims paid triangle is shown in Table 3.16, with the accident year column being labelled "Cohort", to indicate that in theory other definitions than accident year could also be used to construct the data triangle. All figures are in thousands unless stated otherwise.[8] For the purpose of this worked example, the cohorts have been labelled 2005 to 2014 inclusive and the development periods from 0 to 9. The indexing of the development periods starts at 0 rather than 1 to be consistent with the generic mathematical notation used for data triangles in this book, as explained in Section 2.5.6. The indexing for the development periods that is used in any particular reserving exercise will be based, for example, on the practice used at the relevant insurer; it does not affect the underlying details of the application of the CL or other reserving methods.

The data triangle used in the explanation of the triangle concept in Section 2.3 and in the initial explanation of the CL method in Section 3.2.2 relates only to the 2009 to 2014 cohorts of Table 3.16.

Table 3.16 *Chain Ladder example: Taylor and Ashe dataset – Cumulative claims paid*

Cohort	\multicolumn{10}{c}{Development year}									
	0	1	2	3	4	5	6	7	8	9
2005	358	1,125	1,735	2,218	2,746	3,320	3,466	3,606	3,834	3,901
2006	352	1,236	2,170	3,353	3,799	4,120	4,648	4,914	5,339	
2007	291	1,292	2,219	3,235	3,986	4,133	4,629	4,909		
2008	311	1,419	2,195	3,757	4,030	4,382	4,588			
2009	443	1,136	2,128	2,898	3,403	3,873				
2010	396	1,333	2,181	2,986	3,692					
2011	441	1,288	2,420	3,483						
2012	359	1,421	2,864							
2013	377	1,363								
2014	344									

Table 3.17 *CL example for Taylor and Ashe data: Individual link ratios*

Cohort	\multicolumn{9}{c}{Development year}								
	0	1	2	3	4	5	6	7	8
2005	3.143	1.543	1.278	1.238	1.209	1.044	1.040	1.063	1.018
2006	3.511	1.755	1.545	1.133	1.084	1.128	1.057	1.086	
2007	4.448	1.717	1.458	1.232	1.037	1.120	1.061		
2008	4.568	1.547	1.712	1.073	1.087	1.047			
2009	2.564	1.873	1.362	1.174	1.138				
2010	3.366	1.636	1.369	1.236					
2011	2.923	1.878	1.439						
2012	3.953	2.016							
2013	3.619								

---

[7] The data in this source is separated into number of claims finalised and payment per claim finalised. The triangle used in the examples here is derived from the product of these items to give a triangle of claim amount finalised – that is, a claim amount settled triangle. This is treated as a claims paid triangle, without loss of generality.

[8] The calculations have been applied to the original data in units, rather than in thousands, and so some of the results shown here will not agree exactly with those using the data in thousands.

First, the individual link ratios are calculated, as shown in Table 3.17, with the entries in each column representing the link ratios from the development year for that column to the next development year, for each cohort.

In practice, these will usually be examined to consider whether there are any obvious outliers that might justify the exclusion of the underlying data before moving to the next stage of deriving the link ratio estimators. This can also be done graphically, bearing in mind that each individual link ratio is equivalent to the slope of the claims development pattern for the relevant cohort in the corresponding development period. Some specialist reserving software has automated procedures (which may be user-adjustable) that can be used to exclude certain obviously very high or low individual link ratios. Where any such exclusions are used in practice, in some cases further investigation of the cause of any obvious outliers may sometimes need to be investigated, as well as exploring the impact on the link ratio estimators (derived in the next stage of the process) and hence on the results of retaining or excluding such outliers. Exclusion of data in respect of certain cohorts and development periods may in some cases be inappropriate if it leads to removal of a genuine feature of the claims development that may affect other cohorts and hence is something that needs to be allowed for in deriving the reserve estimates. This stage of the process may therefore require careful judgement by the practitioner. For the purposes of this worked example, no exclusions will be applied.

Following the consideration of whether any data is to be excluded from the remaining stages, a selection of different estimators are derived, as shown in Table 3.18. This shows the Column Sum Average (CSA) and the Arithmetic Average (AM) – first across all cohorts and then using the most recent three cohorts in each ratio, denoted by CSA3 and AM3. Table 3.18 shows both the incremental estimators (i.e. just the factor between successive columns) and the cumulative estimators (i.e. the factor between a particular column and ultimate).

Table 3.18 *CL example for Taylor and Ashe data: Link ratio estimators*

	Link ratio estimator							
	CSA		CSA3		AM		AM3	
Period	Incr'l	Cum've	Incr'l	Cum've	Incr'l	Cum've	Incr'l	Cum've
$0 \to 1$	3.491	14.447	3.460	14.108	3.566	14.817	3.498	14.314
$1 \to 2$	1.747	4.139	1.847	4.077	1.746	4.155	1.843	4.092
$2 \to 3$	1.457	2.369	1.392	2.208	1.452	2.380	1.390	2.220
$3 \to 4$	1.174	1.625	1.154	1.586	1.181	1.639	1.161	1.597
$4 \to 5$	1.104	1.384	1.085	1.375	1.111	1.388	1.088	1.375
$5 \to 6$	1.086	1.254	1.097	1.267	1.085	1.249	1.098	1.265
$6 \to 7$	1.054	1.155	1.054	1.155	1.053	1.151	1.053	1.151
$7 \to 8$	1.077	1.096	1.077	1.096	1.075	1.094	1.075	1.094
$8 \to 9$	1.018	1.018	1.018	1.018	1.018	1.018	1.018	1.018
$9 \to$ Ult	1.000	1.000	1.000	1.000	1.000	1.000	1.000	1.000

The tail factor is initially assumed to be 1. Since the details of the calculation used for the CSA estimator have already been covered in the example in Section 3.2.2 (albeit on a

Table 3.19 *CL example for Taylor and Ashe data: Results*

Cohort	Age	Ultimate Claims				Reserve			
		CSA	CSA3	AM	AM3	CSA	CSA3	AM	AM3
2005	9	3,901	3,901	3,901	3,901	0	0	0	0
2006	8	5,434	5,434	5,434	5,434	95	95	95	95
2007	7	5,379	5,379	5,370	5,370	470	470	461	461
2008	6	5,298	5,298	5,283	5,283	710	710	695	695
2009	5	4,858	4,908	4,838	4,899	985	1,034	965	1,026
2010	4	5,111	5,075	5,125	5,078	1,419	1,383	1,433	1,386
2011	3	5,661	5,525	5,710	5,563	2,178	2,042	2,227	2,079
2012	2	6,785	6,325	6,818	6,359	3,920	3,460	3,954	3,494
2013	1	5,642	5,558	5,664	5,578	4,279	4,195	4,301	4,215
2014	0	4,970	4,853	5,097	4,924	4,626	4,509	4,753	4,580
Total		53,039	52,256	53,241	52,389	18,681	17,898	18,883	18,031

smaller dataset) this is not repeated here. Example calculations for the other incremental estimators are as follows:

$$\textbf{CSA3 for } 3 \to 4 = [4{,}030 + 3{,}403 + 3{,}692]/[3{,}757 + 2{,}898 + 2{,}986] = 1.154$$

$$\textbf{AM for } 3 \to 4 = \text{Average } (1.238, 1.133, \ldots, 1.174, 1.236) = 1.181$$

$$\textbf{AM3 for } 3 \to 4 = \text{Average } (1.073, 1.174, 1.236) = 1.161.$$

The cumulative estimators are then derived by multiplying the factors at and after each column, to derive the estimator to ultimate for that column. For example, the cumulative factor for the AM estimator from $6 \to$ Ult is $1.053 \times 1.075 \times 1.018 = 1.151$.

The estimated ultimate claims for each type of estimator can then be determined for each cohort by first determining the "age" of the cohort, and then mutliplying the claims from the leading diagonal of the triangle for that cohort by the relevant cumulative estimator at that development age. The results are shown in Table 3.19, which also shows the implied reserves. So, for 2011, which is at development point 3, the estimated ultimate claims using CSA3 will be $3{,}483 \times 1.586 = 5{,}525$. The reserve will be $5{,}525 - 3{,}483 = 2{,}042$.

In practice, at this stage, if a decision had not been made earlier in the application of the CL method regarding which type of estimator to use, then a judgement would need to be made, perhaps partly based on using the types of graph explained in Section 3.2.7. For the purpose of this worked example, it is assumed that the CSA estimator has been selected.

Before the next step (determining if a tail factor is required) is described, it is helpful to illustrate how the CL method can be used to derive the full projected triangle and hence estimates of the future incremental claim movements. These are shown in Tables 3.20 and 3.21 respectively. The first of these shows the cumulative data triangle above the stepped diagonal line, and the projected cumulative data below it, using the same approach explained for the simplified example in Section 3.2.2. The second table shows the incremental version of this triangle, which can be used to derive the total projected incremental claims movement in each future period. As this example uses a paid claims triangle, this movement is effectively the projected cash flows in each future period. If the reserves were being discounted for the time value of money, then these cash flows would be a key input to the relevant calculation, as discussed further in Section 6.7. The values are derived by summing across each diagonal, and are shown in the final column of the table. Hence, the value of 5,227 shown in this column is the sum of the values in the first diagonal below the stepped line (i.e. $95 + 376 + 247 + \cdots$), with subsequent values being the sum across each of the remaining diagonals. The sum of the cash flow values in all diagonals is the total estimated future claims (prior to allowance for any tail factor) and is equal to the reserve of 18,681.

Table 3.20 *CL worked example: Full projected triangle. CSA estimator. No tail factor*

	Development Year									
Cohort	0	1	2	3	4	5	6	7	8	9
2005	358	1,125	1,735	2,218	2,746	3,320	3,466	3,606	3,834	3,901
2006	352	1,236	2,170	3,353	3,799	4,120	4,648	4,914	5,339	5,434
2007	291	1,292	2,219	3,235	3,986	4,133	4,629	4,909	5,285	5,379
2008	311	1,419	2,195	3,757	4,030	4,382	4,588	4,835	5,206	5,298
2009	443	1,136	2,128	2,898	3,403	3,873	4,207	4,434	4,774	4,858
2010	396	1,333	2,181	2,986	3,692	4,075	4,427	4,665	5,022	5,111
2011	441	1,288	2,420	3,483	4,089	4,513	4,903	5,167	5,562	5,661
2012	359	1,421	2,864	4,175	4,901	5,409	5,876	6,193	6,667	6,785
2013	377	1,363	2,382	3,472	4,075	4,498	4,887	5,150	5,544	5,642
2014	344	1,201	2,098	3,058	3,590	3,962	4,304	4,536	4,883	4,970

Table 3.21 *CL worked example: Incremental projected triangle and projected cash flows*

	Development Year										
Cohort	0	1	2	3	4	5	6	7	8	9	Cash flow
2005	358	767	611	483	527	574	146	140	227	68	–
2006	352	884	934	1,183	446	321	528	266	425	95	5,227
2007	291	1,002	926	1,017	751	147	496	280	376	94	4,179
2008	311	1,108	776	1,562	272	352	206	247	370	92	3,132
2009	443	693	992	769	505	471	334	227	339	85	2,127
2010	396	937	847	805	706	383	352	238	357	89	1,562
2011	441	848	1,131	1,063	606	425	389	264	396	99	1,178
2012	359	1,062	1,443	1,310	726	509	467	317	474	118	744
2013	377	987	1,019	1,090	604	423	388	263	394	98	446
2014	344	857	897	960	532	373	342	232	347	87	87
									Total		18,681

Curve Fitting for Worked Example

Although it might be thought to be reasonably clear from observing the pattern of estimators in Table 3.18 that a tail factor is required, using graphs can also be helpful. In
particular, viewing the Scaled Paid graph shown in Figure 3.7 does seem to suggest that the
development for the oldest cohort is still continuing slightly upwards, and so a small tail
factor may be necessary.

If Exponential and Inverse Power curves are fitted to the Incremental CSA estimators
then the fitted values are as shown in Table 3.22.

This shows the results of fitting both curves using linear regression, by applying the
transformations given in Section 3.2.6. The results of using a Least Squares method for
fitting the Exponential curve are also shown (with zero weight applied to the first three link
ratio estimators to illustrate how this can influence the curve fitting).

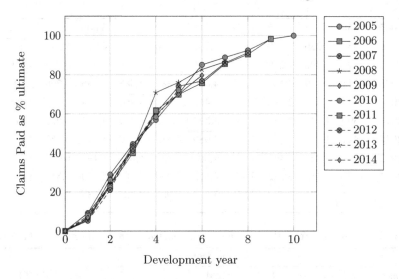

Figure 3.7 CL worked example: Scaled Paid graph

The final row in this table is the tail factor for each curve. Whilst the Exponential
curve using both fitting methods has virtually converged to an incremental factor of 1 by
development point 16, the Inverse Power curve has not; however, it has been capped at this
point to derive the tail factor that is shown, for the purpose of this example.

These results were produced using the R software, and example code for doing this
is given in Appendix B.2.1. The linear regression results in particular can also easily be
produced using a spreadsheet.

A graph can be drawn of the fitted curves, as shown in Figure 3.8. This covers the
development points from 1 to 14, which results in it being difficult to observe how well the
different curves fit the data in the tail. Hence, Figure 3.9 shows the same data, but just for
the tail area of the development. This approach of "zooming" in on user-defined areas of the
graph is possible in some of the proprietary reserving software packages that are available.

Table 3.22 *CL worked example: Curve fitting*

Period	Actual	Exp'l	Inv Pwr	Expo LS fit
0 → 1	3.491	2.366	4.023	1.441
1 → 2	1.747	1.807	1.735	1.318
2 → 3	1.457	1.477	1.322	1.229
3 → 4	1.174	1.281	1.179	1.165
4 → 5	1.104	1.166	1.114	1.119
5 → 6	1.086	1.098	1.078	1.086
6 → 7	1.054	1.058	1.057	1.062
7 → 8	1.077	1.034	1.044	1.045
8 → 9	1.018	1.020	1.034	1.032
9 → 10		1.012	1.028	1.023
10 → 11		1.007	1.023	1.017
11 → 12		1.004	1.019	1.012
12 → 13		1.002	1.016	1.009
13 → 14		1.001	1.014	1.006
14 → 15		1.001	1.012	1.005
15 → 16		1.001	1.011	1.003
9 → Ult		1.029	1.129	1.077

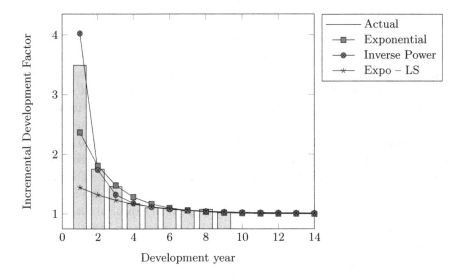

Figure 3.8 CL worked example: Fitted curves

A visual inspection of Figure 3.9 suggests that whilst the Inverse Power curve appears to provide a better fit between development points 5 and 9, it perhaps does not tail off fast enough, and so the Exponential might be preferred. The Least Squares fitting method was applied excluding the first three link ratio estimators from the fitting process.[9] This has the effect, as seen in Figure 3.9, of flattening the Exponential curve in the tail area, and so

---

[9] See the R code in Appendix B.2.1 for how this was implemented.

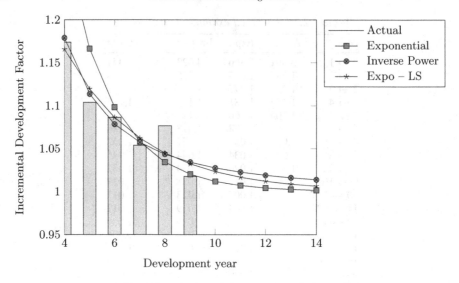

Figure 3.9 CL worked example: Fitted curves tail area

producing a higher tail factor than the linear regression fit, which included all estimators in the fitting process. As well as visual aids such as these graphs, more formalised statistical goodness-of-fit measures can be used, and will often be produced automatically by the software that is used to fit the curves to the link ratios. For example, the R software code in Appendix B.2.1 produces goodness-of-fit measures. Whilst these measures are useful for comparing different curves, the selection of a curve (and hence tail factor and resulting ultimate claims) may also need to place reliance on other criteria, as described in this section.

For example, to help choose between the ultimate claims implied by the different tail factors, the alternative results using the different curves can be analysed using Scaled Paid development graphs. For example, Figure 3.10 shows the development data for all cohorts, scaled to the ultimates derived by using the Linear Regression Exponential tail factor (from Table 3.22) of 1.029. Hence, the horizontal line at 100 on this graph effectively represents these ultimates. If instead, the development were scaled to the Inverse Power ultimate, this line would appear as shown on this graph at approximately 110.[10] One reading of this graph would be that the Inverse Power ultimate is too high, unless the claims development is expected to reverse its recent pattern of stabilising, whereas the Exponential Ultimate does not appear unreasonable. To keep the graph simple, the ultimate derived from the Exponential tail using a Least Squares fit is not shown, but it should be obvious that it would be between the two ultimates shown in Figure 3.10.

As noted in Section 3.2.4, regardless of whether a tail factor is required, curve fitting can also be used to smooth the link ratio estimators. This can be useful if the data is sparse at certain development points or there are other anomalies in the data that are deemed not representative of the overall development pattern. For this example, Table 3.18 shows a

---

[10] 110 is derived from the ratio of the tail factors of 1.129 and 1.029. Of course, 110 would become 100 if the data was scaled to the Inverse Power ultimate, but the gap between the development data and the ultimate would be exactly the same, hence this particular graph can also be used to judge the reasonableness of the Inverse Power ultimate.

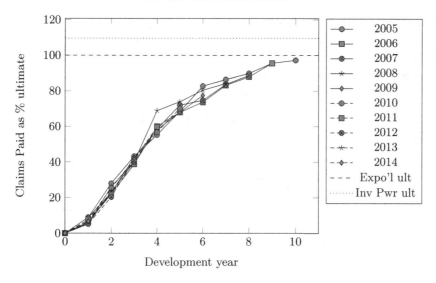

Figure 3.10 CL worked example: Scaled paid graph after adding tail factor

reducing pattern of estimators up to development point 7, but the next estimator (i.e. $7 \to 8$) is then higher. This is also clearly seen graphically in Figure 3.9.[11] In practice, the cause of this may need to be considered further, by investigating the underlying claim movements, and bearing in mind the fact that this estimator is based only on two cohorts and hence is subject to some uncertainty. If this was judged by the practitioner to be a genuine outlier that was distorting the results, then one option would be to replace the actual incremental CSA link ratio estimator from $7 \to 8$, of 1.077, with, say, the Exponential curve value of 1.034. The cumulative link ratio estimators would then be recalculated using this value. To illustrate the impact in this example of using a tail factor and of smoothing the estimators in this way, Table 3.23 shows the implied cumulative link ratio estimators and % developed using the following three options:

Table 3.23 *CL worked example: Alternative options for tail factor and smoothing*

		CSA		CSA+TF		CSA+TF+Smooth	
Cohort	Age	Estim'r	% dev'd	Estim'r	% dev'd	Estim'r	% dev'd
2005	9	1.00	100%	1.029	97%	1.029	97%
2006	8	1.02	98%	1.047	95%	1.047	95%
2007	7	1.10	91%	1.127	89%	1.083	92%
2008	6	1.15	87%	1.188	84%	1.141	88%
2009	5	1.25	80%	1.291	77%	1.240	81%
2010	4	1.38	72%	1.425	70%	1.369	73%
2011	3	1.63	62%	1.672	60%	1.607	62%
2012	2	2.37	42%	2.437	41%	2.341	43%
2013	1	4.14	24%	4.259	23%	4.091	24%
2014	0	14.45	7%	14.866	7%	14.281	7%

---

[11] Note that the ratio $7 \to 8$ is shown in this graph at the histogram bar labelled 8.

1. Original CSA estimators without tail factor or smoothing.
2. CSA estimators with Exponential tail factor (using regression) of 1.029 and no smoothing of development pattern.
3. CSA estimators with Exponential tail factor (using regression) of 1.029 and Exponential curve used at for $7 \rightarrow 8$, to smooth the pattern.

The resulting ultimate claims and reserves using these three approaches are then shown in Table 3.24.

Table 3.24 *CL worked example: Ultimate claims and Reserves for different options*

Cohort	Ultimate claims			Reserve		
	CSA	CSA+TF	CSA+TF +Smooth	CSA	CSA+TF	CSA+TF +Smooth
2005	3,901	4,015	4,015	0	113	113
2006	5,434	5,591	5,591	95	252	252
2007	5,379	5,535	5,317	470	625	408
2008	5,298	5,452	5,237	710	863	649
2009	4,858	4,999	4,803	985	1,126	929
2010	5,111	5,259	5,053	1,419	1,568	1,361
2011	5,661	5,825	5,596	2,178	2,342	2,113
2012	6,785	6,982	6,707	3,920	4,117	3,843
2013	5,642	5,806	5,578	4,279	4,443	4,214
2014	4,970	5,114	4,913	4,626	4,770	4,569
Total	53,039	54,577	52,810	18,681	20,219	18,452

A brief commentary in relation to Tables 3.23 and 3.24 follows:

- The CSA estimator column ("Estim'r") refers to the cumulative estimators and is the same as that shown in Table 3.18, except that the ordering is by cohort. The % developed is just the inverse of the cumulative estimator.
- The CSA with tail factor estimator (which is labelled as "CSA+TF" in the table) can be derived by applying the tail factor of 1.029 to the CSA estimator in the previous columns.
- The CSA with tail factor and smoothing estimator (which is labelled as "CSA+TF+ Smooth" in the table) can be derived by multiplying the previous estimator (i.e. CSA with tail factor) by the ratio of the original development factor $7 \rightarrow 8$ by the fitted curve value. This is $1.034/1.077 = 96\%$. It applies only for cohorts of age 7 or less only.
- The ultimate claims for the three different options shown in Table 3.24 are then derived by multiplying the leading diagonal for each cohort by the estimator, or dividing it by the % developed.
- As expected, the ultimate claims with a tail factor are equal to those in the previous column multiplied by the tail factor.
- The ultimate claims with tail factor and smoothing are 96% of those without smoothing for cohorts 2007 onwards, which are the only ones impacted by the amended link ratio estimator at $7 \rightarrow 8$.

- The reserves are derived by subtracting the claims to date from each ultimate. The gearing effect that this causes as a result of the tail factor is clear from the results, with the reserves (across all cohorts combined) including a tail factor being 8.2% higher than the reserve without a tail factor, compared with 2.9% for the ultimate claims. Similarly, there is a gearing impact on the reserves of using the smoothing estimator, with the reserves being nearly 9% lower than without the smoothing effect, compared to less than 4% lower on an ultimate basis.

In practice, a choice would need to be made between the three options for the ultimate claims, using a combination of numerical and graphical diagnostics. The choice could potentially vary by cohort and may involve using weighted averages across the different options. The results from using other reserving methods may also be factored into the final selections. Results selection is covered more generally in Chapter 5.

3.2.9 Inflation-adjusted Chain Ladder Method

This method does not represent a distinct approach, but rather an adaptation of the standard CL method. Although its name refers to inflation, in theory it can be used to adjust for any calendar period effect. For brevity it is referred to here as the ILCL method, but is intended to relate to any deterministic CL method for which explicit allowance is made for any type of calendar period effects, using the approach described here.

One of the limitations of the CL method, as noted in Section 3.2.5, is that its application can in some circumstances extrapolate calendar period trends in an inappropriate way. Where those trends can be estimated and used to produce an adjusted data triangle, the ILCL method then involves applying the standard CL method to the adjusted data, and modifying the implied reserve to make an allowance for assumed future calendar period trends. Implicit within this process is an assumption that all payments within a particular calendar period are subject to the same inflationary effects. When applying the method to a claims paid data triangle, the following steps are involved:

1. Derive the incremental data triangle.
2. Convert all diagonals so that they are on a common calendar period basis, usually the latest calendar period, which is represented by the leading diagonal. In other words, the past data is adjusted so that it is expressed in current values. So, for example, if the calendar period trend is estimated to be, say, an increase of 10% between the penultimate and latest diagonal, then all incremental values in the penultimate diagonal are increased by 10%. In practice, a suitable historical inflation index might be used. This produces an adjusted data triangle.
3. Re-accumulate the adjusted data triangle.
4. Apply the standard CL method to the adjusted data triangle. All of the considerations mentioned in the description of the standard CL method, including those in Section 3.2.10, apply equally well here.

5. Apply the CL pattern resulting from Step 4 to the adjusted data to derive the estimated future cash flow pattern. If a tail factor has been used, then an assumption will need to be made regarding the period over which the data develops beyond the data triangle.
6. Make an assumption for future calendar period trends.
7. Apply those trends to the cash flow derived from Step 5, to derive the calendar period trend-adjusted cash flows.
8. Sum the adjusted cash flows to produce the estimated reserve, using the assumed future calendar period trend.

A worked example is not included here, as the calculations should be self-explanatory from this description, and the application of the CL method to the adjusted data just follows the earlier worked example. In theory, the ILCL could be applied to incurred claims data, but there will be complications related to the fact that it may not be clear to what extent future inflationary or other calendar period trends have already been factored into the case reserve element.

In many situations, it can be difficult to identify historical calendar period trends due to inflation or other effects, because there will be many factors affecting the development of each cohort, which confound these trends. In some cases it is possible to use specific inflation indices to adjust the data, but unless there is firm evidence that these are appropriate and not confounded by other effects, the results of applying the ILCL method may not be obviously more reliable than using the standard CL method, perhaps with some specific adjustments to allow for material calendar period effects. Testing the impact of alternative assumptions of historical calendar period trends can help understand the sensitivity of results to these assumptions. Brydon and Verrall (2009) provides further discussion of calendar period effects and the CL method. Calendar period trends can also be allowed for in the context of stochastic claims reserving methods, which are covered in Chapter 4.

3.2.10 A Generalised Procedure for Applying the Chain Ladder Method in Practice

Because of its widespread and longstanding usage, the practical application of the CL method involves many variations. Nevertheless, it is possible to describe an overall procedure for its implementation, which can then be adapted for particular circumstances. The focus here is on the application to claim amount triangles, although many of the points will also apply when claim number triangles are being used.

It is assumed that the initial background and data related elements of the reserving process have already been carried out (as described in Section 1.2, with additional details given in Chapter 5). Those initial steps should, for example, have determined the reserving categories to be used and, for each such category whether any individual large claims need to be removed from the data, or whether any other adjustments need to be made. A decision will also have been made as to whether the CL method is to be applied to gross or net of reinsurance data, or to the reinsurance data itself, or to a combination of these options. For the purpose of this procedure, it is assumed that gross data is being used, but the procedure is broadly similar for all these data types. It is also assumed that the objective is to produce a single point estimate for the reserves, rather than a range, although the procedure below can also be used to produce a range of alternative estimates, which can assist in the quantification and communication of uncertainty. After the initial background and data steps, the procedure continues for each reserving category as follows:

1. Determine if the CL method is suitable for application to Paid or Incurred data or both. This could involve, for example, reviewing the limitations and assumptions listed in Section 3.2.5 and determining if any of them are relevant, and if so whether they can be overcome through adjustments to the standard procedure. The description here assumes the method is being applied to both types of data.

2. Even though the application of the CL method to Paid or Incurred claims data does not require premium or exposure data, these are useful for carrying out simple diagnostic checks at the end of the procedure, and so this data may be collected at this stage if possible. In some cases,[12] to derive the correct premium exposure figure for the later cohorts, it may be appropriate to apply the CL method to the premium development triangles, and then to compare the resulting estimated ultimate premiums with those derived from underwriting or finance sources. For the purpose of the remaining steps in this procedure it is assumed that an appropriate premium measure has been collated that is consistent with the cohort definition being used for the claims (e.g. if the cohort is accident year, then earned premiums are required).

3. For each of Paid or Incurred, continue as below.

4. Review the development numerically and/or graphically to assess the overall pattern, determine whether a tail factor is required, decide which is the most appropriate development frequency to use and identify any obvious data anomalies. The cohort definition will usually have been made at an earlier stage of the process.

5. Raise any queries arising from the previous step with the relevant staff (e.g. claims, underwriting and finance).

6. Based on the findings in the two previous steps, decide if a single CL pattern is appropriate, or whether there is more than one distinct pattern which would justify applying the CL method to specific groups of cohorts. The remaining steps assume that only a single pattern is required.

7. Also based on Steps 4 and 5, decide if any individual cells in the data triangle need excluding from the overall process and decide whether there are any cohorts where the CL method is not likely to produce reasonable results. For example, for some classes of business, this may apply to any cohorts that are less than, say, two to three years developed. If this is the case, then aspects such as the consideration of the exclusion of certain data points and the selection of link ratio estimators may only need to focus on the parts of the triangle that will impact on the results for the cohorts where the CL method is going to be used. However, even where it is decided that the CL method will not be used for some cohorts, it is usual to apply the CL method to all cohorts, but in the final results selection replace the CL ultimates for the relevant cohorts with those derived from applying other methods.

---

[12] For example, see the relevant London Market sections in Chapter 7.

8. Apply the CL method to the data triangle, possibly using more than one option for the estimators and for the tail factor estimation (e.g. if curve fitting is used, then alternative results using different types of curve might be produced). In some cases, alternative results might also be obtained by including or excluding individual cells.
9. This will typically produce a small number of alternative estimated ultimate values for each cohort, for one or both of the Paid and Incurred data triangles.
10. The next stage is to decide which of these, or what combination thereof, should be selected as the final selected CL method ultimate for each cohort. The range of alternative estimates may also be retained, however, for use in subsequent analysis and communication of uncertainty, as discussed in Section 5.9.
11. This selection process will involve using a range of diagnostic approaches, including analysis of numerical summary tables including ULRs, a graphical review of the results (using, for example, the types of graph outlined in Section 3.2.7), comparison with any ultimates estimated at previous as-at dates or estimated by other parties involved in the reserving exercise (e.g. external actuaries) and, if available, comparison of actual development with any previously estimated expected development (for each of Paid and Incurred, if relevant).

The above procedure represents an outline of one possible approach for implementing the CL. Practitioners will design their own approach, depending on the circumstances and features of the business being analysed, with many bespoke adjustments being used in practice. Examples of some of these are referred to in Chapter 7.

3.3 The Initial Expected Loss Ratio Method

Although this approach, abbreviated by IELR, is described as a reserving method and is used quite widely in practice, it merely involves selecting an assumption for the ultimate loss ratio (ULR)[13] for a cohort. The ultimate claims are then calculated by multiplying the ULR by the relevant premium figure. The IELR method may also be referred to as the "Expected claim ratio" method or just the "Loss ratio" method.

The ULR could be derived from external benchmark sources, underyling pricing analyses or a business plan assumption, or a combination thereof. In some cases, the IELR assumption for one or more recent cohorts could alternatively be based on the ULRs for earlier cohorts, which themselves could have been derived using other reserving methods. If this approach is used, then the ULRs for the earlier cohorts should preferably be adjusted so that they are appropriate as a basis for selecting the IELR for the recent cohort(s). This adjustment would typically allow for factors such as claims inflation, premium rates and any changes in the business between cohorts that might affect the loss ratios. The process can also be used for determining the prior assumption in the Bornhuetter–Ferguson method. This is described in Section 3.4 in the context of that method.

Unlike the Chain Ladder method, the ultimate claims derived using the IELR method for a particular cohort are independent of the observed claims data for that cohort. Typical

---

[13] A ULR is a ratio of ultimate claims to premium.

situations where it is used in practice include for more recent cohorts, particularly for longer-tailed classes of business, or where there is very limited or unstable development data, making many other methods unreliable. Some companies have a reserving policy or philosophy for one or more reserving categories, whereby they use the IELR method for a specified period of development, only adjusting it in this period if the claims experience suggests a materially different ULR. After that period, they switch to claim-based methods, such as the Chain Ladder.

A worked example of the IELR method is given in the Bornhuetter–Ferguson worked example, in Section 3.4.3.

The Expected Method

The IELR method is sometimes described as being a type of "Expected method". This is effectively a generalised version of the IELR method, whereby some form of initial estimate is determined, but that estimate can relate not just to claim amounts, but to other items such as claim numbers (or counts). In addition, when used on claim amounts, the derivation of the ultimate in the Expected method does not need to be based on using a ULR – any suitable approach can be used. For example, it could be based on multiplying an exposure measure by a selected factor, or by multiplying an expected number of claims by an expected average severity. When used with claim numbers, the ultimate in the Expected method could, for example, be based on multiplying an exposure measure by an estimated frequency. Further discussion of the Expected method is given in Friedland (2013).

In the context of reserving methods applied to claim amounts, the most common type of Expected method is believed to be the IELR method.

3.4 The Bornhuetter–Ferguson Method

This method, abbreviated here by BF, represents a blend between the purely claim-based approach of the CL method and the IELR method, which is independent of the claims experience. It is widely used in practice, particularly for more recent and hence less developed cohorts. The name of the method derives from the surnames of the authors who originally suggested it, in Bornhuetter and Ferguson (1972).

Once the assumptions for the method have been chosen, it is very straightforward to apply. It can be applied to both Claims Paid and Claims Incurred data, and to other data types such as claim numbers. The description here focuses on its application to claim amount data, but very similar principles apply for other data types. When applied to Claims Incurred data, three pieces of information are needed to derive the IBNR for a single cohort – the IELR, the ultimate premiums and the assumed % developed. When used in a BF context, the IELR is usually referred to as the "Prior Loss Ratio", "Prior ULR" or just "Prior", and can be derived using similar approaches to that given in the description of the IELR method. The concept of the % developed is explained in Section 3.2.2. It can be derived from a number of sources, most commonly from applying the CL method to the relevant data triangle or from a benchmark source, or a combination of these using the experience and judgement of the practitioner.

When applied to Claims Incurred, to derive the estimated ultimate claims using the BF method, the following steps are involved for an individual cohort:

1. Calculate the ultimate claims according to the prior. This is just the IELR multiplied by the premium, and is referred to here as the Prior Ultimate Claims.
2. Calculate the estimated "future" % developed, which will be 100% less the % developed. This is sometimes also referred to as the " expected % undeveloped" or the "expected % unobserved".
3. Derive the estimated IBNR as the future % developed from Step 2 multiplied by the Prior Ultimate Claims from Step 1.
4. Derive the Incurred BF ultimate claims by adding the Incurred Claims to date to the IBNR from Step 3.

This can be written as

$$\text{BF Ult(Inc)} = \text{Claims Incurred} + [1-\% \text{ devel'd}] \times [\text{Ult using prior}].$$

In other words, the BF ultimate is the Claims Incurred plus an estimated future proportion of the ultimate based on the prior assumption. When applied to Claims Paid, exactly the same steps are involved, except that the % developed must be appropriate for use with the Claims Paid, rather than Claims Incurred (e.g. it can be derived by applying a CL method to a Claims Paid triangle). The IBNR will instead then be the Claims Reserve and this will be added to the Claims – Paid in the final step to derive the Paid BF Ultimate claims. The $[1 - \%$ devel'd] component can be referred to as the "% Unreported" or "% Unpaid" for the Incurred and Paid BF methods respectively, so that, for example, the Paid BF may be written as

$$\text{BF Ult(Paid)} = \text{Claims Paid} + [\% \text{ Unpaid}] \times [\text{Ult using prior}].$$

The above description of the BF method demonstrates that it is effectively an approach which derives an estimate of the "future" claims (IBNR or Reserve, depending on whether Incurred or Paid data respectively is being used) that is independent of the claims to date. The ultimate claims under the BF method are of course dependent on the claims to date, but the element that is being estimated (i.e. the reserve or IBNR) is not. This means, for example, that if there has been one or more large claims which have caused the paid or incurred claims to be relatively high, then the IBNR or reserve produced by the BF method will not be impacted by the existence of these claims in the data. In contrast, the reserves derived from development factor methods (i.e. Chain Ladder type approaches) will be impacted by such claims. Hence, unlike with development factor methods, with the BF method there is therefore no need to consider whether the amounts relating to such large claims should be excluded from the incurred claims data before carrying out the required calculations. However, consideration may still be needed regarding whether the BF reserve makes appropriate allowance for both the known large claims as well as for other claims.

An alternative, but equivalent description of the ultimate derived using a BF method involves this being seen as a weighted average between an ultimate based on using the % developed assumption and one based on using the prior assumption, with the weight being the % developed. The ultimate based on using the % developed assumption for an Incurred BF method is just the Incurred Claims divided by the % developed. This description of the BF method can be written as

$$\text{BF Ult} = [\text{Ult from \% dev method}] \times [\text{Weight}]$$
$$+ [1 - \text{Weight}] \times [\text{Ult from using prior}].$$

This has an intuitive logic to it, since the more developed the cohort is, the higher the % developed assumption will be and hence the higher the weight, and so the more the BF ultimate will be influenced by the ultimate derived from the data alone, and less from the prior assumption. When the cohort is much less developed, the opposite will be true, with the BF ultimate being strongly influenced by the prior assumption. The equivalence of the two alternative descriptions is shown in the mathematical formulation of the method given in Section 3.4.1.

The weighted average version of the BF formula can be rearranged as

$$\text{BF Ult} = [\text{Ult from \% dev method}]$$
$$+ [1 - \text{Weight}] \times [\text{Ult from prior} - \text{Ult from \% dev'd}].$$

If the weight (which is the % developed) is less than one, which it almost certainly will be, then inspection of this version of the BF formula demonstrates that if the ultimate implied by the BF prior is larger than the ultimate implied by the % developed, then the BF ultimate will be larger than the ultimate derived from the % developed (i.e. the CL ultimate). In other words, the BF method will produce a higher ULR than the CL method (implied by the % developed), if the prior loss ratio assumption is higher than the CL ULR (and vice versa if the prior is lower). Where the % developed is greater than one (i.e. the cumulative development factor is less than one), this implies that the BF method is being used for a cohort that is expected to show a reduction in the relevant claims between its current position and ultimate. Whilst the BF method can theoretically be applied in such situations, its use is not thought to be widespread, with alternative, development-based methods, such as the CL, usually being preferred for the relevant cohorts. This topic is discussed further in Friedland (2013, Section 17.5) and Gluck (1997).

For some reserving categories, in the early stages of the development the claims from different cohorts can be quite variable, even if they all eventually reach similar ultimate loss ratios. If a Chain Ladder or development factor type approach is applied at these early stages, then this variable development will potentially have an undue influence on the implied ultimate values. In other words, particularly good (i.e. low) claims experience may produce unreasonably low estimated ultimate values, and particularly bad (i.e. high) claims experience the opposite. The BF method overcomes this limitation by placing weight on the prior assumption, but the reasonableness of the estimated ultimate is then critically dependent on the appropriateness of the prior. As the claims experience develops, it may become apparent that the prior assumption is not appropriate, and it will either need to be revised, or an alternative reserving method that relies more on the observed claims experience could be used. In certain reserving contexts, some practitioners have a rule of thumb which they use to decide when to use a BF method and when to use a Chain Ladder type approach. For example, this might be based on selecting a particular % developed (e.g. 40%, for illustrative purposes only), so that when the claims are estimated to be less developed than the chosen value, a BF method will be used and then after that a Chain Ladder method will be used. There are many situations where this approach might not be appropriate, and it will always be necessary to use judgement for each reserving category to decide which method(s) should be used for each cohort. This is discussed further in Section 5.7, which covers results selection.

The prior in the BF method does not necessarily need to be based on selecting a prior ULR – any suitable approach to deriving a prior ultimate claim value, such as referred to in the Expected method description in Section 3.3 is equally valid. Furthermore, the BF method can also be applied when estimating ultimate values for other items by replacing the claim amounts in the calculations described above with, for example, claim numbers, claim frequency or loss ratios.

A typical use of the BF method in conjunction with the CL method, applied to incurred claims data for an individual reserving category would involve first applying the CL method to the data to produce an initial set of ultimate claims for all cohorts, referred to here as the "CL ultimates". Then, for each cohort where a BF ultimate is required, proceeding as follows:

1. **Derive % developed assumption.** This is initially produced directly from the CL ultimates and may need some amendment using judgement, perhaps based on a review of the development factors or of the relevant development graphs. External benchmark sources may also be taken into account.
2. **Calculate Ultimate Loss Ratios using the CL ultimates.** These are called the CL ULRs here, and are derived for all cohorts in the data triangle.
3. **Adjust ULRs for use as BF prior.** This will usually involve adjusting the CL ULRs for all cohorts (except those where a BF ultimate is being derived) to allow for claims inflation and premium rate changes, and in some cases changes in any other relevant factors, such as business mix or legislation, or the impact of large losses or catastrophes. This is designed to make them appropriate for use as priors for the cohorts where a BF ultimate is required. The calculations involved are explained in the example that follows.
4. **Select final BF prior.** This can be based on either the adjusted BF priors from the previous step or from other external or business plan assumptions, or some combination thereof. Where the adjusted BF priors are used, a suitable group of cohorts needs to be chosen for any averaging that is done, along with suitable weights for each cohort and/or allowance for trends.

5. **Apply the BF method.** This is done either on a loss ratio or claim amount basis.
6. **Select final ultimate.** This can be based on either the BF ultimate or the CL ultimate, or a combination thereof.

This is a generalised procedure that can usually only be used for datasets that have a reasonable number of cohorts, otherwise there will not be sufficient history to derive the average priors from the CL ultimates, for use in the BF method. The procedure will inevitably have numerous adjustments in practice, depending on the circumstances. For example, in the final step, other methods may also be used and the results factored into the final selected ultimate for each cohort. This procedure is very similar to that used in the Cape Cod method to derive the prior assumption, as described in Section 3.6.

When using a BF method, the derived ultimate for an individual cohort may be materially higher or lower than that produced by one or more other reserving methods (e.g. Chain Ladder-type approaches). This can either be for a very good reason (e.g. unusually low or high claims experience to date causing these other methods to produce potentially unreasonable results), but in some circumstances it might be at least partly caused by an inappropriate prior assumption. If so, reconsideration of the basis for that assumption or appropriate allowance in the final result selection process may be necessary.

As for the CL method, stochastic versions of the BF method have been suggested, as referred to in Section 4.7.2, but they are not thought to be used very much in practice at present, compared to the deterministic version. The Bayesian approach described in Section 4.5 also provides a stochastic equivalent version of the BF method.

The Iterated BF Method

A development of the standard BF approach involves first applying the BF method in the normal way, and then using the resultant BF ultimate (or reserve) as a prior in a further application of the BF method. This can, if required, be repeated any number of times, with the estimated ultimate eventually equalling the CL ultimate. This approach has been suggested (in slightly different ways) by a number of different authors, as discussed further in Section 3.4.1, which also describes the mathematical formulation. As well as "Iterated BF", it has also been referred to as the "Benktander method" and the "Benktander–Hovinen" method.

The Iterated BF method can also be described using the alternative weighted average format for the BF method described in Section 3.4. First recall that this format involves the weight assigned to the BF prior as being the future % developed, with the CL ultimate having a weight equal to one minus the future % developed. The Iterated BF method involves using the same approach, but with the first iteration using the square of the future % developed in this same weighted average. Subsequent iterations use powers of 3, 4 and so on. This is explained further in the mathematical formulation of the method, as well as in the worked example of the Iterated BF method given at the end of Section 3.4.3.In practice, there will often not be any obvious basis for choosing between a standard BF method and an Iterated BF method. Furthermore, if an Iterated BF method is used it will also not be obvious how many iterations to select. For this reason, for cohorts where it is deemed appropriate to derive ultimates using both a CL and BF method, an alternative approach to the Iterative BF method is to derive the estimated ultimate values using a CL and BF method, and then combine the results in some way. For example, if the results from the two methods are relatively close, then there will be little to choose between them, and thus no need to consider whether the BF method needs to be iterated at all. If they are further apart, and one is not obviously preferable to the other, then judgement can be used to select a weighted average, rather than using the Iterated BF method. If necessary, the ultimate produced by this alternative approach can be used as the ultimate in a BF calculation, along with the original BF prior, to derive the implied weight that is being assigned to the CL ultimate and the BF prior. An example of the Iterated BF approach, along with this alternative approach is given in the BF and IELR worked example in Section 3.4.3. Finally, it is worth noting that the Iterated BF method is not thought to be widely used in practice, whereas the approach of making a judgement-based weighted selection or choice between a BF and CL ultimate for individual cohorts is probably more common.

3.4.1 Mathematical Formulation of the BF Method

As for the Chain Ladder model, a mathematical formulation[14] of the BF method can help understand what lies behind the approach, although as should be apparent from the introductory paragraphs on the BF method, such a formulation is not necessary for understanding and applying the method in practice. In Section 3.2.6, $\hat{C}_{i,J}$ was used to denote the estimated ultimate claims. To clarify that this is based on using the Chain Ladder method and to distinguish it from the ultimate claims derived using the BF method, the notation is changed here to $\hat{C}_{i,J}^{CL}$, so that, as per (3.6),

$$\hat{C}_{i,J}^{CL} = C_{i,I-i}\hat{f}_{I-i}\cdots\hat{f}_{J-1} = \frac{C_{i,I-i}}{\hat{\alpha}_{I-i}}. \tag{3.14}$$

Similarly, $\hat{R}_{i,J}$ from (3.7) is replaced by $\hat{R}_{i,J}^{CL}$.

The prior assumption used in the BF method for cohort i is denoted by U_i^{BF}, which could be derived, for example, by multiplying an assumed prior ULR by the relevant premium figure. The ultimate claims estimate derived using the BF method is denoted by $\hat{C}_{i,J}^{BF}$, with

$$\hat{C}_{i,J}^{BF} = C_{i,I-i} + U_i^{BF}\left(1 - \frac{1}{\hat{f}_{I-i}\cdots\hat{f}_{J-1}}\right)$$

$$= C_{i,I-i} + U_i^{BF}(1 - \hat{\alpha}_{I-i}). \tag{3.15}$$

The $(1 - \hat{\alpha}_{I-i})$ component of the formula in the second row effectively represents the estimated future proportion of the ultimate claims (or future % developed), so that the BF formula here is equivalent to the non-mathematical explanation of the BF method given in the introduction, namely that the BF ultimate is simply the claims to date plus the estimated

---

[14] Note that the formulation given here is intended to be an algebraic representation of the BF method, rather than a description of the method in a statistical or stochastic modelling context, which is discussed further in Section 4.7.2.

future proportion of the prior assumption. Given that the reserve (or IBNR if the method is being used with incurred claims rather than paid claims) will be calculated by subtracting $C_{i,I-i}$ from the ultimate, it should be clear that the BF reserve is independent of the claims paid/incurred to date. In mathematical terms, this is

$$\hat{R}_{i,J}^{\text{BF}} = U_i^{\text{BF}}(1 - \hat{\alpha}_{I-i}). \tag{3.16}$$

The approach to thinking of the BF method as a weighted average of an ultimate using the CL method and a prior assumption is expressed mathematically by first rearranging (3.14) to give

$$C_{i,I-i} = \hat{C}_{i,J}^{\text{CL}} \hat{\alpha}_{I-i}. \tag{3.17}$$

Then, putting this in (3.15) gives:

$$\hat{C}_{i,J}^{\text{BF}} = \hat{C}_{i,J}^{\text{CL}} \hat{\alpha}_{I-i} + U_i^{\text{BF}}(1 - \hat{\alpha}_{I-i}). \tag{3.18}$$

This has the weighted average format discussed in the introduction to the BF method, with the weight being the assumed proportion developed for the cohort. So, as this proportion increases, more weight is given to the ultimate derived from a chain ladder approach, and less to the prior assumption of the BF method. By formulating the method in this way it can be regarded as a type of simple Bayesian or credibility method, since it takes into account prior information and combines it with the known data to produce the estimated ultimate claims – i.e. the posterior in Bayesian terminology.

As mentioned in the general description of the BF method, it will give a higher or lower ultimate than the Chain Ladder ultimate, depending on the relationship between the Chain Ladder ultimate and the BF prior assumption. Specifically, it is straightforward to see from (3.18) that if $U_i^{\text{BF}} > \hat{C}_{i,J}^{\text{CL}}$ then the BF ultimate will be higher than the Chain Ladder ultimate.[15] In other words, if the BF prior assumption is greater than the pure Chain Ladder ultimate implied by the % developed assumption, then the BF ultimate will always be higher than the Chain Ladder ultimate (and the opposite if it is lower).

As noted in the description of the method at the start of this section, the BF method can be formulated using claim amounts, numbers or loss ratios. Mathematically, when expressed in loss ratio terms, the formulae are the same in substance, with all terms in (3.15) and (3.16) simply being divided by the appropriate premium figure corresponding to the claims cohort.

3.4.2 Mathematical Formulation of the Iterated BF Method

This variation on the standard BF method, introduced earlier in this section, can be formulated mathematically by first defining the future % developed as

$$\hat{\beta}_{I-i} = 1 - \hat{\alpha}_{I-i}.$$

---

[15] As long as $\hat{\alpha}_{I-i}$ is less than one, which, as noted in Section 3.4, is usually the case where the BF method is used.

Then, the weighted average version of the standard BF method given in (3.18) becomes

$$\hat{C}_{i,J}^{\text{BF}} = \hat{C}_{i,J}^{\text{CL}}(1 - \hat{\beta}_{I-i}) + \hat{\beta}_{I-i} U_i^{\text{BF}}. \tag{3.19}$$

The first iteration of the Iterated BF method puts $\hat{C}_{i,J}^{\text{BF}}$ as the prior in another BF calculation, to give

$$\hat{C}_{i,J}^{\text{BF2}} = \hat{C}_{i,J}^{\text{CL}}(1 - \hat{\beta}_{I-i}) + \hat{\beta}_{I-i} \left[\hat{C}_{i,J}^{\text{CL}}(1 - \hat{\beta}_{I-i}) + \hat{\beta}_{I-i} U_i^{\text{BF}} \right], \tag{3.20}$$

which can be simplified to

$$\begin{aligned} \hat{C}_{i,J}^{\text{BF2}} &= \hat{C}_{i,J}^{\text{CL}}(1 - \hat{\beta}_{I-i})(1 + \hat{\beta}_{I-i}) + \hat{\beta}_{I-i}^2 U_i^{\text{BF}} \\ &= \hat{C}_{i,J}^{\text{CL}}(1 - \hat{\beta}_{I-i}^2) + \hat{\beta}_{I-i}^2 U_i^{\text{BF}}. \end{aligned} \tag{3.21}$$

Hence, the weight assigned to the BF prior in this formula is the square of what it was in the original BF formula, so that for any value for the future % developed assumption (i.e. for $\hat{\beta}_{I-i}$) the Iterated BF will assign less weight to the original BF prior, and more to the CL ultimate.

The iteration can be continued, and it is relatively straightforward to show that, at the mth iteration, the formula for the ultimate will be

$$\hat{C}_{i,J}^{\text{BF}m} = \hat{C}_{i,J}^{\text{CL}}(1 - \hat{\beta}_{I-i}^m) + \hat{\beta}_{I-i}^m U_i^{\text{BF}}. \tag{3.22}$$

This will assign less and less weight to the original BF prior as the number of iterations increases, at any particular value for the future % developed. Similarly, as the number of iterations increases, the BF prior will have relatively low weight for successively higher values (and above) for the future % developed. In other words, where several iterations are used, even for cohorts which are at an early stage of development (so that the future % developed is high), the BF prior will be assigned relatively low weight in the iterated BF method.

As mentioned in Section 3.4, the Iterated BF method is also sometimes referred to as the Benktander method and the Benktander–Hovinen method. In particular, Mack (2000) used this name, citing Benktander (1976) as the source of the method. Mack (2000) also states that the method was described independently by two other authors, in Hovinen (1981) and Neuhaus (1992). Merz and Wüthrich (2008c) refers to the method as the Benktander–Hovinen method.

A number of sources analyse the Iterated BF method in a credibility context. They do so by considering the approach as a weighted average of the CL and BF results, with the weight being selected so that it satisfies certain optimality criteria. These sources include Neuhaus (1992), Mack (2000) and Hürlimann (2009).

3.4.3 Worked Example of the IELR and BF Methods

This section provides an example of the IELR and BF methods, first assuming that the BF priors have already been selected. Then, an example of an approach to deriving the priors for recent cohorts is given, using the results obtained by applying the CL method to earlier cohorts (as referred to in the general description of the BF method). Finally, an example of the Iterated BF method is given. All the examples use the same data as for the CL worked example given in Section 3.2.8.

Table 3.25 *BF and IELR worked example*

Cohort	Claims Paid	Prem	Prior ULR	Prior Ult	% dev'd	CL Ult	Paid LR	BF Ult	BF ULR	CL ULR
2005	3,901	5,333	75%	4,000	100%	3,901	73%	3,901	73%	73%
2006	5,339	7,333	75%	5,500	98%	5,434	73%	5,435	74%	74%
2007	4,909	6,875	80%	5,500	91%	5,379	71%	5,389	78%	78%
2008	4,588	6,875	80%	5,500	87%	5,298	67%	5,325	77%	77%
2009	3,873	6,111	90%	5,500	80%	4,858	63%	4,988	82%	79%
2010	3,692	5,914	93%	5,500	72%	5,111	62%	5,219	88%	86%
2011	3,483	6,316	95%	6,000	62%	5,661	55%	5,791	92%	90%
2012	2,864	6,316	95%	6,000	42%	6,785	45%	6,331	100%	107%
2013	1,363	6,000	100%	6,000	24%	5,642	23%	5,914	99%	94%
2014	344	5,714	105%	6,000	7%	4,970	6%	5,929	104%	87%

The basic calculations are straightforward, with the key figures being given in Table 3.25. The BF and IELR methods are applied to all cohorts for illustrative purposes only. Although in practice the results from these two methods might be used more in the recent cohorts, there is no reason why in theory the calculations cannot be done on all cohorts. The specific cohorts for which the results from these methods are taken into account in the final results selection process will be a matter for judgement.

In this example, the Chain Ladder ultimate (and hence the % developed) is taken from the CSA estimator without a tail factor, as given in the CL worked example in Section 3.2.8. The BF prior amounts are the same as those used in Verrall (2007), which are then also used in the worked example for the Bayesian version of the BF method given in Chapter 4. The premiums shown in Table 3.25 are values chosen solely for the purpose of this worked example, so as to produce broadly plausible ULRs.

If it assumed that the IELR assumption is equal to the prior ULR, then the IELR ultimate is equal to the prior ultimate. There are no further calculations required to apply the IELR method, and so the remainder of this example focuses solely on the BF method.

An example of the calculation of the BF ultimate can be seen by considering cohort 2012 in Table 3.25. Applying the description of the method outlined in Section 3.4, or the mathematical formulation in (3.15), gives:[16]

$$
\text{BF Ult} = \underbrace{2,864}_{\text{Paid to date}} + \underbrace{(1 - 42.22\%)}_{1 - \% \text{ dev'd}} \times \underbrace{6,000}_{\text{BF prior}}
$$
$$
= 6,331.
$$

---

[16] Full values for percentages etc. are shown here, whereas the table just shows rounded figures, for ease of presentation.

It is also straightforward to show that this is equivalent to using the weighted average description of the BF method given in Section 3.15 and the mathematical description of it given by (3.18), so that

$$
\begin{aligned}
\text{BF Ult} = \;& \underbrace{6{,}785}_{\text{Ult using \% dev'd}} \times \underbrace{42.22\%}_{\text{\% dev'd weight}} \\
&+ \underbrace{(1 - 42.22\%)}_{1 - \text{\% dev'd weight}} \times \underbrace{6{,}000}_{\text{Ult using prior}} \\
= \;& 6{,}331.
\end{aligned}
$$

It is quite common for the BF prior to be quoted in loss ratio terms. For the 2012 cohort, using the relevant figures from Table 3.25, the first version of the calculation of the ultimate then becomes

$$
\begin{aligned}
\text{BF ULR} &= \underbrace{45.35\%}_{\text{Paid LR}} + \underbrace{(1 - 42.22\%)}_{1 - \text{\% dev'd}} \times \underbrace{95\%}_{\text{BF prior}} \\
&= 100.25\%.
\end{aligned}
$$

For all other cohorts, the calculations follow exactly the same approach. It is worth noting, however, that the 2012 cohort is the only one in this example where the BF ultimate is less than the ultimate implied by the % developed (i.e. the CL ultimate). This is because it is the only cohort where the BF prior is less than the CL ultimate (i.e. it is 6,000, which is less than the CL ultimate of 6,785). This is the generic rule for the BF method, as referred to in both the general and mathematical descriptions of the BF method given in Sections 3.4 and 3.4.1 respectively.

Worked Example of Derivation of BF Priors using the CL Method

When using the BF method, one approach to deriving the BF prior assumptions for the more recent, less developed cohorts is to base these on the estimated ultimate values for the older, more developed cohorts. The resulting prior estimates can then be compared against priors that have been derived from sources external to the data, where available, and then a final prior chosen. This prior can be used in both the BF method and/or as an assumption for the IELR method. The procedure to estimate the prior in this way was described towards the end of Section 3.4, and is now demonstrated for this worked example.

The calculations are summarised in Table 3.26. This table includes an index for premiums and claims, with values chosen solely for the purpose of this example. In practice, the premium index would typically be derived based on the insurer's knowledge of how premium rates have moved over time and the claims index would be chosen to represent expected past

Table 3.26 *BF method worked example showing derivation of priors from older cohorts*

Cohort	Index Premium	Index Claims	Adjusted to 13 Premium	Adjusted to 13 Claims	CL ULR	Prior for: 2013	Prior for: 2014
2005	100	100	115%	122%	73%	78%	76%
2006	103	102	112%	120%	74%	79%	78%
2007	105	105	110%	116%	78%	83%	81%
2008	107	107	107%	114%	77%	82%	80%
2009	105	107	110%	114%	79%	83%	81%
2010	100	109	115%	112%	86%	84%	83%
2011	105	110	110%	111%	90%	91%	89%
2012	95	120	121%	102%	107%	90%	89%
2013	115	122			94%		
2014	120	125			87%		
					Avg (all yrs)	84%	82%
					Avg (last 3)	88%	87%
					Selected	90%	89%
					BF % dev'd	24.16%	6.92%
					BF ULR	90.98%	88.86%

changes in claims inflation across the cohorts.[17] The reliability of a premium rate index can be affected by several factors, such as changes in terms and conditions, the degree to which the index takes appropriate account of the position of the relevant business in the underwriting cycle, and the balance between renewal and new business in the portfolio. The practitioner may therefore take these and other factors into account when selecting the final prior loss ratio in the BF method to use for each relevant cohort.

It is assumed in this example that the 2012 and prior cohorts are being used to estimate the prior loss ratio assumptions for use in the application of the BF method to the 2013 and 2014 cohorts. The unadjusted Chain Ladder ULR shown in the "CL ULR" column is the same as the ULR given in the final column of Table 3.25. These are adjusted so that they represent estimates of the loss ratio for 2013 and 2014, allowing for premium and claims inflation. By way of example, for the 2008 cohort:

- First, consider the values in the "Adjusted to 13" columns of Table 3.26, starting with the premiums. Here, the premium index implies an adjustment factor to 2013 of 107%, derived from the ratio of the index in 2013 to the index in 2008 = 115/107.

- Similarly, the claim index implies an adjustment factor to 2013 of 114%, derived from the ratio of the index in 2013 to the index in 2008 = 122/107.

- The 2008 CL ULR of 77% then translates to a 2013 equivalent ULR by multiplying it by the ratio of claim and premium adjustment factors, that is, by 114/107 = 106.5, so that it becomes 77% × 106.5 = 82%. In other words, the numerator of the CL ULR (i.e. the

---

[17] Since claims inflation is a calendar period effect, the index to be used to convert one cohort to the equivalent basis as another needs to allow for the impact of inflation between cohorts, which will be influenced by the estimated claims settlement pattern, for example.

claims) are inflated by the claims index so that they represent 2013 equivalent values, and the denominator (i.e. the premiums) are similarly inflated by the premium index.

- To translate the 2013 equivalent ULR to a 2014 equivalent, the premium and claim indices from 2013 to 2014 are used. Hence, for 2008, the 2013 equivalent ULR of 82% is multiplied by 125/122 (= 1.025) and divided by 120/115 (1.043), i.e. by 0.98, so that it becomes 80%. In other words, because premium inflation of 4.3% between 2013 and 2014 has exceeded claims inflation of 2.5%, the 2014 equivalent ULRs are all slightly lower than the 2013 equivalent.

These calculations are done for all 2012 and prior cohorts. Various weighted or unweighted averages can be taken of these in order to derive suitable priors for 2013 and 2014 cohorts, and then a final selection for the priors can be made, based on consideration of trends and other relevant factors. In this case, the simple average over all cohorts and over the most recent three cohorts has been calculated, as shown in Table 3.26. Then, selected priors of 90% and 89% for 2013 and 2014 respectively are chosen, which in this case is based on placing relatively more weight on the higher ULRs that appear to occur in the two most recent cohorts.

In practice, these would be compared with any priors that are available from other sources (e.g. business plan or pricing analyses), before a final selection is made. In this example, the priors shown in Table 3.25 are intended to be from such a source, and at 100% and 105% for 2013 and 2014 respectively, are somewhat higher than the priors produced by adjusting the 2012 and prior CL ULRs for premium rates and claims inflation. Hence, a choice would need to be made between the sets of priors, partly depending on the assumed credibility of the other sources. In this case, to conclude this part of the worked example showing the BF calculations derived from the CL ULRs, the BF ULRs are calculated using these priors (rather than the priors from other sources) and the CL % developed derived previously, as shown in Table 3.26. The resulting BF ULRs are, for 2013 and 2014 respectively, lower and higher than the CL ULRs, reflecting the differing relationship between the priors and the CL ULRs for the two years (i.e. the prior is lower than the CL ULR for 2013, and the opposite for 2014).

Worked Example of Iterated BF Method

Continuing with the example used for the standard BF method applied to the 2012 cohort, the BF ultimate of 6,331 is used as a prior in a further application of the BF method to derive the Iterated BF ultimate, denoted here by BF*:

$$
\text{BF* Ult} = \underbrace{2,864}_{\text{Paid to date}} + \underbrace{(1 - 42.22\%)}_{\text{1-\% dev'd}} \times \underbrace{6,331}_{\text{BF iterated prior}}
$$

$$
= 6,523.
$$

Alternatively, the weighted average approach for the BF method can be used, by first calculating the future % developed as 57.78%, and then the calculation becomes

$$BF^* \, Ult = \underbrace{6{,}785}_{\text{Ult using \% dev'd}} \times \underbrace{1 - (57.78\%)^2}_{\text{Iterated BF weight for CL Ult}}$$

$$+ \underbrace{(57.78\%)^2}_{\text{Iterated BF weight for prior}} \times \underbrace{6{,}000}_{\text{Ult using prior}}$$

$$= 6{,}523.$$

Further iterations either involve repeating the first version above, using the new prior at each successive stage, or just using powers of 3, 4, etc. in the weighted average version. Using the latter approach it is easier to see how many iterations are needed before the weight assigned to the prior is so low that the ultimate is the same as from the CL method. In this case, if 15 iterations are used, so that the weight is 57.78%[15], the ultimate becomes 6,785 – i.e. the CL ultimate.

As mentioned in the description of the Iterated BF method, in practice when selecting final results it may not be obvious whether the reserves implied by its application are preferable in some sense, compared with the "pure" BF method.[18] In this case, the first iteration of the method produces an ultimate (of 6,523), which is, as expected, between the pure BF ultimate (of 6,331) and the CL ultimate of 6,785. Further iterations will progressively bring the ultimate closer to the CL ultimate. After the first iteration, in this case the ultimate is very close to the simple average of the ultimates from the two methods, which is 6,558. Hence, if for example, a review of the implied ULRs and of the graphs of the different ultimates and/or any other suitable diagnostics suggested that there was little to choose between the pure BF ultimate and the CL ultimate (i.e. each seemed reasonably plausible, with neither being obviously better) then taking an average would be one possible approach.

3.5 Frequency–Severity Methods

3.5.1 Introduction

This section covers a group of methods, abbreviated here as "FS", that differ from many of the other methods described in this chapter, in that they use both claim amount and claim count or number data, rather than just claim amount data. As well as producing an estimate of the ultimate claims, and hence reserve, these methods typically will also provide additional useful information for key items such as claims frequency, claim number reporting and settlement patterns and average costs per claim. This can be beneficial in both a reserving context and more widely in areas such as pricing and claims management. Even in situations where purely claim-amount-based methods are deemed sufficient, FS methods can add value in this way.

---

[18] Note, however, that the references mentioned in Section 3.4.2 give some examples of possible approaches to producing an optimal weighting between the ultimates produced by the BF and CL methods.

FS methods can either use raw claim numbers or claim frequency, which is the number of claims per unit of exposure, where exposure is some form of measure which is believed to be a key driver for the number of claims. For example, for property-related insurance, sum insured might be used. For some classes, premium might be used, which, although it may not always be the best measure of true exposure, may be the only metric that is readily available.

In relation to claims severity, this is usually intended to mean the average cost per claim, which explains why FS methods are also sometimes referred to as Average Cost Per Claim (or ACPC) methods, or simply as Average Cost methods. In some countries and contexts, these methods are probably the most widely used deterministic methods other than Chain Ladder, IELR and BF applied to purely claim amount triangles. For example, the results from the survey at ASTIN Committee (2016) suggest that Average Cost methods were the fourth most used deterministic methods amongst survey respondents as a whole, although there were significant variations between countries.

In general terms, FS methods tend to be used more in a direct insurance context than a reinsurance context, where claim count data is not always readily available. In addition, for both direct and reinsurance business, even if claim count data is available, if the proportion covered by the insurer for some contracts and policies is less than 100%, then interpretation of average costs can be difficult. In such cases, unless the proportion itself is also included as a parameter in the reserving method, the results produced by FS methods may be unreliable.

The overall approach of all FS methods is to estimate the number of claims and average severity for each cohort, and then multiply these together to derive the ultimate claim amount, and hence reserve. The estimation of both claim numbers and severity in many FS methods involves applying one of the commonly used reserving methods (such as the Chain Ladder) to each of the respective data triangles. Hence, several of the general points made about the CL method in Section 3.2 also apply to FS methods. In particular, the same limitations and assumptions as for purely claim-amount-based methods apply, as described in Section 3.2.5. In addition, the use of graphs as a diagnostic tool is equally applicable for FS methods as it is for claim amount-based methods.

FS methods can be particularly useful for the more recent cohorts, where a purely amount-based projection can be more uncertain. The reported claim number development data for these cohorts can effectively act as a leading indicator of claims trends, since it will typically reach a stable level more quickly than the corresponding claim amount data, for which the development will be subject to potential variability until all claim amounts are finalised. This is particularly the case for claims with an injury or other liability-related element. Combining the projected ultimate number of claims derived from the reported development data with a projected average cost can be a useful way of estimating an ultimate for the more recent cohorts. This FS-based ultimate can either be used on its own, or, where a % developed can be derived from the data or from an external source (or both), then it can be used as a prior ultimate in a BF method applied to claim amount data.

Before describing a number of different FS methods, the next section identifies some general considerations that arise when collating data and applying such methods.

3.5.2 Data Considerations for Frequency–Severity Methods

The use of claim number data inevitably raises some additional data considerations, beyond that required for reserving methods involving only claim amount data. These will vary depending on the context, but some examples are as follows:

- **Types of claim number triangles:** There are two main types of claim number triangles that are typically collated – claims settled (or "closed") and claims reported. As would be expected, the latter usually has a much faster development pattern than for the settled claims, with the difference being particularly marked for claims with a liability element. Hence, it will usually be used in preference to claim number settled triangles to estimate the ultimate number of claims for each cohort. Number of claims outstanding can also be collated in triangle format, being the difference between the number of claims reported and settled at each development point. A further claim number definition that might be encountered is "finalised" claims. Where this is used, it will most likely be in a situation where the term "settled" refers to an intermediate stage before full closure or finalisation of the claim, when perhaps the indemnity element of the claim has been paid, with the claim being referred to as being only truly closed or finalised when all remaining fees etc. have been paid. In some situations, however, "finalised" can also be another term for "settled" or "closed". If there is any doubt about the definition used in any data to be used for reserving purposes, then the obvious response is to seek clarification, wherever possible.
- **Relationship between reported and settled claims:** If both number of claim reported and settled data triangles are available, then before any reserving methods are applied to the data, the ratio of these two triangles can be examined to look for any trends or patterns in the data. Such trends can not only inform the application of FS methods, but also other claim amount-based methods. For example, an adjustment to the standard application of one or more such methods might be needed if this analysis suggests a change across the cohorts in the proportion of reported claims that are settled at a particular development point.
- **Types of claim amount triangles:** The definition used for the claim amount triangle will be partly dependent on the definition used for the claim number triangle, as well as the requirements of the particular FS method being used. A typical claims paid amount triangle will include payments relating to claims that have fully settled (or have been finalised) and claims that are only partially settled. It will also include amounts related to claims that involve single payment amounts, and others that involve multiple partial payments. Hence, if claims paid triangles are used in FS methods, the definition used for the claim number triangles will affect whether there is appropriate correspondence between the claim amount and claim number triangles. This issue is considered further in the paragraph at the end of this section, after the identification of the other key data considerations.
- **Cohort definitions:** FS methods are usually described as being applied to data that is organised by accident or underwriting period cohorts. However, as for other reserving

methods, they can also be applied to data organised by report period cohorts. This will then produce an estimate of the reserve in respect of reported claims only (i.e. it will exclude any allowance for "pure" IBNR). When applied to claim settled data that is organised by report period, FS methods potentially provide a way of assessing the level of redundancy or deficit in the case reserves for reported claims, as estimated by the claims department. In fact, one of the FS methods described below, the Fisher–Lange method (FS Method 4), was designed specifically for this purpose.

- **Multiple claimants:** If a claim involves more than one claimant (e.g. a motor accident involving several persons) then these can either all be included under a single claim reference for the purpose of the claim number data triangles, or alternatively each claimant can be counted separately for this purpose. It is possible for the number of claimants per claim to change over time for a particular class of business, so that even if the average claim per claimant is not changing, the overall average cost per claim (for all claimants related to a single claim) may change. Hence, understanding any changes in the average number of claimants per claim and average cost per claimant over time can sometimes be important in identifying the underlying trend in claim costs.

- **Zero claims:** If some claims settle or close with zero cost (e.g. they are deemed invalid claims) then a decision needs to be made as to how these are treated in the relevant data triangles. In particular, average settled costs can either include or exclude the effect of zero (or "nil") claims – i.e. the average can either be only for claims settled at non-zero cost or for all claims, including those that settle at zero cost. One approach is to treat the settlement of a claim at zero cost as a reduction of 1 in the incremental claim number settled and reported triangles. If this is done, then projecting the cumulative version of these triangles to ultimate (e.g. using the CL method) will effectively then result in an estimate of the ultimate number of non-zero claims. To ensure consistency when applying FS methods, this will then need to be used with a corresponding average claim amount that excludes the effect of zero claims. Other approaches are possible in practice, and so the practitioner will need to determine how zero claims have been treated in the relevant data, if this is not already known.

- **Reopened claims:** Depending on how these are treated in the data, they can cause apparently anomalous movements in claim number development, and possibly also in average costs. If reopening of claims is prevalent, then it may be necessary to analyse closure and reopening separately, to prevent reopening distorting the application of FS methods to the claims data.

- **Grouping similar types of claims:** In common with purely amount-based reserving methods, as well as grouping by reserving category (e.g. class of business) it can also be appropriate to analyse different types of claim separately when applying FS methods. This is because both claim frequency and average cost can vary considerably between claim types within a single class of business.

- **Grouping by claim size:** In FS methods, average claim size becomes more prominent compared to amount-only methods, and hence it can be beneficial to subdivide the analysis by claim size, particularly if the trends are thought to vary by claim size (e.g.

large claims often take longer to settle). This topic is discussed further in general terms in Section 6.9.5, and in relation to UK Personal Motor in Section 7.2.3.

- **Treatment of excess/deductibles:** A decision will need to be made as to whether the claim count data includes claims that are known about, but are below the relevant excess/deductible point. These are in effect zero claims unless they subsequently reach the deductible. An additional practical aspect related to excess/deductible points is where they have changed over time. This can cause the average cost of claims to also change, and so can impact on the use of FS methods. For example, consider a situation where most of the liability relates to a small number of recent cohorts, but where the relevant policies have different excesses to those for earlier cohorts. One approach to dealing with such a change is to construct "as-if" data triangles, whereby the claim amount data for all cohorts is recast using the excesses for the recent cohorts. The FS methods are then applied using this as-if data to derive reserve estimates for the recent cohorts. This is an example of allowing for changes in business mix, as discussed further in more general terms in Section 6.9.2.

- **Treatment of expenses/fees:** Many claims involve both an indemnity element and an expense element. If the claim amount data includes all claims with either or both of these elements, then it is important that the claim count data also uses this definition (i.e. it should, for example, include claims in the count data even if they only have an expense element). In other words, there should be a consistent definition used in the claim amount and count data, otherwise any derived averages may be misleading.

- **Impact of seasonality:** For some types of business there can be a difference between the claim number settlement patterns and average cost per claim, according to the accident period. For example, Friedland (2010) gives an example (in Chapter 11) of Canadian motor insurance data with half-yearly cohorts, where the claim number settlement pattern appears to be slower for claims in the second half of the year than for those in the first half. Various possible reasons are given in Friedland (2010) for this. These include the impact of winter weather on driving conditions, perhaps causing more claims, combined with the fact that the second half-year has two winter months at the end of the period (November and December), whereas the first half-year has two winter months at the beginning (January and February). Hence, there is more time within the first half-yearly cohort to settle these claims within the cohort. Such a feature might need to be allowed for in the selection of the cohort length and in the application of the FS method to the data (e.g. with different development patterns being selected for different cohort periods, to allow for the seasonality). Similar considerations regarding seasonality may also apply for some classes of business when using claim-amount based reserving methods.

- **Gross/net of reinsurance:** As with other reserving methods, FS methods can in theory be applied to data that is gross or net of reinsurance. However, care is needed when using net of reinsurance data, particularly where the reinsurance programme has changed over time. This is discussed further in Section 6.2.1.

- **Operational time:** This refers to a particular way in which data can be collated, which instead of using time as the development dimension, the cumulative proportion settled is used. The concept is explained further in Section 3.5.4, before the FS methods worked examples.

The way in which these and other data considerations will impact on the application of FS methods in practice will depend, for example, on the specific method(s) being used, the data availability and the context for the reserving exercise.

Relationship Between Claim Number and Amount Triangles

When deriving the average cost of claims, a cumulative or incremental claim amount triangle is divided by the corresponding claim number triangle. This raises the question of correspondence between the number and amount triangles – i.e. whether the individual claims that make up the claim amount total in each development period for each cohort relate to the same claims for that cohort in the corresponding period of the claim number triangle.

For methods relying on settled claim averages, one approach is to use the claims paid triangle, along with the number of settled claims triangle. This should have appropriate correspondence if all claims settle as single payments, with no partial payments. However, where this is not the case the derived averages will be distorted. For example, the partial payments in respect of claims not yet settled will be included in the claims paid triangle, but will have no corresponding entry in the number of settled claims triangle. Furthermore, partial payments related to individual settled claims will potentially occur in more than one development period of the amount triangles. If the impact of partial payments is thought to be material, then before using methods involving average settled claim costs, it may be necessary to consider whether claim amount triangles can be created only in respect of settled claims, rather than simply using the claims paid triangles.

If settled only triangles are created, then a further decision will need to be made as to whether all partial payments related to settled claims are allocated to the development period in which the claim is finally settled, or left in the development period in which the payment is made. The former approach will produce a more accurate representation of the true total average cost per settled claim.

For methods relying on incremental or cumulative average reported (i.e. incurred) claims, one approach is to use an incurred claim amount triangle in conjunction with a reported number triangle. If the cumulative incurred amount triangle is divided by the cumulative number reported triangle, to derive the cumulative average incurred claims, then there will be correspondence between these two triangles; it will show the average incurred value of the claims reported up to each development point for each cohort. However, on an incremental basis, there will not be correspondence between the two triangles, as movements in claims incurred in any particular development period may relate to claims reported in earlier periods as well as in that period.

3.5.3 *Types of Frequency–Severity Method*

Although all FS methods have the same basic component of combining an estimated number (or frequency) of claims with an estimated average severity, there are a number of distinct types of method that can be used in practice. Five different FS methods are described in this section, in order to provide examples of the different types of method. Where relevant, reference is also made to some of the variations that are possible. The five methods are:

1. **FS Method 1: Cumulative Averages:** This is the simplest FS method described here and relies on using cumulative data triangles to estimate the ultimate number of claims and average costs. A number of slight variations are possible, some of which are described in the relevant section.
2. **FS Method 2: Payments Per Claim Finalised ("PPCF"):** This method involves estimating the number of future settled claims and average cost at each duration. Inflation can be allowed for, as well as any dependency between average cost and delay to settlement.
3. **FS Method 3: Payments Per Claim Incurred ("PPCI"):** This method involves estimating the payments per ultimate number of claims (incurred) at each duration. Inflation can be allowed for, as well as any relationship between duration and size of PPCI.
4. **FS Method 4: Fisher–Lange:** This is a type of FS method which is similar to the PPCF method, except that is specifically designed to be used with data where the cohort is a reporting period.
5. **FS Method 5: Payments Per Active Claim ("PPAC"):** This method is deigned to be used where claim payments are made up of a series of regular amounts, rather than lump sum payments. It involves estimating the number of active claims for each cohort in each future development period, and then combining this with an estimate of the average cost in these periods.

Worked examples are given in Section 3.5.5 for the first three methods to aid the explanation of these methods.

FS Method 1 is the simplest of the five methods. FS Methods 2 and 3 require similar data to the first method, but are slightly more complicated to implement in practice, with more assumptions being required. FS Method 2 in particular is an example of a method that allows for the relationship between average cost of settled claim and delay to settlement. FS Method 3 has a similar feature, except that it uses the concept of payments per ultimate number of claims incurred (or reported). FS Method 4 can be a useful approach where an estimate is required of the reasonableness of the case reserves for reported claims only. The first four methods may not work well if claimants receive a series of regular amounts, rather than one or more lump sum payments; FS Method 5 is designed to be used in situations where there are these regular payments.

The choice of method, and the specific version of it that is used in any particular situation, will depend partly on the data available and on the context and objective for the reserving exercise. Other types of FS method (and variations thereof) beyond those described here may be used in practice and hence this section is not intended to provide a description of all possible FS methods.

For each method, where settled averages are used, for the purpose of the description given here, it is assumed that the claims paid data triangle can be used. If the distortions caused by partial payments are material, then, as discussed in the previous section, it may be necessary to replace this triangle with a settled only claim amount triangle.

Many of the data considerations mentioned in the previous section, such as carrying out the analysis by claim size, apply to each of these methods, and they are not repeated here. Unless otherwise stated, it is assumed that the data is organised on either an accident or underwriting year (or policy year) basis. Where the methods involve projecting data triangles to ultimate, for example using the CL method, then unless stated otherwise, an allowance for a tail factor may need to be made if the relevant claims data is not fully developed.

Like all reserving methods, FS methods will rarely be used in practice in isolation. For example, if the particular FS method that is chosen uses only claims paid or settled amount data (as many do) then a comparison of the derived reserve estimate for each cohort with the latest case reserves would often be appropriate, along with the corresponding estimate derived by applying other purely amount-based methods to paid and/or incurred data.

A further group of methods which encompass some types of FS methods are those applied using operational time. Due to the distinct features of these methods, they are described in a separate section – Section 3.5.4, and a worked example of an operational time method is given in Section 3.5.5.

FS Method 1 – Cumulative Averages

This is an example of one of the simpler types of FS method. It involves using, for example, the CL method applied to the number of claims reported triangle to estimate the ultimate number of claims for each cohort. The average cost per claim is derived by dividing the cumulative paid claim amount at the end of each development period by the corresponding cumulative number of claims settled. This triangle of averages is then projected to ultimate, using, for example, the CL method. The resulting ultimate average settled amount and number reported are then multiplied together to derive the estimated ultimate claim amount for each cohort.

Other variations on this type of FS method are possible. For example, if for some reason the number of claims reported triangle is not available, then the number of claims settled triangle could be used instead to derive the estimated ultimate number of claims. However, this may in some cases produce unreliable estimates, particularly for the more recent cohorts which may have a relatively small proportion of claims settled in the early development periods. As an alternative to using the settled average triangle, the cumulative incurred claims triangle could be divided by the cumulative reported claim number triangle. The resulting average incurred claim triangle could then be projected to ultimate using, for example, the CL method. This would then be combined with the estimated ultimate number of reported claims to derive the estimated ultimate claim amount. A further variation for this type of FS method would involve adjusting the claim amount data triangle for inflation (before calculating the cumulative averages), in a similar way as for with the inflation-adjusted chain ladder, described in Section 3.2.9.

The number of reported claims triangle may exhibit relatively short-tail and stable development, compared to, for example, the incurred claims amount triangle. However, there will still usually be an element of uncertainty in the more recent cohorts. One way of addressing this is to calculate the claim frequencies for all cohorts, using the estimated ultimate number of claims expressed as a proportion of a suitable exposure measure. The trend in these frequencies can then be examined, and averages taken (excluding the recent cohorts), in order to derive a selected claim frequency for the more recent cohorts. This is then multiplied by the known exposure for these cohorts to derive the estimated ultimate number of claims.

Assuming that a reported claim number triangle is available with sufficient and reasonably stable development history for an appropriately homogeneous reserving category, then in theory this type of FS method should in many circumstances produce a reasonable estimate of the ultimate number of claims for each cohort. There will usually be more uncertainty associated with the estimation of the average claim amount for each cohort (on an ultimate basis), as the cumulative settled or reported average may not necessarily show a stable development over time. If this is the case, then one or more of the other FS methods described in this section may be a possible alternative.

FS Method 2 – Disposal Rates and Delay-dependent Inflation-adjusted Severities

When the average cost per claim is believed to be related to the delay to settlement, this method is an improvement on the first FS method, as this relationship is explicitly allowed for by estimating the settlement pattern and the average cost at each duration. Assumptions regarding historical and future claims inflation (or other calendar period trends) can also be allowed for in this method, if required. This is assumed to be the case in the description of the method given here; the relevant steps can be omitted if inflation is not being allowed for explicitly. The method also introduces the concept of "disposal rates", in the context of claim number settlement patterns – or alternatively "closure rates". Given a selected ultimate number of claims per cohort, disposal rates allow the estimated number settled in each development period to be derived. Disposal rates are typically defined as the number of claims settled (or "disposed of") in a period divided by the number of claims outstanding at the beginning of that period. However, depending on the length of the development period, it is possible for a proportion of the claims settled in the period to have been reported in that period, and so some versions of the calculation of disposal rates take this into account by adding a proportion of those reported in the period to the denominator of the ratio. The worked example for this method in Section 3.5.5 explains the disposal rates concept further.

The steps involved in FS Method 2 are summarised below:

1. Estimate the ultimate number of claims per cohort, using similar approaches as for the first type of method (e.g. typically using number of claims reported triangles).
2. If the previous step has not used the claim number settlement triangle, then project this triangle to ultimate (e.g. using the CL method) and derive a claim number settlement pattern.

3. Calculate the disposal rates from the settlement pattern from the previous step. The details of the calculations involved are most easily understood by example, as shown in Section 3.5.5.

4. Use the selected disposal rates and the projected ultimate number of claims from the first step to derive the estimated future number of claims settled at each duration for each cohort. These settled claim numbers will enable allowance to be made, where appropriate, for the relationship between delay to settlement and average settled value, as explained in the remaining steps.

5. Derive the incremental average settled amount for each cell in the triangle (i.e. for each cohort in each development period) by dividing the incremental settled (or paid) claim amount triangle by the incremental settled claim number triangle. Depending on the context, this may show higher averages for later development periods, as larger claims may take longer to settle.

6. Make an assumption regarding historical inflation, which is designed to represent how the average settled values have changed across calendar periods, due to the impact of claims inflation. This may be based partly on market knowledge and experience, and partly on known historical inflation indices (e.g. for wages, prices or from claims inflation market studies, where available). This inflation represents calendar period inflation, rather than the relationship between delay and average claim size. It will usually be formed either as a single value or as an index across calendar periods.

7. Make an assumption regarding future calendar period inflation, based on a combination of judgement, market studies or projections of future inflation and examination of inflationary trends shown in the selected historical inflation index.

8. Using the selected historical inflation assumption, adjust the average settled claims in each development period for all cohorts, to express all figures in latest calendar period values (or to another calendar period if preferred, with the relevant step to add future inflation being adjusted accordingly).

9. Select an average settled claim amount at each duration, using the data triangle from the previous step. This will require judgement to be exercised, particularly in the later development periods, where the data may be more volatile, and where it may be necessary, for example, to derive an overall average for all development periods after a certain point, rather than individually.

10. The average values in the previous step are all expressed in monetary terms of the latest calendar period, which enables averages to be taken across the cohorts. To use these averages to derive the reserves for each cohort, future calendar period inflation needs to be allowed for. This step is most easily seen in the worked example later in this section. This produces estimates of the future average settled claim amounts for each cohort in each future development period, expressed in monetary terms appropriate for the relevant calendar periods.

11. The averages from the previous step are then multiplied by the corresponding estimated future number of settled claims from Step 4, to derive the total estimated settled claim amounts in each future development period for each cohort.

12. The future values from the previous step are then combined by cohort to derive the estimated reserve for each cohort.

The above description is similar to a method referred to as "Payments Per Claim Finalised", or PPCF, where the term "finalised" is as explained in Section 3.5.2. Further details of the PPCF method are given in Chapter 16 of Hart et al. (2007).

FS Method 3 – Payments Per Claim Incurred

This method, abbreviated by PPCI, is one of a number of reserving methods that have been referred to by some authors as being "Australian methods" (e.g. see Claughton, 2013), due to their use by actuaries and other reserving practitioners in that country. Another such method is the PPCF method. The reference to "per Claim Incurred" in the name of this method is synonymous with "per Claim Reported", rather than being a reference to claim incurred amounts, which are not utilised in this method; instead claim payments are used. The method involves the following steps, which are based on Claughton (2013):

1. Estimate the ultimate number of claims based on the reported claim number triangle, using, for example, the CL method. As for some of the other FS methods, the settled claim number triangle could also in theory be used, but in most cases it is likely to produce less reliable results than those based on the reported numbers, particularly for the more recent cohorts.
2. The claim amount part of this method is represented by the incremental claims paid data triangle. Where possible, this triangle is adjusted for inflation, to express all values in monetary terms of the latest calendar period. This is done by making an assumption for past inflation, based on knowledge of the factors affecting claim values over the relevant period. This can be done in a similar way as for the PPCF type method (i.e. FS Method 2), but applied to claim amounts rather than averages. It is comparable to the adjustment for inflation used in the Inflation Linked Chain Ladder method, described in Section 3.2.9.
3. Using either the unadjusted or inflation-adjusted incremental claims paid triangle, divide all values in each cohort by the estimated ultimate number of claims for that cohort derived from Step 1, to produce the incremental PPCI triangle. This effectively adjusts the claim amount data for exposure and creates a triangle of payments-per-claim-incurred – hence the name of the method. Note that these values are not the average amount settled per claim in each development period (as for the PPCF method), but rather an expression of the level of claims paid for each cohort in units represented by the number of claims reported (i.e. incurred) for that cohort. In effect, this is the average cost per claim per unit of exposure, where the unit is the total number of claims reported. Other exposure measures can also be used, including premiums or, where relevant, sums insured, wages etc.
4. Using the PPCI triangle from the previous step, select a suitable average over the cohorts of the incremental PPCI for each development period. This approach assumes that the average PPCI in each development period is consistent across the cohorts, although

it may also be appropriate to allow for any observed trends across the cohorts when selecting the average PPCI that will be used to derive estimated future claims. If the claims are still developing for the oldest cohorts in this triangle, then an allowance will need to be made for an average PPCI in the tail.

5. For each cohort, the selected averages in each future development period derived in the previous step are then multiplied by the estimated ultimate number of claims incurred (i.e. reported) to derive the estimated future paid amounts. If the claim amount data was adjusted for inflation into latest calendar period terms, then, as with FS Method 2, an allowance for future inflation will need to be built in at this stage.

6. These future amounts are then accumulated to derive the estimated reserve for each cohort.

Hart et al. (2007) provides a detailed description and worked example of the PPCI method.

FS Method 4 – Fisher–Lange

This method, abbreviated here as "FL", was originally formulated in Fisher and Lange (1973), and is a type of FS method that is specifically designed to be used with data where the cohort is a reporting period. In other words, the development relates to all claims reported in each cohort, so that the projection to ultimate (using either the FL method or any other method) will produce an estimated ultimate only in respect of claims reported at the valuation date. For reserve booking purposes, an additional amount will be needed in respect of IBNR claims.

Although the description of the other FS methods assumes that an accident or underwriting cohort is being used, in theory, any FS (or other triangle-based reserving method) can be applied to report cohort data to produce an estimate of the reserve for reported, but still outstanding claims at a particular valuation date, which can then be compared directly with any corresponding case-by-case (or just "case") estimates made by the claims department for individual reported claims. This will therefore give an estimate of the redundancy or deficiency in the case estimates made by the claims department. This is different to using data organised by accident cohort, for example, where the estimated reserves will include an implicit allowance for IBNR claims. Since the total case estimates will not include such an allowance, this will need to be taken into account when comparing them with a reserve derived from accident cohort data.

The original description of the FL method, given in Fisher and Lange (1973), involves first using exponential regression to estimate the average cost of future settled claims for each cohort and future development period, allowing for inflation. Disposal rates are then used to estimate the number of claims settled in future development periods, possibly allowing for any observed change in speed of settlement. These results are then combined to provide an estimate of the ultimate average incurred claim, which can then be compared with that implied by the known settled amounts together with the case estimates made by the claims department. Hence, an estimate of the surplus or deficit in the reported case estimates can be derived.

From this brief description, it can be seen that the FL method is similar to the other FS methods described here that use incremental claim settled/finalised number and average amount data – for example, it can be viewed as a type of PPCF method (i.e. FS Method 2). The detailed descriptions of these other FS methods can easily be adapted to apply them to report cohort data, in which case there is no need to estimate the ultimate number of claims, as this will be known. Further details of the FL method are given in Fisher and Lange (1973), and that paper is discussed in Skurnick (1973a).

FS Method 5 – Payments Per Active Claim

All of the other FS methods described here are designed for use where the majority of claim payments are made in lump sum amounts. When the payments are made as a series of regular amounts, either for short periods or indefinitely until, for example, death or retirement, these methods may not work very well, and so an alternative approach is needed.

The Payments Per Active Claim, or "PPAC" method, is an example of such a method. It is another example of one of the so-called "Australian methods", as referred to in the description of the PPCI method (i.e. FS Method 3). It can also be referred to as the "Payment per Claim Handled" method. Examples of the classes of business where a PPAC-type approach might be used are Medical Expenses, Creditor, Workers' Compensation and Unemployment Insurance. A brief summary of the method is given below, following Claughton (2013).

The claim number data that is collated for the PPAC method will be the number of claimants for each cohort whose claims are "active" in each development period (i.e. received a payment in that period). Hence, this will eventually reduce to zero once all the claimants for that cohort have ceased to have active claims. The relationship between the number of active claimants in successive development periods in this claim number triangle is often referred to as the "continuance rate" – that is, the proportion of active claimants in one period that are still active in the next period. The triangle is projected to ultimate, so as to derive the estimated number of active claims in each future development period for each cohort. Allowance may need to be made at this stage if there has been a change across the cohorts in the average length of the period over which regular payments are made to claimants.

The average cost per active claim is derived by dividing the total costs for active claims in each cohort/development period by the number of active claims in that cell. In a similar way as for the delay-dependent PPCF method (FS Method 2), the averages are usually adjusted for inflation and then an average cost per active claim is selected for each development period. These are then multiplied by the projected number of active claims in each future cell to derive the estimated future claim amounts for each cohort/development period, and hence the reserve.

Implementations of a PPAC-type approach in practice can include an allowance for seasonality, legislative changes and superimposed inflation, and a separate analysis by different claim types. Further details of the PPAC method are given in Claughton (2013).

3.5.4 Operational Time

This concept was referred to briefly in the earlier section on data considerations for FS methods. This subsection explains it further and summarises briefly its application to FS methods. Operational time is also used in at least one stochastic reserving method, as referred to in Section 4.7.13.

Operational time refers to the cumulative proportion of claims (by number) that are settled by a particular development time (i.e. as measured since the start of the cohort). It therefore ranges from zero to one. It is similar to the % developed concept, as described in the description of the Chain Ladder method in Section 3.2.2, applied to the claim number settled triangle.

The concept of operational time is believed to have been originally introduced in Reid (1978). Some types of FS method explicitly incorporate a relationship between development time and average cost per claim (e.g. the PPCF method – FS Method 2). However, an alternative form of this relationship might be more appropriate in some circumstances – namely that the average cost per claim is related to the order in which claims are settled, rather than development time. If claims for each cohort are grouped according to operational time intervals, then (ignoring inflation) FS methods can be used by estimating the average cost of settled claims in each such interval for each cohort. It is therefore assumed that the size of claims is related to the point in operational time when they settle, rather than the point in development time when they settle. If full claims transaction data is available, then it can be used to derive average costs for selected operational time intervals for each cohort. This is done by ordering the settled claims for each cohort according to their settlement date, and then calculating the average cost of those claims for the first, say, 5% in the ordered list, the next 10% and so on. A triangle of average costs (possibly adjusted for inflation) can then be constructed with operational time as the column dimension, and this can then be used in FS methods in exactly the same way as if it were a normal claims development triangle. If claims transaction data is not available, then approximations can be made using the usual claims development triangles to derive operational time triangles.

The potential advantage of using operational time, rather than the much more commonly used development time, can be seen by considering the situation where there has been a change in settlement rates for one or more reserving categories – i.e. the proportion settled by development time may have changed across the cohorts. This can be difficult to allow for in FS methods that explicitly model average cost by development time. An alternative is to model the average cost by operational time and review whether there is more stability across the cohorts. For example, where there has been a delay in the claims department in settling claims for a particular cohort or group of cohorts (e.g. due to staff shortages perhaps) then this may affect the observed average cost of settled claims by development time for those cohorts compared to other cohorts. Unless allowance is made for this in the application of the relevant FS method, the results may be incorrectly impacted, particularly if the delay itself does not change the actual cost of settling claims. However, when measured by operational time, there may be no impact on the average costs for the relevant operational time intervals, and hence there will be no need to make any specific allowance in the application of the method.

Operational time can be used in a range of different FS methods. A worked example of using it for FS Method 3 (PPCI) is given in Section 3.5.5. Hart et al. (2007) also gives an example of using operational time for a PPCI method and Bain (2003) provides a simplified example to help explain the concept.

3.5.5 Worked Examples of Frequency–Severity Methods

This section describes the application of a number of different FS methods to the same data used for the worked examples of the CL and BF methods given earlier in this chapter. For those methods, only claim amount data was required, but for the FS worked examples, claim number data is also required. The original source for the data used in the CL and BF worked examples, Taylor and Ashe (1983), contains both claim amount (paid) and number data, with the latter relating to claims "finalised". For the purposes of these examples, it is assumed that "finalised" is equivalent to "settled". It is also assumed that the claim amount paid data can be used to derive average settled claim amounts, and that in doing so there are no material distortions caused by either zero claims or partial payments. Any references to settled claim amount data in these examples are to this claims paid amount data. For simplicity, no tail factor is used for projecting either the claim number or amount data triangles.

The original data source does not contain any data relating to the number of reported claims, and so a suitable data triangle has been created, solely for the purpose of these worked examples. The relevant data triangles are shown in Tables 3.27 to 3.29 inclusive. These do not include a claims incurred triangle, which was not provided in Taylor and Ashe (1983); in practice this would usually be available and would be used to validate any projections based only on settled claims data, and could also be used to apply FS methods involving incurred data. The premium data used in this example is the same as that used in the BF method worked example in Section 3.4.3.

Table 3.27 *FS example: Incremental Settled Claim Amounts*

Cohort	Development year									
	0	1	2	3	4	5	6	7	8	9
2005	358	767	611	483	527	574	146	140	227	68
2006	352	884	934	1183	446	321	528	266	425	
2007	291	1002	926	1017	751	147	496	280		
2008	311	1108	776	1562	272	352	206			
2009	443	693	992	769	505	471				
2010	396	937	847	805	706					
2011	441	848	1131	1063						
2012	359	1062	1443							
2013	377	987								
2014	344									

Table 3.28 *FS example: Incremental Settled Claim Numbers*

Cohort	\multicolumn{10}{c}{Development year}
	0	1	2	3	4	5	6	7	8	9
2005	40	124	157	93	141	22	14	10	3	2
2006	37	186	130	239	61	26	23	6	6	
2007	35	158	243	153	48	26	14	5		
2008	41	155	218	100	67	17	6			
2009	30	187	166	120	55	13				
2010	33	121	204	87	37					
2011	32	115	146	103						
2012	43	111	83							
2013	17	92								
2014	22									

Table 3.29 *FS example: Incremental Number of Reported claims*

Cohort	\multicolumn{10}{c}{Development year}
	0	1	2	3	4	5	6	7	8	9
2005	539	6	25	30	6	0	0	0	0	0
2006	631	29	21	36	0	0	0	0	0	
2007	580	62	27	14	7	0	0	0		
2008	480	61	31	31	6	6	0			
2009	469	72	29	24	0	0				
2010	453	41	11	10	0					
2011	412	44	14	5						
2012	323	48	16							
2013	301	36								
2014	305									

Table 3.30 shows the average settled claim amount in each cohort/development year. This provides some indication that the larger claims settle later than the smaller claims, as the higher averages are mostly in the later development periods, but it is worth bearing in mind that claims inflation can materially affect tables such as this.

The first three FS methods described in Section 3.5.3 are applied to this data, as well as a version of the PPCI method (Method 3), using operational time. All these methods require

Table 3.30 *FS example: Average incremental settled claim amount*

Cohort	\multicolumn{10}{c}{Development year}
	0	1	2	3	4	5	6	7	8	9
2005	8.95	6.19	3.89	5.19	3.74	26.11	10.45	14.00	75.74	33.97
2006	9.52	4.75	7.18	4.95	7.31	12.35	22.95	44.36	70.84	
2007	8.30	6.34	3.81	6.64	15.64	5.65	35.43	56.08		
2008	7.58	7.15	3.56	15.62	4.07	20.71	34.38			
2009	14.77	3.71	5.98	6.41	9.18	36.20				
2010	12.00	7.74	4.15	9.25	19.08					
2011	13.78	7.37	7.75	10.32						
2012	8.36	9.56	17.39							
2013	22.16	10.72								
2014	15.64									

an estimate of the ultimate number of claims, so the calculations begin by projecting the claim number triangles. In this case, both the settled and reported triangles are projected using a standard CL method, using the CSA estimators. The results are shown in Table 3.31. The detailed CL method calculations are not shown here; suffice it to say that all cohorts were used in the calculation of the CSA estimators, with no exclusions or tail factor being used. The selected ultimate number of claims has been chosen as being equal to the ultimate derived from the reported claim number triangle.

Table 3.31 *FS example: Estimated ultimate number of claims*

	Latest no. claims		CL ultimate from:			
Cohort	Settled	Reported	Settled	Reported	Avg	Selected
2005	606	606	606	606	606	606
2006	714	717	716	717	717	717
2007	682	690	689	690	690	690
2008	604	615	617	615	616	615
2009	571	594	596	594	595	594
2010	482	515	521	516	519	516
2011	396	475	485	478	482	478
2012	237	387	392	404	398	404
2013	109	337	348	366	357	366
2014	22	305	355	363	359	363
Total	4,423	5,241	5,325	5,349	5,337	5,349

Graphs of the development of settled and reported number of claims are shown in Figures 3.11 and 3.12 respectively. These use the respective CL ultimates derived from each data triangle, as shown in Table 3.31, with the data at each development point for each cohort

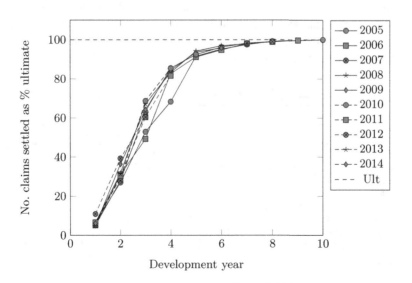

Figure 3.11 FS worked example: No. claims settled as % ultimate

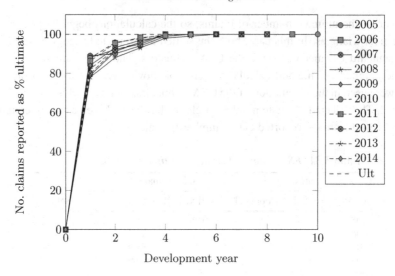

Figure 3.12 FS worked example: No. claims reported as % ultimate

being scaled to its estimated ultimate (i.e. the graphs are derived in the same way as the scaled graphs referred to in the CL method description in Section 3.2.7). Both show reasonably stable development across the cohorts, and as expected the development pattern for the number of reported claims is noticeably shorter than for settled claims.

Having selected the estimated ultimate number of claims for each cohort, it is useful to review the claim number settlement pattern across the cohorts, otherwise referred to as the claim closure pattern – calculated as the incremental number of settled claims in each cohort/development year divided by the selected ultimate number of claims for that cohort. The results are shown numerically in Table 3.32 and graphically in Figure 3.13. Inspection of both of these does not suggest any obvious trend towards slower or faster settlement patterns across the cohorts. An alternative formulation that can be used to examine claim

Table 3.32 *FS example: Incremental no. claims settled as % ultimate*

Cohort	Development Year									
	0	1	2	3	4	5	6	7	8	9
2005	7%	20%	26%	15%	23%	4%	2%	2%	0%	0%
2006	5%	26%	18%	33%	9%	4%	3%	1%	1%	
2007	5%	23%	35%	22%	7%	4%	2%	1%		
2008	7%	25%	35%	16%	11%	3%	1%			
2009	5%	31%	28%	20%	9%	2%				
2010	6%	23%	40%	17%	7%					
2011	7%	24%	31%	22%						
2012	11%	27%	21%							
2013	5%	25%								
2014	6%									

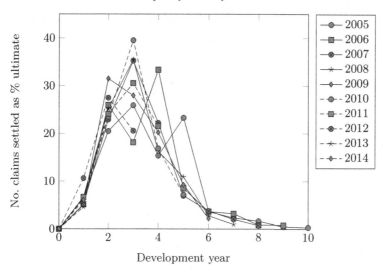

Figure 3.13 FS worked example: Incremental No. claims settled as % ultimate

number settlement patterns is to use the ratio of the cumulative number of settled claims to the cumulative number reported for each cohort/development year – i.e. the proportion of claims closed/settled of those that have been reported so far.

The remainder of this section explains the application of each selected FS method to the relevant data.

FS Method 1 (Cumulative Averages) – Worked Example

This FS method is the simplest of those explained in Section 3.5.3. The first version of it uses the reported number and claims paid data triangles. It should be straightforward to see how the example could be adapted to allow for the other variations of this method described in Section 3.5.3.

The ultimate number of claims is derived by applying the CL method to the reported number triangle, which produces the ultimate values shown in the relevant column in Table 3.31, with an estimated ultimate number of claims across all cohorts of 5,349.

The next step when applying this FS method is to derive the cumulative average claim amount data triangle and then to estimate the ultimate averages for each cohort. This triangle is shown in Table 3.33.

This triangle of cumulative average settled claim amounts does not appear particularly stable, with the values in the first few development periods being quite volatile and many cohorts having link ratios less than 1. Overall, the average settled claim does not appear to have stabilised across the development periods, even for the older cohorts. This might suggest that using this data to derive an estimate of the ultimate average claim for each cohort could be unreliable, but the remaining steps will be applied anyway, to demonstrate

Table 3.33 *FS Method Type 1 example: Cumulative average settled claim amounts*

Cohort	Development year									
	0	1	2	3	4	5	6	7	8	9
2005	8.95	6.86	5.41	5.36	4.95	5.75	5.87	6.00	6.35	6.44
2006	9.52	5.54	6.15	5.66	5.82	6.07	6.62	6.94	7.48	
2007	8.30	6.70	5.09	5.49	6.26	6.23	6.84	7.20		
2008	7.58	7.24	5.30	7.31	6.94	7.33	7.60			
2009	14.77	5.24	5.56	5.76	6.10	6.78				
2010	12.00	8.66	6.09	6.71	7.66					
2011	13.78	8.77	8.26	8.80						
2012	8.36	9.23	12.09							
2013	22.16	12.51								
2014	15.64									

the overall approach. Applying the CL Method to this data triangle, using all cohort CSA estimators and no tail factor, gives the ultimate values shown in Table 3.34.

Table 3.34 *FS Method Type 1 example: Results summary*

Cohort	Ultimate			Claims		CL method
	Number	Avg settled	Total	To date	Reserve	reserve
2005	606	6.44	3,901	3,901	0	–
2006	717	7.59	5,439	5,339	99	95
2007	690	7.80	5,382	4,909	473	470
2008	615	8.58	5,276	4,588	688	710
2009	594	8.12	4,826	3,873	953	985
2010	516	9.82	5,066	3,692	1,374	1,419
2011	478	11.72	5,600	3,483	2,117	2,178
2012	404	17.35	7,008	2,864	4,143	3,920
2013	366	16.63	6,086	1,363	4,723	4,279
2014	363	13.95	5,064	344	4,720	4,626
Total	5,349		53,648		19,290	18,681

The ultimate in the "Total" column in this table is the estimated ultimate number of claims multiplied by the estimated ultimate average settled claim amount. The reserve is then this ultimate less the claims paid to date, shown in the next column in the table. The reserve from the CL method applied to the claim amount data, as per Section 3.2.8, is also given in the table, for comparative purposes. This shows that, despite the apparent volatility of the average settled triangle, the results for FS Method 1 are reasonably close to the standard CL method.

One of the variations of this method described in Section 3.5.3 is to use claim frequency. For this example, premiums will be used – specifically, the same as those used in the worked example for the BF and IELR methods given in Section 3.4.3. These are shown in Table 3.35. In practice, the premiums would be adjusted for rate changes, so that they are expressed in constant values (e.g. of the latest cohort), but for this example it will be assumed that rates have, on average, been flat (i.e. unchanged) over the relevant period. The premiums may also need adjusting for other factors (such as business mix) that may be expected to impact upon claims frequency, but again that will be ignored for the purpose of this example. Having collated the exposure data, the next step in this method is to derive the claim frequency for each cohort, using the estimated ultimate number of claims given in the first part of this example. The resulting claim frequency is shown in Table 3.35.

Table 3.35 *FS Method Type 1 example: Calculation of claims frequency*

Cohort	Premium	Ult number	Frequency	Ultimate
2005	5,333	606	11.36%	
2006	7,333	717	9.78%	
2007	6,875	690	10.04%	
2008	6,875	615	8.95%	
2009	6,111	594	9.72%	
2010	5,914	516	8.73%	
2011	6,316	478	7.57%	
2012	6,316	404	6.40%	
2013	6,000	366	6.10%	420
2014	5,714	363	6.35%	400
		Avg – all	8.50%	
		to 2010	9.76%	
		to 2011	9.45%	
		to 2012	9.07%	
		Selected 13&14	7.00%	

For this variation of the method, it is assumed that the data is being used to derive estimated ultimate claims for only the latest two cohorts. This situation might arise in practice, if, for example, claim amount-based methods are deemed to produce reasonable estimates for the earlier cohorts, but an alternative to such methods is required for the more recent (and hence less developed) cohorts.

Having derived the frequencies for each cohort, consideration would be given as to whether they need adjusting to allow for any known legislative or other trends, before they can be used as a basis for estimating the frequency for the latest two cohorts. In this case, it is assumed that no such adjustment is necessary.

Various averages of the claims frequency are taken, as shown in Table 3.35. The averages to consider should not include 2013 and 2014, since this method uses the frequency for the more developed cohorts as a basis for deriving the frequency for these recent cohorts. Although the average for 2012 and prior appears to be around 9%, there is also some evidence of an overall fall in frequency across the cohorts, and so a value of 7% is selected for 2013 and 2014. In practice, this selection may be informed by a discussion with relevant claims and underwriting staff to understand the reasons for the apparent trend. Having selected this value, the estimated ultimate number of claims is then derived by multiplying it by the premiums for each cohort, giving the figures of 420 and 400 for 2013 and 2014 respectively, as shown in the table. These are significantly higher than the corresponding ultimates produced by projecting both the claim number settled and claim number reported triangles, as given in Table 3.31. If such a difference were to arise with a real dataset, this would be investigated further before proceeding.

The next step in the application of this version of the method is to derive the average claim amounts, which for this method can be based on projecting to ultimate either the cumulative average settled or reported data triangle, or both. In this example, there is no incurred data triangle, and so as with the example of the first version of this method, the ultimate is based on projecting just the cumulative settled average claim amount data triangle (i.e. Table 3.33), to give the same ultimate averages as for that example. These are shown in the "Avg settled" column in Table 3.34.

The final stage of this variation of the method can now be applied by multiplying the estimated ultimate number and average together for each cohort, as shown in Table 3.36. The estimated ultimates for 2013 and 2014 derived using this method can be described as those implied by using the claim frequency observed for the more developed years, combined with an estimated average cost per settled claim for each year. These ultimates could either be used directly, or as priors in a BF method.

Table 3.36 *FS Method Type 1 example (using frequency): Results summary*

Cohort	Ultimate			
	No.	Avg	Amount	Reserve
2013	420	16.63	6,984	5,620
2014	400	13.95	5,580	5,236

The higher reserve value for these two cohorts, compared to the first version of this method, is just a function of the higher number of ultimate claims that was implied by using the selected claim frequency, as noted earlier.

The average values selected for the two most recent cohorts could also have been based on using the averages from the more developed cohorts, perhaps allowing for inflation. This type of approach is described, for example, in Chapter 15 of Friedland (2013).

FS Method 2 (Delay-dependent Severity or PPCF) – Worked Example

This method involves multiplying the estimated number of claims settled in each future development period by the estimated average cost per settled claim in that period. It is therefore different from the FS methods used in the first worked example, as it uses average costs per claim that are dependent on the delay to settlement. It also allows an explicit assumption for historical and future calendar period inflation to be used.

For the number of claims, the example begins by using the same estimated ultimates as for the Cumulative Averages FS method (FS Method 1) example, which were based on projecting the claim number reported triangle. The next step is to estimate the number settled in each future development period. To do this, a settlement pattern needs to be selected.

As the settlement pattern, shown in Table 3.32 and Figure 3.13, appears to be similar across the cohorts, a single pattern can be used to estimate the future number settled in each future development period. This pattern can be derived in a number of ways – for example, by taking averages of the values in each column in Table 3.32, by using the pattern directly from the CL method applied to the claim number settlement triangle, or by inferring it from the relationship between the settled to date and the selected ultimate number of claims. In this case, the CL pattern derived from the claim number settled data will be used to determine the disposal rates, from which the estimated number of future settled claims in each development period can be calculated. In doing so, it is assumed that there is no need to adjust the disposal rates to allow for the proportion of claims that might be reported and settled in the same development period, as referred to in the description of FS Method 2.

The calculation of the disposal rates is shown in Table 3.37.

Table 3.37 *FS Method Type 2 Example: Calculation of disposal rates*

	Development year									
	0	1	2	3	4	5	6	7	8	9
CL–Cum(%)	6.2	31.3	60.5	81.6	92.6	95.8	97.9	99.0	99.7	100.0
CL–Inc(%)	6.2	25.1	29.1	21.2	10.9	3.2	2.2	1.0	0.7	0.3
Selected(%)	6.0	25.0	29.0	21.0	11.0	3.0	2.0	1.0	1.0	1.0
OS @ start(%)	100.0	94.0	69.0	40.0	19.0	8.0	5.0	3.0	2.0	1.0
Disposal rate(%)	6.0	26.6	42.0	52.5	57.9	37.5	40.0	33.3	50.0	100.0

The calculations are explained as follows:

- The cumulative pattern derived from the CL method (labelled "CL–Cum" in the table) can be calculated using the relationship between the selected ultimate number of claims, derived from projecting the settlement triangle, and the settled to date. For example, to derive the value at development year 2, the ultimate settled values in Table 3.31 for 2012 cohort are used (since the latest development point for this year is 2), giving $237/392 = 60.5\%$. In practice, a curve could be fitted to these values if required.
- The incremental pattern, labelled "CL–Inc" in the table, is then derived by subtracting successive cumulative values, and then a selected incremental pattern is chosen (which sums to 100%), as shown in the next row of the table.
- To derive the disposal rates, the proportion of claims outstanding at the start of each development period (i.e. the future proportions settled) is first calculated, by subtracting from 100% the total incremental % values for earlier development points. For example, at development period 3, this is $100\% - [6\% + 25\% + 29\%] = 40\%$.
- The disposal rate in each development period is then the incremental % divided by the proportion outstanding at the start of that development period. For example, at development period 3, this will be $21\%/40\% = 52.5\%$. This represents the number of claims settled in this period, expressed as a proportion of the number of claims that are estimated to be outstanding at the end of the previous development period (i.e. the estimated number of future settled claims).

Table 3.38 shows the results of using this pattern to derive the estimated number of claims settled in each future development period.

Table 3.38 *FS Method Type 2 Example: Projection of future incremental settled claim numbers*

Cohort	0	1	2	3	4	5	6	7	8	9	Total	Future
				Development year								
2005	40	124	157	93	141	22	14	10	3	2	606	–
2006	37	186	130	239	61	26	23	6	6	3	717	3
2007	35	158	243	153	48	26	14	5	4	4	690	8
2008	41	155	218	100	67	17	6	4	4	4	615	11
2009	30	187	166	120	55	13	9	5	5	5	594	23
2010	33	121	204	87	37	13	9	4	4	4	516	34
2011	32	115	146	103	47	13	9	4	4	4	478	82
2012	43	111	83	88	46	13	8	4	4	4	404	167
2013	17	92	108	78	41	11	7	4	4	4	366	257
2014	22	91	105	76	40	11	7	4	4	4	363	341
										Total	5,349	926

The estimated future settled number of claims, shown below the diagonal in Table 3.38, are then derived using these disposal rates, by successively determining the number outstanding at each development point, moving from left to right in each row. The ultimate number of claims input to this process is the estimate selected at the start of the example – i.e. that based on projecting the reported number of claims (rather than the ultimate implied by projecting the settled number triangle). For example, for the 2012 cohort, the number settled in each development period is calculated as shown in Table 3.39.

Table 3.39 *FS Method Type 2 example: Disposal rate calculations for 2012 cohort*

Period	No. outstanding at start	Disposal rate	Estimated settled in period
3	167.0	52.5%	87.7
4	79.3	57.9%	45.9
5	33.4	37.5%	12.5
6	20.9	40.0%	8.4
7	12.5	33.3%	4.2
8	8.4	50.0%	4.2
9	4.2	100.0%	4.2
		Total	167.0

The first row in this table is calculated by starting with the total estimated future number of settled claims for this cohort, which is 167 (calculated from the ultimate of 404 less settled to date of 237) and then proceeding as follows:

$$167.0 \quad \times \quad 52.5\% \quad = 87.7.$$

No. outstanding claims   Disposal rate period 3

The next row is then

$$[\quad 167.0 \quad - \quad 87.7 \quad] \times \quad 57.9\%$$

No. OS at start   No. settled in previous period   Disposal rate period 4

$$= 79.3 \times 57.9\% = 45.9.$$

The remaining rows continue in a similar way, and comparable calculations are done for all other cohorts, resulting in the values below the diagonal in Table 3.38 (which are rounded to whole units, for ease of presentation).

An alternative to the disposal rate approach could in theory have been used here, by applying the selected cumulative settlement pattern to the estimated ultimate number of claims, and then calculating the resulting incremental values. Although the total number of future settled claims using this approach would be the same, the incremental pattern will be different to that derived using the disposal rate approach, as the settlement pattern was not itself used to derive the ultimate number of claims (the reported data triangle was used in this case). The disposal rate approach has the advantage that it uses only the future part of the settlement pattern (for each cohort) to derive the estimated future incremental number of claims, and hence any difference between that pattern and the one implied by the selected ultimates will not affect the estimated future incremental proportion settled. The next step is to derive the estimated average future settled claim amounts, allowing for inflation, in each future cell of the triangle. As explained in the description of this method in Section 3.5.3, the first step in allowing for inflation is to select suitable historical calendar period inflation assumptions. For the purposes of this example, for simplicity, it will be assumed that a single assumption is appropriate, with a value of 13% per annum.[19] In practice, this would need to be selected based on an understanding of the inflationary impact on average costs per claim for the relevant reserving category and claim type, taking into account market knowledge where appropriate. In some cases, it might be possible to fit a model to the averages in order to estimate historical inflation, but there will always be an element of judgement here, and sensitivity testing using different inflation assumptions may be helpful.

The historical incremental average settled claim amounts given in Table 3.30 can then be adjusted using this inflation rate, to express them all in current calendar period values; the resulting triangle is shown in Table 3.40.

For example, the value for 2005 cohort in development period 1 is calculated as follows (with any difference to that shown in the table being due to rounding):

---

[19] The particular value used here is not of critical importance from the perspective of demonstrating the mechanics of this FS method. However, the value of 13% was derived using a relatively simple approach of estimating the inflation implied by the ultimate average settled amounts in FS Method 1.

Table 3.40 *FS Method 2 example: Average inflation-adjusted settled claim amounts (2014 values)*

Cohort	Development year									
	0	1	2	3	4	5	6	7	8	9
2005	26.87	16.44	9.15	10.81	6.89	42.57	15.08	17.87	85.59	33.97
2006	25.30	11.18	14.96	9.12	11.91	17.81	29.30	50.13	70.84	
2007	19.53	13.20	7.02	10.83	22.57	7.22	40.03	56.08		
2008	15.77	13.17	5.81	22.54	5.19	23.40	34.38			
2009	27.22	6.04	8.62	8.19	10.37	36.20				
2010	19.57	11.17	5.30	10.46	19.08					
2011	19.88	9.41	8.76	10.32						
2012	10.67	10.81	17.39							
2013	25.04	10.72								
2014	15.64									
Average	20.55	11.35	9.63	11.75	12.67	25.44	29.70	41.36	78.22	33.97
RawAvg	12.10	7.06	6.71	8.34	9.84	20.20	25.80	38.15	73.29	33.97
Selected	20.00	11.50	9.50	12.00	12.00	25.00	30.00	30.00	30.00	30.00

$$\underbrace{6.19}_{\text{Average from Table 3.30}} \times \underbrace{(1 + 13\%)^8}_{\text{Assumed annual inflation for 8 years}} = 16.44.$$

Then, one of the key features of this FS method can be implemented by selecting future delay-dependent estimated average future settled claim amounts. Since all the values are now expressed in 2014 calendar period values, averages can be taken across the cohorts for each development period. The calculations are shown below the triangle in Table 3.40.

Simple averages of each column are taken, and the pre-inflation adjusted values are also shown (labelled "RawAvg" in the table). A selected average at each delay is then chosen, with those in the later development periods being derived by considering the combined average across the relevant periods, to smooth the volatility (for example, the average across all values in development periods 5 onwards, excluding period 8 values, is approximately 30). Judgement is needed at this critical stage, allowing for any prior knowledge or experience of average settled claims costs, including how they vary by delay.

These estimated future average settled amounts are all expressed in current calendar year values. To use them for the future settled averages, future inflation needs to be added. For this simple example, this has been assumed to be the same as the past inflation, at 13%. This is applied across the calendar periods to derive the values given below the diagonal in Table 3.41.

This table also has the historical average settled claim amounts shown above the diagonal (before they are adjusted for inflation). Thus, for example, the value for 2013 cohort in development period 4 is the assumed value for that period in 2014 calendar year values

Table 3.41 *FS Method 2 example: Future average settled claim amounts*

Cohort	Development year									
	0	1	2	3	4	5	6	7	8	9
2005	8.95	6.19	3.89	5.19	3.74	26.11	10.45	14.00	75.74	33.97
2006	9.52	4.75	7.18	4.95	7.31	12.35	22.95	44.36	70.84	33.90
2007	8.30	6.34	3.81	6.64	15.64	5.65	35.43	56.08	33.90	38.31
2008	7.58	7.15	3.56	15.62	4.07	20.71	34.38	33.90	38.31	43.29
2009	14.77	3.71	5.98	6.41	9.18	36.20	33.90	38.31	43.29	48.91
2010	12.00	7.74	4.15	9.25	19.08	28.25	38.31	43.29	48.91	55.27
2011	13.78	7.37	7.75	10.32	13.56	31.92	43.29	48.91	55.27	62.46
2012	8.36	9.56	17.39	13.56	15.32	36.07	48.91	55.27	62.46	70.58
2013	22.16	10.72	10.74	15.32	17.31	40.76	55.27	62.46	70.58	79.75
2014	15.64	13.00	12.13	17.31	19.57	46.06	62.46	70.58	79.75	90.12

(12.00) increased by three year's inflation at 13% (i.e. 1.13^3), to give 17.31. The averages shown below the diagonal in this table therefore allow for both future inflation and the assumed relationship between delay and average settled claim cost. Judgement could be used at this stage to make appropriate adjustments to derive the final selected average future settled amounts. In this example, the values will be left unchanged. In practice, it can be useful to test the impact on the estimated reserve of alternative assumptions, and to factor in any knowledge of the remaining outstanding claims (e.g. for some cohorts, there may be relatively few outstanding claims, where there may be good knowledge of their likely settlement values). Furthermore, if the analysis is being done on net of reinsurance data, then if there is excess of loss reinsurance in place, this can be factored into the selected averages.

The final step in this FS method is to multiply the estimated future average claim amounts by the estimated future number of settled claims in each cell, to derive the future settled claim amounts. The calculations are summarised in Table 3.42. For example, for 2013, development period 4, the value of 709 is derived by multiplying the projected number of settled claims of 41 from the corresponding cell in Table 3.38 by the estimated average settled amount of 17.31 from that cell in Table 3.41 (with any differences due to rounding).

Table 3.42 also shows the reserve implied by this method, together with that using the standard CL method.

FS Method 3 (PPCI) – worked example

This method begins by selecting the ultimate number of claims for each cohort, for which the same values as for the other methods will be used – that is, those based on applying the CL method to the reported claim number triangle. This ultimate represents the number of claims "incurred" for each cohort, which are the values referred to in the "per claims incurred" element of the PPCI method.

Table 3.42 *FS Method 2 example: Results*

Cohort	0	1	2	3	4	5	6	7	8	9	Res've	CL Res
2005	358	767	611	483	527	574	146	140	227	68	–	–
2006	352	884	934	1,183	446	321	528	266	425	102	102	95
2007	291	1,002	926	1,017	751	147	496	280	136	153	289	470
2008	311	1,108	776	1,562	272	352	206	124	140	159	423	710
2009	443	693	992	769	505	471	312	176	199	225	912	985
2010	396	937	847	805	706	360	326	184	208	235	1,313	1,419
2011	441	848	1,131	1,063	644	413	374	211	239	270	2,150	2,178
2012	359	1,062	1,443	1,189	704	452	408	231	261	295	3,539	3,920
2013	377	987	1,160	1,199	709	455	412	233	263	297	4,727	4,279
2014	344	1,179	1,276	1,319	781	501	453	256	289	327	6,382	4,626
										Total	19,837	18,681

The next step is to derive the relevant incremental claim amount triangle, from which the PPCI values can be calculated. This triangle can either be the incremental claims paid/settled triangle, or alternatively that triangle adjusted for historical inflation. In this example, it will be adjusted for inflation using the same constant rate of 13% used in the PPCF (FS Method 2) example. The resulting triangle is shown in Table 3.43.

Table 3.43 *FS Method Type 3 Example: Incremental claims paid adjusted for inflation*

Cohort	0	1	2	3	4	5	6	7	8	9
2005	1,075	2,039	1,436	1,005	972	937	211	179	257	68
2006	936	2,080	1,944	2,180	727	463	674	301	425	
2007	683	2,086	1,706	1,658	1,083	188	560	280		
2008	647	2,042	1,266	2,254	348	398	206			
2009	816	1,130	1,431	983	570	471				
2010	646	1,352	1,082	910	706					
2011	636	1,082	1,278	1,063						
2012	459	1,200	1,443							
2013	426	987								
2014	344									

This table is derived by inflation adjusting the incremental data shown in Table 3.27. For example, the value for 2010 cohort in delay 2 of 1,082 is calculated as the original value in Table 3.27 of 847, inflated for two years at 13% (with differences to the value in the table being due to rounding).

Having derived this, the inflation-adjusted PPCI triangle can be created by dividing the values for each cohort in Table 3.43 by the corresponding estimated ultimate number of

Table 3.44 *FS Method Type 3 Example: Calculation of inflation-adjusted PPCI values*

Cohort	Development year									
	0	1	2	3	4	5	6	7	8	9
2005	1.77	3.36	2.37	1.66	1.60	1.55	0.35	0.29	0.42	0.11
2006	1.31	2.90	2.71	3.04	1.01	0.65	0.94	0.42	0.59	
2007	0.99	3.02	2.47	2.40	1.57	0.27	0.81	0.41		
2008	1.05	3.32	2.06	3.67	0.57	0.65	0.34			
2009	1.37	1.90	2.41	1.65	0.96	0.79				
2010	1.25	2.62	2.10	1.76	1.37					
2011	1.33	2.26	2.67	2.22						
2012	1.14	2.97	3.57							
2013	1.16	2.70								
2014	0.95									
Average	1.23	2.78	2.55	2.34	1.18	0.78	0.61	0.37	0.51	0.11
Avg last 4	1.14	2.64	2.69	2.33	1.12	0.59	0.61	0.37	0.51	0.11
Avg last 2	1.06	2.83	3.12	1.99	1.16	0.72	0.57	0.41	0.51	0.11
Selected	1.15	2.65	2.70	2.35	1.16	0.50	0.50	0.40	0.40	0.40

Table 3.45 *FS Method 3: PPCI – Past and Future PPCI values adjusted for inflation*

Cohort	Development year									
	0	1	2	3	4	5	6	7	8	9
2005	0.59	1.27	1.01	0.80	0.87	0.95	0.24	0.23	0.37	0.11
2006	0.49	1.23	1.30	1.65	0.62	0.45	0.74	0.37	0.59	0.45
2007	0.42	1.45	1.34	1.47	1.09	0.21	0.72	0.41	0.45	0.51
2008	0.51	1.80	1.26	2.54	0.44	0.57	0.34	0.45	0.51	0.58
2009	0.75	1.17	1.67	1.30	0.85	0.79	0.57	0.51	0.58	0.65
2010	0.77	1.82	1.64	1.56	1.37	0.57	0.64	0.58	0.65	0.74
2011	0.92	1.77	2.37	2.22	1.31	0.64	0.72	0.65	0.74	0.83
2012	0.89	2.63	3.57	2.66	1.48	0.72	0.82	0.74	0.83	0.94
2013	1.03	2.70	3.05	3.00	1.67	0.82	0.92	0.83	0.94	1.06
2014	0.95	2.99	3.45	3.39	1.89	0.92	1.04	0.94	1.06	1.20

claims taken from the Selected Ultimate column in Table 3.31. This produces the triangle shown in Table 3.44.

Then, for each development period, various averages can be taken, and judgement applied based on experience and any prior knowledge of PPCI amounts to choose a selected value, as shown at the foot of this table. These values are all in 2014 calendar year terms, so to use these selections to derive the reserve estimates for each cohort, they need to have future inflation added, which is assumed to be 13% as for FS Method 2. This produces the results below the diagonal in Table 3.45. The values above the diagonal are the original PPCI values, before adjustment for inflation.

The final step in the PPCI method is to multiply the estimated future PPCI values in each cohort by the corresponding estimated ultimate number of claims. This gives the results shown in Table 3.46.

For example, the value of 598 in 2012 cohort, delay 4, is calculated as the PPCI for this cell of 1.48 (from Table 3.45) multiplied by the ultimate number of claims for this cohort of 404 (with any differences due to rounding).

Table 3.46 also shows the reserve using this method, derived by summing the future values. The reserve using the CL Method is also shown, for comparison purposes.

Table 3.46 *FS Method 3: PPCI – Calculation of estimated future claims and reserve*

Cohort	0	1	2	3	4	5	6	7	8	9	Res've	CL Res
2005	358	767	611	483	527	574	146	140	227	68	–	–
2006	352	884	934	1,183	446	321	528	266	359	324	324	95
2007	291	1,002	926	1,017	751	147	496	280	312	352	664	470
2008	311	1,108	776	1,562	272	352	206	278	314	355	1,153	710
2009	443	693	992	769	505	471	336	303	343	387	1,369	985
2010	396	937	847	805	706	292	329	298	337	380	1,636	1,419
2011	441	848	1,131	1,063	627	305	345	312	352	398	2,339	2,178
2012	359	1,062	1,443	1,073	598	291	329	298	336	380	3,306	3,920
2013	377	987	1,117	1,098	613	298	337	305	344	389	4,501	4,279
2014	344	1,087	1,251	1,231	687	334	378	342	386	436	5,045	4,626
											20,338	18,681

Operational Time with FS Method 3 (PPCI) – Worked Example

For this example, it is assumed that it has not been possible to derive the operational time values by ordering the individual claims according to their settlement date. Instead, a more approximate approach is used, based on the available triangle data.

The first stage of this example is to derive the operational time values, which, as explained in Section 3.5.4 represent the proportion settled (by number) at a particular development point. The cumulative settled number triangle and the estimated ultimate number of claims are therefore needed for this calculation, with the ultimate being the same as used in the FS Method 3 worked example. These are shown in Table 3.47.

A triangle showing the cumulative operational time values can then be calculated, as shown in Table 3.48.

For example, the operational time for 2009 at delay 2 of 64.48% is the cumulative number settled of 383 taken from the corresponding cell in Table 3.48 divided by the estimated ultimate number of claims for this cohort of 594.

The next step is to select operational time values for use in the subsequent calculations. Table 3.48 shows two alternative values – the latest value (i.e. the value for the most recent

Table 3.47 *FS Method 3 PPCI in operational time – worked example: Cumulative settled claim numbers and estimated ultimate*

Cohort	Development year										Ult'
	0	1	2	3	4	5	6	7	8	9	
2005	40	164	321	414	555	577	591	601	604	606	606
2006	37	223	353	592	653	679	702	708	714		717
2007	35	193	436	589	637	663	677	682			690
2008	41	196	414	514	581	598	604				615
2009	30	217	383	503	558	571					594
2010	33	154	358	445	482						516
2011	32	147	293	396							478
2012	43	154	237								404
2013	17	109									366
2014	22										363

Table 3.48 *FS Method 3 PPCI in operational time – worked example: Calculation of operational time (%)*

Cohort	Calcs	Development year									
		0	1	2	3	4	5	6	7	8	9
2005	0.00	6.60	27.06	52.97	68.32	91.58	95.21	97.52	99.17	99.67	100.00
2006	5.16	5.16	31.10	49.23	82.57	91.07	94.70	97.91	98.74	99.58	
2007	5.07	5.07	27.97	63.19	85.36	92.32	96.09	98.12	98.84		
2008	0.00	6.67	31.87	67.32	83.58	94.47	97.24	98.21			
2009	5.05	5.05	36.53	64.48	84.68	93.94	96.13				
2010	0.00	6.40	29.84	69.38	86.24	93.41					
2011	0.00	6.69	30.75	61.30	82.85						
2012	0.00	10.64	38.12	58.66							
2013	4.64	4.64	29.78								
2014	0.00	6.06									
Latest		6.06	29.78	58.66	82.85	93.41	96.13	98.21	98.84	99.58	100.00
Avg		6.30	31.45	60.82	81.94	92.80	95.87	97.94	98.92	99.63	100.00
Select		6.06	29.78	58.66	82.85	93.41	96.13	98.21	98.84	99.58	100.00

cohort at each development point) and the average across the cohorts. In this case, the latest value is chosen as the selected value.

Having selected the operational time values at each development point (on a cumulative basis here), the cumulative PPCI at those operational time values need to be estimated. This starts with the inflation-adjusted PPCI used in the FS Method 3 worked example, derived by accumulating the triangle given in Table 3.44, to produce the triangle in Table 3.49.

This has the cumulative PPCI values at each development point, which need to be translated to an operational time basis. The calculations are summarised in Table 3.50.

Each of the values in Table 3.50 are derived using linear interpolation of the original PPCI values in development time. The first column requires a suitable lower value to be used, as shown in the "Calcs" column of Tables 3.48 and 3.49. For example, consider 2006,

Table 3.49 *FS Method 3 PPCI in operational time – worked example: Cumulative inflation adjusted PPCI*

Cohort	Calcs	Development year								
		0	1	2	3	4	5	6	7	8
2005	0.00	1.77	5.14	7.51	9.17	10.77	12.32	12.66	12.96	13.38
2006	1.31	1.31	4.21	6.92	9.96	10.97	11.62	12.56	12.98	13.57
2007	0.99	0.99	4.01	6.49	8.89	10.46	10.73	11.54	11.95	
2008	0.00	1.05	4.37	6.43	10.10	10.66	11.31	11.64		
2009	1.37	1.37	3.28	5.69	7.34	8.30	9.09			
2010	0.00	1.25	3.87	5.97	7.73	9.10				
2011	0.00	1.33	3.60	6.27	8.49					
2012	0.00	1.14	4.11	7.68						
2013	1.16	1.16	3.86							
2014	0.00	0.95								

Table 3.50 *FS Method 3 PPCI in operational time – worked example: Cumulative inflation adjusted PPCI at selected operational time values*

Cohort	Selected operational time values(%)									
	6.06	29.78	58.66	82.85	93.41	96.13	98.21	98.84	99.58	100.00
2005	1.63	5.39	8.12	10.17	11.55	12.45	12.79	12.90	13.31	13.50
2006	1.41	4.06	7.78	9.99	11.39	12.04	12.71	13.05	13.57	
2007	1.12	4.14	6.17	8.62	10.54	10.75	11.60	11.95		
2008	0.96	4.10	5.93	9.93	10.61	11.05	11.64			
2009	1.44	2.87	5.19	7.19	8.25	9.09				
2010	1.19	3.86	5.40	7.38	9.10					
2011	1.20	3.50	6.04	8.49						
2012	0.65	3.20	7.68							
2013	1.31	3.86								
2014	0.95									

delay 0. The "target" operational time value is 6.06% for this delay, whereas the cumulative PPCI are only known at the operational time values shown in Table 3.48, as summarised in Table 3.51.

Table 3.51 *FS Method 3 PPCI in operational time – worked example: Calculation example for 2006 cohort at delay 0*

Operational time value	Cumulative PPCI
5.16%	1.31
31.10%	4.21

Hence, the estimated cumulative PPCI for 2006 at operational time value of 6.06% is calculated using linear interpolation:

$$1.31 + \left[\frac{6.06\% - 5.16\%}{31.10\% - 5.16\%} \right] \times (4.21 - 1.31) = 1.41.$$

All other values in Table 3.50 are calculated in a similar way, with the linear interpolation being applied to the left or right of each cell, according to whether the "target" operational time value is below or above the corresponding operational time value in the relevant cell.

The remaining stages follow those of the PPCI method in development time. First the incremental PPCI are derived and selected values chosen at each development period. The results are given in Table 3.52.

Table 3.52 *FS Method 3 PPCI in operational time – worked example: Incremental PPCI at selected operational time values*

Cohort	Selected operational time values(%)									
	6.06	29.78	58.66	82.85	93.41	96.13	98.21	98.84	99.58	100.00
2005	1.63	3.76	2.74	2.04	1.38	0.91	0.33	0.11	0.41	0.19
2006	1.41	2.65	3.72	2.21	1.40	0.65	0.67	0.34	0.52	
2007	1.12	3.02	2.03	2.45	1.92	0.21	0.85	0.35		
2008	0.96	3.14	1.83	4.00	0.68	0.44	0.59			
2009	1.44	1.43	2.32	2.01	1.06	0.85				
2010	1.19	2.68	1.54	1.98	1.72					
2011	1.20	2.30	2.54	2.46						
2012	0.65	2.56	4.47							
2013	1.31	2.54								
2014	0.95									
Avg	1.18	2.68	2.65	2.45	1.36	0.61	0.61	0.27	0.47	0.19
Avg last 4	1.03	2.52	2.72	2.61	1.34	0.54	0.61	0.27	0.47	0.19
Ag last 2	1.13	2.55	3.50	2.22	1.39	0.64	0.72	0.34	0.47	0.19
Select	1.03	2.50	2.70	2.50	1.35	0.50	0.40	0.40	0.40	0.40

The final two stages involve completing the future incremental PPCI values. To do this, future inflation is included at 13% per annum as with the PPCI method using development time (as shown in Table 3.53) and then these values are multiplied by the corresponding estimated ultimate number of claims. This gives the estimated future claims and hence the reserve, as shown in Table 3.54. This table also shows, in the final column, the PPCI (FS Method 3) reserve using development time.

For this example, the results are very similar for the PPCI method using development time and operational time (a total reserve of 20,338 vs 20,558), but where there have been material changes in settlement speed this may not be the case.

In some implementations of operational time for FS methods the calculations can be more complicated than shown in this example, depending on how the "target" operational time intervals relate to the known operational time values for each cohort and development point. An example of such complications is given in the PPCI worked example in Chapter 16 of Hart et al. (2007). However, as mentioned in Section 3.5.4, these complications can

Table 3.53 *FS Method Type 4 PPCI in operational time – worked example: Future PPCI at selected operational time values*

Cohort	Selected operational time values(%)									
	6.06	29.78	58.66	82.85	93.41	96.13	98.21	98.84	99.58	100.00
2005	1.63	3.76	2.74	2.04	1.38	0.91	0.33	0.11	0.41	0.19
2006	1.41	2.65	3.72	2.21	1.40	0.65	0.67	0.34	0.52	0.45
2007	1.12	3.02	2.03	2.45	1.92	0.21	0.85	0.35	0.45	0.51
2008	0.96	3.14	1.83	4.00	0.68	0.44	0.59	0.45	0.51	0.58
2009	1.44	1.43	2.32	2.01	1.06	0.85	0.45	0.51	0.58	0.65
2010	1.19	2.68	1.54	1.98	1.72	0.57	0.51	0.58	0.65	0.74
2011	1.20	2.30	2.54	2.46	1.53	0.64	0.58	0.65	0.74	0.83
2012	0.65	2.56	4.47	2.83	1.72	0.72	0.65	0.74	0.83	0.94
2013	1.31	2.54	3.05	3.19	1.95	0.82	0.74	0.83	0.94	1.06
2014	0.95	2.83	3.45	3.61	2.20	0.92	0.83	0.94	1.06	1.20

Table 3.54 *FS Method 3 PPCI in operational time – worked example: Future claims paid and reserves*

Cohort	Development year										Res've	PPCIRes
	0	1	2	3	4	5	6	7	8	9		
2005	358	767	611	483	527	574	146	140	227	68	–	–
2006	352	884	934	1,183	446	321	528	266	359	324	324	324
2007	291	1,002	926	1,017	751	147	496	280	312	352	664	664
2008	311	1,108	776	1,562	272	352	206	278	314	355	1,153	1,153
2009	443	693	992	769	505	471	268	303	343	387	1,302	1,369
2010	396	937	847	805	706	292	264	298	337	380	1,570	1,636
2011	441	848	1,131	1,063	729	305	276	312	352	398	2,372	2,339
2012	359	1,062	1,443	1,141	696	291	263	298	336	380	3,407	3,306
2013	377	987	1,117	1,168	713	298	270	305	344	389	4,604	4,501
2014	344	1,025	1,251	1,309	799	334	302	342	386	436	5,160	5,045
											20,558	20,338

be overcome if it is possible to derive average claim amounts at specific operational time values, using the underlying claims data. This ought to be possible in many cases, and will usually be preferable to the more approximate approach described here.

3.6 The Cape Cod Method

This method can be regarded as a subset of the BF method, which involves a particular approach for deriving the prior assumption. It gets its name from the location of a summer school devoted to reinsurance reserving held in August 1983, at which the method was discussed. It is also known as the "Stanard–Bühlmann" method, since, as noted in Stanard (1985), it is described in Stanard (1980) and Bühlmann (1983). It is abbreviated here as "CC". It was included as a distinct deterministic method in the survey at ASTIN Committee (2016), and although it is not one of the widely used reserving methods, some respondents indicated that it was a method they used, either as one of the main reserving methods or for peer review purposes, for example.

The approach in the CC method to deriving the prior assumption in the BF formula involves first calculating an average expected loss ratio across all cohorts, using the total claims to date in the numerator and an estimate of the so-called "used-up" premium in the denominator. The latter is derived, for each cohort, by multiplying the assumed proportion developed from the CL method (derived from the claims, not the premium) by the relevant premium amount for the cohort. In doing this calculation, where possible the claims are usually adjusted so that they are all on the same basis – i.e. they are adjusted for claims inflation and any other differences between the cohorts related to claims trends (e.g. legislative reforms). Similarly, where possible the premiums are adjusted for rate changes using what are described as "on-level" factors in some contexts.[20] These "trend" adjustments are effectively the same as those made in the calculation of the priors in the version of the BF method which uses adjusted ultimates derived from the CL method, as described in Section 3.4.[21] The usual approach is to adjust the claims and premiums so that they are all expressed in terms of values appropriate for the latest cohort.

Once this overall average loss ratio has been determined, to use it in the CC method for a particular cohort, it is first adjusted so that it is expressed in values appropriate for that cohort, using the claim and premium adjustment factors referred to above. It is then used as the prior assumption in the standard BF formula for each cohort. The method can be applied using either paid claims or incurred claims, the only variation being the assumed proportion developed.

The validity of the results derived using the CC method is critically dependent on whether the approach to deriving the average loss ratio is appropriate, and whether it is suitable for use as a prior for a particular cohort. As shown in the mathematical formulation of the CC method in the next section, the average expected loss ratio (i.e. the prior) can be shown to be a weighted average of the ultimate loss ratios derived from the CL method, with the weight for each cohort being the product of the premiums for that cohort and the percentage developed (as estimated by the CL method). These weights have some intuitive logic, since the cohorts which have higher premium and which are more developed are assigned more weight in the calculation of the average loss ratio. However, whether this average loss ratio is appropriate to use as a prior in the CC method for a particular cohort will depend on several factors, including whether the business mix has changed and whether there is any trend in the loss ratios across cohorts. If the average loss ratio is not

[20] For example, this term is commonly used in the US. For further details, see Chapter 12 of Friedland (2013).

[21] In fact, the only difference between the two approaches for deriving the priors is that the CC method uses a form of weighted average across the cohorts. This can be seen in the second worked example later in this section.

representative of the possible performance of a particular cohort, then the CC method is not likely to produce a reasonable estimate of the reserve for that cohort. Furthermore, since the CC method simply involves using the average loss ratio as the prior in the BF formula, the reasonableness of the ultimate claims (and hence reserve) derived using the CC method will depend on the same considerations as for the BF method, as referred to in Section 3.4.

The mechanics of the method, both with and without trend adjustments, can be seen in the worked example given later in this section.

Modifications to the standard CC method include using an alternative exposure measure rather than premiums in the formula, and using only a subset of the cohorts to derive the prior assumption. There is also a so-called "generalised" form of the method, which is explained in the next subsection.

3.6.1 Generalised Cape Cod Method

In the CC method described above, other than adjusting for trend, the same average loss ratio is used as the prior in the BF calculation for each cohort. An alternative approach to this was suggested in Gluck (1997), which is referred to as the Generalised Cape Cod method (abbreviated as the "GCC method" here). For the purpose of this description of the method, the cohort for which the prior (and hence ultimate) is being derived is referred to as the Target Cohort.

The GCC method involves deriving a different trended prior for each Target Cohort, by assigning a weight to each cohort in the calculation of the average loss ratio that depends on the proximity of that cohort to the Target Cohort. The idea is to assign more weight in the average to cohorts that are close to the Target Cohort and less to those that are further away. This assumes that cohorts which are closer to the Target Cohort are more likely to be similar in terms of the relevance of their loss ratios to that of the Target Cohort (e.g. due to claims and premium trends being more similar) than those which are further away. This may not necessarily be the case, of course, in particular situations, and in theory other weighting approaches could easily be used.

Some texts (e.g. Lyons et al., 2002 and Friedland, 2013) refer to the distance between the Target Cohort and the other cohorts as the "trending period". The weight suggested in Gluck (1997) is expressed as a selected parameter value (between 0 and 1) raised to the power of the absolute trending period. For example, if the parameter value is, say, 0.5, then when calculating the average loss ratio for a particular Target Cohort, the values for the Target Cohort itself are assigned a weight of 1 (from 0.5^0), the values for the cohorts immediately before and after the Target Cohort are assigned a weight of 0.5 (from 0.5^1) and so on. This is an exponentially decaying weight, and the parameter is therefore often referred to as the decay factor. It is selected based upon the degree of confidence (i.e. reliability) that there is assumed to be in the CL ULR for the Target Cohort itself; where the confidence is high, then a parameter closer to 0 would be more appropriate, since this will assign high relative weight to the Target Cohort and low relative weight to the other cohorts. The opposite is true where degree of confidence is low – e.g. as may be the case with some long tail classes of business. As pointed out in Friedland (2013), as the parameter approaches 0, the GCC method becomes equivalent to the CL method for that cohort, and when it approaches 1, the GCC method becomes equivalent to the CC method.

The mechanics of employing the GCC method in practice are straightforward, and can be seen in the worked example that follows, and in the mathematical formulation of the method. In effect, it represents a formalised alternative way of deriving the prior assumption for a particular cohort in the BF formula. As for the BF and CC methods, judgement will still be needed to assess whether this prior is appropriate for use in deriving an estimated ultimate, and hence reserve, for that cohort. In some cases, the average loss ratio procedure will need bespoke adjustments to allow for features of the data that are not incorporated in the standard GCC procedure (e.g. removal of individual cohorts from the calculation of the average loss ratio due to genuinely exceptional claims experience or known data errors).

When the % developed is used in the weights to derive the average loss ratio in the GCC method (and the CC method), it is effectively being used as a proxy for the inverse of the variance for each cohort – i.e. less developed cohorts, which will have low % developed are expected to have more variance and so are assigned less weight in the formula. Other approaches to allowing for the relative variance between the cohorts are also possible, such as the "variance functions" described in Gluck (1997). Lyons et al. (2002) provides a worked example of the GCC method applied to incurred claims, where the variance function is defined as the % developed derived from CL method applied to the paid claims. Regardless of what approach is used, the critical issue will always be whether the resulting prior assumption is appropriate to use for the relevant cohort, which will require judgement to be exercised.

3.6.2 Mathematical Formulation of CC Method

This formulation assumes that premiums are used as the exposure measure. First, assume that neither the premiums or claims are adjusted for any trends such as inflation and rate changes etc. As for the mathematical formulation of the CL and BF methods, the % developed derived from the CL method that is applicable for cohort i at its latest development point $I - i$ is denoted by $\hat{\alpha}_{I-i}$. The premium for cohort i is denoted by π_i. The average loss ratio used in the CC method, denoted by $\hat{\kappa}$, is determined as follows:

$$\hat{\kappa} = \frac{\sum\limits_{i=0}^{I} C_{i,I-i}}{\sum\limits_{i=0}^{I} \hat{\alpha}_{I-i}\pi_i}. \tag{3.23}$$

This is used to calculate the prior assumption for the ultimate claims for cohort i as $\hat{\kappa}\pi_i$, which is then used in exactly the same way as for the BF method to derive the estimated reserve for cohort i by

$$\hat{R}_{i,J}^{CC} = \hat{\kappa}\pi_i(1 - \hat{\alpha}_{I-i}). \tag{3.24}$$

Similarly, the ultimate claims for cohort i under the CC method is

$$\hat{C}_{i,J}^{CC} = C_{i,I-i} + \hat{\kappa}\pi_i(1 - \hat{\alpha}_{I-i}).$$
(3.25)

Equation (3.23) can be written as

$$\hat{\kappa} = \frac{\sum\limits_{i=0}^{I} \pi_i \frac{C_{i,I-i}}{\hat{C}_{i,J}^{CL}} \frac{\hat{C}_{i,J}^{CL}}{\pi_i}}{\sum\limits_{i=0}^{I} \hat{\alpha}_{I-i}\pi_i} = \frac{\sum\limits_{i=0}^{I} \hat{\alpha}_{I-i}\pi_i \frac{\hat{C}_{i,J}^{CL}}{\pi_i}}{\sum\limits_{i=0}^{I} \hat{\alpha}_{I-i}\pi_i}.$$
(3.26)

This can be seen as a weighted average of ultimate loss ratios derived from the CL method, with the weight for each cohort being the product of the premiums for that cohort and the percentage developed (as estimated by the chain ladder). These weights are intuitively reasonable, since the cohorts which have higher premium and which are more developed are assigned more weight in the calculation of the average loss ratio.

If the premiums and claims are adjusted for rate changes and inflation, then the adjustments are applied to the relevant figures in the numerator and denominator in the calculation of $\hat{\kappa}$. If the adjusted claims for cohort i are denoted by $C_{i,I-i}^*$ and the adjusted premiums by π_i^*, then the adjusted average loss ratio, denoted by $\hat{\kappa}^*$, becomes

$$\hat{\kappa}^* = \frac{\sum\limits_{i=0}^{I} C_{i,I-i}^*}{\sum\limits_{i=0}^{I} \hat{\alpha}_{I-i}\pi_i^*}.$$
(3.27)

As in the unadjusted case, denoting the adjusted ultimate claims for cohort i by $\hat{C}_{i,J}^{CL*}$, this can be rewritten as

$$\hat{\kappa}^* = \frac{\sum\limits_{i=0}^{I} \hat{\alpha}_{I-i}\pi_i^* \frac{\hat{C}_{i,J}^{CL*}}{\pi_i^*}}{\sum\limits_{i=0}^{I} \hat{\alpha}_{I-i}\pi_i^*}.$$
(3.28)

When this adjusted loss ratio assumption is used in the CC method as a prior in the BF formula for a particular cohort, it is adjusted to remove the premium and claims trend factors so that it is expressed in values appropriate for that cohort. Similarly, if an alternative exposure measure is used rather than premiums, this is used instead of $\hat{\pi}_i$ in the formula. This formulation of the CC method is effectively the same as that presented in Section 6 of Lyons et al. (2002).

Mathematical Formulation of the Generalised Cape Cod Method

If claims and premiums are not trend adjusted, the average loss ratio for cohort i in the GCC method is defined as

$$\hat{\kappa}_i^{GCC} = \frac{\sum\limits_{j=0}^{I} \hat{\alpha}_{I-j}\pi_j \frac{\hat{C}_{j,J}^{CL}}{\pi_j} D^{|i-j|}}{\sum\limits_{j=0}^{I} \hat{\alpha}_{I-j}\pi_j D^{|i-j|}}.$$
(3.29)

$D^{|i-j|}$ is the decay factor (in the range [0,1]), where D is selected based on the considerations given in the description of the GCC method earlier in this section. Note that this average loss ratio formula differs from the equivalent formula in the CC method in that $\hat{\kappa}$ is specific to each cohort i, since the decay factor depends on the distance between the cohort for which the GCC ultimate is being derived and the other cohorts in the formula. Hence the notation $\hat{\kappa}_i^{GCC}$.

For each cohort, this average loss ratio is then used as the prior in the standard BF formula. The average loss ratio formula is effectively the same in the trend-adjusted version, but with $C_{i,J}^{CL}$ and π_i being replaced by the trend-adjusted values.

As for the standard CC method, the GCC ULR for each cohort is calculated using the BF formula with the average loss ratio, as defined above, used as the prior, either with or without trend adjustments. This is demonstrated further in the worked example of the GCC method given later in this section.

3.6.3 *Worked Example of the Cape Cod Method*

This worked example has two applications of the CC method – first without applying trend adjustments to the CL method ULRs and then with these adjustments. In practice, it will generally be preferable to make the trend adjustments, but in some cases the relevant data is not available, in which case the results will have certain limitations. This comment applies equally well for the BF method if the prior for a particular cohort is derived from estimated ULRs for earlier cohorts, as described in Section 3.4.

Example Without Trend Adjustments

Using the same data as for the CL and BF worked examples, the calculations for the CC method, without any trend adjustment to the CL method ULRs, are shown in Table 3.55. The CC weight in this table is the Premium multiplied by the % developed, which is then used to derive the weighted average ULR as follows:

$$\text{Weighted Avg ULR} = \underbrace{34,358}_{\text{Sum of weight} \times \text{ULR}} / \underbrace{42,310}_{\text{Sum of CC weight}}$$

$$= 81.21\%.$$

The next step in the CC method is to use this ULR as the prior in the standard BF formula. The resulting "CC ULR" is shown in the table.

For example, for the 2013 cohort, the CC ULR is calculated as

$$\text{CC ULR} = \underbrace{22.72\%}_{\text{Paid LR}} + \underbrace{(1 - 24.16\%)}_{1 - \% \text{ dev'd}} \times \underbrace{81.21\%}_{\text{CC weighted avg ULR}}$$

$$= 84.31\%.$$

Table 3.55 *Cape Cod method: Worked example*

Cohort	Paid	Prem	Paid LR%	% dev'd	CC Wt	CL ULR%	Wt× ULR	CC ULR%	BF ULR%
2005	3,901	5,333	73.15	100.00	5,333	73.15	3,901	73.15	73.15
2006	5,339	7,333	72.81	98.26	7,206	74.10	5,339	74.22	74.11
2007	4,909	6,875	71.41	91.27	6,275	78.24	4,909	78.50	78.39
2008	4,588	6,875	66.74	86.61	5,954	77.06	4,588	77.62	77.45
2009	3,873	6,111	63.38	79.73	4,872	79.50	3,873	79.84	81.63
2010	3,692	5,914	62.42	72.23	4,272	86.43	3,692	84.98	88.25
2011	3,483	6,316	55.15	61.53	3,886	89.63	3,483	86.39	91.70
2012	2,864	6,316	45.35	42.22	2,666	107.43	2,864	92.28	100.25
2013	1,363	6,000	22.72	24.16	1,450	94.04	1,363	84.31	98.56
2014	344	5,714	6.02	6.92	396	86.97	344	81.61	103.75
Total	34,358	62,788			42,310		34,358		

Table 3.55 also shows the BF ULR in the final column, using the "external" priors, as shown in the BF method worked example in Section 3.4.3. For each cohort, the BF ULR is higher or lower than the corresponding CC ULR, according to whether the prior ULR (shown in Table 3.25) is correspondingly higher or lower than the CC Weighted Avg ULR of 81.21%. So, for 2013, the BF ULR is higher than the CC ULR, as the BF prior is 100%, whereas for 2006, for example, the BF ULR is lower, as the BF prior is 75%.

Example With Trend Adjustments

This example is now repeated, but with trend adjustments being made to the premiums and claims so that each cohort is expressed in 2014 value terms. The same adjustment, or "trend" factors are used as those in the BF worked example, where the CL ultimates were used as a basis for the prior assumptions, as described in Section 3.4.3. The relevant indices for premiums and claims are shown in Table 3.56, in the columns labelled "ULR adj to 14(%)". Although the claims index here represents claims inflation, it could equally well incorporate other factors such as legislative changes which might have affected the claims development across the cohorts. The objective in adjusting the premiums and the claims is to put each cohort's ULR on the same effective basis, before deriving a weighted average to use as the basis for the prior assumption in the CC method. In some cases, it may be difficult to derive the relevant adjustment factors, in which case the approach follows that outlined in the worked example without trend adjustments.

The calculations needed to derive the results using the trend-adjusted CC method are given in Tables 3.56 and 3.57 – labelled Part 1 and Part 2 of the calculations respectively. Further points to note in relation to the calculations are as follows:

• The premium index given in the column labelled "Prem" under "ULR adj to 14(%)" in Part 1 of the calculations is used to adjust the premiums so that they are all in 2014

Table 3.56 *Cape Cod method worked example: Trend-adjusted calculations – Part 1*

Cohort	Paid	Prem	PaidLR(%)	ULR adj to 14(%)		% dev'd	AdjPrem
				Prem	Claims		
2005	3,901	5,333	73.15	120.00	125.00	100.00	6,400
2006	5,339	7,333	74.10	116.50	122.55	98.26	8,544
2007	4,909	6,875	78.24	114.29	119.05	91.27	7,857
2008	4,588	6,875	77.06	112.15	116.82	86.61	7,710
2009	3,873	6,111	79.50	114.29	116.82	79.73	6,984
2010	3,692	5,914	86.43	120.00	114.68	72.23	7,097
2011	3,483	6,316	89.63	114.29	113.64	61.53	7,218
2012	2,864	6,316	107.43	126.32	104.17	42.22	7,978
2013	1,363	6,000	94.04	104.35	102.46	24.16	6,261
2014	344	5,714	86.97	100.00	100.00	6.92	5,714
Total	34,358	62,788					

Table 3.57 *Cape Cod method worked example: Trend adjusted calculations – Part 2*

Cohort	CC weight	CL ULR(%)	AdjCL ULR(%)	Wt× AdjULR	CC prior(%)	CC ULR(%)
2005	6,400	73.15	76.20	4,877	78.41	73.15
2006	8,395	74.10	77.94	6,543	77.65	74.16
2007	7,171	78.24	81.50	5,844	78.41	78.25
2008	6,678	77.06	80.27	5,360	78.41	77.24
2009	5,568	79.50	81.26	4,525	79.90	79.58
2010	5,126	86.43	82.59	4,234	85.46	86.16
2011	4,441	89.63	89.12	3,958	82.14	86.75
2012	3,368	107.43	88.59	2,984	99.04	102.58
2013	1,513	94.04	92.34	1,397	83.18	85.80
2014	396	86.97	86.97	344	81.67	82.04
Total	49,056			40,066		
	CC prior(%)	81.67				

equivalent values, with the resulting adjusted premiums being given in the "AdjPrem" column.

- As with the example without trend adjustments, the CC weight for each cohort shown in Part 2 of the calculations is then equal to the adjusted premiums multiplied by the % developed.
- The CL method ULRs are then adjusted so that they are all in 2014 equivalent values. For example, the 2013 cohort CL ULR becomes

$$\text{Adj 2013 ULR} = \underbrace{94.04\%}_{\text{CL ULR}} \times \underbrace{102.46\%}_{\text{Clms adj factor}} / \underbrace{104.35\%}_{\text{Prem adj factor}}$$

$$= 92.34\%.$$

This produces the values shown in the "AdjCL ULR(%)" column of Part 2 of the calculations, which are identical to those in the BF worked example, where the BF priors were derived by adjusting the earlier cohorts CL ULRs (as per the 2014 Prior column in Table 3.26), which demonstrates the close relationship between these two approaches.

- The average loss ratio is then derived in exactly the same way as without trend adjustments, by dividing the sum of the product of the CC weights and the adjusted ULRs by the sum of the CC weights. This gives 81.67% as shown in Part 2 of the calculations. This is derived using all the cohorts, but in practice, a subset might be chosen, for example by excluding some of the more recent cohorts for which the CL ULRs might be unreliable.

- Since this value is in 2014 value terms, to use it as a prior in the application of the BF formula to each cohort for the CC method it is first translated back to the terms of each cohort. So, for example, to derive the prior for 2013 cohort, 81.67% becomes

$$2013 \text{ Prior} = \underbrace{81.67\%}_{\text{CC avg LR}} \times \underbrace{104.35\%}_{\text{Prem adj factor}} / \underbrace{102.46\%}_{\text{Clms adj factor}}$$

$$= 83.18\%.$$

- Finally, the resulting CC priors are used in the standard BF formula to derive the CC ULRs shown in the final column of Table 3.57. The calculations are not explained further here, as they follow the standard BF approach. In practice, in some cases the selection of results might only use the CC method for the more recent cohorts, but the results are shown here for all cohorts for completeness.

3.6.4 Worked Example of the Generalised Cape Cod Method

This worked example uses the trend-adjusted version of this method, applied to the same data as for the CC method, and just for the 2014 cohort. The calculations involve first deriving the adjusted premiums, as per the CC worked example, with the results as shown in Table 3.56, so they are not repeated here. The remaining calculations for 2014 cohort are summarised in Table 3.58.

The decay factor is selected as 0.75. Because this example is intended to apply for the 2014 cohort, that cohort has a factor of 1.0 and then the factor for the earlier cohorts are successively reduced by a multiplicative factor of 0.75. This gives progressively less weight to cohorts that are further away from 2014. The CC weight is the adjusted premiums, taken from the last column in Table 3.56, multiplied by the decay factor for each cohort. The rest of the calculations for deriving the GCC prior and applying the BF formula follow the same approach as for trend-adjusted CC method, with the result being a GCC prior of 84.63% and a GCC ULR of 84.79% for 2014, as shown in Table 3.58.

Table 3.59 shows the effect of using alternative decay factors.

Table 3.58 *Generalised Cape Cod method worked example for 2014: Using trend-adjusted claims and premiums*

Cohort	Decay	CC weight	CL ULR(%)	AdjCL ULR(%)	Weight× AdjULR
2005	0.075	481	73.15	76.20	366
2006	0.100	840	74.10	77.94	655
2007	0.133	957	78.24	81.50	780
2008	0.178	1,188	77.06	80.27	954
2009	0.237	1,321	79.50	81.26	1,074
2010	0.316	1,622	86.43	82.59	1,340
2011	0.422	1,874	89.63	89.12	1,670
2012	0.563	1,895	107.43	88.59	1,678
2013	0.750	1,135	94.04	92.34	1,048
2014	1.000	396	86.97	86.97	344
Total		11,708			9,909
			CC prior 14(%)	84.63	
			CC ULR 14(%)	78.83	

Table 3.59 *Generalised Cape Cod worked example with alternative decay factors*

Decay	GCC prior	GCC ULR
0.7500	84.63%	84.79%
0.5000	87.83%	87.77%
0.0001	86.97%	86.97%
0.9999	81.67%	82.04%

As expected, when the decay factor is very close to zero, both the prior and ULR are the same as the CL ULR for 2014 of 86.97%, since the only cohort contributing to the average loss ratio (i.e. the prior assumption) is the 2014 cohort itself. At the opposite extreme, when the decay factor is close to 1, the prior and ULR are the same as with the standard CC method, at 81.67% and 82.04% respectively. With intermediate decay factors, the effect will depend on the relative trend-adjusted CL ULRs across the cohorts. So, with a decay factor of 0.5, Table 3.59 shows that the resulting prior and ULR are higher than all the other corresponding values in the table; this is due to the relatively higher weights that the higher ULRs in cohorts 2011 to 2013 inclusive are assigned when this factor is used. As commented in the description of the GCC method, in practice, whatever decay factor is chosen, the resulting prior assumption will inevitably need to be scrutinised to assess whether it is appropriate for the particular cohort(s).

3.7 The Separation Method

A version of this method was originally formulated in Verbeek (1972), for use with claim number triangles. Subsequently, Taylor (1977) adapted it for use on claim amount triangles and referred to it as the Separation method. It is the latter formulation that is described here. The motivation for the method, as described in Taylor (1977), is to improve upon the CL method, when there are factors affecting the calendar period dimension of the data triangle which may make results produced by the CL method unreliable. It is not thought to be used very much in practice at the current time – for example, it was not included as a separately identified method in the survey at ASTIN Committee (2016). However, it is included here partly because it has some features that are distinct from other deterministic methods, and partly because it has received some attention in the literature in relation to the popular bootstrapping procedure which is used in the context of stochastic methods (for further details, see the Separation method paragraphs in Section 4.4.8).

The data used for the method is the incremental claims paid (or settled or finalised) data triangle and some form of exposure measure for each cohort. The exposure measure which is typically used is a claim number related item – for example, the estimated ultimate number of claims. As for the Frequency–Severity methods described in Section 3.5, this ultimate can be derived by applying the CL method to either the claim number settled or reported triangle. Each cohort in the incremental claims triangle is then divided by the corresponding exposure measure. The resulting triangle is equivalent to the PPCI triangle described in Section 3.5.3. It is then assumed that each cell in the resulting triangle is represented by two parameters – one relating to the development period (i.e. the columns in the triangle) and one relating to calendar period effects (i.e. relating to the diagonals in the triangle). The second of these parameters is usually assumed to be claims escalation or inflation, but in practice it can represent the combined effect of any calendar period factors or drivers of claims experience. The implementation of the Separation method involves applying a procedure to the data to "separate" out the two sets of parameters.

The details of how to apply the method in practice are most easily explained by introducing its mathematical formulation, although the method can also be understood by following the worked example at the end of this section.

3.7.1 Mathematical Formulation of the Separation Method

The following notation is used, which with some minor differences to ensure consistency with the remainder of this book, follows that given in Björkwall et al. (2010):

$X_{i,j}$ ~ incremental claims for cohort i, development period j,

$\hat{N}_i$ ~ estimated ultimate number of claims for cohort i,

r_j ~ parameter representing development period j, and

λ_k ~ parameter representing calendar period k.

It is assumed, as for the mathematical formulation of other methods in this book, that the data triangle is of dimension $I \times I$, although as for the other methods, it can be applied to triangles which do not have the same number of rows and columns. It is also assumed that no tail factor is required, but this could at least in theory be relaxed, with some modification to the mathematical formulation.

The Separation method described here uses the incremental claims data for each cohort, expressed in units of exposure defined by the estimated ultimate number of claims for that cohort. This is represented by $s_{i,j}$, defined as follows:

$$s_{i,j} = \frac{X_{i,j}}{\hat{N}_i}.$$

It is assumed that the observed $s_{i,j}$ values can be estimated using the following:

$$E[\hat{s}_{i,j}] = r_j \lambda_{i+j}.$$

This is shown in triangle format in Table 3.60.

Table 3.60 *Mathematical notation for the Separation method*

Cohort	Development period							
	0	1	2	3	4	. . .	$I-1$	I
0	$r_0\lambda_0$	$r_1\lambda_1$	$r_2\lambda_2$	$r_3\lambda_3$	$r_4\lambda_4$	. . .	$r_{I-1}\lambda_{I-1}$	$r_I\lambda_I$
1	$r_0\lambda_1$	$r_1\lambda_2$	$r_2\lambda_3$	$r_3\lambda_4$	$r_4\lambda_5$	. . .	$r_{I-1}\lambda_I$	
2	$r_0\lambda_2$	$r_1\lambda_3$	$r_2\lambda_4$	$r_3\lambda_5$	$r_4\lambda_6$	. . .		
3	$r_0\lambda_3$	$r_1\lambda_4$	$r_2\lambda_5$	$r_3\lambda_6$	$r_4\lambda_7$	. .		
4	$r_0\lambda_4$	$r_1\lambda_5$	$r_2\lambda_6$	$r_3\lambda_7$	$r_4\lambda_8$	.		
.		.	.	.	.			
.			.	.	.			
.			.	.				
$I-1$	$r_0\lambda_{I-1}$	$r_1\lambda_I$						
I	$r_0\lambda_I$							

It is assumed that

$$\sum_{j=0}^{I} r_j = 1. \tag{3.30}$$

The implementation of the Separation method in practice involves using a recursive procedure to derive estimates of the parameters in Table 3.60, which are denoted by $\hat{r}_j$ and $\hat{\lambda}_k$. To complete the lower right-hand part of the triangle an assumption needs to be made about the future values of $\hat{\lambda}_k$ based on analysing the pattern of estimated $\hat{\lambda}_k$, combined with any knowledge of likely calendar period (e.g. inflationary) effects in future. This will then give future estimated values, expressed in exposure units, $\hat{N}_i$, so that the reserve can be calculated by multiplying these future values for each cohort by $\hat{N}_i$.

The recursive procedure involves using the diagonal and column marginal sum equations derived from Table 3.60 and the observed data $s_{i,j}$, which are defined as follows. First, the equation for diagonal k is

$$\sum_{i=0}^{k} s_{i,k-i} = \hat{\lambda}_k \sum_{j=0}^{k} \hat{r}_j, \tag{3.31}$$

and for column j is

$$\sum_{i=0}^{I-j} s_{i,j} = \hat{r}_j \sum_{k=j}^{I} \hat{\lambda}_k. \tag{3.32}$$

These can be rearranged to give

$$\hat{\lambda}_k = \frac{\displaystyle\sum_{i=0}^{k} s_{i,k-i}}{\displaystyle\sum_{j=0}^{k} \hat{r}_j} = \frac{\displaystyle\sum_{i=0}^{k} s_{i,k-i}}{1 - \displaystyle\sum_{j=k+1}^{I} \hat{r}_j}, \tag{3.33}$$

and

$$\hat{r}_j = \frac{\displaystyle\sum_{i=0}^{I-j} s_{i,j}}{\displaystyle\sum_{k=j}^{I} \hat{\lambda}_k}. \tag{3.34}$$

The recursive procedure starts by using the constraint (3.30) in (3.33) to find $\hat{\lambda}_I$, as

$$\hat{\lambda}_I = \sum_{j=0}^{I} s_{j,I-j}.$$

In other words, $\hat{\lambda}_I$ is just the sum of the values in the leading diagonal of the triangle containing $s_{i,j}$.

The calculated value of $\hat{\lambda}_I$ is then used in (3.34) to find $\hat{r}_I$ from

$$\hat{r}_I = \frac{s_{0,I}}{\hat{\lambda}_I}.$$

This can then be used in (3.33) to derive $\hat{\lambda}_{I-1}$, which in turn is then used in (3.34) to find $\hat{r}_{I-1}$. This recursive procedure continues until all values of $\hat{\lambda}_k$ and $\hat{r}_j$ have been found. Then, the expected claim amounts in the known part of the triangle are estimated using

$$\hat{m}_{i,j} = \hat{N}_i \hat{r}_j \hat{\lambda}_{i+j}.$$

For the future part of the triangle, $\hat{\lambda}_k$ needs to be estimated where $k > I$. If it is assumed that $\hat{\lambda}_k$ changes in future at a rate of $K\%$, then the future values of $\hat{\lambda}_k$ can be estimated using

$$\hat{\lambda}_{k+t} = \hat{\lambda}_I \left(1 + \frac{K}{100}\right)^t.$$

The term K can be estimated by reviewing the ratio of the successive values of $\hat{\lambda}_k$, where $K \leq I$, or by using exponential regression on these values, as well as factoring in any expectation for future calendar period effects (e.g. related to claims inflationary drivers). If required, K could be chosen to be different across individual future calendar periods.

Having derived the values of $\hat{\lambda}_k$ for all values of k up to $2I - 1$, the future values of $\hat{m}_{i,j}$ can be derived, and hence so can the reserves.

The above formulation of the Separation method was set out in Taylor (1977) for use on claim amount triangles. This was based on the formulation in Verbeek (1972), which was designed for use with claim number triangles, but has the same structure as in Table 3.60. In that formulation, each cell in the triangle represented by Table 3.60 was assumed to follow a Poisson distribution, and then maximum likelihood was used to estimate the $\hat{\lambda}_k$ and $\hat{r}_j$ parameters. Verbeek (1972) then derives recursive formulae which have an identical format to those given earlier in this section. Hence, although maximum likelihood could also be used with the formulation of the Separation method given here, the equivalence of the two approaches makes this unnecessary.

3.7.2 Worked Example of the Separation Method

This example uses the Taylor and Ashe (1983) dataset, as used for the other worked examples in this chapter. Specifically, the claim amount and claim number triangles are the same as those used in the Frequency–Severity method examples in Section 3.5.5. For ease of reference the incremental claim amount settled and the cumulative claim number settled triangles are shown in Tables 3.61 and 3.62 respectively.

Table 3.61 *Separation method example: Incremental settled claim amounts*

Cohort	Development period									
	0	1	2	3	4	5	6	7	8	9
2005	358	767	611	483	527	574	146	140	227	68
2006	352	884	934	1183	446	321	528	266	425	
2007	291	1002	926	1017	751	147	496	280		
2008	311	1108	776	1562	272	352	206			
2009	443	693	992	769	505	471				
2010	396	937	847	805	706					
2011	441	848	1131	1063						
2012	359	1062	1443							
2013	377	987								
2014	344									

Table 3.62 also shows the estimated ultimate number of settled claims in the final column. This is derived from applying the CL method to the cumulative triangle, and is the same as the ultimate number of settled claims used in the FS examples in Section 3.5.5. This ultimate could also be based on using the number of claims reported triangle, but in the context of the Separation method worked example, it is assumed that only the settled claim number triangle is available.

The data triangle on which the recursive Separation method procedure is applied is derived by dividing the values for each cohort in the incremental settled amount triangle

Table 3.62 *Separation method example: Cumulative settled claim numbers and estimated ultimate*

Cohort	\multicolumn{10}{c}{Development period}										Ult
	0	1	2	3	4	5	6	7	8	9	
2005	40	164	321	414	555	577	591	601	604	606	606
2006	37	223	353	592	653	679	702	708	714		716
2007	35	193	436	589	637	663	677	682			689
2008	41	196	414	514	581	598	604				617
2009	30	217	383	503	558	571					596
2010	33	154	358	445	482						521
2011	32	147	293	396							485
2012	43	154	237								392
2013	17	109									348
2014	22										355

by the corresponding estimated ultimate number of settled claims. The resulting triangle is shown in Table 3.63.

Table 3.63 *Separation method example: Settled claim amount per unit of exposure*

Cohort	\multicolumn{10}{c}{Development period}									
	0	1	2	3	4	5	6	7	8	9
2005	0.59	1.27	1.01	0.80	0.87	0.95	0.24	0.23	0.37	0.11
2006	0.49	1.23	1.30	1.65	0.62	0.45	0.74	0.37	0.59	
2007	0.42	1.45	1.34	1.48	1.09	0.21	0.72	0.41		
2008	0.50	1.80	1.26	2.53	0.44	0.57	0.33			
2009	0.74	1.16	1.66	1.29	0.85	0.79				
2010	0.76	1.80	1.63	1.55	1.36					
2011	0.91	1.75	2.33	2.19						
2012	0.92	2.71	3.68							
2013	1.08	2.84								
2014	0.97									

For example, for cohort 2012, development period 2, the value of 3.68 is calculated by dividing the value of 1,443 (from the corresponding cell in Table 3.61) by 392 (the ultimate number of settled claims for 2012 given in Table 3.62). The "unit of exposure" in the title of Table 3.63 is the ultimate number of claims for each cohort.

Before explaining the recursive calculations, it is helpful to show the triangle of parameters that need to be estimated, which are given in Table 3.64. For ease of presentation, the years have been relabelled from 0 to 9.

The recursive calculations require the column and diagonal sums of Table 3.63. These are shown in Table 3.65, which also shows the other values derived using the recursive procedure, that are explained below.

Any differences between the values that are derived if the figures quoted below are used are due to the underlying calculations having been done on figures with additional decimal places.

Table 3.64 *Separation method example: Notation for parameters to be estimated*

					Development period					
Cohort	0	1	2	3	4	5	6	7	8	9
0	$r_0\lambda_0$	$r_1\lambda_1$	$r_2\lambda_2$	$r_3\lambda_3$	$r_4\lambda_4$	$r_5\lambda_5$	$r_6\lambda_6$	$r_7\lambda_7$	$r_8\lambda_8$	$r_9\lambda_9$
1	$r_0\lambda_1$	$r_1\lambda_2$	$r_2\lambda_3$	$r_3\lambda_4$	$r_4\lambda_5$	$r_5\lambda_6$	$r_6\lambda_7$	$r_7\lambda_8$	$r_8\lambda_9$	
2	$r_0\lambda_2$	$r_1\lambda_3$	$r_2\lambda_4$	$r_3\lambda_5$	$r_4\lambda_6$	$r_5\lambda_7$	$r_6\lambda_8$	$r_7\lambda_9$		
3	$r_0\lambda_3$	$r_1\lambda_4$	$r_2\lambda_5$	$r_3\lambda_6$	$r_4\lambda_7$	$r_5\lambda_8$	$r_6\lambda_9$			
4	$r_0\lambda_4$	$r_1\lambda_5$	$r_2\lambda_6$	$r_3\lambda_7$	$r_4\lambda_8$	$r_5\lambda_9$				
5	$r_0\lambda_5$	$r_1\lambda_6$	$r_2\lambda_7$	$r_3\lambda_8$	$r_4\lambda_9$					
6	$r_0\lambda_6$	$r_1\lambda_7$	$r_2\lambda_8$	$r_3\lambda_9$						
7	$r_0\lambda_7$	$r_1\lambda_8$	$r_2\lambda_9$							
8	$r_0\lambda_8$	$r_1\lambda_9$								
9	$r_0\lambda_9$									

Table 3.65 *Separation method example: Recursive calculations summary*

Cohort	Diag Sum	Col	Col Sum	$\hat\lambda$ param	Cum've $\hat\lambda$	$\hat r$ param	Cum've $\hat r$	$\hat\lambda$ Change
2005	0.59	0	7.39	6.283	78.62	0.094	1.000	
2006	1.76	1	16.00	5.574	72.33	0.221	0.906	−11.29%
2007	2.66	2	14.22	5.041	66.76	0.213	0.685	−9.55%
2008	4.06	3	11.49	5.681	61.72	0.186	0.472	12.69%
2009	6.41	4	5.23	7.933	56.04	0.093	0.286	39.64%
2010	6.23	5	2.97	7.164	48.10	0.062	0.192	−9.70%
2011	8.68	6	2.03	9.450	40.94	0.050	0.131	31.92%
2012	7.21	7	1.01	7.576	31.49	0.032	0.081	−19.83%
2013	10.55	8	0.97	10.644	23.91	0.040	0.049	40.49%
2014	13.27	9	0.11	13.271	13.27	0.008	0.008	24.69%
							Avg	11.01%

- The column and diagonal sums should be self-explanatory. The year for the diagonal sum represents the calendar year for that diagonal.

- The estimates of the λ parameters, denoted by $\hat\lambda$, represent the calendar period effects and the estimates of the r parameters, denoted by $\hat r$, represent the development period effects. The first $\hat\lambda$ parameter that is derived is $\hat\lambda_9$ – that is, the parameter in the latest calendar period or leading diagonal. Because the $\hat r$ parameters must sum to 1, $\hat\lambda_9$ is just equal to the sum of the values in this diagonal – 13.27 in this example.

- This value, and the column 9 sum of 0.11 can then be used to derive $\hat r_9$ as $0.11/13.271 = 0.008$.

- The recursive procedure now moves to the next diagonal to derive $\hat\lambda_8$. Consideration of the parameters in this diagonal in Table 3.64 shows that their sum can be expressed as $\hat\lambda_8(1 - \hat r_9)$, since the $\hat r$ parameters must sum to 1. Hence, the derived value of $\hat r_9$ can be used along with the diagonal sum of 10.55 to derive $\hat\lambda_8$ as $10.55/(1 - 0.008) = 10.644$, as shown in Table 3.65.

- Now the column 8 sum of 0.97 can be used, along with the sum of $\hat{\lambda}_8$ and $\hat{\lambda}_9$ (shown in the "Cum've $\hat{\lambda}$" column of Table 3.65) to derive $\hat{r}_8$ as $0.97/23.91 = 0.04$.
- This continues backwards through the diagonals, and across from right to left for the column sums, to derive all parameter values, as shown in Table 3.65.

To derive the estimated reserve using the Separation method, the future values of $\hat{\lambda}$ first need to be estimated. One way of doing this is to examine the ratio of successive values of the known $\hat{\lambda}$ parameters, and select a future value based on suitable averages or observed trends. These ratios are shown in Table 3.65 and are very volatile in this example.[22] A simple average gives 11.01%, which would be one possible selection for future values of $\hat{\lambda}$. Another approach would be to use exponential regression through the known $\hat{\lambda}$ values, which when applied here produces a value of 9.33%. For the purposes of this example, the simple average of 11.01% will be used.

Table 3.66 *Separation Method example: Calculation of reserves*

| Cohort | Development period | | | | | | | | | | Future | No. | Res |
	0	1	2	3	4	5	6	7	8	9			
2005	0.59	1.27	1.01	0.80	0.87	0.95	0.24	0.23	0.37	0.11	–	–	–
2006	0.49	1.23	1.30	1.65	0.62	0.45	0.74	0.37	0.59	0.12	0.12	606	89
2007	0.42	1.45	1.34	1.48	1.09	0.21	0.72	0.41	0.60	0.14	0.73	716	506
2008	0.50	1.80	1.26	2.53	0.44	0.57	0.33	0.47	0.66	0.15	1.29	689	794
2009	0.74	1.16	1.66	1.29	0.85	0.79	0.73	0.52	0.74	0.17	2.16	617	1,288
2010	0.76	1.80	1.63	1.55	1.36	0.91	0.81	0.58	0.82	0.19	3.31	596	1,723
2011	0.91	1.75	2.33	2.19	1.37	1.01	0.90	0.65	0.91	0.21	5.05	521	2,448
2012	0.92	2.71	3.68	2.74	1.53	1.12	1.00	0.72	1.01	0.23	8.34	485	3,270
2013	1.08	2.84	3.14	3.04	1.69	1.24	1.11	0.80	1.12	0.26	12.40	392	4,314
2014	0.97	3.26	3.48	3.38	1.88	1.38	1.23	0.88	1.24	0.29	17.02	348	6,043
												Sum	20,476

The calculations used to derive the reserve are summarised in Table 3.66. The values below the leading diagonal use the estimated $\hat{r}$ values along with $\hat{\lambda}$ values inflated by the required number of years. So, for example, the value of 0.72 for 2012 in development period 7 is calculated as follows:

$$\underbrace{13.271 \times}_{\hat{\lambda}_9} \quad \underbrace{(1.1101)^5}_{\text{5 calendar years' growth}} \times \underbrace{0.032}_{\hat{r}_7} = 0.72.$$

[22] In practice, such a volatile pattern would be investigated further and might suggest that the calendar period trend that the Separation method is capturing is unreliable.

The "Future" column in this table shows the sum of the individual values for each cohort after the leading diagonal. This sum is then multiplied by the estimated ultimate number of claims to give the reserve. Another reference source, Björkwall et al. (2010), uses the same data as this example and shows the same results as those in Table 3.66, as well as those using the $\hat{\lambda}$ growth assumption of 9.33%. This reference also explores the use of bootstrapping the Separation method, to produce an estimate of the full distribution of reserves, which is also discussed in Section 4.4.8.

3.8 Other Deterministic Methods

This section briefly describes a small selection of other deterministic reserving methods. None of these are thought to be widely used in practice at present, although there may be specific contexts in which some of them are the preferred method. This selection of methods is not intended to be complete, as there are many methods that have been developed for particular circumstances, such as for unusual classes of business where the established methods may not work well. Furthermore, in some individual practical situations, bespoke methods may be developed which may be entirely new or represent variations on those described in this section or the preceding sections. Where relevant, references are provided for the methods described here, which the interested reader can use to obtain further details if required.

3.8.1 Munich Chain Ladder (MCL) Method

This method was originally described in Quarg and Mack (2004), and was reprinted in Quarg and Mack (2008). Its key feature is that it is designed to be applied to both claims paid and claims incurred data triangles within a single method. This is intended to address the relatively common situation that can arise whereby the independent application of the standard CL method to these two triangle types can lead to potentially significantly different estimated ultimate claims for one or more cohorts.

This may be overcome in practice by selecting results derived from the paid triangle for some cohorts and the incurred triangle for others (or a weighting between the two), as explained further in Section 3.2.5. The MCL method effectively provides a procedure for more formally blending these two sets of ultimates, by explicitly taking into account the observed historical correlation between the paid and incurred data triangles. It can be viewed either as a deterministic reserving method or within the context of a stochastic model, as described briefly in Section 4.7.13. This section focuses on the deterministic version.

The method relies on using the standard chain ladder applied to each of the paid and incurred data triangles, except that the development factors are adjusted to take into account the correlations between these two triangles. This means that individual cohorts can have different development factors at each development point, whereas the standard chain ladder (in its unadjusted form) will use the same development factors for each cohort. The method builds on two key observations made in Quarg and Mack (2004):

1. If an individual cohort has a particularly low or high current ratio of paid to incurred claims relative to the average current ratio across all relevant cohorts, then when the paid and incurred claims are projected independently using the standard CL method (without any ad-hoc adjustments), this relativity will apply on an ultimate basis. This observation is shown mathematically in Quarg and Mack (2004). The implication of this is that, for some cohorts, independently projecting paid and incurred claims using a standard CL method will result in estimated ultimate claims which imply a ratio of paid to incurred claims that is potentially materially different to 100%, which must mean that one or both of the independent projections is invalid.

2. There can be older cohorts where the claims development has finished, but where at some point in their development they have had paid to incurred ratios that were above or below the average across cohorts at the relevant development point, but where they have still ultimately reached a ratio of 100%, which they all must do eventually.

The implication of these two observations is that where cohorts have at some point in their development shown a relatively low paid to incurred ratio, then at some subsequent point in the development there must have been relatively high paid development factors, or relatively low incurred development factors, or both. Otherwise, the ultimate paid to incurred ratio would not have reached 100% at ultimate. The opposite is true where relatively high paid to incurred ratios are observed. In other words, there is a correlation structure within any particular pair of paid and incurred data triangles that is not modelled in an independent chain ladder projection of these triangles. The MCL method seeks to allow for this correlation explicitly, by adjusting the standard CL method development factors so that where the observed paid to incurred ratio is relatively low, an above average paid development and/or a below average incurred development factor is used (and vice-versa where the observed ratio is above average).

The way in which the standard development factors are adjusted in the MCL method is described further in Quarg and Mack (2004), but in summary it involves assuming that:

- the paid development factors have a linear dependency on the preceding incurred to paid ratios (e.g. where the former are relatively low, they will usually have been preceded by relatively low incurred to paid ratios); and
- the incurred development factors have a linear dependency on the preceding paid to incurred ratios (e.g. where the former are relatively low, they will usually have been preceded by relatively low paid to incurred ratios).

The MCL method builds on this by using standardised residuals, rather than the raw data, which allows all development periods to be used in determining the way in which the standard CL development factors are adjusted. A summary of the procedure for deriving estimated ultimate claims using the MCL method is as follows (assuming tail factors are not needed):

1. Apply the standard column sum average CL method to each of the paid and incurred data triangles. This produces both the individual development factors and the average CSA development factors.

2. Derive triangles representing the ratio of the paid to incurred and incurred to paid data, and calculate the average ratio at each development point for these two triangles (using the ratio of the sum of the respective columns).

3. Derive standard deviation parameters at each development period for each of the paid and incurred data triangles, using the same approach as used with the Mack method, which is described in Section 4.2.

4. Derive standard deviation parameters at each development point for each of the ratio triangles at Step 2, using a similar formulation as per the previous step.

5. Derive four triangles of standardised residuals, using these standard deviations – in respect of the paid claims development factors, incurred claims development factors, paid to incurred ratios and incurred to paid ratios.

6. Plot two graphs as follows: The first is a plot the residuals of the paid development factors (*y*-axis) against the residuals of the preceding incurred to paid ratios (*x*-axis) and the second is a plot the residuals of the incurred development factors (*y*-axis) against the residuals of the preceding paid to incurred ratios (*x*-axis). Because of the standardisation of the residuals, the plots can be drawn across all development periods combined to analyse the relationship between the development ratios and the paid to incurred (or vice versa) ratios.

7. Fit a regression line through the origin for each of these two graphs and identify the slope of each line.

8. Using recursive formulae given in Quarg and Mack (2004), derive adjusted paid and incurred development factors, using the slope of the respective line from the previous step as the correlation parameter. The formulae are designed to adjust the development factors to take into account the relative level of the ratio of incurred to paid claims (or vice versa).

9. Apply these adjusted development factors to derive estimated ultimate claims based on each of the paid and incurred data. These ultimates will not be identical, but if the MCL method has worked, they should be a lot closer compared to an independent chain ladder projection of the two triangles.

10. Select final estimated ultimate claims for each cohort, using judgement and relevant diagnostics or analysis (e.g. blending in results from other methods for one or more cohorts, as appropriate).

The MCL method was one of the deterministic methods separately identified in ASTIN Committee (2016). This source suggests that, although the method does not appear to be in widespread use at present, it does appear to be used in some countries (e.g. Austria, Norway, Portugal, Turkey and Ukraine, where respondents to the survey indicated that it was one of either the main or peer review deterministic methods that is used). Full details of the method, including its mathematical formulation and a worked example, are given in Quarg and Mack (2004, 2008). Forray (2011) and Wright (2012) provide a useful summary and worked examples. Francis (2010) similarly provides an introduction to the method, but is also accompanied by a spreadsheet that produces results for the same example as in Quarg and Mack (2004). Finally, it is worth noting that the MCL method has been implemented in the R software via the "MunichChainLadder" function in the "ChainLadder" package. See Gesmann et al. (2014) for further details.

3.8.2 Projected Case Estimates ("PCE") Method

This is an example of a so-called "Australian" reserving method, as referred to in the description of Frequency–Severity methods in Section 3.5. The summary of the method given here is based on Claughton (2013), in which it is noted that the method is a commonly used alternative in Australia to the chain ladder, applied to incurred claims data.

The PCE method involves first deriving the following ratios or factors, for data in each cell of a triangle:

1. The Paid claims in an interval divided by the case estimates (i.e. the outstanding claims) at the start of that interval. These are referred to as the Pay Out (or "PO") factors.
2. The sum of the case estimates at the end of the interval and the paid claims in that interval, divided by the case estimates at the start of the interval. These are referred to as the Case Estimate Development (or "CED") factors.

Then, averages of the PO and CED factors are derived for each development period, across a suitable number of cohorts, based on judgement. The data is then projected forward in an iterative way using a chain-ladder type approach, which will produce a cash flow projection of paid claims, the total of which will be the estimated reserve. For example, to project forward a cohort that is currently developed to period 4, the case estimates at the end of period 4 (i.e. effectively at the start of period 5) are multiplied by each of the relevant PO and CED factors to derive an estimate of the paid claims in period 5 and the paid in period 5 plus the case estimates at the end of period 5, respectively. The latter is subtracted from the former to produce an estimate of the case estimates at the end of period 5. The process continues with these new case estimates being multiplied by the next PO and CED factors, and stops once the development is complete.

For further details of the PCE method, see Claughton (2013) and Hart et al. (2007).

3.8.3 Granular Reserving Methods

All of the methods described in this chapter rely on using some form of aggregated claims data. An obvious criticism of any reserving method that relies on such data is that it ignores the detailed underlying individual claims data, and hence may mask some important trends or patterns in this data. Efforts have been made to address this, by attempting to develop models that apply directly to the individual claims data. They are referred to in various ways, including "individual claim models", "granular reserving", "micro-level reserving" or "triangle-free reserving". Their common feature is that they tend to be stochastic in nature, whereby some form of statistical model is fitted to the data, rather than being deterministic-type reserving methods. Their principal application is to estimate the distribution of reserves, as per stochastic methods based on aggregate data.

Whilst it is possible that granular reserving methods will become more widely used for the purpose of estimating the distribution of reserves, their use for deriving a point estimate of the reserves (e.g. for financial reporting purposes) seems a long way off at present. They are described further in the stochastic methods chapter, in Section 4.7.13.

4

Stochastic Reserving Methods

4.1 Introduction

Stochastic reserving methods provide both a point estimate (e.g. a best estimate) of the future claims and an estimate of the variability surrounding that point estimate. Some of these methods also provide an estimate of the full distribution of future claims and their associated cash flows. In most practical reserving situations, if a point estimate for the reserves is required (e.g. as part of a regular reserving exercise for financial reporting purposes) this will often be determined using a combination of deterministic, as opposed to stochastic, reserving methods. Although there is no reason why stochastic methods cannot be used to derive the point estimate, in practice this is not usually the case, at least at present. Despite this, the majority of insurers will be making some use of stochastic reserving methods, in one form or another, and hence an understanding of the different methods available and their limitations is vital for most insurers.

This chapter provides a description of several stochastic reserving methods, including those that are commonly used in practice (at least in the UK in 2017), along with worked examples of these methods. There is also a brief description of other less commonly used methods and to some more recently developed approaches. Although some details of the mathematical formulation of a number of methods is included, this chapter is not intended to represent a thorough analysis of the underlying theoretical basis for the methods that are described here, but rather an overview of some of the commonly used methods. Readers who require more details of the underlying theory will find many sources in the extensive body of material that is available on stochastic reserving. Several of these sources are referred to in this chapter.

Most of the stochastic claims reserving techniques that have been developed are designed to be used where the data triangle structure is an appropriate one to use in order to estimate reserves. As discussed in Section 2.5, there are some instances where the triangle format is not the most appropriate format to use. Furthermore, there are also many reserving contexts where, whilst the triangle format might be the most appropriate format to use for analysing the data, there may be very good reasons (such as insufficient development history) that make it difficult for standard deterministic, let alone stochastic methods, to be applied

146

successfully. In these situations, other approaches need to be used, which will depend on the nature of the business or claim category being modelled.

The theory and practice of stochastic reserving continues to evolve, with new methods and approaches being developed from time to time. The practitioner can use various sources to keep up to date, including those available online which may be updated when new approaches are developed. One such source is Merz and Wüthrich (2015b), which provides a mathematical description of several stochastic methods, building on the same authors' book, Merz and Wüthrich (2008c).

4.1.1 Remainder of this Chapter

The rest of this introduction covers some background topics on stochastic reserving, including a review of the different uses for and types of stochastic methods, the stochastic representation of the Chain Ladder method and the time horizon for stochastic methods.

After this section, the next five sections explain the Mack, ODP, Bootstrap, a Bayesian-type method and finally the Merz–Wüthrich method. Several other methods are then described in less detail, followed by sections dealing with measuring reserve risk over a one-year time horizon (and extensions thereof) and combining results across reserving categories. The descriptions of each method, including the worked examples, are designed to enable some of the theory behind them to be understood, as well explaining the steps involved in applying the method in practice to a dataset. Some of the generic issues that need to be considered when they are applied in practice, such as testing the appropriateness of a method and the validity of the results, are considered in the final section of this chapter, Section 4.10, which considers stochastic reserving in practice.

Readers who only want an introduction to stochastic methods should read the uses and types of method sections, perhaps followed by a review of the worked examples for the Mack method and the Bootstrap procedure, as these are two of the most commonly used stochastic approaches. Readers who are not interested in the mathematics of the methods can skip those sections, as in most cases the other sections will provide a sufficient explanation of the method.

4.1.2 Uses for Stochastic Methods

The two main contexts in which stochastic reserving methods are used in practice are as follows:

- **Assessing the uncertainty surrounding a proposed reserve figure.** When reserving for financial reporting purposes, although the key requirement is usually to establish a single value for the reserves to be booked, understanding the uncertainty surrounding that figure can also be important. For example, if management wish to book a reserve figure in excess of an actuarial best estimate, then establishing an approximate indication of the "strength" of this proposed figure can be helpful, for example, to assist with ensuring it is consistent with any stated reserving policy and/or with the strength at the previous period end. Stochastic methods can be used, for example, to give an indication of the percentile of the proposed booked reserve figure, which can be used to measure the reserve strength. This helps to address the question that might be asked by, for example, senior management – "What is the chance of the actual claims being above (or below) the proposed reserve?" Assessment of reserve uncertainty may involve using stochastic methods as well as other approaches; estimating and communicating uncertainty is discussed in more detail in Section 5.9.

- **Assessment of reserving risk in the context of determination of capital requirements.** A key component of any risk-based capital regulatory regime is the consideration of the risk associated with the reserves, for which stochastic reserving is an essential tool. Depending on the context, the requirement can either be over an ultimate or one-year time horizon, as discussed further in Section 4.1.6. This chapter focuses on the use of stochastic reserving methods to estimate reserve uncertainty, rather than the details of how reserving risk is allowed for within a capital modelling context.

In addition to these two key contexts, stochastic methods can be used in a wide range of other situations, whenever an assessment of the uncertainty surrounding future claims outgo is required. These include direct and reinsurance pricing, commutations, M&A transactions, reasonableness of reserves for tax purposes, business planning, portfolio transfers and valuation. Furthermore, solvency regulations and financial reporting standards can potentially have an impact on the use of such methods. For example, the definition of best estimate under Solvency II includes reference to the use of probability-weighted cash flows, which might lead insurers affected by these regulations to make greater use of stochastic methods. This is explained further in to Section 6.8. International Financial Reporting Standards (IFRS) for insurance business may at some point in future require risk margins to be calculated,[1] in which case stochastic methods could also be an important tool in this context.

In reports that describe reserving work, some professional actuarial standards require at least a discussion, if not a quantification of reserve uncertainty, even when the scope of work relates only to a point estimate. Stochastic methods can therefore also be helpful in this context.

Regardless of the context for the use of stochastic methods, or the particular methods being used, there will inevitably be a considerable degree of judgement involved in the application and interpretation of the results. This is perhaps even greater than for deterministic methods. The use of benchmarking and other comparatives, such as scenario testing, to assist with interpreting the results from stochastic methods is discussed further in Section 4.10.

4.1.3 Types of Uncertainty / Errors

Before describing the different types of stochastic methods, it is helpful to consider the types of uncertainty or errors that exist when estimating future claims outgo. From the perspective

---

[1] Specifically, IFRS 17 Insurance Contracts, as referred to in Section 6.8.

of the management or Board of an insurance entity, in a reserving context, perhaps the key type of uncertainty that concerns them is that surrounding the difference between an estimate made now of the future claims outgo (i.e. the reserve figure) and the actual, but currently unknown future outgo. In other words, how wrong could the reserve estimate ultimately be that has been established now? Management might also be concerned with the uncertainty related to the timing of the outgo, but the uncertainty around the amount usually receives more focus, so that will be discussed first. The uncertainty around the amount is referred to as the "Total Uncertainty" below.

One simple way of assessing the Total Uncertainty is to look at how wrong the estimates made in the past have been relative to the corresponding actual claims outgo that subsequently transpired. If there is sufficient historical data, this can be a useful diagnostic tool, but it will not usually give the granularity required for estimating the Total Uncertainty in practice, and relies on the assumption that the future uncertainty is captured fully in the historical uncertainty, which often may not be the case. Hence, in most situations, to estimate the Total Uncertainty it is necessary to apply some form of statistical model or other procedure to the available data. If the model is inherently stochastic in nature, it can be fitted to the data and then used to produce an estimate of at least part of the Total Uncertainty. When this approach is used, there will be three different components of the Total Uncertainty:

- The model being used could be "wrong" in the sense that it doesn't accurately reflect the real world claims process that is being analysed. This is usually called "Model error".
- The parameters that are selected when the model is fitted to the data could be wrong, because they need to be estimated and hence could be different to the "true" but unobservable parameters. This is usually called "Parameter error" or "Estimation error".
- Even if the model is correct and the parameters are known, so that there is no uncertainty from these sources, there will still be residual uncertainty caused by the random nature of the underlying claims process. This is usually called "Stochastic error" or "Process error".

Some descriptions of the sources of risk in reserving include a fourth category, systemic risk, which is described as the additional uncertainty that arises from unforseen trends or shifts away from the current claims environment. In the breakdown above, this is assumed to be part of the model error.

Some authors have also used a different breakdown of Total Uncertainty. For example, Marshall et al. (2008) define two sources of risk – *independent* risk, which represents those risks arising due to the randomness in the insurance process, and *systemic* risk, which represents those risks which are common across reserving categories. They further subdivide independent risk into parameter and process risk, which effectively map to parameter and process error in the breakdown of Total Uncertainty given above. For systemic risk, they subdivide this into Internal and External systemic risk, with the former being related to the inability of the chosen model to represent the real world claims process, and the latter being related to future systemic changes in claims cost that are external to the claims modelling process. In the breakdown above of Total Uncertainty, model error is therefore effectively the same as systemic risk from Marshall et al. (2008), as long as it is deemed to include the potential impact of future systemic claims cost trends, which is logical since that is one aspect of the real world claims process that needs to be modelled.

Thus, although management might wish to focus on the Total Uncertainty, in practice, in order to estimate this, it is usually necessary to do so by breaking it down into various components, as described above. When communicating results, the uncertainty from each of these components can then be aggregated to produce an estimate of the Total Uncertainty.

When a given statistical model is fitted to a reserving dataset or triangle, it produces fitted data points to the historical and known data. The model is then used to forecast or "predict" the future, beyond the domain, or "out of the sample", of the known dataset – i.e. to estimate the future claims outgo. There will be uncertainty around this prediction, called "Prediction" error, which is made up of the second and third of the types of error listed above. Instead of "error", "risk" or "variance" is sometimes used, so that (at least approximately):

$$\text{Prediction Variance} = \text{Estimation Variance} + \text{Process Variance}.$$

In a statistical modelling context, estimation error refers to the uncertainty involved in estimating the expected or "mean" outcome. Process error still remains, however, since even if the mean were known, there is still uncertainty surrounding whether the outcome will be equal to the mean or not. In most applications of stochastic reserving methods, the prediction error and possibly also the distribution of the future claims will be estimated.

The other component of the Total Uncertainty, model error, will not usually be included implicitly or explicitly in the results that arise from applying most stochastic reserving methods. Therefore, the practitioner may need to consider what allowance, if any, should be made for model error in interpreting and using the results. Model error is very difficult to determine in practice, and may not always receive detailed consideration. Approaches to determining the potential impact of model error include reviewing the sensitivity of results to alternative underlying models, taking weighted results from different models (with due consideration of dependencies), identification of any potential future claims trends not adequately captured by the data, back-testing of models and comparison of results against scenario testing approaches.

Model error can also be regarded as being made up of different types of internal and external systemic risk. An example breakdown of these is given in Table 4.1, which is based on Marshall et al. (2008).

This can be helpful in considering what types of risk are not captured by stochastic reserving methods. For specific reserving contexts, some of the items in Table 4.1 will either be irrelevant or will already have been captured by the stochastic model, and there may be others not listed here. Marshall et al. (2008) also provides some suggestions as to how to adjust estimates of reserve uncertainty (e.g. by uplifting the Coefficients of Variation) to allow for the various categories of risk listed in Table 4.1, which may not already be captured by applying stochastic methods to the data.

Table 4.1 *Types of systemic risk*

#	Category	Error/risk type	Description
1	Internal	Specification	Failure to build a model that is fully representative of underlying process that gives rise to claims, due, for example, to complexity or lack of knowledge of that process.
2		Parameter selection	Failure to build a model that fully captures all the underlying predictors or parameters of claims cost outcome and trends.
3		Data	Error arising from poor quality or insufficient data, including due to lack of knowledge of the underlying business being modelled.
4	External	Economic and social	Normal inflation and other social/environmental trends.
5		Legislative, political and claim inflation	Known or unknown changes to legislative and political environment specific to the portfolio being analysed. Similar to "superimposed inflation".
6		Claim management process change	Changes in claim reporting, payment, estimation or finalisation.
7		Expense	Uncertainty related to cost of managing run-off of earned and unearned element of claims, where relevant.
8		Event	Uncertainty related to natural or made-made events.
9		Latent claim	Uncertainty related to sources of claim that is currently not considered to be covered by the issued policies.
10		Recovery	Uncertainty related to reinsurance or other types of recovery such as subrogation.

Regardless of the way in which risks are categorised, the important point is to consider what allowance, if any, needs to be made for categories of risk that are not captured adequately by applying the selected stochastic methods to the data. The term "Standard Error" is commonly used by reserving practitioners to describe a measure of the uncertainty surrounding a point estimate of claims reserves. In a purist statistical sense, the Standard Error is more appropriately another term for only the estimation error component of the prediction error. Hence, it may be preferable to use the term "prediction error" or "prediction variance" to describe a measure of the uncertainty surrounding a reserve estimate, which incorporates uncertainty related to both process and estimation sources (but not usually to sources of model error). This terminology has an intuitive logic, since reserving is analogous to a forecasting or prediction exercise, and stochastic reserving assists with estimating the uncertainty surrounding that forecast, taking account of both the uncertainty related to estimating the parameters of the model (i.e. parameter error) and the inherent uncertainty related to the data itself (i.e. process error). However, in part for consistency with other sources, in the description of some of the stochastic methods in this chapter (e.g. Mack's method), the term Standard Error is used to mean prediction error. It should be obvious from the context whether this is referring to the full prediction error or to just the estimation error component.

Mathematical Representation of Prediction Uncertainty

In statistical applications, a commonly used measure for assessing the prediction variance is termed the "Mean-Squared Error of Prediction", or "MSEP". Another abbreviation that might also be used in this context is "RMSEP", which refers to the Root Mean Squared Error of Prediction, which is just the square root of the MSEP, and hence is equivalent to the prediction error. Consider a variable X, which represents a generic and unknown future value that is being estimated or "predicted" (e.g. the future incremental claims in a particular cell of a triangle or the total reserve for an individual cohort). In a reserving context, X is usually estimated based on the observed past data, which is denoted by D_I. If the estimator or predictor of X is denoted by $\hat{X}$, then the so-called "conditional MSEP" of this is defined by

$$\text{MSEP}_{X|D_I}(\hat{X}) = E\left[(\hat{X} - X)^2|D_I\right]. \tag{4.1}$$

An approximate formula for (4.1) can be derived, based on certain assumptions, which breaks down the MSEP into components for process and parameter estimation variance, as follows:

$$\text{MSEP}_{X|D_I}(\hat{X}) \approx \underbrace{\text{Var}(X)}_{\text{Process variance}} + \underbrace{\text{Var}(\hat{X})}_{\text{Parameter estimation variance}}. \tag{4.2}$$

This formula is explored further in Appendix A. The approach used to derive the MSEP in practice will vary according to which stochastic reserving method or procedure is being used. For example, if a simulation-type procedure is being used (e.g. bootstrapping, as explained later in this chapter), as long as the procedure takes into account both parameter and process uncertainty, then the MSEP can be estimated by calculating the variance of the simulated forecast.

4.1.4 Types of Stochastic Methods

There are several types of stochastic method, although relatively few that are currently used commonly in practice. The different types of method can be categorised in various ways, but one of the key attributes is whether the method is analytical or simulation based. The former is characterised by some form of mathematical or statistical model which is fitted to the data and then used to produce estimates of the prediction error. In contrast, whilst most simulation-based approaches also involve specifying some form of underlying model, the estimates of prediction error are produced using stochastic simulation methods, rather than via mathematical formulae. Simulation-based methods have the additional benefit that

Table 4.2 *Selected characteristics of stochastic reserving methods*

Method	Type	Distribution	Horizon
Mack	Analytical	N	Ult
ODP	Analytical	N	Ult
ODP Bootstrap[a]	Simulation	Y	Both
Bayesian ODNB MCMC[bc]	Simulation	Y	Ult
Stochastic BF	Analytical	N	Ult
Log-linear regression[d]	Analytical	N	Ult
ODNB[b]	Analytical	N	Ult
PTF[e]	Analytical	N	Ult
Wright's Hoerl curve	Analytical	N	Ult
Merz–Wüthrich	Analytical	N	1-year
Actuary-in-the-box	Simulation	Y	1-year

[a] See main body of text for reference to bootstrapping other models, such as Mack.
[b] ODNB stands for Over-dispersed Negative Binomial.
[c] See main body of text for reference to other Bayesian models.
[d] Also called the Lognormal model.
[e] PTF stands for Probabilistic Trend Family.

it is relatively straightforward to use them to produce an estimate of the full distribution of future claims. Commonly used methods include both analytical (e.g. Mack's method) and simulation-based methods (e.g. various forms of the Bootstrap method). Table 4.2 summarises a number of stochastic methods, showing whether they are analytical or simulation based (in the "Type" column), whether they produce the full distribution of future outgo, and also whether they operate on an ultimate or one-year time horizon basis.

Only one bootstrap method is referred to in Table 4.2 – the ODP Bootstrap. In fact, it is possible to bootstrap any suitably defined analytical model, such as Mack's method, so that any such model can in theory then be used to produce a full predictive distribution, as long as an allowance for process variance can also be included. Similarly, for Bayesian methods, although only the Bayesian ODNB model using Markov-Chain Monte Carlo (MCMC) is listed in the table, many other statistical models can be formulated in a Bayesian framework.

Other characteristics could be used to describe stochastic methods, such as ease of use or understanding. These will largely be dependent on the individual practitioner's knowledge and experience, and the software available to them within which the methods can be applied. Ease of use could also be influenced by the specific features or limitations of each method, such as whether the method copes with negative incremental amounts, which could be important in some situations.

Another characteristic that might be relevant is whether the method is in common usage. According to the survey at ASTIN Committee (2016), the respondents (who were from 42 countries) indicated that the bootstrap method (of the CL method)[2] and Mack's method were the most commonly used stochastic methods. Other responses for this characteristic might apply in specific countries, and indeed there may be other methods not listed in Table 4.2 that are in common use elsewhere.

In relation to whether the method produces an estimate of the full distribution or just the mean and variance, even where the latter is the case, it may be possible to use a bootstrapping procedure applied to the relevant method to estimate the full distribution. If for some reason this is not possible, then a very approximate full distribution can be produced by fitting an analytic distribution (e.g. lognormal) using the estimated mean and variance. A simple example of this approach is given in Section 6.9.4. However, such an approach has potentially significant limitations, including the fact that it is very difficult to preserve the dependencies between cohorts.

For methods that are described as producing estimates on an ultimate basis, in many cases it may be possible to extend the methodology so as to produce results on either a single or multi-period basis. The time horizon of the stochastic method is discussed further in Section 4.1.6.

Other characteristics can be used to classify stochastic methods, including for example whether the method is recursive (e.g. Mack, Negative Binomial) or non-recursive (e.g. ODP, log-linear regression). In recursive models, the expectations of observations are conditioned on earlier observations, whereas with non-recursive methods they are not.

It can also be relevant to consider whether the method is a stochastic version of the popular deterministic chain ladder. This might be thought of as a somewhat academic question, since the chain-ladder method was initially developed as a simple numerical algorithm or procedure, without reference to any underlying stochastic model. However, there has been much discussion of this matter in the literature and it is discussed further in Section 4.1.5.

In practice, there will be a range of factors to take into account in selecting a method, and in some situations more than one stochastic method can be applied anyway. Selecting a stochastic reserving method is discussed further in Section 4.10.

Most stochastic methods that have been developed to date are applied independently to each subdivision of the reserving data. In practice, the results usually need to be combined to produce an estimate of the variability across all categories combined. The aggregation of results is discussed in Section 4.9, which also includes a description of some stochastic methods which can be applied to several categories as part of a single integrated framework.

All stochastic methods have certain assumptions underlying them or constraints regarding whether they can be applied to any particular dataset. For example, some methods can be more difficult to apply where there are negative incremental claim amounts. The key assumptions and constraints for each method are discussed in the relevant section of this chapter.

---

[2] This is labelled as the ODP Bootstrap in Table 4.2.

In order to help understand the underlying details of a selection of methods, this chapter provides worked examples for five methods – Mack, ODP, Bootstrap (of ODP and Mack), a Bayesian-type method and the Merz–Wüthrich method. This is intended to help address the commonly stated criticism of the way in which stochastic methods may be used in practice, namely that they are a "black box". These examples are designed to aid the understanding of each method, but they also show how some of the methods can be implemented in practice, using a spreadsheet for example, without the need for specialist reserving software (although that will often be used in practice). Sufficient detailed tables are given in the examples for many of the methods, so that if the reader wishes to construct their own version of the method in a spreadsheet, they can then check that their implementation produces the same results at each stage of the process.

All of the worked examples in this chapter use the same dataset as used in the worked examples for the deterministic methods in Chapter 3, namely that originally analysed in Taylor and Ashe (1983). This dataset has been referred to in several papers on stochastic reserving.

4.1.5 The Chain Ladder as a Stochastic Model

As noted in Chapter 3, the Chain Ladder (CL) method, in its various forms, is probably the most widely used deterministic claims reserving method, so it is not surprising that there has been considerable discussion in numerous papers regarding equivalent stochastic reserving methods. Equivalence in this context is deemed to mean that the stochastic method produces the same reserve estimate as that produced by a specified type of deterministic CL, when both are applied mechanically, without any judgement or interventions. In practice, of course, in most situations neither the deterministic CL method or any stochastic methods would be applied in a mechanistic way, without any overlaying of judgement or modification to the standard procedure. However, equivalence of the two approaches in the context of their most basic application is still helpful, as explained below.

Prior to Hachemeister and Stanard (1975), the CL method did not have an equivalent stochastic formulation, and was originally developed as a simple mechanical procedure that could be applied without any reference to an underlying mathematical or statistical model. Hence, the purpose of identifying an equivalent stochastic model is not to help understand its original formulation, since this was not stochastic. Rather, it is simply to find stochastic methods that are equivalent to it so that, for example, if the practitioner believes that the deterministic CL method is a suitable method to use to derive the reserve estimate in a particular situation, then they might also conclude that this provides some support for the use of an equivalent stochastic method to derive other diagnostics such as the prediction error of the reserve estimate or the full distribution of the reserves. Even if equivalence is thought to be of benefit in this way, it will not override the need to assess whether a stochastic method is itself a suitable model for a particular dataset. Selection and testing of stochastic methods is discussed in Section 4.10.3.

In practice, at least in a valuation / financial reporting context, best estimates of reserves are commonly obtained using deterministic rather than stochastic methods and several modifications to the standard mechanical procedure are used, with judgement being applied when selecting the individual assumptions within each stage of the procedure and in selecting the final results. It will not usually be possible to replicate all these bespoke adjustments when applying a stochastic method. Hence, even if the deterministic method used is based on the CL method, the best estimate reserve produced by an equivalent stochastic model will not necessarily equal the selected best estimate reserve. In this case, some form of scaling may be required in order to make the variability metrics that are produced from the stochastic method consistent with the selected best estimate. This is discussed further in the context of the bootstrap method in Section 4.4.5 and more generally in Section 4.10.4.

Several stochastic methods have been proposed that have been described as being equivalent to the deterministic CL method. These include:

- **Mack's method**: This analytic method, described in Section 4.2, has been referred to in many sources as the first one to be characterised as a stochastic version of the deterministic CL. Its equivalence is clear since it has the CL method with column sum average link ratio estimators as part of its core specification.
- **The ODP model**: This model, described in Section 4.3, is a type of Generalised Linear Model (GLM) that can be fitted to a data triangle using standard statistical fitting procedures. There are various forms of the ODP model; for example, when used with a single scale parameter and a parameter for each cohort and development period, the fitted model can be formulated so that it produces exactly the same results as the CL method with column sum average link ratio estimators. This equivalence is somewhat surprising when first encountered, particularly since the fitted parameters of the ODP model have no obvious interpretation in the context of a deterministic CL method. The equivalence can be proved from a theoretical standpoint, and is demonstrated in practice for two datasets described in the worked examples in Sections 4.3.7 and 4.3.8.
- **ODP Bootstrap**: This method[3] is described in Section 4.4 and if the right formulation of the ODP model is used, it will also be equivalent to the deterministic CL method, since that is the approach which is "bootstrapped". The equivalence between the ODP model and the CL method as noted above is key, and this is used in the method to make bootstrapping the ODP model relatively straightforward to apply in practice, as described in Section 4.4. This section also explains how it is possible to bootstrap Mack's method, which therefore defines a further approach which is equivalent to the CL method.

Other stochastic models have also been described as equivalent to the deterministic CL method, including the Poisson model (see Section 4.7.6), the Negative Binomial model (see Section 4.7.10), some formulations of the ELRF models (described in Section 4.7.8) and some Bayesian models (for an example see Section 4.5). Hence, there are several stochastic methods that might be deemed equivalent to the CL method. However, although

---

[3] Strictly speaking, this is not a distinct method in itself, but rather a procedure for generating an estimate of the predictive distribution, but its popularity as an overall stochastic approach is in part based on its inherent relationship with the CL method; that is the sense in which it is described here as CL-equivalent.

these will all produce the same reserve estimates (under certain conditions) they will not all necessarily produce the same prediction error, or distribution of reserves.

In most cases, the type of CL method for which equivalence is examined involves using the column sum average (all cohorts) for the link ratio estimators. Where references are made in this chapter to the equivalence of a particular stochastic method to the CL method, then it is this form that is intended, unless stated otherwise. Equivalance to CL methods using other types of estimators does not appear to have been examined as extensively in the literature, although, as explained in Section 3.2.6, the CL method can be defined within the statistical framework of linear regression, and by varying the "weighting" parameter different types of link ratio are produced. This is also discussed in Barnett and Zehnwirth (2000).

There has been much debate in the literature regarding which stochastic method is the "best" or most appropriate representation of the deterministic CL method. In particular, Mack and Venter (2000) and Verrall (2000b) discuss this in relation to the Mack method and the ODP model. In addition, Taylor (2011) discusses other criteria that might be used to judge the equivalence of stochastic methods with the Chain Ladder, including whether the method produces biased estimators and whether the variance of the estimators is low or high. Overall, there does not seem to be a consensus over which stochastic model can claim to be the "best" representation of the CL method.

From a practitioner's perspective, this debate is, in any case, often not of paramount importance, and there may be other equally or more important factors to consider when selecting a stochastic reserving method. This is covered further in Section 4.10, which discusses stochastic reserving in practice, including selection and testing of methods.

4.1.6 Time Horizon for Stochastic Methods

Most of the methods described in this chapter are designed to estimate the uncertainty in the future claims, with a time horizon measured between the latest diagonal of the triangle and the ultimate (or "ultimo") position. In other words, they assist in estimating the uncertainty in the estimated ultimate claims, or equivalently in the total future claims outgo (i.e. the reserve). However, a number of stochastic approaches have been developed which estimate the uncertainty over a one-year time horizon. The need for this has arisen principally from the European Solvency II regulations, which require sufficient capital to be held to cover the risk that emerges over a one-year time horizon. Measuring risk over a shorter period, such as one-year, as opposed to an ultimate time horizon is also of interest more generally, as it can assist with short-term planning for example. Methods that estimate uncertainty over an ultimate time horizon can also be adapted so that the uncertainty is subdivided into that which emerges in each future year – effectively as a series of one-year views.

In the context of reserving, a one-year view of risk can be seen as the extent to which a best estimate of reserves made using information available at the start of the year can vary between then and when a further corresponding estimate is made at the end of the year. In other words, it is a measure of how much an estimate of ultimate claims (on an undiscounted basis) might vary between the start and end of the year, allowing for claims development and emergence of new information during the year. The difference between the estimated ultimate claims at the start and end of the year is commonly referred to as the "Claims Development Result", or "CDR". At the start of the year, if the undiscounted reserves are established on a best estimate basis, then the CDR is expected to be zero – i.e. the estimated ultimate claims are not expected to change over the year, but in practice there is risk / uncertainty around this outcome of zero, due for example to claim movements being greater or less than expected. Measuring the uncertainty in the CDR, and putting aside capital to cover that risk to a defined level of confidence, will mean that the insurer has enough capital at the start of the year to cover that uncertainty, to the defined level of confidence.

When an insurer is summarising its financial result for the previous accounting year, there will normally be an item referred to as the "Prior Year Development", or "PYD". If it is positive, then the current year financial result is improved due to prior year movements, and if it is negative, then the opposite is the case. The CDR can be thought of as a prospective view of the PYD. As for the CDR, if the reserves have been established on a best estimate basis, the PYD is expected to be zero. In practice, there can be unexpected claim developments during the year, so that at the end of the year, the PYD may be non-zero. Furthermore, if the reserves are established on a basis that is designed to produce a reserve in excess of a best estimate, as may be the case for financial reporting purposes for some insurers, the expected CDR will be positive. This does not mean that the PYD will definitely be positive, as there could be unforseen adverse claims movements in the year. Most of the technical work around estimating the one-year reserve risk assumes that reserves are established on a best estimate basis, with any additional margin being dealt with separately.

Unless otherwise stated, the methods described in this chapter are designed to produce estimates of reserve uncertainty to ultimate. Methods that measure reserve risk over a one-year time horizon are discussed further in Section 4.6, which describes one particular one-year method – the Merz–Wüthrich method – and in Section 4.8, which describes other approaches. Both of these sections also discuss the concept of measuring reserve risk over a multi-year time horizon, so that the total future uncertainty can be broken down into the emergence over all future periods.

4.2 Mack's Method

4.2.1 Introduction

This method was originally described in Mack (1993a). It consists of formulae for deriving the mean reserve and prediction error, with the underlying reserving method being the standard column sum Chain Ladder approach.

It is said to be "distribution-free", as no assumptions are made regarding the distribution of the underlying data. Although the method only produces an estimate of the mean reserve and the prediction error, the full distribution of claims can be estimated by applying a bootstrap procedure to Mack's method, as described in Section 4.4.3. Alternatively, if this is not possible, then an assumed distribution (e.g. Lognormal) can in theory be fitted to the mean and variance, but as described in Section 6.9.4 this approach has potentially significant limitations.

As will be seen in the worked example in this section, the method can be implemented easily in a spreadsheet, without the need for specialist reserving software or extensive computer programming. Because it is analytical in nature, there is also no need for simulations to be carried out.

In simple terms, the method involves applying the standard column-sum CL method to the data (with or without a tail factor) and then implementing formulae to calculate the prediction error for individual cohorts and in total. Since it is based on the CL method, the validity of the results produced by the relevant formulae includes reliance on the same underlying assumptions as for that method, as described in Section 3.2. Specific tests that can be used to assess the validity of the assumptions underlying the Mack method are discussed further below, in Section 4.2.7.

The detail of the method is best explained by describing it in mathematical terms, which follows, although an understanding of this is not a prerequisite for implementing it in practice. Readers who do not wish to study the mathematical formulation can gain a good understanding of how to apply the method by reviewing the worked example in Section 4.2.3.

4.2.2 Mathematical Formulation of Mack's Method

Using the notation[4] given in Section 3.2.6 for the basic CL method, the method is defined in mathematical terms, assuming an $I \times J$ triangle, initially with no tail factor being required. Later sections explain how a tail factor can be allowed for in the model.

1. Specify development factors, $f_0, f_1, \ldots, f_{J-1} > 0$ such that, for all $0 \le i \le I$ and $1 \le j \le J$,

$$E(C_{i,j}|C_{i,0}, C_{i,1}, \ldots, C_{i,j-1}) = f_{j-1}C_{i,j-1}. \tag{4.3}$$

   This is the same as the mathematical formulation of the CL method, given in Section 3.2.6.

2. Assume that the cohorts are independent. In mathematical notation this means that

$$\{C_{i,0}, C_{i,1}, \ldots, C_{i,J}\}, \{C_{k,0}, C_{k,1}, \ldots, C_{k,J}\}, i \neq k, \text{ are independent.} \tag{4.4}$$

   This independence assumption between the cohorts applies to a number of other stochastic methods. A specific test for this is explained in Section 4.2.7.

3. Specify variance parameters, $\sigma_0^2, \sigma_1^2, \ldots, \sigma_{J-1}^2 > 0$ such that, for all $0 \le i \le I$ and $1 \le j \le J$,

$$\text{Var}(C_{i,j}|C_{i,0}, C_{i,1}, \ldots, C_{i,j-1}) = \sigma_{j-1}^2 C_{i,j-1}. \tag{4.5}$$

---

[4] The notation used here is consistent with that used in Merz and Wüthrich (2008c), which is slightly different to that given in Mack's original paper (Mack, 1993a).

4. The development factors are estimated using the column sum average link ratio estimators[5] as in the standard CL method in (3.2):

$$\hat{f}_j = \frac{\sum\limits_{i=0}^{I-j-1} C_{i,j+1}}{\sum\limits_{i=0}^{I-j-1} C_{i,j}}, \quad \text{where } 0 \le j \le J - 1. \tag{4.6}$$

As per (3.4), this can be rewritten as a weighted average of the individual development factors:

$$\hat{f}_j = \frac{\sum\limits_{i=0}^{I-j-1} C_{i,j} F_{i,j}}{\sum\limits_{i=0}^{I-j-1} C_{i,j}}, \tag{4.7}$$

where the individual development factors, $F_{i,j}$, are defined as in (3.3). That is,

$$F_{i,j} = \frac{C_{i,j+1}}{C_{i,j}}.$$

5. In his original paper (Mack, 1993a), Mack shows that the estimators $\hat{f}_j$ are unbiased and uncorrelated.

6. The variance parameters are estimated using

$$\hat{\sigma}_j^2 = \frac{1}{I - j - 1} \sum\limits_{i=0}^{I-j-1} C_{i,j}(F_{i,j} - \hat{f}_j)^2. \tag{4.8}$$

This can be thought of as a weighted average of the square of the differences between the individual development factors and the column sum average link ratio estimators, and in that sense it has some intuitive logic. Mack (1993a) shows that $\hat{\sigma}_j^2$ is an unbiased estimator of σ_j^2.

It should be noted that in (4.8) the index for j is only valid for values up to $J - 2$. For index value $J - 1$, if $\hat{f}_{J-1} = 1$ and if the development can be assumed to be complete after $J + 1$ periods, then it is appropriate to use $\hat{\sigma}_{J-1} = 0$. If not, $\hat{\sigma}_{J-1}$ needs to be estimated. The successive values of $\hat{\sigma}_j$ are usually decreasing, so in estimating $\hat{\sigma}_{J-1}$, it is appropriate to maintain this pattern. Hence, if the two values immediately preceding $\hat{\sigma}_{J-1}$ are decreasing (i.e. $\hat{\sigma}_{J-2} < \hat{\sigma}_{J-3}$) then, linear interpolation can be used for example to estimate $\hat{\sigma}_{J-1}$. In this case,

$$\hat{\sigma}_{J-1}^2 = \left(\frac{\hat{\sigma}_{J-2}^2}{\hat{\sigma}_{J-3}^2}\right)\hat{\sigma}_{J-2}^2 = \frac{\hat{\sigma}_{J-2}^4}{\hat{\sigma}_{J-3}^2}. \tag{4.9}$$

---

[5] This was the original formulation of Mack's method given in Mack (1993a). Subsequently a more generalised formulation was given in Mack (1999), which allows other definitions of the development factors to be used, as discussed in Section 4.2.9.

However, when $\hat{\sigma}_{J-2} > \hat{\sigma}_{J-3}$ and assuming it is appropriate for the decreasing pattern to continue, this cannot be done. An alternative approach is to just put $\hat{\sigma}_{J-1} = \hat{\sigma}_{J-3}$. Mack (1993a) summarises this approach in the following:

$$\hat{\sigma}_{J-1}^2 = \min(\hat{\sigma}_{J-2}^4/\hat{\sigma}_{J-3}^2, \min(\hat{\sigma}_{J-3}^2, \hat{\sigma}_{J-2}^2)). \tag{4.10}$$

Other extrapolation methods (such as log-linear regression fitted across all or a subset of the σ_j^2 values) can also be used, and some software packages allow the user to select which method to use, or to input their own value (e.g. "MackChainLadder" in the "ChainLadder" package in R allows formula (4.10), log-linear regression or a user-defined value to be selected).

7. As in the standard Chain Ladder method described in Section 3.2.6, if no tail factor is required,[6] the ultimate claims, $C_{i,J}$ and the reserve (when using a claims paid triangle), R_i for cohort i are estimated using the following:

$$\hat{C}_{i,J} = C_{i,I-i}\hat{f}_{I-i} \cdots \hat{f}_{J-1} \text{ for } i + j > I, \tag{4.11}$$

and

$$\hat{R}_i = C_{i,I-i}\left(\hat{f}_{I-i} \cdots \hat{f}_{J-1} - 1\right). \tag{4.12}$$

8. The key component of the Mack method is the formula for determining the prediction error, or Mean-Squared Error of Prediction ("MSEP") of the estimated ultimate claims, $\hat{C}_{i,J}$. This is defined in the model in a similar way as (4.1):

$$\text{MSEP}(\hat{C}_{i,J}) = E\left[(\hat{C}_{i,J} - C_{i,J})^2|D_I\right], \tag{4.13}$$

where D_I is the historical data triangle, as defined as in (2.2) – that is

$$D_I = \{C_{i,j} : i + j \le I; 0 \le j \le J\}.$$

Because D_I relates only to the known historical data and is therefore fixed, by including the conditioning on D_I, the formula for $\text{MSEP}(\hat{C}_{i,J})$ represents the average deviation between $\hat{C}_{i,J}$ and $C_{i,J}$ due to future randomness only.

9. As explained in Appendix A, $\text{MSEP}(\hat{C}_{i,J})$ can be rewritten as

$$\text{MSEP}(\hat{C}_{i,J}) = \underbrace{\text{Var}(C_{i,J}|D_I)}_{\text{Process error}} + \underbrace{\left[(\hat{C}_{i,J} - E(C_{i,J}|D_I))^2\right]}_{\text{Estimation error}}. \tag{4.14}$$

In other words, the prediction error is the sum of the process error (the first term) and the parameter (i.e. estimation) error (the second term).

10. Using the definition of the reserve in (4.12) it is clear that

$$\begin{aligned}\text{MSEP}(\hat{R}_i) &= E((\hat{R}_i - R_i)^2|D_I) \\ &= E((\hat{C}_{i,J} - C_{i,J})^2|D_I) \\ &= \text{MSEP}(\hat{C}_{i,J}). \end{aligned} \tag{4.15}$$

---

[6] See Section 4.2.6 for details of how Mack's method can be adapted to include a tail factor.

In other words, the mean-squared error of the reserve for each cohort is the same as the mean squared error of the ultimate claims for that cohort.[7] This makes sense, as the only difference between the ultimate claims and the reserve is the inclusion of the known paid claims amount in the former, which of course is fixed and hence is not subject to any variation.

11. Mack (1993a) shows that for $i \in \{1, 2, \ldots, I\}$, $\text{MSEP}(\hat{R}_{i,J})$ can be estimated using

$$\widehat{\text{MSEP}}(\hat{R}_{i,J}) = \widehat{\text{MSEP}}(\hat{C}_{i,J})$$

$$= \hat{C}_{i,J}^2 \sum_{j=I-i}^{J-1} \frac{\hat{\sigma}_j^2}{\hat{f}_j^2} \left(\frac{1}{\hat{C}_{i,j}} + \frac{1}{\sum_{k=0}^{I-j-1} C_{k,j}} \right),$$

where the leading diagonal is given by $\hat{C}_{i,I-i} = C_{i,I-i}$.

The summation in the brackets represents the sum of column j excluding the leading diagonal, and is sometimes written as[8]

$$S_j^I = \sum_{i=0}^{I-j-1} C_{i,j}, \tag{4.16}$$

so that

$$\widehat{\text{MSEP}}(\hat{R}_{i,J}) = \hat{C}_{i,J}^2 \sum_{j=I-i}^{J-1} \frac{\hat{\sigma}_j^2}{\hat{f}_j^2} \left(\frac{1}{\hat{C}_{i,j}} + \frac{1}{S_j^I} \right). \tag{4.17}$$

12. This can be subdivided into process and estimation error, as per (4.14) as follows:

$$\widehat{\text{MSEP}}(\hat{R}_{i,J}) = \underbrace{\hat{C}_{i,J}^2 \sum_{j=I-i}^{J-1} \frac{\hat{\sigma}_j^2}{\hat{f}_j^2} \frac{1}{\hat{C}_{i,j}}}_{\text{Process error}} + \underbrace{\hat{C}_{i,J}^2 \sum_{j=I-i}^{J-1} \frac{\hat{\sigma}_j^2}{\hat{f}_j^2} \frac{1}{S_j^I}}_{\text{Estimation error}}. \tag{4.18}$$

13. The final component of Mack's method involves defining the formula for the mean squared error of the total reserve, across all cohorts combined. That is, determining $\text{MSEP}(\hat{R})$, where $\hat{R} = \hat{R}_1 + \hat{R}_2 + \cdots + \hat{R}_I$. In doing so, the MSEPs for each cohort cannot simply be added together as they are correlated through using common $\hat{f}_j$ and $\hat{\sigma}_j^2$. Instead Mack (1993a) defines the following formula to estimate $\text{MSEP}(\hat{R})$:

---

[7] Note that, theoretically, in a general stochastic modelling context, the prediction error of the reserves is only the same as the prediction error of the ultimate claims if the estimated ultimate claims in the stochastic model are conditioned on the (known) claims to date.

[8] The relevance of the I superscript notation can be understood by referring to the description of the Merz–Wüthrich method given in Section 4.6.

$$\widehat{\mathrm{MSEP}}(\hat{R}) = \sum_{i=1}^{I} \left\{ \widehat{\mathrm{MSEP}}(\hat{R}_i) + \hat{C}_{i,J} \left(\sum_{l=i+1}^{I} \hat{C}_{l,J} \right) \sum_{k=I-i}^{J-1} \frac{2\hat{\sigma}_k^2 / \hat{f}_k^2}{\sum\limits_{m=1}^{I-k} C_{m,k}} \right\}. \tag{4.19}$$

14. To facilitate the calculation of MSEP($\hat{R}$) in practice, and to help understand the relationship between Mack's method and the Merz–Wüthrich method, as described in Section 4.6, this can be expressed in a slightly different way, as per Estimator 3.16 in Merz and Wüthrich (2008c), as follows:

$$\widehat{\mathrm{MSEP}}(\hat{R}) = \sum_{i=1}^{I} \widehat{\mathrm{MSEP}}(\hat{R}_i) + 2 \sum_{1 \le i < k \le I} \hat{C}_{i,J} \hat{C}_{k,J} \sum_{j=I-i}^{J-1} \frac{\hat{\sigma}_j^2}{\hat{f}_j^2} \frac{1}{S_j^I}. \tag{4.20}$$

The second term is effectively the covariance between the different cohorts. In this term, the second summation is the same as the summation in the Estimation error component of the individual cohort MSEP, as in (4.18). It will be helpful in the example that follows if this summation is defined as follows:

$$\hat{\Lambda}_i = \sum_{j=I-i}^{J-1} \frac{\hat{\sigma}_j^2}{\hat{f}_j^2} \frac{1}{S_j^I}, \tag{4.21}$$

so that

$$\widehat{\mathrm{MSEP}}(\hat{R}) = \sum_{i=1}^{I} \widehat{\mathrm{MSEP}}(\hat{R}_i) + 2 \sum_{1 \le i < k \le I} \hat{C}_{i,J} \hat{C}_{k,J} \hat{\Lambda}_i. \tag{4.22}$$

Section 7.4 of Taylor (2000) provides a detailed mathematical derivation of the key formulae in Mack's method – i.e. (4.8), (4.17) and (4.22). More recently, Röhr (2016) showed that an alternative approach to deriving the parameter estimation error for both single and combined cohorts, involving the use of "error propagation methods", can be used to derive the same MSEP formulae as above.

The above formulation is consistent with the original description of Mack's method in Mack (1993a). It is possible to express Mack's method as a Generalised Linear Model, fitted to the development ratios, as explained further in Section 4.4.3.

The assumptions underlying the Mack method are encompassed in points 1 to 3 above. Tests can be used to assess the validity of these, as explained further in Section 4.2.7. For the purpose of the worked example that follows, it is assumed that these are valid.

4.2.3 Worked Example of Mack's Method

The same Taylor and Ashe (1983) dataset that is used for the worked examples of the deterministic methods in Chapter 3 is also used for many of the examples for the stochastic methods covered in this book, including Mack's method.[9] This data is quoted in many papers on stochastic claims reserving, such as England and Verrall (2006).

The example starts with a cumulative claims paid data triangle, as shown in Table 4.3. All figures are in thousands unless stated otherwise.

Table 4.3 *Mack example: Taylor and Ashe dataset – claims paid*

Year	Development year									
	0	1	2	3	4	5	6	7	8	9
0	358	1,125	1,735	2,218	2,746	3,320	3,466	3,606	3,834	3,901
1	352	1,236	2,170	3,353	3,799	4,120	4,648	4,914	5,339	
2	291	1,292	2,219	3,235	3,986	4,133	4,629	4,909		
3	311	1,419	2,195	3,757	4,030	4,382	4,588			
4	443	1,136	2,128	2,898	3,403	3,873				
5	396	1,333	2,181	2,986	3,692					
6	441	1,288	2,420	3,483						
7	359	1,421	2,864							
8	377	1,363								
9	344									

The results of applying Mack's method to this dataset can be produced using the "MackChainLadder" function within the "ChainLadder" package in R (see R Core Team, 2013 and Gesmann et al., 2014 for further details). The relevant R code is in Appendix B.3.1. The results are shown in Table 4.4. The R code in the Appendix also includes code to produce diagnostic charts, including residual plots, as explained in Section 4.10.3. As mentioned in Section 4.1.3, for consistency with other sources, the term "Standard Error" is used in Table 4.4 and in other results tables for the Mack worked examples. It is equivalent to "prediction error".

The cohorts and development periods in this example are years, which both take values from 0 to 9, so that, using the notation of Section 4.2.2, $I = J = 9$. For the purposes of this example, it is assumed that no tail factor is required. The addition of a tail factor in Mack's method is explored further in Section 4.2.6.

These results are the same as those shown in the original Mack paper, Mack (1993a). In particular, the CVs shown in Table 4.4 are the same, ignoring rounding, as those shown under the "Chain Ladder" column in Table 3 of Mack (1993a). Furthermore, the reserve for each year is the same as that derived using the deterministic column sum average CL method (as shown in the results for the CL worked example given in Table 3.19), which is simply a result of the fact that the Chain Ladder development factors are used in the formulation of Mack's method (as per (4.6)).

This worked example will show that equations (4.18) and (4.20) are relatively straightforward to implement using a matrix-type layout in a spreadsheet. The section after this

---

[9] Note that the cohorts are numbered from 0 to 9 (rather than 2005 to 2014 as in the worked examples for the deterministic methods) for consistency with the labelling convention used in the mathematical formulation of the stochastic methods in this chapter. This has no effect on the calculations.

Table 4.4 *Mack example: Results using R*

Yr	Latest	Ultimate	IBNR	Standard Error Process	Standard Error Est'n	Standard Error Total	CV
0	3,901	3,901	0	–	–	–	–
1	5,339	5,434	95	49	58	76	80%
2	4,909	5,379	470	91	81	122	26%
3	4,588	5,298	710	103	85	134	19%
4	3,873	4,858	985	228	128	261	27%
5	3,692	5,111	1,419	367	186	411	29%
6	3,483	5,661	2,178	500	248	558	26%
7	2,864	6,785	3,920	786	386	875	22%
8	1,363	5,642	4,279	896	376	971	23%
9	344	4,970	4,626	1,285	456	1,363	30%
Total	34,358	53,039	18,681	1,878	1,569	2,447	13%

worked example also shows how the formulae for Mack's method can be expressed recursively, which facilitates, for example, addition of a tail factor. A further worked example then shows how these recursive formulae, with and without a tail factor, can be implemented in practice.

The example begins with the calculation of the individual development factors – that is, the $F_{i,j}$ factors defined in (3.3) as part of the CL method description, and used in Mack's method, formula (4.8). These are shown in Table 4.5.

Table 4.5 *Mack example: Development factors ($F_{i,j}$)*

Year	Development year 0	1	2	3	4	5	6	7	8
0	3.143	1.543	1.278	1.238	1.209	1.044	1.040	1.063	1.018
1	3.511	1.755	1.545	1.133	1.084	1.128	1.057	1.086	
2	4.448	1.717	1.458	1.232	1.037	1.120	1.061		
3	4.568	1.547	1.712	1.073	1.087	1.047			
4	2.564	1.873	1.362	1.174	1.138				
5	3.366	1.636	1.369	1.236					
6	2.923	1.878	1.439						
7	3.953	2.016							
8	3.619								

The estimators, $\hat{f}_j$, are also needed, which are simply the column sum average (CSA) Chain Ladder development factors. These are shown in Table 4.6. "Sum 1" and "Sum 2" in this table represent the numerator and denominator respectively of the formula for $\hat{f}_j$, (4.6). This table also shows the cumulative development factors (i.e. the factor from time j to J) and the implied percentage developed at each development point (i.e. the inverse of the cumulative development factors). These last two items are not required for this example, but they will be referred to in later examples. The values in this table are the same as those for the CL method worked example in Chapter 3, as shown in Table 3.18.

Table 4.6 *Mack example: Chain Ladder development factors*

	Development year								
	0	1	2	3	4	5	6	7	8
Sum 1	11,615	17,912	21,931	21,655	19,828	17,331	13,430	9,173	3,901
Sum 2 $(= S_j^I)$	3,327	10,251	15,048	18,448	17,963	15,955	12,743	8,520	3,834
Ratio $(= \hat{f}_j)$	3.491	1.747	1.457	1.174	1.104	1.086	1.054	1.077	1.018
Cum've	14.447	4.139	2.369	1.625	1.384	1.254	1.155	1.096	1.018
% dev'd	7%	24%	42%	62%	72%	80%	87%	91%	98%

These can then be used to calculate the projected triangle, as shown in Table 4.7, which will be needed to calculate the process and parameter error later in the example. The values below the "stepped" line are simply the projected cumulative values derived using the Chain Ladder factors. This is same as the projected triangle for the CL worked example in Chapter 3 (Table 3.20), but is repeated here for ease of reference.

The variance parameter terms, that is, $C_{i,j}(F_{i,j} - \hat{f}_j)^2$ from (4.8), are then calculated, as shown in Table 4.8.

So, for example, the value[10] of 1,211 for cohort 2, delay 1 is

$$C_{2,1}(F_{2,1} - \hat{f}_1)^2 = 1,292,306 * (1.71672 - 1.74733)^2.$$

The $C_{i,j}$ value is taken from cell (2, 1) (multiplied by 1000) in Table 4.7. Note that the top right-hand value will always be zero, as there is only one row contributing to the chain ladder development factor.

Table 4.7 *Mack example: Actual and projected data triangle*

Year	Development year									
	0	1	2	3	4	5	6	7	8	9
0	358	1,125	1,735	2,218	2,746	3,320	3,466	3,606	3,834	3,901
1	352	1,236	2,170	3,353	3,799	4,120	4,648	4,914	5,339	5,434
2	291	1,292	2,219	3,235	3,986	4,133	4,629	4,909	5,285	5,379
3	311	1,419	2,195	3,757	4,030	4,382	4,588	4,835	5,206	5,298
4	443	1,136	2,128	2,898	3,403	3,873	4,207	4,434	4,774	4,858
5	396	1,333	2,181	2,986	3,692	4,075	4,427	4,665	5,022	5,111
6	441	1,288	2,420	3,483	4,089	4,513	4,903	5,167	5,562	5,661
7	359	1,421	2,864	4,175	4,901	5,409	5,876	6,193	6,667	6,785
8	377	1,363	2,382	3,472	4,075	4,498	4,887	5,150	5,544	5,642
9	344	1,201	2,098	3,058	3,590	3,962	4,304	4,536	4,883	4,970

---

[10] The cumulative data item is shown in full, rather than in thousands, and sufficient decimal places are shown so that the calculated value equals 1,211, as shown in Table 4.8.

Table 4.8 *Mack example: Variance parameter terms*

Year	\multicolumn{9}{c}{Development year}
	0	1	2	3	4	5	6	7	8
0	43,189	47,051	55,673	9,049	30,492	5,910	632	662	0
1	140	82	16,756	5,616	1,420	7,211	54	486	
2	266,530	1,211	2	10,968	17,874	4,705	208		
3	360,548	56,914	142,030	38,584	1,092	6,731			
4	380,334	17,933	19,561	0	4,048				
5	6,191	16,621	16,984	11,697					
6	142,127	22,032	786						
7	76,955	102,314							
8	6,227								

Summing the columns in Table 4.8 and dividing by the number of values in each column (less 1), as in (4.8), enables the $\hat{\sigma}_j^2$ parameters to be derived, as shown in Table 4.9.

Table 4.9 *Mack example: Variance parameters –* $\hat{\sigma}_j^2$

\multicolumn{9}{c}{Development year}
0	1	2	3	4	5	6	7	8
160,280	37,737	41,965	15,183	13,731	8,186	447	1,147	447

As noted in Section 4.2.2, since $\hat{f}_8 \neq 1$, it is necessary to estimate $\hat{\sigma}_8^2$. If the two values immediately preceding $\hat{\sigma}_8^2$ were decreasing, then linear interpolation could be used to estimate $\hat{\sigma}_8^2$. However, since they are not, instead $\hat{\sigma}_6^2$ is used, i.e. 447,[11] as shown in Table 4.9.

Mack Worked Example: Calculation of Process Error

The next step is to calculate the Process error, using the first part of equation (4.18). The workings and results are shown in Tables 4.10 and 4.11.

The values[12] below the stepped line in Table 4.10 are the items in the first summation of (4.18). The "Sum" column in Table 4.11 adds these values across the columns for each row, and the next two columns show the estimated ultimate for that row (i.e. $\hat{C}_{i,J}$) and then the product of the sum and the square of the ultimate. The final column shows the Process Error (i.e. the square root of the previous column).

Mack Worked Example: Calculation of Estimation Error

The next step is to calculate the Estimation error for each year individually, using the second part of equation (4.18). The workings and results are shown in Tables 4.12 and 4.13.

The values below the stepped line in Table 4.12 are the items in the second summation of (4.18). As for the Process error, these values refer back to the previous column values.

---

[11] The same result of 447 is produced if (4.10) is used.

[12] The items in each cell of Table 4.10 refer back to the previous columns for the $\hat{\sigma}_j$ and $\hat{f}_j$ values due to the formulation at (4.5).

Table 4.10 *Mack example: Process error results (×1,000) – Part 1*

Year	Development year									
	0	1	2	3	4	5	6	7	8	9
1										0.081
2									0.202	0.082
3								0.088	0.205	0.083
4							1.791	0.096	0.223	0.090
5						3.053	1.702	0.091	0.212	0.086
6					3.163	2.756	1.537	0.082	0.192	0.078
7				6.897	2.639	2.300	1.282	0.068	0.160	0.065
8			9.066	8.294	3.174	2.765	1.542	0.082	0.192	0.078
9	38.239	10.293	9.416	3.603	3.140	1.751	0.093	0.218	0.088
Total	38.239	19.359	24.607	12.580	14.014	9.606	0.600	1.604	0.730	

Table 4.11 *Mack example: Process error results – Part 2*

Year	Sum[a]	Ult($\hat{C}_{i,J}$)	Product	Proc' Error[a]
1	8.076E-05	5,434	2.385E+09	49
2	2.832E-04	5,379	8.195E+09	91
3	3.752E-04	5,298	1.053E+10	103
4	2.200E-03	4,858	5.193E+10	228
5	5.144E-03	5,111	1.344E+11	367
6	7.808E-03	5,661	2.502E+11	500
7	1.341E-02	6,785	6.174E+11	786
8	2.519E-02	5,642	8.020E+11	896
9	6.684E-02	4,970	1.651E+12	1,285
Total	1.213E-01	N/A	3.528E+12	1,878

[a] Ultimate and Proc' Error values are in thousands. All other values are in units.

For example, the value of 0.627 in cell (7, 5) of the table is calculated as follows (multiplied by 1,000 as shown in the table):

$$\frac{\hat{\sigma}_4^2}{\hat{f}_4^2} \frac{1}{S_4^I} = \frac{13,731}{1.104^2} \frac{1}{17,963,259}$$

$$= 0.000627.$$

The "Sum" column in Table 4.13 adds these values across the columns for each row, to give the $\hat{\Lambda}_i$ values defined in (4.21). The next two columns show the estimated ultimate for that row (i.e. $\hat{C}_{i,J}$) and then the product of the sum and the square of the ultimate. The final column shows the Estimation Error (i.e. the square root of the previous column).

Table 4.12 *Mack example: Estimation error results (×1,000) – Part 1*

					Development year					
Year	0	1	2	3	4	5	6	7	8	9
1										0.112
2									0.116	0.112
3								0.032	0.116	0.112
4							0.435	0.032	0.116	0.112
5						0.627	0.435	0.032	0.116	0.112
6					0.597	0.627	0.435	0.032	0.116	0.112
7				1.313	0.597	0.627	0.435	0.032	0.116	0.112
8			1.206	1.313	0.597	0.627	0.435	0.032	0.116	0.112
9		3.953	1.206	1.313	0.597	0.627	0.435	0.032	0.116	0.112

Table 4.13 *Mack example: Estimation error results ('000) – Part 2*

Year	Sum[a]	Ult($\hat{C}_{i,J}$)	Product	Est Error[a]
1	1.125E-04	5,434	3.321E+09	58
2	2.287E-04	5,379	6.616E+09	81
3	2.602E-04	5,298	7.304E+09	85
4	6.950E-04	4,858	1.640E+10	128
5	1.322E-03	5,111	3.455E+10	186
6	1.920E-03	5,661	6.152E+10	248
7	3.233E-03	6,785	1.488E+11	386
8	4.438E-03	5,642	1.413E+11	376
9	8.392E-03	4,970	2.073E+11	455

[a] Ultimate and Est Error values are in thousands. All other values are in units.

Unlike the Process error, for the Estimation error the value across all years combined cannot simply be derived by summing the values across each year, due to the covariance term in (4.20). The calculations for this term are shown in Table 4.14. Setting out the calculation in this way facilitates the calculation of this term using a spreadsheet.

For example, the value of 38,136 in cell (5, 8) in Table 4.14 is calculated as follows (with decimal places being shown, rather than whole units as per the tables):

$$\hat{C}_{5,J}\hat{C}_{8,J}\hat{\Lambda}_5 = 5,111.171 \times 5,642.266 \times 1.322403/1,000.$$

$\hat{\Lambda}_5$ is taken from Row 5 of the "Sum" column in Table 4.13.

The covariance term in (4.20) is then calculated by summing the values in the triangle part of Table 4.14 and multiplying by two. Combining the square root of this (which is 1,354k),

Table 4.14 *Mack example: Calculation of covariance term*

Year	0	1	2	3	4	5	6	7	8	9	Ult[a]
						Development year					
1		3,287	3,238	2,969	3,124	3,460	4,147	3,448	3,037	5,434
2			6,516	5,976	6,287	6,963	8,345	6,940	6,113	5,379
3				6,698	7,047	7,804	9,354	7,779	6,852	5,298
4					17,258	19,114	22,909	19,052	16,781	4,858
5						38,261	45,859	38,136	33,591	5,111
6							73,730	61,314	54,007	5,661
7								123,751	109,002	6,785
8									124,456	5,642
9										4,970
Ult[a]		5,434	5,379	5,298	4,858	5,111	5,661	6,785	5,642	4,970

[a] Ultimate values are in thousands. All other values are in millions.

with the Process and Estimation error values in Tables 4.11 and 4.13 respectively gives the results summary shown in Table 4.15.

The Standard Error in the Total row for each of Process and Estimation error is simply the square root of the sum of squares of the values in each column. The Standard Error in the Total column is similarly calculated as the square root of the sum of squares of the Process and Estimation errors in each row. As expected, the results in Table 4.15 are the same as the results produced from the "MackChainLadder" package in R, shown in Table 4.4.

4.2.4 Recursive Formulation of Mack's Method

The MSEP formulae for Mack's method, (4.18) and (4.19), can be written recursively, so as to provide an alternative means of implementing the model in practice. Mack (1999)

Table 4.15 *Mack example: Results summary split by process and estimation error*

Year	Process	Estimation	Total
		Standard error ('000)	
1	49	58	76
2	91	81	122
3	103	85	134
4	228	128	261
5	367	186	411
6	500	248	558
7	786	386	875
8	896	376	971
9	1,285	455	1,363
Cov		1,354	
Total	1,878	1,569	2,447

gives recursive formulae both for single cohort years and for combined cohort years, but not subdivided by process and estimation components. Papers by other authors provide such formulae – specifically Murphy (2007) gives recursive formulae for single cohort years and Bardis et al. (2012) gives recursive formulae for cohort years combined. Buchwalder et al. (2006) also gives recursive formulae for Mack's method. This section explains these formulae, which are then used in the worked example in Section 4.2.5.

Mack Recursive Formulae for a Single Cohort

First, for a single cohort, i, the recursive formulae for each of process and estimation risk build up across the development periods, starting on the left-hand side of the triangle. Again, working with an $I \times J$ triangle, the formulae, based on those given in Murphy (2007), are as follows:[13]

$$\overline{\text{ProcessRisk}}(i, k)$$

$$= \begin{cases} \hat{f}_{k-1}^2 \overline{\text{ProcessRisk}}(i, k - 1) + \hat{C}_{i,k-1}\hat{\sigma}_{k-1}^2, & \text{for } k > I - i, \\ C_{i,I-i}\hat{\sigma}_{k-1}^2, & \text{for } k = I - i, \end{cases} \tag{4.23}$$

$$\overline{\text{EstRisk}}(i, k)$$

$$= \begin{cases} \hat{f}_{k-1}^2 \overline{\text{EstRisk}}(i, k - 1) + \hat{C}_{i,k-1}^2 \widehat{\text{Var}}(\hat{f}_{k-1}), & \text{for } k > I - i, \\ C_{i,I-i}^2 \widehat{\text{Var}}(\hat{f}_{k-1}), & \text{for } k = I - i, \end{cases} \tag{4.24}$$

where

$$\widehat{\text{Var}}(\hat{f}_k) = \frac{\hat{\sigma}_k^2}{\sum_{i=0}^{I-k-1} C_{i,k}}. \tag{4.25}$$

Using the S_j^I notation of (4.16) given in Section 4.2.2, the right-hand side of the first row of (4.24) becomes

$$\hat{f}_{k-1}^2 \overline{\text{EstRisk}}(i, k - 1) + \hat{C}_{i,k-1}^2 \frac{\hat{\sigma}_{k-1}^2}{S_{k-1}^I}. \tag{4.26}$$

The above recursive formulae have been expressed partly in words to help explain their structure, with "EstRisk" and "ProcessRisk" being equivalent to Estimation variance and Process variance respectively. They can equally well be expressed in full mathematical notation. For example, following the notation of Buchwalder et al. (2006) and Bardis et al. (2012), where the process risk is denoted by Γ^2 and the estimation risk is denoted by Δ^2, the recursive formula (4.23) for a single cohort's process risk becomes

$$\Gamma_{i,k}^2 = \hat{f}_{k-1}\Gamma_{i,k-1}^2 + \hat{C}_{i,k-1}\hat{\sigma}_{k-1}^2, \tag{4.27}$$

---

[13] Note that, for parameter risk, the formula given here is consistent with the original formulation of Mack's method and therefore excludes the third cross-product term given in Murphy (2007). Further details are given in Section 4.2.8.

and the recursive formula (4.26) for a single cohort's estimation risk becomes

$$\Delta_{i,k}^2 = \hat{f}_{k-1}^2 \Delta_{i,k-1}^2 + \hat{C}_{i,k-1}^2 \frac{\hat{\sigma}_{k-1}^2}{S_{k-1}^I}. \tag{4.28}$$

These are equivalent to formula (4.9) and (4.27) respectively in Buchwalder et al. (2006).

Mack Recursive Formulae Across all Cohorts Combined

For the MSEP across all cohorts combined, first define $\hat{Y}_j$ as the estimated ultimate claims at delay j in respect of cohorts that have not yet reached development period j in their development. That is,

$$\hat{Y}_j = \sum_{i=I-j+1}^{I} \hat{C}_{i,j}, \text{ where } j > 0.$$

Then the process risk of $\hat{Y}_j$ will just be the sum of the process risk across cohorts, as an underlying assumption of the model is that cohorts are independent. Thus

$$\widehat{\text{ProcessRisk}}(\hat{Y}_j) = \sum_{i=I-j+1}^{I} \widehat{\text{ProcessRisk}}(i, j). \tag{4.29}$$

So, once the recursive formula (4.23) has been used to determine the process risk for each cohort at delay J (which for current purposes is at ultimate as it is assumed that there is no tail factor), then the process risk for the total reserve across all cohorts combined will be

$$\widehat{\text{ProcessRisk}}(\hat{Y}_J) = \sum_{i=1}^{I} \widehat{\text{ProcessRisk}}(i, J). \tag{4.30}$$

For the estimation risk across all cohorts combined, as noted earlier, the estimation risk for each cohort cannot simply be added together, since there is dependency through use of common parameters. The recursive formula begins on the left-hand side of the triangle, with $j = 1$. In this case,

$$\widehat{\text{EstRisk}}(\hat{Y}_1) = C_{I,0}^2 \widehat{\text{Var}}(\hat{f}_1).$$

Then, using Formula (12) from Bardis et al. (2012), but without the final cross-product term[14] the recursive formula for estimation risk across all cohorts combined, for $j = 2, 3, \ldots, J$, is

$$\widehat{\text{EstRisk}}(\hat{Y}_j) = (\hat{Y}_{j-1} + C_{I-j+1,j-1})^2 \widehat{\text{Var}}(\hat{f}_{j-1})$$
$$+ \hat{f}_{j-1}^2 \widehat{\text{EstRisk}}(\hat{Y}_{j-1}). \tag{4.31}$$

When the j index has reached the value J, the above formula will represent the parameter risk across all cohorts combined.

---

[14] See Section 4.2.8 for further details. Also note that Bardis et al. (2012) uses $\hat{X}_j$ rather than $\hat{Y}_j$, which is used here to avoid confusion with the notation used elsewhere in this book for incremental claims.

To derive the estimation and process components of the Standard Error, square roots of the relevant components are taken. The total Standard Error is the square root of the sum of the process and estimation risk components. Or, more simply:

$$\text{SE}(\hat{R}_{i,n}) = \sqrt{\widehat{\text{ProcessRisk}}(i, J) + \widehat{\text{EstRisk}}(i, J)}, \tag{4.32}$$

and

$$\text{SE}(\hat{R}) = \sqrt{\widehat{\text{ProcessRisk}}(\hat{Y}_J) + \widehat{\text{EstRisk}}(\hat{Y}_J)}. \tag{4.33}$$

Table 4.16 *Mack recursive example: Process risk results ('000m)*

Cohort	0	1	2	3	4	5	6	7	8	9
						Development year				
1										2.4
2									5.6	8.2
3								2.0	7.9	10.5
4							31.7	37.1	48.1	51.9
5						50.7	93.2	105.5	127.6	134.4
6					52.9	120.6	179.2	201.2	239.2	250.2
7				120.2	229.0	346.3	453.0	505.7	593.2	617.4
8			51.4	209.2	341.0	471.5	593.2	661.0	772.0	802.0
9		55.1	213.7	541.9	793.1	1,015.6	1,230.9	1,369.0	1,591.8	1,650.9
Total		55.1	265.1	871.3	1,416.0	2,004.7	2,581.1	2,881.5	3,385.3	3,528.0

4.2.5 Worked Example of Mack's Method – Using Recursive Formulae

This section shows how Mack's method can be implemented using the recursive formulae given in Section 4.2.4. Although this example uses the same dataset that has already been used in the first worked example of Mack's method, given in Section 4.2.3, this alternative implementation is helpful since it can be easily adapted to incorporate a tail factor, as explained in Section 4.2.6.

Mack Recursive Example: Calculation of Process Risk for a Single Cohort

The process risk is calculated using the recursive formula (4.23). The results are shown in Table 4.16 (with values being shown in units of thousand million ('000m), for ease of presentation).

The first entry in each row in this table is taken from the second part of (4.23) when $k = I - i$. For example, when $i = 7$:

Table 4.17 *Mack recursive example: Calculation of variance of development factors*

Item[a]				Development year					
	0	1	2	3	4	5	6	7	8
$\hat{\sigma}_k^2$	160.280	37.737	41.965	15.183	13.731	8.186	0.447	1.147	0.447
S_k^I	3,327	10,251	15,048	18,448	17,963	15,955	12,743	8,520	3,834
$\widehat{\text{Var}}(\hat{f}_k)$	48.17	3.68	2.79	0.82	0.76	0.51	0.04	0.13	0.12
$\widehat{\text{SE}}(\hat{f}_k)$	0.219	0.061	0.053	0.029	0.028	0.023	0.006	0.012	0.011
$\hat{f}_k$	3.491	1.747	1.457	1.174	1.104	1.086	1.054	1.077	1.018
CV(%)	6.288	3.472	3.623	2.444	2.505	2.085	0.562	1.078	1.061

[a] For ease of presentation, $\hat{\sigma}_k^2$ and S_k^I are shown in '000, $\widehat{\text{Var}}(\hat{f}_k)$ is shown multiplied by 1,000 and the other items are in units.

$$\text{ProcessRisk}(7,3) = C_{7,2}\hat{\sigma}_2^2 = 2{,}864{,}498 \times 41{,}965/1{,}000\text{m} = 120.2.$$

The subsequent entries in each row are then calculated using the first part of (4.23). For example, when $i = 7$, the first recursively calculated value is as follows (with any differences to the respective values shown in the tables being due to rounding):

$$\text{ProcessRisk}(7,4) = \hat{f}_3^2\,\text{ProcessRisk}(7,3) + \hat{D}_{8,4}\hat{\sigma}_4^2$$
$$= 1.174^2 \times 120.2 + 4{,}174{,}756 \times 15{,}183/1{,}000\text{m}$$
$$= 229.0.$$

This is continued from left to right across each row until the value in the final column is reached, being the process risk for the reserve for that cohort.

Mack Recursive Example: Calculation of Estimation Risk for a Single Cohort

To apply the recursive formula (4.24), it is first necessary to calculate the variance of the development factors, $\widehat{\text{Var}}(\hat{f}_k)$, using (4.25). This is given in Table 4.17.

For example, for $\widehat{\text{Var}}(\hat{f}_2)$, the denominator is the sum of column 2 in the original triangle, excluding the leading diagonal (i.e. S_2^I) This is the same as "Sum 2" from column "2" in Table 4.6 – i.e. 15,048. $\widehat{\text{Var}}(\hat{f}_2)$ is then $41.695/15{,}048 = 0.002\,789$ (shown multiplied by 1,000 in the table for ease of presentation). The coefficient of variation (CV) of each development factor is also shown for information purposes (and they are used in the worked example incorporating a tail factor in Section 4.2.6). The results of then applying the recursive formula are given in Table 4.18.

The first entry in each row in this table is taken from the second part of (4.24) when $k = I - i$. For example, when $i = 7$:

Table 4.18 *Mack example: Estimation risk results ('000m)*

Cohort	0	1	2	3	4	5	6	7	8	9
					Development year					
1										3.3
2									3.2	6.6
3								0.7	4.0	7.3
4							7.7	9.2	13.3	16.4
5						10.4	20.8	23.8	30.5	34.5
6					10.0	24.9	39.9	45.1	55.9	61.5
7				22.9	45.9	74.3	102.6	115.2	138.7	148.8
8			6.8	30.4	51.8	75.7	99.8	111.6	133.0	141.3
9		5.7	22.7	60.5	91.1	120.8	150.6	168.0	197.4	207.3
Total		5.7	41.6	238.8	494.7	919.2	1,440.4	1,638.4	2,113.1	2,460.3

$$\widehat{\text{EstRisk}}(7,3) = C_{7,2}^2 \widehat{\text{Var}}(\hat{f}_2)$$
$$= 2{,}864{,}498^2 \times 0.002789$$
$$= 22.9 \times 1{,}000\text{m}.$$

The subsequent entries in each row are then calculated using the first part of (4.24). For example, when $i = 7$, the first recursively calculated value is

$$\widehat{\text{EstRisk}}(7,4) = \hat{f}_3^2 \widehat{\text{EstRisk}}(7,3) + \hat{C}_{7,3}^2 \widehat{\text{Var}}(\hat{f}_3)$$
$$= 1.174^2 \times 22.9 \times 1{,}000\text{m} + 4{,}174{,}756^2 \times 0.00082302$$
$$= 45.9 \times 1{,}000\text{m}.$$

As for the process risk, this is continued across each row until the value in the final column is reached, being the parameter risk for the reserve for that cohort.

Mack Example: Calculation of Process and Parameter Risk for all Cohorts Combined

For the process risk, this can be obtained by summing across cohorts, as per (4.30). The total row in Table 4.16 shows these results, with the final entry in column 9 being the process risk for the total reserve, that is $3{,}528 \times 1{,}000\text{m}$.

For the parameter risk across all cohorts combined, the recursive formula (4.31) is used. The results are shown in the total row of Table 4.18, where the formula is applied recursively across this row from left to right. The first entry (denoted by $\widehat{\text{EstRisk}}(\hat{Y}_1)$) is simply the parameter risk for delay 1, i.e. $5.70 \times 1,000$m. For the next entry, at delay 2, from (4.31):

$$\widehat{\text{EstRisk}}(\hat{Y}_2) = (\hat{Y}_1 + C_{8,1})^2 \widehat{\text{Var}}(\hat{f}_1)$$
$$+ \hat{f}_1^2 \widehat{\text{EstRisk}}(\hat{Y}_1).$$

From (4.29), $\hat{Y}_1 = \hat{C}_{9,1} = 1,200,818$ (= 1,201k as shown in Table 4.7), so that

$$\widehat{\text{EstRisk}}(\hat{Y}_2) = (1,200,818 + 1,363,294)^2 \times 0.060673$$
$$+ 1.7473^2 \times 5.70 \times 1,000\text{m}$$
$$= 41.6 \times 1,000\text{m}.$$

This continues recursively across the columns, with the successive $\hat{Y}_j$ just being the sum of the amounts below the stepped line in column j of Table 4.7. The final value, of 2,460 is the estimation risk for the total reserve across all cohorts (in '000m).

The total Standard Error can then be derived, using (4.32) and (4.33), which gives the results (in thousands) in Table 4.19.

The results for the individual cohorts for each of the process and estimation risk are taken from the final columns of Tables 4.16 and 4.18 respectively. For example, for cohort no. 6, the process element of the Standard Error is $[\sqrt{250.2 \times 1,000\text{m}}]/1,000 = 500$ and the estimation element is $[\sqrt{61.5 \times 1,000\text{m}}]/1,000 = 248$. The total Standard Error is the square root of the sum of squares of the parameter and process elements. As expected, the results using the recursive formulae, shown in Table 4.19, are identical to both the results produced using R, shown in Table 4.4, and those given in the first worked example, shown in Table 4.15.

Table 4.19 *Mack recursive example: Results showing Process and Estimation Error*

Cohort	Standard Error ('000)		
	Process	Estimation	Total
1	49	58	76
2	91	81	122
3	103	85	134
4	228	128	261
5	367	186	411
6	500	248	558
7	786	386	875
8	896	376	971
9	1,285	455	1,363
Total	1,878	1,569	2,447

4.2.6 Adapting Mack's Method to Include a Tail Factor

The mathematical formulation of Mack's method given in Section 4.2.2 and the worked examples given in Sections 4.2.3 and 4.2.5 assume that the triangle to which the model is being applied is "complete" – i.e. no tail factor needs to be estimated. In practice, there will be many situations where this is not the case, and hence where the basic formulation of Mack's method will need amending to incorporate a tail factor, and its associated variability. Like most practical reserving exercises, there will likely be an element of judgement involved in the selection of the tail factor, as explained in Section 3.2.4.

Once the tail factor has been selected, a very simple way to allow for it in the determination of the Standard Error is to assume that the CV derived without the tail factor also applies to the reserve including the tail factor (either applied per cohort or in total). This simplistic approach ignores any additional volatility that might be present in the tail of the triangle, and so other approaches are sometimes preferable.

The inclusion of a tail factor in Mack's method is considered in Mack (1999), but the recursive formulae presented in Section 4.2.4 can also be adapted to incorporate a tail factor.

If the selected tail factor for an $I \times J$ triangle is denoted by $\hat{f}_J$ and "ult" is used to denote values at ultimate[15] (i.e. after allowance for the tail factor), then it is straightforward to extend the recursive formulae (4.23), (4.24), (4.30) and (4.31). First, for individual cohorts:

$$\widetilde{\text{ProcessRisk}}(i, \text{ult})$$
$$= \begin{cases} \hat{f}_J^2 \widetilde{\text{ProcessRisk}}(i, J) + \hat{C}_{i,J} \hat{\sigma}_J^2, & \text{for } i > 0, \\ C_{i,I-i} \hat{\sigma}_J^2, & \text{for } i = 0, \end{cases} \tag{4.34}$$

and

$$\widetilde{\text{EstRisk}}(i, \text{ult}) = \begin{cases} \hat{f}_J^2 \widetilde{\text{EstRisk}}(i, J) + \hat{C}_{i,J}^2 \widehat{\text{Var}}(\hat{f}_J), & \text{for } i > 0, \\ C_{i,I-i}^2 \widehat{\text{Var}}(\hat{f}_J), & \text{for } i = 0. \end{cases} \tag{4.35}$$

Next, across all cohorts combined:

$$\widetilde{\text{ProcessRisk}}(\hat{Y}_{\text{ult}}) = \sum_{i=0}^{I} \widetilde{\text{ProcessRisk}}(i, \text{ult}), \tag{4.36}$$

and

$$\widetilde{\text{EstRisk}}(\hat{Y}_{\text{ult}}) = (\hat{X}_J + C_{0,J})^2 \widehat{\text{Var}}(\hat{f}_J) + \hat{f}_J^2 \widetilde{\text{EstRisk}}(\hat{Y}_J). \tag{4.37}$$

Hence, in order to apply these formulae, estimates of $\hat{\sigma}_J^2$ and $\widehat{\text{Var}}(\hat{f}_J)$ need to be derived. The latter represents an estimate of the variability in the tail factor. This can be estimated using judgement, perhaps based on consideration of the likely variability in the tail part of the claims development for the class of business being analysed. Alternatively, analytical methods could be used, such as log-linear regression through the previously estimated

---

[15] $\hat{f}_{\text{ult}}$ would be more logical, but the recursive formula are easier to extend to ultimate if the index value of J is used.

$\widehat{\mathrm{Var}}(\hat{f}_j)$ values. Similarly, $\hat{\sigma}_j^2$ can also be estimated by analytical methods, such as log-linear regression. Alternatively, Mack (1999) suggests another approach for $\hat{\sigma}_j^2$, derived from using judgement to select the random error between the individual development factors for each cohort and the selected tail factor.

To show how a tail factor can be incorporated into Mack's method, a tail factor can be added to the recursive version of the worked example given in Section 4.2.5. Viewing the Taylor and Ashe (1983) data graphically, as shown in Figure 4.1, suggests that a tail factor is probably necessary, as the oldest cohort is still developing at the latest development point; without this the reserves could be under-estimated.

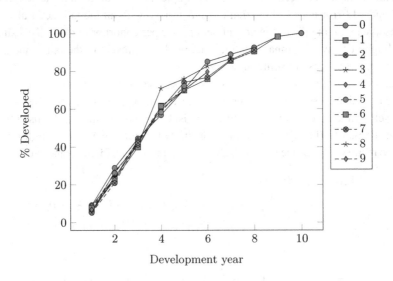

Figure 4.1 Taylor and Ashe data: Claims Paid as % of CL ultimate

As noted in Chapter 3, various methods can be used to estimate the tail factor, but for the purposes of this example, log-linear regression is used,[16] which suggests a tail factor of 1.029.[17] For $\widehat{\mathrm{Var}}(\hat{f}_J)$, from Table 4.17 it can be seen that the coefficient of variation for the value $j = 7$ and $j = 8$ is around 1%, so this could be used to derive $\widehat{\mathrm{Var}}(\hat{f}_J)$. This gives $\widehat{\mathrm{Var}}(\hat{f}_J) = (1\% \times 1.029)^2 = 0.00010588$. For $\hat{\sigma}_j^2$, the average $\hat{\sigma}$ value at $j = 7$ and $j = 8$ is 27.5 (calculated using the relevant values from Table 4.17 as the average of $\sqrt{1,147}$ and $\sqrt{447}$), so this could be used for $\hat{\sigma}_J$.

Putting these values into the recursive formulae (4.34), (4.35), (4.36) and (4.37) gives the results (in thousands) shown in Table 4.20. Note that an additional year (i.e. Cohort) "0" is included in the results, since the inclusion of a tail factor means that this year now has a reserve, with associated Standard Error.

---

[16] For example, using the LINEST function in MS Excel applied to the factors in Table 4.6 produces a cumulative tail factor (i.e. from delay 9 to ultimate) value of 1.029.

[17] This is the same as the tail factor derived for the exponential curve in the worked example of the CL method in Section 3.2.8.

Table 4.20 *Mack example: Results with a tail factor*

| Cohort | Ultimate | Reserve | Standard Error of reserve | | | CV |
			Process	Estimation	Total	
0	4,015	113	54	40	68	60%
1	5,591	252	81	82	115	46%
2	5,535	625	113	100	151	24%
3	5,452	863	123	103	161	19%
4	4,999	1,126	242	141	280	25%
5	5,259	1,568	382	198	431	27%
6	5,825	2,342	519	262	581	25%
7	6,982	4,117	812	403	906	22%
8	5,806	4,443	924	391	1,003	23%
9	5,114	4,770	1,324	471	1,405	29%
Total	54,577	20,219	1,943	1,704	2,584	13%

The total estimated ultimate claims, as expected, exceed the ultimate value without a tail factor, shown in Table 4.4, by 2.9% (i.e. the selected tail factor), whereas the total reserve exceeds the equivalent reserve figure by more than 8%, indicating the geared effect of the tail factor (as discussed in the Chain Ladder worked example in Chapter 3). The CV varies a little by cohort, but is the same overall, indicating that in this case the simplistic approach of using the CV derived from the model excluding a tail factor, referred to at the start of this section, would have produced broadly comparable results, at least at an aggregate level.

These results were produced by applying the recursive formulae in a spreadsheet, but the same results can be produced using, for example, "MackChainLadder" in the "ChainLadder" package in R. This package can also be used to derive estimates of $\widehat{\text{Var}}(\hat{f}_J)$ and $\hat{\sigma}_J^2$, using log-linear regression. The results using tail factors derived in this way, rather than using those based on judgement, are very similar to those shown in Table 4.20. The relevant R code using the values of $\widehat{\text{Var}}(\hat{f}_J)$ and $\hat{\sigma}_J^2$ based on judgement, and using log-linear regression is in Appendix B.3.1.

4.2.7 Testing the Validity of Assumptions Underlying Mack's Method

As explained in Mack (1997), there are three key assumptions that underpin Mack's method. These are equivalent to the first three items in the mathematical formulation of Mack's method given in Section 4.2.2:

1. There are development factors $f_0, f_1, \ldots, f_{J-1} > 0$ such that, for all $0 \le i \le I$ and $1 \le j \le J$,

$$E(C_{i,j}|C_{i,0}, C_{i,1}, \ldots, C_{i,j-1}) = f_{j-1}C_{i,j-1}. \tag{4.38}$$

Mack (1997) shows that this implies that the sequential development factors $C_{i,j}/C_{i,j-1}$ and $C_{i,j+1}/C_{i,j}$ are not correlated.

2. The cohorts are independent. In mathematical notation this means that

$$\{C_{i,0}, C_{i,1}, \ldots, C_{i,J}\}, \{C_{k,0}, C_{k,1}, \ldots, C_{k,J}\}, i \neq k, \text{ are independent.} \qquad (4.39)$$

3. There are variance parameters, $\sigma_0^2, \sigma_1^2, \ldots, \sigma_{J-1}^2 > 0$ such that, for all $0 \leq i \leq I$ and $1 \leq j \leq J$,

$$\mathrm{Var}(C_{i,j} | C_{i,0}, C_{i,1}, \ldots, C_{i,j-1}) = \sigma_{j-1}^2 C_{i,j-1}. \qquad (4.40)$$

Tests that can be used to assess the validity of these assumptions are summarised below. Further details, including worked examples, are given in Mack (1997). Carrato et al. (2015) also summarises these tests and shows some worked examples of their practical application.

Test for Uncorrelated Development Factors

This test looks at the degree of correlation between successive development factors across the whole triangle, rather than separately between successive pairs.

The idea is to first rank the development factors in each column (leaving out the last one of the first column in each pair of columns). Then, add the squared differences of each adjacent rank columns of equal length, derive Spearman's rank correlation coefficients, and finally an overall weighted average of these. The result is then tested against a 50% confidence limit, assuming a standard Normal distribution. As explained in Appendix G of Mack (1997), 50% is used here (rather than, say, 95% which is often used in statistical tests such as this), as the test is intended to be approximate in nature and to only identify correlations in a substantial part of the triangle.

In mathematical terms, using the same triangle notation as for the mathematical formulation of Mack's method, the Spearman rank correlation coefficient, T_j, for a single adjacent pair of columns is defined as

$$T_j = 1 - 6 \sum_{i=1}^{I-j} \frac{(r_{i,j} - s_{i,j})^2}{[(I-j)^3 - I + j]}, \qquad (4.41)$$

where $r_{i,j}$ is the rank for the development factor in cohort i, development period j and $s_{i,j}$ is the rank for the development factor in cohort i, development period $j - 1$.

These are then combined into a single statistic, T, using

$$T = \sum_{j=2}^{I-2} \frac{I-j-1}{(I-2)(I-3)/2} T_j. \qquad (4.42)$$

As explained further in Mack (1997), the 50% confidence interval for T, assuming a Normal distribution, is

$$-\frac{0.67}{\sqrt{(I-2)(I-3)/2}} \leq T \leq +\frac{0.67}{\sqrt{(I-2)(I-3)/2}}.$$

Hence, if the calculated value for T lies outside this range, then the assumption that the development factors are uncorrelated appears invalidated according to this test, and hence it may not be advisable to use the Mack method for the relevant dataset.

Test for Independence of Cohorts

There can be a number of different reasons why this assumption is invalid, including for example the existence of catastrophe-type claims or calendar period effects that affect more than one cohort. In relation to the former, where such claims are present, this will usually be known and hence a test procedure is not usually necessary. Furthermore, it may be possible to remove the relevant claims data from the triangle and deal with these claims separately. In relation to calendar-year effects, one approach to testing for these is to use residual plots, which are relevant for many stochastic methods, as explained in Section 4.10.3. These can be used by the practitioner, in conjunction with general reasoning, based on their knowledge of the underlying business and claims trends. However, another approach that can be used, and which is specifically applicable for Mack's method, is the test procedure described in Appendix H of Mack (1997). In summary, this involves the following:

1. For each development period, order the development factors and divide them into two groups – one that is lower than the median of the factors (the "Small" group) and one that is larger (the "Large" group). If the number of factors is odd, then discard the factor that is equal to the median. Hence, every development factor that is not eliminated has a 50% chance of belonging to either the Small or Large group.
2. For each diagonal, count the number of development factors categorised as Small and Large. Intuitively, if there were no calendar year-trends, then the number of Small and Large factors in each diagonal should be approximately the same size.
3. Since every development factor that is not eliminated has a 50% chance of belonging to either the Small or Large group, this means that the number of Large factors in each diagonal follows a binomial distribution with n = the number of entries in the diagonal and $p = 0.5$.
4. Once the binomial probabilities have been calculated it is possible to derive the distribution of a further statistic, namely the minimum of the number of Large and Small factors in each diagonal. Then, each realisation of this statistic in a dataset can be compared with this distribution to test whether each diagonal contains significantly more Large or Small factors, thus indicating a possible calendar period effect.

Test for Variance Parameters

The equivalence of the Chain Ladder to a linear regression on the data in successive pairs of columns is explained in Section 3.2.6. Mack (1997) builds on this to suggest examining residual plots of the weighted residuals, using a form consistent with that implied by the Mack variance assumption at (4.40). If the variance assumption is valid, then these residual plots should show a random pattern. Further details are given in Mack (1997).

Carrato et al. (2015) also suggests that the behaviour of the variance parameters, σ^2, can be used to test the validity of the assumption. In particular, they suggest that a graph of these parameters against development period should show an approximate exponential decay pattern to zero (assuming zero tail factor) and that unusually large kinks in the pattern may indicate that the assumption is at least partly invalid.

Further diagnostic tests that apply generally to stochastic methods are described in Section 4.10.3.

4.2.8 Adding a Cross-product Term to the Formulae for Mack's Method

Other CL-based stochastic models have been defined that are closely related to Mack's method, but which differ in their formulation and resulting formula for the estimation error. The reserve estimate in these models is the same as in Mack's method. For example, in Buchwalder et al. (2005, 2006) a time-series model is defined which is equivalent to the CL model, and has the same process error as Mack's method, but which includes an additional term in the estimation error component of the MSEP. This model is abbreviated as "BBMW" here and in other sources, after the authors of these papers. The same results are produced by another model defined in Murphy (2007) and Bardis et al. (2012). Merz and Wüthrich (2008c) gives a thorough mathematical analysis of Mack's method, and demonstrates that the formula for estimation (i.e. parameter) error in Mack's method can be shown to be a linear approximation to the model described as "Approach 3" in Section 3.2.3 of Merz and Wüthrich (2008c). This linear approximation is also discussed in Buchwalder et al. (2006).

The key difference in the implementation of these models and of Mack's method is the additional term in the estimation error, which Murphy (2007) refers to as the "cross-product term", which is also used here.

As per Murphy (2007), the recursive formula for the estimation (i.e. parameter) risk for a single cohort, equivalent to (4.24), including the cross-product term is

$$\widehat{\text{EstRisk}}(i, k) = \hat{f}_{k-1}^2 \widehat{\text{EstRisk}}(i, k - 1) + \hat{C}_{i,k-1}^2 \widehat{\text{Var}}(\hat{f}_{k-1})$$
$$+ \widehat{\text{Var}}(\hat{f}_{k-1})\widehat{\text{EstRisk}}(i, k - 1). \tag{4.43}$$

This applies for values of $k > I - i$, as the formula is the same as in (4.24) when $k = I - i$.

As per Formula (12) from Bardis et al. (2012), the recursive formula for the parameter risk for combined cohorts, equivalent to (4.31), including the cross-product term is

$$\widehat{\text{EstRisk}}(\hat{X}_j) = (\hat{X}_{j-1} + C_{I-j+1,j-1})^2 \widehat{\text{Var}}(\hat{f}_{j-1})$$
$$+ \hat{f}_{j-1}^2 \widehat{\text{EstRisk}}(\hat{X}_{j-1})$$
$$+ \widehat{\text{Var}}(\hat{f}_{j-1})\widehat{\text{EstRisk}}(\hat{X}_{j-1}). \tag{4.44}$$

This applies for $j = 2, 3, \ldots, J$, as the formula is the same as in (4.31) when $j = 1$.

Implementing these recursive formulae in practice is no more difficult than doing so without the cross-product term. The impact on the MSEP of the inclusion of the cross-product term in the formulation will vary, depending on the data. When the triangle is reasonably stable, the increase in the MSEP is not usually material. If the term is included when the BBMW method is applied to the Taylor and Ashe (1983) data used in the example in Section 4.2.5, then the results, derived by implementing the amended recursive formulae in a spreadsheet, are as shown in Table 4.21.

These results are the same[18] as those in the columns labelled "Murphy Formula" in Murphy (2007:9). They are also the same as the results on the "BBMW" row of Table 5 in Buchwalder et al. (2006), demonstrating the equivalence of the BBMW model and the model in Murphy (2007). Finally, the results are the same as the "Ultimate" row in Merz and Wüthrich (2007b), which also includes results using a model incorporating their "expected claims-development result" approach (as described in Section 4.6). The reserve is not shown in Table 4.21, as it is the same as for Mack's method without the cross-product term, given in Table 4.4.

Table 4.21 *Mack example: Results incorporating the cross-product term*

Cohort	Process	Standard Error Estimation	Total	CV
1	49	58	76	80%
2	91	81	122	26%
3	103	85	134	19%
4	228	128	261	27%
5	367	186	411	29%
6	500	248	558	26%
7	786	386	875	22%
8	896	376	971	23%
9	1,285	456	1,363	29%
Total	1,878	1,569	2,448	13%

Hence, the process error component of the results are the same as those in Mack's method in Table 4.15, as expected, and there is a negligible difference in the parameter error results compared to the results without using the cross-product term. For other datasets, particularly those where the development is relatively volatile, this may not be the case. The "MackChainLadder" function in the "ChainLadder" package in R makes it straightforward to test the effect in practice on any dataset, since there is an option to include the cross-product term. The relevant R code is in Appendix B.3.1.

There has been some debate in the literature regarding whether the BBMW model incorporating the additional cross-product term is an appropriate model or not. For example, Mack et al. (2006) argue that it will overstate the estimation error, with the authors of Buchwalder et al. (2006) responding that their approach is based on using a so-called "conditional resampling" approach. In subsequent work, some authors (e.g. Liu and Verrall, 2009b) have chosen to use both approaches, and others (e.g. Venter, 2006) have compared the two approaches with other methods.

Whether Mack's method or the BBMW model incorporating the cross-product term is "better" will depend on the circumstances, and judgement will need to be used to determine which model is producing a more reasonable indication of the variability in the future claims outgo. In some cases, the differences between the results according to each

---

[18] The exact total parameter error across all cohorts combined is 1,569,349 – the results shown in Table 4.21 are in thousands for ease of presentation.

model may be negligible anyway. As with any model selection for both deterministic and stochastic claims reserving purposes, applying more than one model and then comparing the results may be helpful when there is no clear indication as to which model is "best".

4.2.9 Using Different Types of Development Factors

The mathematical formulation of the Mack method given in Section 4.2.2 follows that of the original Mack (1993a) paper. Subsequently, in Mack (1999), a more generalised formula for the development factors was given, whereby (4.7) is written as

$$\hat{f}_j = \frac{\sum\limits_{i=0}^{I-j-1} w_{i,j} C_{i,j}^\alpha F_{i,j}}{\sum\limits_{i=0}^{I-j-1} w_{i,j} C_{i,j}^\alpha}, \tag{4.45}$$

where $\alpha = 0, 1$ or 2 and $w_{i,j} \in [0, 1]$.

When $\alpha = 1$ and $w_{i,j} = 1$, the formula is the column sum average, as per the original formulation of Mack's method. When $\alpha = 0$, the development factor becomes the simple average of the individual cohort development factors. When $\alpha = 2$, the formula becomes the result of an ordinary regression of $C_{i,j+1}$ against $C_{i,j}$ with an intercept of 0. The weights, $w_{i,j}$ allow for the possibility of assigning more or less weight to individual development factors.

With this formulation, the variance parameter formula changes from (4.8) to

$$\hat{\sigma}_j^2 = \frac{1}{I-j-1} \sum\limits_{i=0}^{I-j-1} w_{i,j} C_{i,j}^\alpha (F_{i,j} - \hat{f}_j)^2. \tag{4.46}$$

Mack (1999) gives recursive formulae for the MSEP of the estimated ultimate claims for single cohorts and for all cohorts combined, using the more generalised form for the development factors. As for the original form of Mack's method, these formulae can be broken down into process and estimation error, as per the formulae in Section 4.2.4. In fact, if $\widehat{\mathrm{Var}}(\hat{f}_k)$ is changed from (4.25) to

$$\widehat{\mathrm{Var}}(\hat{f}_k) = \frac{\hat{\sigma}_k^2}{\sum\limits_{i=0}^{I-k-1} w_{i,k} C_{i,k}^\alpha}, \tag{4.47}$$

then the recursive formulae in Section 4.2.4 for both single and combined cohorts can be used without modification. Similarly, when a tail factor is required, the recursive formulae given in Section 4.2.6 can also be used for the more generalised form of Mack's method. Finally, if a cross-product term is also included in the model formulation, then the recursive formulae in Section 4.2.8 can be used.

All of these alternative versions of the Mack method can be implemented in practice using the "MackChainLadder" function in the "ChainLadder" R package. Example code, applied to the Taylor and Ashe (1983) data, is given in Appendix B.3.1. The results are not shown here, but the CV is very similar for all the different options shown in the Appendix, at approximately 13% across all cohorts combined, as per the results for the original Mack method approach given in Table 4.4. For other datasets, the results could vary between the different options, perhaps significantly.

4.3 The Over-dispersed Poisson ("ODP") Model

4.3.1 Introduction

The ODP model is a form of statistical model that, when used in the context of claims reserving, is fitted to the incremental claims in a data triangle. It can be thought of as a generalisation of the more straightforward Poisson model, which is described in Section 4.7.6. The ODP model was initially proposed in Renshaw and Verrall (1994) and subsequently in Renshaw and Verrall (1998). It is often referred to in papers and articles covering different types of stochastic claims reserving methods. When it is formulated with a single scale parameter and with separate parameters for each cohort and development period, then like Mack's method it produces results that are equivalent to the column sum weighted average deterministic CL method. It provides an estimate of the mean reserve and of the variance around that reserve, but not the full distribution. It is formulated as a Generalised Linear Model (or "GLM"), which are a family of models that are a flexible generalisation of ordinary linear regression. GLMs can be fitted to a data triangle using standard statistical software, as will be demonstrated in the worked example later in this section. In common with some other stochastic models, it does have constraints regarding the nature of the data to which it can be fitted successfully, as explained later in this section. Despite these constraints, the ODP model is an important stochastic method in claims reserving, partly because it provides the statistical basis for a very commonly used method for estimating reserve uncertainty – the ODP Bootstrap method – which is described in Section 4.4.

4.3.2 Mathematical Formulation of the ODP Model

Using the notation from Section 2.5.6, where the incremental claims for cohort i in development period j are denoted by $X_{i,j}$, then it is first useful to note the distributional assumptions for the mean and variance in GLMs can be expressed as follows:

$$E[X_{i,j}] = m_{i,j} \text{ and } \text{Var}[X_{i,j}] = \frac{\phi_j V[m_{i,j}]}{w_{i,j}}. \tag{4.48}$$

The term $V[m_{i,j}]$ is the so-called *variance function*, which is a function of the mean. ϕ_j is a scale parameter (that may or may not vary by the index j) and $w_{i,j}$ are the weights assigned to $X_{i,j}$.

For the ODP model, $X_{i,j}$ are assumed to be independent over-dispersed Poisson random variables, with all the weights being 1, a single scale parameter,[19] ϕ, is used and $V[m_{i,j}] = m_{i,j}$. Based on this, the mean and variance of the ODP model, using the GLM formulation in (4.48), are defined under the original multiplicative form of the model as follows:[20]

$$\text{E}[X_{i,j}] = m_{i,j} = x_i y_j \text{ and } \text{Var}[X_{i,j}] = \phi\text{E}[X_{i,j}] = \phi x_i y_j, \qquad (4.49)$$

where $i + j \leq I; 0 \leq j \leq J$ and $\sum_{k=0}^{J-1} y_k = 1$.

In this formulation, x_i represent the expected "ultimate" claims up to the latest development point of the triangle, and y_j represent the proportion of this ultimate that is expected to emerge in development period j. In this way, the ODP model can be seen as having a "cross-classified" structure, and hence is sometimes described as being part of the family of Cross-Classified models, which also includes a more general form of model referred to as Tweedie models, as mentioned in Section 4.7.11. Cross-Classified models are described further in Merz and Wüthrich (2015b).

For estimation purposes, it is usual to express the ODP model mean in a linear form, by taking logarithms. This leads to the following alternative formulation of the mean and variance, which is the one most often quoted when referring to the ODP model:

$$\text{E}[X_{i,j}] = m_{i,j} \text{ and } \text{Var}[X_{i,j}] = \phi\text{E}[X_{i,j}] = \phi m_{i,j}, \qquad (4.50)$$

$$\log m_{i,j} = \eta_{i,j},$$

$$\eta_{i,j} = c + \alpha_i + \beta_j, \quad \text{where } \alpha_0 = \beta_0 = 0.$$

Using standard GLM terminology, this defines a GLM with a logarithmic link function, with the variance being proportional to the mean. Readers who are unfamiliar with GLMs can find details in many statistical textbooks, including McCullagh and Nelder (1989) and Dobson (2002). The parameter ϕ is an unknown scale parameter that is estimated as part of the fitting procedure. It is usually greater than one and hence is the source of the "over-dispersion" in the ODP model, as the variance is therefore larger than the mean. The formulation of the model given here uses a constant scale parameter; this can be modified so that, for example, there is a different scale parameter, ϕ_j, for each development period. This is discussed further in Section 4.4.3, in the context of bootstrapping the ODP model.

The ODP model is one type of a more generic model from the GLM family where $\text{Var}[X_{i,j}] = \phi m_{i,j}^z$, so that when $z = 1$ this becomes the ODP model. When $z = 0$ the model is a Gaussian model, when $z = 2$ it is a Gamma model and when $z = 3$ it is an Inverse Gaussian model. These other models are described in Section 4.7.11.

In the ODP model, the incremental claims, $X_{i,j}$, in each cell of the triangle is assumed to be distributed as independent over-dispersed Poisson variables with the above means

---

[19] Note that non-constant scale parameters can also be used and may be more appropriate in some circumstances, as discussed in Section 4.4.3 in the context of bootstrapping the ODP model.

[20] In some papers that describe the ODP model, $C_{i,j}$ is used to denote the incremental claims, but $X_{i,j}$ is used here to be consistent with the notation used in other parts of the book.

and variances. The constraint that $\alpha_0 = \beta_0 = 0$ is used[21] to avoid the model being over-parameterised. This will mean that the fitted values in the top right and bottom left cells of the triangle will be equal to the actual values.

The α_i represent factors associated with the rows of the triangle and the β_j represent factors associated with the columns of the triangle, with c being a "constant" parameter that applies to all cells. Hence, for a simple 3×3 triangle, the model has the structure shown in Table 4.22.

Table 4.22 *Structure for the ODP model*

Cohort	Development period		
	0	1	2
0	c	$c + \beta_1$	$c + \beta_2$
1	$c + \alpha_1$	$c + \alpha_1 + \beta_1$	
2	$c + \alpha_2$		

Thus, the model assumes the same development pattern, defined by the β_j parameters, is appropriate for all cohorts. Because the link function is logarithmic, once the parameters, c, α_i and β_j, have been fitted to the data, the fitted incremental values will be derived through $e^{\eta_{i,j}}$ (i.e. $e^{c+\alpha_i+\beta_j}$). In that sense, the parameters of the model do not have values that have any obvious intuitive interpretation. However, due to the way in which the ODP model is formulated, it can be shown (both in practice and theory) to produce exactly the deterministic chain ladder reserve estimates (subject to the constraint that the sum of each column in the triangle is positive), with the advantage that it also can be used to produce estimates of the variability around those estimates. This partly explains why the ODP model is sometimes referred to as a stochastic version of the chain ladder method, although it is not the only model which has been given that label, as discussed in Section 4.1.5.

The worked example given later in this section and the application of the method to a larger dataset both demonstrate the equivalence of the stochastic ODP model and the deterministic chain ladder method. A theoretical explanation of the equivalence is given, for example, in Renshaw and Verrall (1998).

Although the ODP model will still work when there are a small number of negative incremental values in the triangle, it is subject to the constraint that the sum of each column and row in the triangle must be positive. Furthermore, the fitted values will always be positive, and hence if the underlying claims process exhibits material reductions in claims during some development periods (as can occur with some classes of business for claims incurred data and for others for claims paid data, for example due to salvage and subrogration recoveries being received) then the model may not be suitable. For this, and other reasons, it is not always possible to fit the ODP model to a particular dataset.

---

[21] Note that in many formulations of the ODP model the index value begins at 1, so this constraint becomes $\alpha_1 = \beta_1 = 0$, but here the notation begins at zero, as explained in Section 2.5.6.

This represents a practical disadvantage of the method, but since it underpins the ODP Bootstrap approach (described in Section 4.4) it is explored in detail in the remainder of this section.

Fitting the model to a particular dataset and then producing reserve estimates, along with the associated prediction error, involves the following key high-level stages:

1. Fit the model parameters using standard GLM model fitting procedures, and test the goodness of fit.
2. Use the parameters to determine the estimated future claims and hence reserves.
3. Determine the prediction errors in the individual "future" cells of the triangle
4. Determine the prediction error of the reserves for each cohort and in total.

For consistency with other sources, where the term "Standard Error" or its abbreviation "SE" is used in the worked examples below, it is intended to be the same as "prediction error".

Each of these stages is discussed further below.

4.3.3 Stage 1: Fitting the ODP Model

Because the model is formulated as a GLM, it can be fitted to claims triangles relatively easily using a number of widely available software packages. They include both commercially available software and other freely available software on the internet.

The model can also be fitted by understanding the process required to fit the parameters and then programming this in a suitable software package. Although this approach is likely to enable a more complete understanding of the ODP model, it is time consuming and would not be a viable option for many. Hence, many practitioners who wish to explore the use of an ODP model would often make use of an existing software package. However, since it is beneficial to have some understanding of the process that is being used in the software packages, rather than simply regarding them as "black boxes", the remainder of this section provides a high-level summary of one way in which an ODP model can be fitted, together with a worked example. Other explanations of how to fit the parameters of an ODP model are also possible – for example, Verrall (2010b) shows how it can be done using MS Excel. Readers who do not wish to understand the theory can just follow the worked example in this section, if required.

The method described here is an implementation of maximum likelihood estimation known as the *Fisher scoring algorithm*. It can be shown to be equivalent to an iterative weighted least squares method, solved using a Newton–Raphson type approach. This is the approach used by many of the software packages that incorporate GLM fitting procedures. To explain the implementation, the model is formulated in matrix format:

$$\log(E[X]) = \Gamma\beta, \tag{4.51}$$

where Γ is known as the *design matrix*.[22] This structure is easily understood if the 3×3 matrix example shown in Table 4.22 continues to be used. Then:[23]

$$X = \begin{bmatrix} X_{0,0} \\ X_{1,0} \\ X_{2,0} \\ X_{0,1} \\ X_{1,1} \\ X_{0,2} \end{bmatrix}; \Gamma = \begin{bmatrix} 1 & 0 & 0 & 0 & 0 \\ 1 & 1 & 0 & 0 & 0 \\ 1 & 0 & 1 & 0 & 0 \\ 1 & 0 & 0 & 1 & 0 \\ 1 & 1 & 0 & 1 & 0 \\ 1 & 0 & 0 & 0 & 1 \end{bmatrix}; \beta = \begin{bmatrix} c \\ \alpha_1 \\ \alpha_2 \\ \beta_1 \\ \beta_2 \end{bmatrix}. \tag{4.52}$$

The iterative procedure used to fit the β parameters to the data makes use of a *weight* matrix, W, which, in the case of a Poisson model, has diagonal elements equal to the fitted values at each successive stage of the iteration, that is

$$W = \begin{bmatrix} e^c & 0 & 0 & 0 & 0 & 0 \\ 0 & e^{c+\alpha_1} & 0 & 0 & 0 & 0 \\ 0 & 0 & e^{c+\alpha_2} & 0 & 0 & 0 \\ 0 & 0 & 0 & e^{c+\beta_1} & 0 & 0 \\ 0 & 0 & 0 & 0 & e^{c+\alpha_1+\beta_1} & 0 \\ 0 & 0 & 0 & 0 & 0 & e^{c+\beta_2} \end{bmatrix}. \tag{4.53}$$

The general matrix formula for the procedure, which can be found in a number of textbooks on GLMs, is as follows:

$$\hat{\beta} = (\Gamma^T W \Gamma)^{-1} \Gamma^T W z, \tag{4.54}$$

where

$$z = \Gamma\beta + \frac{X - e^{\Gamma\beta}}{e^{\Gamma\beta}}. \tag{4.55}$$

The W and β matrices on the right-hand side are taken from each successive stage of the iteration, and (4.54) above generates the values for the next stage of the iteration.

If the ith row of the matrices $\Gamma\beta, X$ and z are denoted by b_i, x_i and z_i respectively, then

$$z_i = b_i + \frac{x_i - e^{x_i}}{e^{x_i}}. \tag{4.56}$$

In the simple 3×3 example, the initial values for the cells in W matrix will be as follows:

$$W(0) = \begin{bmatrix} X_{0,0} & 0 & 0 & 0 & 0 & 0 \\ 0 & X_{1,0} & 0 & 0 & 0 & 0 \\ 0 & 0 & X_{2,0} & 0 & 0 & 0 \\ 0 & 0 & 0 & X_{0,1} & 0 & 0 \\ 0 & 0 & 0 & 0 & X_{1,1} & 0 \\ 0 & 0 & 0 & 0 & 0 & X_{0,2} \end{bmatrix}. \tag{4.57}$$

[22] A commonly used statistical notation for the design matrix is X, but this represents the incremental claims used in the notation in this book, so Γ is used instead, which is consistent with some other sources, such as Merz and Wüthrich (2008c).

[23] The bold β notation simply refers to the matrix shown here, comprising the c, α and β parameters. The use of bold text is standard matrix notation and should not be confused with the β_1 and β_2 parameters, which are only two elements of the β parameter matrix.

The estimated values for the $\hat{\beta}$ parameters in the first iteration are given by a first estimation of the fitted data, which in this model will be the log of the actual data, that is:[24]

$$\beta(0) = \begin{bmatrix} \log(X_{0,0}) \\ \log(X_{1,0}) \\ \log(X_{2,0}) \\ \log(X_{0,1}) \\ \log(X_{1,1}) \\ \log(X_{0,2}) \end{bmatrix}. \tag{4.58}$$

Putting these $\beta(0)$ values into (4.55) or (4.56) will then give $z(0) = \beta(0)$.

The initial values for W and β are then used in (4.54) to produce the values for the fitted parameters, β, in the next stage of the iteration. The values on the diagonal of W are then set equal to the fitted values implied by these fitted parameters. This procedure continues until successive values of β are sufficiently close to each other.

Further detail of the generalised statistical procedure, using similar matrix notation to above (except with the design matrix being denoted by X rather than Γ), can be found in Sections 4.3 and 4.4 of Dobson (2002). It is also consistent with the generalised procedure outlined in Section 2.5 of McCullagh and Nelder (1989). These sources will help interested readers to understand the rationale for the key equations, (4.54) and (4.55), that define this procedure, but it is sufficient for the purpose of explaining the ODP model as a reserving method to be aware that these formulae are consistent with established statistical theory.

Once the model has been fitted, then as with any GLM model, some assessment of how well the model fits the data would normally be carried out at this stage. In summary, this can be assessed using, for example, residual plots – overall and by cohort, development and calendar period, residual Normality plots, Box-Whisker plots to check for outliers and standard statistical goodness-of-fit tests such as Akaike Information Criterion and Bayesian Information Criterion. For further details of the selection and testing of stochastic methods in general see Section 4.10.3, and for validation and communication of results from stochastic methods see Section 4.10.4.

Before moving on to the next stage of the process, it is helpful to define a matrix referred to in statistical modelling as the "hat" matrix. Although this matrix is not required for the description here of fitting the ODP model, it becomes relevant when describing one formulation of the bootstrap method applied to the ODP model, as given in Section 4.4. The hat matrix is described in detail in statistical textbooks[25] and is defined, in a GLM context, as follows:

$$H = \Gamma(\Gamma^T W \Gamma)^{-1} \Gamma^T W. \tag{4.59}$$

The description of the procedure now continues with the next stage of the ODP model fitting process.

---

[24] For ease of presentation, the ^ notation for the iterative estimated values for β have been omitted, as the fact that they are iterative estimated values should be clear from the (0) notation, which shows the iteration number.

[25] For example, see Davison and Hinkley (1997) for further details in the context of bootstrapping statistical models.

4.3.4 Stage 2: Determining the Fitted Values and Hence the Estimated Reserves

Once the fitted parameters, $\hat{\beta}$ have been determined, they are combined for each cell of the triangle, using the structure in Table 4.22 and then the historical fitted values are obtained using

$$\hat{X}_{i,j} = \hat{m}_{i,j} = \exp(\hat{\eta}_{i,j}) = e^{\hat{c} + \hat{\alpha}_i + \hat{\beta}_j}. \tag{4.60}$$

The estimated future incremental values are determined in a similar way and the estimated reserve, denoted by $\hat{R}_i$, for cohort i ($i > 0$) for the standard claims paid triangle referred to in Section 2.5.6 with $I + 1$ rows and $J + 1$ columns is then derived, as follows:

$$\hat{R}_i = \sum_{k=I-i+1}^{J} \hat{X}_{i,k}. \tag{4.61}$$

The total reserve across all cohorts, denoted by $\hat{R}$, is

$$\hat{R} = \sum_{i=1}^{I} \hat{R}_i. \tag{4.62}$$

4.3.5 Stage 3: Determining the Estimated Prediction Error in the Individual Future Incremental Values

The previous stage has involved using the ODP model to predict a value in the future part of the triangle. The next stage is to estimate the uncertainty, or error, around this prediction, to determine an estimate of the reserve uncertainty, which is the key purpose of fitting a stochastic model. As explained in Section 4.1.3, the commonly used statistical approach to estimating this prediction uncertainty is to use the variance of the prediction, also known as the Mean-Squared Error of Prediction ("MSEP"). The prediction error is the square-root of the MSEP, and is usually referred to as the Root-Mean-Squared Error of Prediction, or "RMSEP". For a single future incremental payment, $X_{i,j}$, the MSEP is defined as follows:

$$\text{MSEP}[\hat{X}_{i,j}] = \text{E}[(X_{i,j} - \hat{X}_{i,j})^2]. \tag{4.63}$$

Replacing $(X_{i,j} - \hat{X}_{i,j})$ in (4.63) with

$$(X_{i,j} - \text{E}[X_{i,j}]) - (\hat{X}_{i,j} - \text{E}[X_{i,j}]),$$

and expanding, gives

$$\text{MSEP}[\hat{X}_{i,j}] = \text{E}[(X_{i,j} - \text{E}[X_{i,j}])^2] - \text{E}[(\hat{X}_{i,j} - \text{E}[X_{i,j}])^2]$$
$$- 2\text{E}[(X_{i,j} - \text{E}[X_{i,j}])(\hat{X}_{i,j} - \text{E}[X_{i,j}])]. \tag{4.64}$$

If it is assumed that future claims are independent of past claims, and since $\hat{X}_{i,j}$ is based only on past claims for the ODP model, the third term in (4.64) will be zero. For the first term, note that

$$\text{E}[(X_{i,j} - \text{E}[X_{i,j}])^2] = \text{Var}[X_{i,j}].$$

Then, from equation (4.50):

$$\text{Var}[X_{i,j}] = \phi m_{i,j}. \tag{4.65}$$

For the second term, Renshaw (1994b) shows that

$$\text{E}[(\hat{X}_{i,j} - \text{E}[X_{i,j}])^2] \approx \text{Var}[\hat{\eta}_{i,j}] m_{i,j}^2. \tag{4.66}$$

Hence, combining (4.65) and (4.66):

$$\text{MSEP}[\hat{X}_{i,j}] \approx \phi m_{i,j} + \text{Var}[\hat{\eta}_{i,j}] m_{i,j}^2. \tag{4.67}$$

This can be thought of as

$$\underbrace{\text{MSEP}[\hat{X}_{i,j}]}_{\text{Prediction variance}} = \underbrace{\text{Var}[X_{i,j}]}_{\text{Process variance}} + \underbrace{\text{Var}[\hat{X}_{i,j}]}_{\text{Estimation variance}} . \tag{4.68}$$

This has the same form as the generic equation (4.2) given in the introduction to stochastic methods at Section 4.1.3. Further details are given in Appendix A.3. The terms in (4.67) will be replaced by their estimates, so that the MSEP is estimated using[26]

$$\widehat{\text{MSEP}}[\hat{X}_{i,j}] = \hat{\phi} \hat{m}_{i,j} + \text{Var}[\hat{\eta}_{i,j}] \hat{m}_{i,j}^2. \tag{4.69}$$

In the ODP model, to derive $\hat{\phi}$, the so-called "Pearson" residuals are used, as follows:

$$\hat{\phi} = \frac{1}{n-p} \sum \left(\frac{X_{i,j} - \hat{m}_{i,j}}{\sqrt{\hat{m}_{i,j}}} \right)^2, \tag{4.70}$$

where n is the number of data points and p is the number of parameters, so that $n - p$ is the number of degrees of freedom. For the application of the ODP model to a triangle with $I + 1$ origin periods and $J + 1$ development periods, for ease of notation, $(I + 1) = (J + 1) = k$ is used, so that $n = \frac{1}{2}k(k + 1)$ and $p = 2k - 1$.

England and Verrall (2006) show, in the Appendix to their paper, how the formula for the scale parameter, (4.70), can be reformulated as

$$\hat{\phi} = \frac{n}{n-p} \times \frac{\sum \left(\frac{X_{i,j} - \hat{m}_{i,j}}{\sqrt{\hat{m}_{i,j}}} \right)^2}{n}$$

$$= \frac{\sum \left[\left(\frac{n}{n-p} \right)^{\frac{1}{2}} \left(\frac{X_{i,j} - \hat{m}_{i,j}}{\sqrt{\hat{m}_{i,j}}} \right) \right]^2}{n}. \tag{4.71}$$

This may at first appear like an unnecessary complication, but its relevance will become clear in Section 4.4.3 where bootstrapping the ODP model is considered.[27]

Hence, in this stage an approximate formula for the MSEP of the individual incremental future values has been defined – which can be of interest when looking at the uncertainty

---

[26] The hat notation is dropped in subsequent paragraphs, for ease of presentation.

[27] Specifically, this formulation is used to derive the scale parameter in the ODP Bootstrap worked example in Section 4.4.4 and it is also in used in the description of the ODP Bootstrap using non-constant scale parameters given in Section 4.4.3.

of future cash flows. In practice, this needs to be extended to estimate the MSEP of the sum of the estimated future incremental values – i.e. the reserves. This is the next and final stage of the process of implementing the ODP model.

4.3.6 Stage 4: Determining the Prediction Error of the Reserves Using the ODP Model

This stage of the process moves beyond that normally found in typical GLM applications. The aim is to estimate $\text{MSEP}(\hat{R}_i)$ and $\text{MSEP}(\hat{R})$. Various papers give the following formulae for estimates of these items, including Renshaw (1994b) and England and Verrall (2002):

$$\text{MSEP}[\hat{R}_i] \approx \sum_{j \in \Delta_i} \hat{\phi}\hat{m}_{ij} + \sum_{j \in \Delta_i} \hat{m}_{ij}^2 \text{Var}[\hat{\eta}_{ij}]$$

$$+ 2 \sum_{\substack{j_1, j_2 \in \Delta_i \\ j_2 > j_1}} \hat{m}_{ij_1} \hat{m}_{ij_2} \text{Cov}[\hat{\eta}_{ij_1}, \hat{\eta}_{ij_2}], \tag{4.72}$$

where Δ_i represents the cells in cohort i in the data triangle corresponding to the future estimated values (i.e. those in the lower right-hand part of the triangle). Similarly, Δ is used to represent these cells across all cohorts.

The same papers show that

$$\text{MSEP}[\hat{R}] \approx \sum_{i,j \in \Delta} \hat{\phi}\hat{m}_{ij} + \sum_{i,j \in \Delta} \hat{m}_{ij}^2 \text{Var}[\hat{\eta}_{ij}]$$

$$+ 2 \sum_{\substack{i_1 j_1 \in \Delta \\ i_2 j_2 \in \Delta \\ i_1 j_1 \neq i_2 j_2}} \hat{m}_{i_1 j_1} \hat{m}_{i_2 j_2} \text{Cov}[\hat{\eta}_{i_1 j_1}, \hat{\eta}_{i_2 j_2}]. \tag{4.73}$$

The first summation term in both of these formula represents the process variance and the remaining two summation terms represent the estimation variance. The formulae are relatively difficult to implement in practice, due to the need to select very carefully the values over which to carry out the summations. Fortunately, they are much easier to understand and apply in practice if they are expressed in matrix format, by building on the matrix formulation of the ODP model set out in Section 4.3.3. This will become particularly apparent in the worked example that follows this section.

This formulation starts with the matrix formula for the estimated variance/covariance of the fitted parameters. This is

$$\hat{\Sigma} = \hat{\phi}(\mathbf{\Gamma}^T \mathbf{W} \mathbf{\Gamma})^{-1}. \tag{4.74}$$

The $\mathbf{\Gamma}$ and $\mathbf{W}$ matrices here are the design and weight matrices as defined in Section 4.3.3. For the 3×3 triangle example, which has five ODP model parameters, this matrix will be a 5×5 matrix, with the diagonal elements representing the estimated variance of the fitted parameters.

In the MSEP equations (4.72) and (4.73) the i and j values are only those that are valid for the future part of the triangle. To derive these formulae in matrix format, it is necessary to define the *future design matrix*, Γ_F. For the simple 3×3 triangle in the example above, this will be

$$\Gamma_F = \begin{array}{c} \\ C_{1,2} \\ C_{2,1} \\ C_{2,2} \end{array} \begin{array}{cccccc} \text{cells} & c & \alpha_1 & \alpha_2 & \beta_1 & \beta_2 \\ \left[\begin{array}{ccccc} 1 & 1 & 0 & 0 & 1 \\ 1 & 0 & 1 & 1 & 0 \\ 1 & 0 & 1 & 0 & 1 \end{array}\right] \end{array}. \tag{4.75}$$

It is then straightforward to see that the matrix product $\Gamma_F \beta$ will produce the parameters for the future values in the correct structure (by extending the structure in Table 4.22 into the future part of the triangle). Defining $\Gamma_F \beta$ in this way ensures that the awkward summation limits in the above formulae are handled correctly.

The MSEP equations contain terms relating to the variance and covariance of the so-called *linear predictors*, $\hat{\eta}_{ij}$. In matrix terms, the variance / covariance matrix of the linear predictors in the future part of the triangle will be

$$N = \Gamma_F \hat{\Sigma} (\Gamma_F)^T. \tag{4.76}$$

For the simple 3×3 triangle case, N will also be a 3×3 matrix, as there are only three future values in the triangle (assuming no development tail).

To determine the product of the $\hat{m}_{ij}$ with $\mathrm{Var}[\hat{\eta}_{ij}]$ and $\mathrm{Cov}[\hat{\eta}_{ij_1}, \hat{\eta}_{ij_2}]$ etc. in the second and third terms of the MSEP equations (4.72) and (4.73) a diagonal matrix of the fitted values in the future part of the triangle is formed. In the 3×3 example, this will be

$$\mathbf{diag}(\hat{m}) = \begin{bmatrix} \hat{m}_{1,2} & 0 & 0 \\ 0 & \hat{m}_{2,1} & 0 \\ 0 & 0 & \hat{m}_{2,2} \end{bmatrix}. \tag{4.77}$$

That is,

$$\mathbf{diag}(\hat{m}) = \begin{bmatrix} e^{\hat{c}+\hat{\alpha}_1+\hat{\beta}_2} & 0 & 0 \\ 0 & e^{\hat{c}+\hat{\alpha}_2+\hat{\beta}_1} & 0 \\ 0 & 0 & e^{\hat{c}+\hat{\alpha}_2+\hat{\beta}_2} \end{bmatrix}. \tag{4.78}$$

To derive the required second and third terms in (4.72) and (4.73) matrix multiplication is used, as follows:

$$\mathbf{diag}\,(\hat{m}) N \mathbf{diag}(\hat{m}). \tag{4.79}$$

If this is then added to a further diagonal matrix, $\hat{\phi}\,\mathbf{diag}(\hat{m})$, representing the first term in (4.72) and (4.73), the resulting matrix will contain all the terms needed to derive the values for these formulae, and hence the required MSEP estimates. The way this matrix is used in practice is most easily demonstrated via a worked example, which follows.

4.3.7 *Worked Example of the ODP Model*

This section provides a worked example of the ODP model using a 3×3 matrix. The example uses the same example dataset that is used in the "Simple GLM" spreadsheet that accompanied Shapland and Leong (2010).[28] Although this is a very small data triangle, this helps keep the matrices to a manageable size in the text that follows and so enables the reader to see clearly the practical steps that are involved; the methodology is identical regardless of the size of the triangle.

The example starts with the first stage of the process outlined in Section 4.3.3.

ODP Worked Example: Stage 1 – Fitting the Parameters to the Data

The example data triangle in incremental format, and its logarithms, is shown in Tables 4.23 and 4.24. The data is assumed to be claims paid, although that is not a critical assumption for the analysis.

Table 4.23 *ODP Worked example: Incremental data*

	0	1	2
0	95	55	30
1	115	45	
2	105		

Table 4.24 *ODP Worked example: Log of Incremental data*

	0	1	2
0	4.55	4.01	3.40
1	4.74	3.81	
2	4.65		

The matrices used in the first stage of the iterative fitting procedure are then formed, using (4.52), (4.57) and (4.58). Thus:

$$
X = \begin{bmatrix} 95 \\ 115 \\ 105 \\ 55 \\ 45 \\ 30 \end{bmatrix} ; \quad
W(0) = \begin{bmatrix} 95 & 0 & 0 & 0 & 0 & 0 \\ 0 & 115 & 0 & 0 & 0 & 0 \\ 0 & 0 & 105 & 0 & 0 & 0 \\ 0 & 0 & 0 & 55 & 0 & 0 \\ 0 & 0 & 0 & 0 & 45 & 0 \\ 0 & 0 & 0 & 0 & 0 & 30 \end{bmatrix} ; \quad
\beta(0) = \begin{bmatrix} 4.55 \\ 4.74 \\ 4.65 \\ 4.01 \\ 3.81 \\ 3.40 \end{bmatrix} .
$$

Next, the matrix algebra required to derive $\beta(1)$ using (4.54), is carried out:

---

[28] Note that, although the same data is used in this example, the design matrix in the "Simple GLM" spreadsheet that accompanies Shapland and Leong (2010) needs to be changed to be the same as shown in Section 4.3.3, in order for the results to reconcile. Note also that similar functionality is available in the relevant spreadsheet that accompanies Shapland (2016), which the author of that paper has advised is an update and partial upgrade of Shapland and Leong (2010).

$$\Gamma^T W(0)\Gamma = \begin{bmatrix} 445 & 160 & 105 & 100 & 30 \\ 160 & 160 & 0 & 45 & 0 \\ 105 & 0 & 105 & 0 & 0 \\ 100 & 45 & 0 & 100 & 0 \\ 30 & 0 & 0 & 0 & 30 \end{bmatrix},$$

and its inverse:

$$(\Gamma^T W(0)\Gamma)^{-1} = \begin{bmatrix} 0.009 & -0.007 & -0.009 & -0.005 & -0.009 \\ -0.007 & 0.013 & 0.007 & 0.001 & 0.007 \\ -0.009 & 0.007 & 0.018 & 0.005 & 0.009 \\ -0.005 & 0.001 & 0.005 & 0.015 & 0.005 \\ -0.009 & 0.007 & 0.009 & 0.005 & 0.042 \end{bmatrix}. \qquad (4.80)$$

Since, $z(0) = \beta(0)$:

$$W(0)z(0) = \begin{bmatrix} 432.62 \\ 545.67 \\ 488.67 \\ 220.40 \\ 171.30 \\ 102.04 \end{bmatrix}. \qquad (4.81)$$

Finally, for the first iteration, using (4.54), the three items – (4.80), Γ^T and (4.81) – are multiplied together, which gives

$$\beta(1) = \begin{bmatrix} 4.623 \\ 0.065 \\ 0.031 \\ -0.735 \\ -1.222 \end{bmatrix}.$$

This procedure is repeated for the next iteration, with the $\beta(1)$ parameters above being used in both (4.53) to derive $W(1)$ and in (4.55) to derive $z(1)$. Only two further iterations (i.e. four in total) are needed for the parameters to converge to give the final fitted parameters as follows:

$$\beta(4) = \begin{bmatrix} 4.621 \\ 0.065 \\ 0.033 \\ -0.742 \\ -1.220 \end{bmatrix} \quad \text{or} \quad \begin{array}{l} \hat{c} = 4.621 \\ \hat{\alpha}_1 = 0.065 \\ \hat{\alpha}_2 = 0.033 \\ \hat{\beta}_1 = -0.742 \\ \hat{\beta}_2 = -1.220 \end{array}. \qquad (4.82)$$

Interested readers should be able to produce the same fitted parameters, either by creating their own spreadsheets/software programmes, or using the "Simple GLM" spreadsheets supplied with Shapland and Leong (2010) (but using the above design matrix) or by using one of the many software packages that are available. For example, in the software environment R, if the "glmReserve" function within the "ChainLadder" package is used, then the same fitted parameters can be produced very quickly.

Since this simple example is included merely to show the model fitting process, the assessment of the model fit, as discussed in Section 4.3.3, is not included here.

ODP Worked Example: Stage 2 – Deriving the Estimated Reserves

Having derived the fitted parameters, the fitted values shown in Table 4.25 can be derived, using (4.60), and hence the estimated reserves.

Table 4.25 *ODP example: Fitted values and reserve*

	Development year			Reserve
	0	1	2	
0	101.61	48.39	30.00	0.00
1	108.39	51.61	32.00	32.00
2	105.00	50.00	31.00	81.00
			Total	113.00

This table shows both the past and future fitted values, as well as the estimated reserve. Note that the fitted values in the top right and bottom left cells in the triangle are equal to actual incremental values, which is a function of the constraint specified in (4.50) that $\alpha_0 = \beta_0 = 0$.

If the original incremental data in Table 4.23 is converted to cumulative format and then the standard chain ladder method is applied to it, the results shown in Table 4.26 are produced.

Hence, as expected, the stochastic ODP model reproduces the reserves of the deterministic chain ladder method (using the full column sum average link ratios). This is an entirely hypothetical (and very small) data triangle, but the same equality will also apply to real (and much larger) data triangles.

Table 4.26 *Chain ladder results applied to ODP example data*

	Development year			Reserve
	0	1	2	
0	95.00	150.00	180.00	0.00
1	115.00	160.00	192.00	32.00
2	105.00	155.00	186.00	81.00
Col sum 1	310.00	180.00		
Col sum 2	210.00	150.00		
Ratio	1.48	1.20		

The next two stages of the process can now be implemented, which are combined for the purpose of this example.

ODP Worked Example: Stages 3 and 4 – Estimating the MSEP of the Individual Future Incremental Values and the Reserves

First, the scale parameter, $\hat{\phi}$, is derived using (4.70). This requires the corresponding cells in Tables 4.23 and 4.25 to be used in this formula with $n = 6$ and $p = 5$, which gives $\hat{\phi} = 2.5849$.

Next, the variance/covariance matrix of the fitted parameters, $\hat{\Sigma}$, is derived using (4.74). For this, the weight matrix is needed, which, as per (4.53), will be a 5×5 matrix with the diagonals as the fitted values in the historical part of the triangle. These values are taken from Table 4.25, so that

$$W = \begin{bmatrix} 101.61 & 0 & 0 & 0 & 0 & 0 \\ 0 & 108.39 & 0 & 0 & 0 & 0 \\ 0 & 0 & 105.00 & 0 & 0 & 0 \\ 0 & 0 & 0 & 48.39 & 0 & 0 \\ 0 & 0 & 0 & 0 & 51.61 & 0 \\ 0 & 0 & 0 & 0 & 0 & 30.00 \end{bmatrix}. \qquad (4.83)$$

Then, using Γ from (4.52) this produces

$$\Gamma^T W \Gamma = \begin{bmatrix} 445.00 & 160.00 & 105.00 & 100.00 & 30.00 \\ 160.00 & 160.00 & 0.00 & 51.61 & 0.00 \\ 105.00 & 0.00 & 105.00 & 0.00 & 0.00 \\ 100.00 & 51.61 & 0.00 & 100.00 & 0.00 \\ 30.00 & 0.00 & 0.00 & 0.00 & 30.00 \end{bmatrix}. \qquad (4.84)$$

This is then inverted and multiplied by $\hat{\phi}$ to give

$$\hat{\Sigma} = \hat{\phi}(\Gamma^T W \Gamma)^{-1}$$

$$= \begin{bmatrix} 0.0212 & -0.0172 & -0.0212 & -0.0123 & -0.0212 \\ -0.0172 & 0.0334 & 0.0172 & 0.0000 & 0.0172 \\ -0.0212 & 0.0172 & 0.0458 & 0.0123 & 0.0212 \\ -0.0123 & 0.0000 & 0.0123 & 0.0382 & 0.0123 \\ -0.0212 & 0.0172 & 0.0212 & 0.0123 & 0.1074 \end{bmatrix}. \qquad (4.85)$$

Having determined $\hat{\Sigma}$, using (4.76), the future design matrix, Γ_F, can be taken from (4.75), to calculate the matrix representing the variance/covariance of the linear predictors, as follows:

$$N = \Gamma_F \hat{\Sigma} (\Gamma_F)^T = \begin{array}{c} \\ (1,2) \\ (2,1) \\ (2,2) \end{array} \begin{array}{c} \text{cells} \\ \begin{bmatrix} 0.1196 & 0.0000 & 0.1034 \\ 0.0000 & 0.0628 & 0.0369 \\ 0.1034 & 0.0369 & 0.1320 \end{bmatrix} \end{array}$$

with header cells $(1,2)$ $(2,1)$ $(2,2)$. $\qquad (4.86)$

The cell references shown above are simply to indicate which terms in the relevant formulae, the matrix entries relate to. So, for example, $\text{Cov}(\hat{\eta}_{1,2}, \hat{\eta}_{2,2}) = 0.1034$.

Next, **diag**($\hat{m}$) can be created using the future elements of Table 4.25, as follows:

$$\mathbf{diag}(\hat{m}) = \begin{bmatrix} 32.00 & 0 & 0 \\ 0 & 50.00 & 0 \\ 0 & 0 & 31.00 \end{bmatrix}. \tag{4.87}$$

Then, using (4.79) enables the calculation

$$\mathbf{diag}(\hat{m})\,\mathbf{N}\,\mathbf{diag}(\hat{m}) = \begin{array}{c} \text{cells} \\ (1,2) \\ (2,1) \\ (2,2) \end{array} \begin{array}{ccc} (1,2) & (2,1) & (2,2) \\ \begin{bmatrix} 122.4195 & 0.0000 & 102.5677 \\ 0.0000 & 156.9386 & 57.2364 \\ 102.5677 & 57.2364 & 126.8359 \end{bmatrix} \end{array}. \tag{4.88}$$

This matrix provides the awkward second and third terms in equations (4.72) and (4.73). It is added to $\hat{\phi}\,\mathbf{diag}(\hat{m})$, which is

$$\hat{\phi}\,\mathbf{diag}(\hat{m}) = \begin{bmatrix} 82.7159 & 0 & 0 \\ 0 & 129.244 & 0 \\ 0 & 0 & 80.131 \end{bmatrix},$$

to give a matrix that then provides all three terms in equations (4.72) and (4.73), as follows:

$$\begin{array}{c} \text{cells} \\ (1,2) \\ (2,1) \\ (2,2) \end{array} \begin{array}{ccc} (1,2) & (2,1) & (2,2) \\ \begin{bmatrix} 205.1354 & 0.0000 & 102.5677 \\ 0.0000 & 286.1821 & 57.2364 \\ 102.5677 & 57.2364 & 206.9669 \end{bmatrix} \end{array}. \tag{4.89}$$

The shaded entries shown in the diagonal blocks in this matrix represent the cells to use for the MSEP calculations for cohort 1 (top left-hand block) and cohort 2 (bottom right-hand block), as explained further below. Larger triangles would produce correspondingly larger matrices, but the diagonal blocks would follow a similar pattern for the cohorts.

At this stage of the process, all the calculations have been carried out that are needed to determine the standard errors of the ODP parameters themselves, the MSEP of the future incremental values and the MSEP of the reserve for each cohort and of the total reserve.

The standard errors of the fitted ODP parameters are derived by taking the square root of the diagonal elements of (4.85), as shown in Table 4.27.

Table 4.27 *ODP worked example: Standard error of fitted parameters*

Parameter	Standard Error
c	0.1456
α_1	0.1827
α_2	0.2141
β_1	0.1953
β_2	0.3277

For the MSEP of the future incremental values and the reserves for each cohort in the triangle, the shaded blocks in (4.89) can be used. Hence, for MSEP($\hat{X}_{1,2}$), which is the

MSEP of the reserve for cohort 1 in the triangle, this is just the top left-hand shaded value in this matrix, of 205.13, giving an estimated Standard Error of 14.32.

For the MSEP of the reserve for cohort 2 in the triangle (i.e. MSEP($\hat{R}_2$)), the cells in the next shaded diagonal block of (4.89) are added together to give 607.62, so that the Standard Error is 24.65. Finally, for the MSEP of the total reserve (i.e. MSEP($\hat{R}$)) this is equal to the sum of all values in (4.89), which is 1017.89, so that the Standard Error is 31.90.

The results are summarised in Table 4.28. "SE" is an abbreviation for Standard Error (the same as prediction error here).

Table 4.28 *ODP worked example: Results summary*

Cohort no.	Reserve	SE	CV
1	32	14.32	45%
2	81	24.65	30%
Total	113	31.90	28%

These values, along with the fitted parameters and their standard errors, can also be produced using the "glmReserve" function in the "ChainLadder" package in the software environment, R. They can also be produced using other suitable software packages. The results have been derived here entirely independently, using matrix algebra, in order to demonstrate the calculations in a more transparent way than is possible by simply running such software.

4.3.8 Results of Applying ODP Model to Taylor and Ashe (1983) Dataset

In order to compare the ODP model with other stochastic methods, this section shows the results of applying the method to the same dataset used in the worked example for Mack's method – i.e. from Taylor and Ashe (1983). Since the fitting process for the ODP model has already been demonstrated, only the results of applying the ODP model to this dataset are shown. The model is fitted using the "glm.Reserve" function within the "ChainLadder" package in R. The R code is given in Appendix B, which also includes code for displaying residual plots, Q–Q plots and a test for Normality in the residuals, as explained further in Section 4.10.3. This produces the results in Table 4.29. All figures in this and the other data and results tables in this section are in the thousands.

The R code shown in Appendix B also shows how to derive the fitted parameters for the ODP model. These are shown in Table 4.30.

These results and parameter estimates are the same[29] as those shown in a number of papers on stochastic reserving methods, such as England and Verrall (2006). Tables 2 and 3 of that paper show the same results and parameters estimates as those shown here.

Using (4.60) the fitted values for the past data can be derived, as shown in Table 4.31.

---

[29] Noting that the indexing used in this book for the rows and columns of data triangles starts at 0, and in some other sources it starts at 1, so that, for example, α_1 here will be the same as α_2 in these other sources.

Table 4.29 *Summary results for ODP model applied to Taylor and Ashe (1983) dataset*

Cohort	Latest	Latest/Ult	Ultimate	Reserve	PE	CV
1	5,339	98%	5,434	95	110	116%
2	4,909	91%	5,379	470	216	46%
3	4,588	87%	5,298	710	261	37%
4	3,873	80%	4,858	985	304	31%
5	3,692	72%	5,111	1,419	375	26%
6	3,483	62%	5,661	2,178	495	23%
7	2,864	42%	6,785	3,920	790	20%
8	1,363	24%	5,642	4,279	1,047	24%
9	344	7%	4,970	4,626	1,980	43%
Total	30,457	62%	49,137	18,681	2,946	16%

Table 4.30 *Fitted parameters for ODP model applied to Taylor and Ashe (1983) dataset*

Parameter	Notation	Value	SE
(Intercept)	c	12.506	0.173
factor(origin)1	α_1	0.331	0.154
factor(origin)2	α_2	0.321	0.158
factor(origin)3	α_3	0.306	0.161
factor(origin)4	α_4	0.219	0.168
factor(origin)5	α_5	0.270	0.171
factor(origin)6	α_6	0.372	0.174
factor(origin)7	α_7	0.553	0.187
factor(origin)8	α_8	0.369	0.239
factor(origin)9	α_9	0.242	0.428
factor(dev)1	β_1	0.913	0.149
factor(dev)2	β_2	0.959	0.153
factor(dev)3	β_3	1.026	0.157
factor(dev)4	β_4	0.435	0.184
factor(dev)5	β_5	0.080	0.215
factor(dev)6	β_6	−0.006	0.238
factor(dev)7	β_7	−0.394	0.310
factor(dev)8	β_8	0.009	0.320
factor(dev)9	β_9	−1.380	0.897
Scale parameter	ϕ	52,601	

It can then be shown that these fitted values are the same as those produced using the simple, deterministic column sum average CL method. To do this, the column sum average CL factors are first calculated from the cumulative data triangle shown at the start of this subsection. These are shown in Table 4.32. They are the same as the development factors for the Mack method, as given in Table 4.6, but are repeated here for ease of reference.

These factors are then applied recursively backwards from the leading diagonal of the cumulative data triangle to generate the "fitted" data, as shown in Table 4.33.

For example, Cohort 7 at development point 1 is $1,639 = 2,864/1.747$ and at development point 0 it is $470 = 1,639/3.491$.

The incremental fitted triangle can then be derived from this, which can be seen to be the same as Table 4.31, and hence is not shown again.

Table 4.31 *Fitted values for ODP model applied to Taylor and Ashe (1983) dataset*

Cohort	\multicolumn{10}{c}{Development year}
	0	1	2	3	4	5	6	7	8	9
0	270	673	704	753	417	293	268	182	273	68
1	376	937	981	1,049	581	407	374	254	380	
2	372	927	971	1,039	575	403	370	251		
3	367	913	957	1,023	567	397	364			
4	336	838	877	938	520	364				
5	354	881	923	987	547					
6	392	976	1,022	1,093						
7	470	1,170	1,225							
8	391	973								
9	344									

Table 4.32 *ODP example: Chain Ladder development factors*

Colsum	\multicolumn{9}{c}{Factor for development years}
	0–1	1–2	2–3	3–4	4–5	5–6	6–7	7–8	8–9
Sum 1	11,615	17,912	21,931	21,655	19,828	17,331	13,430	9,173	3,901
Sum 2	3,327	10,251	15,048	18,448	17,963	15,955	12,743	8,520	3,834
Ratio	3.491	1.747	1.457	1.174	1.104	1.086	1.054	1.077	1.018

If the chain ladder factors shown in Table 4.32 are applied to the cumulative data, this will produce the same reserve as the ODP model (of 18,681). Thus, as for the simple 3×3 example, the equivalence of the ODP model and the simple Chain Ladder method has been demonstrated, at least as regards the fitted model and the estimated reserve. The ODP model has the advantage that it also produces an estimate of the prediction error of the reserves.

Table 4.33 *ODP example: Fitted cumulative data using Chain Ladder factors*

Cohort	\multicolumn{10}{c}{Development year}
	0	1	2	3	4	5	6	7	8	9
0	270	943	1,647	2,401	2,818	3,111	3,379	3,561	3,834	3,901
1	376	1,313	2,294	3,343	3,925	4,332	4,706	4,959	5,339	
2	372	1,300	2,271	3,310	3,885	4,288	4,658	4,909		
3	367	1,280	2,237	3,260	3,827	4,224	4,588			
4	336	1,174	2,051	2,989	3,509	3,873				
5	354	1,235	2,158	3,145	3,692					
6	392	1,368	2,390	3,483						
7	470	1,639	2,864							
8	391	1,363								
9	344									

4.4 The Bootstrap Method

The bootstrap method was first mentioned in a general insurance reserving context as early as 1986 (e.g. see Ashe, 1986), but has only become popular in this context since about the late 1990s and early 2000s. It is now one of the most commonly used methods to estimate reserve uncertainty. For example, the bootstrap of the Chain Ladder method was the most popular stochastic method amongst respondents in the survey at ASTIN Committee (2016). It is based on using a form of Monte Carlo simulation, and can be used to produce an estimate of the distribution of future claims outgo, and the associated cash flows, rather than just the prediction error.

Some reserving practitioners have tended to use the term "bootstrap" to refer to the specific implementation using a Chain Ladder method. However, bootstrapping should be regarded as a numerical procedure or algorithm, rather than a claims reserving method in its own right, and can in theory be applied to any underlying statistical model of the claims process. Hence, whenever bootstrapping is used in a reserving context, it is preferable to specify which claims reserving model or method is being "bootstrapped". For ease of reference, the terms "bootstrap method", "bootstrap model" and "bootstrap procedure", or just "bootstrapping" are used interchangeably here to mean the generic procedure applied to any form of claims reserving model.

In order to understand the use of the method in a general insurance reserving context, there are three building blocks required:

1. The general statistical background to bootstrapping – to understand the basic idea behind it. This is covered in Section 4.4.1.
2. The application of bootstrapping in a statistical modelling context (such as classic linear regression and GLMs). This is covered in Section 4.4.2.
3. The application of bootstrapping when applied in a general insurance reserving context. This is covered in Section 4.4.3 and a numerical example of bootstrapping two different reserving models (ODP and Mack) is given in Section 4.4.4.

Readers who are not interested in the underlying theory can just read the first part of each of Section 4.4.1 and Section 4.4.3, followed by the numerical example in Section 4.4.4.

Additional details and other considerations related to the use of bootstrapping in practice in a reserving context are given after the worked examples, as well as the assessment of bootstrap models and results. The section concludes with a brief consideration of bootstrapping other models not considered elsewhere in this section.

4.4.1 General Statistical Background to Bootstrapping

Bootstrapping is a well-established statistical technique that is used in a wide variety of applications. The first reference in the statistical literature to bootstrapping was in Efron et al. (1979), where the technique was described as a development of another well-established statistical technique known as the "jacknife". There are various forms of bootstrap used in statistics, but the one that tends to be used in a general insurance reserving context is a "resampling" type bootstrap, so that is the one which is covered in this general background section.

The basic idea begins with a sample of data, which has been drawn from a wider population. The goal is to estimate a statistic in that wider population (e.g. the mean of a random variable) and to understand the distribution around that statistic, using the sample data. A standard statistical approach is to make assumptions about the distribution of the random variable in the population, derive the statistic within the sample and then use sampling theory to determine the distribution around that sampling statistic. This approach works well if assumptions can reasonably be made about the underlying distribution in the population.

However, where the distribution is not known, an alternative approach is needed. One such approach that has been shown to work well in certain circumstances is the bootstrap method. This involves repeatedly taking samples, with replacement, from the original sample, and then calculating the relevant statistic on each resample. The results across all of these samples can then be used to draw conclusions about the statistic in the underlying population. What is effectively being done in this procedure is to assume that the relationship between the original sample and the underlying population can be inferred from the relationship between the original sample and the bootstrap samples – or, in other words, *"The population is to the sample as the sample is to the bootstrap"*. In effect, the conclusions drawn from analysis of the simulations in the bootstrap world are assumed to be approximately valid in the real world – which is sometimes referred to as the "plug-in-principle".

The term "bootstrap" is thought to derive from the phrase *"pulling oneself up (above the ground) by one's bootstraps"*.[30] In other words, the derivation is based on an action that does not rely upon the use of external sources – which is analogous to the statistical bootstrap procedure which repeatedly uses just the sample from the population, without reference to any other data. Despite the source of the phrase appearing to relate to something that is impossible to achieve, the statistical procedure is recognised as a very useful tool in many situations, and there is a substantial body of statistical work on the subject.

To further explain the concept, a brief explanation of the basic theory in mathematical terms is given below, followed by a simple numerical example.

Mathematical Description of the Generic Bootstrap Approach

This description is based on Efron and Tibshirani (1993), where further details can be found if required.

- Suppose that there are n independent and identically distributed realisations or values from an unknown distribution, F, that represent a sample, $\mathbf{x} = (x_1, x_2, \ldots, x_n)$.
- Suppose that the objective is to estimate, using $\mathbf{x}$, a parameter θ which is a function of F, say $\theta = t(F)$. For example, θ might simply be the mean or variance of $\mathbf{x}$ or a much more complicated statistic.
- An estimate, $\hat{\theta} = s(\mathbf{x})$, is derived from the sample, and the goal is to understand the accuracy of $\hat{\theta}$. The bootstrap procedure was designed to assist with this, regardless of how complicated $\hat{\theta}$ may be, by allowing items of interest, such as the standard error and confidence interval for $\hat{\theta}$ to be determined.
- The first step in this procedure is to derive a "bootstrap resample" of $\mathbf{x}$ as $\mathbf{x}^* \doteq (x_1^*, x_2^*, \ldots, x_n^*)$, where each x_j^* is a random sample from the population of n values

---

[30] A bootstrap in the context of this phrase being a pair of loops at the top of heavy riding boots.

$\mathbf{x} = (x_1, x_2, \ldots, x_n)$ with replacement. In other words, each x_j can occur more than once in the resample. The asterisk notation is intended to indicate that $\mathbf{x}^*$ is not the actual dataset $\mathbf{x}$ but rather a random, resampled version of it. It is important to stress that the validity of this resampling procedure is critically dependent on the assumption that the x_i values are independent and identically distributed.

- This procedure is repeated B times to form B independent resamples $(x^*)^{(1)}, (x^*)^{(2)}, \ldots, (x^*)^{(B)}$.

- For each kth resample, the bootstrap resampled values for the bootstrap estimator as $\hat{\theta}_k^* = s((x^*)^{(k)})$ are calculated.

- Finally, the Standard Error $\mathrm{SE}_F(\hat{\theta})$ is estimated, using the sample standard deviation of the B replications, as

$$\hat{\mathrm{SE}}_B = \sqrt{\frac{1}{B-1} \sum_{k=1}^{B} (\hat{\theta}_k^* - \hat{\theta}_{\mathrm{mean}}^*(\mathbf{x}))^2},$$

where $\hat{\theta}_{\mathrm{mean}}^*(\mathbf{x}) = \frac{1}{B} \sum_{k=1}^{B} \hat{\theta}_k^*$.

Numerical Example of the Generic Bootstrap Approach

The concept can also be explained by considering a simple example. With thanks to Peter England, this is based on the example given in England (2010).

Assume that an insurer has experienced the number of large claims shown in Table 4.34 for a particular class of business in the last 11 years.

Table 4.34 *Simple bootstrap example: No. large claims per year*

Year	04	05	06	07	08	09	10	11	12	13	14
No. claims	3	7	6	7	5	10	8	4	5	6	5

The goal is to predict the number of claims in 2015, along with the uncertainty surrounding that prediction. This uncertainty can be measured using an estimate of the prediction error, which will need to allow for both estimation and process error. Taking the square root of the approximate formula given in Section 4.1.3, this can be derived using:

> Prediction error $\approx$ (Estimation variance + Process variance)$^{1/2}$.

If an analytic approach is used, and a Poisson distribution is assumed for process error, then the results can be summarised as follows:

- Mean = 6.0; Standard Deviation (of population) σ = 1.859.
- No. of observations (n) = 11.
- Standard Error of the mean = $\sigma / \sqrt{n}$ = 0.560.
- Variance of the mean = 0.560^2 = 0.314.

- So, Estimation variance = 0.314.
- Poisson variance = Poisson mean = 6.
- So, Process variance = 6.0.
- Hence, **Prediction error** = $(0.314 + 6.0)^{1/2} = 2.513$.

If a bootstrap approach is used, then using the observed data for the last 11 years, these are repeatedly sampled from randomly with replacement to produce the "pseudo data", which might give the results shown in Table 4.35, assuming 10,000 versions of the pseudo data are generated.

Table 4.35 *Simple bootstrap example: Sample pseudo data*

Data	Year											Mean
	04	05	06	07	08	09	10	11	12	13	14	
Observed data	3	7	6	7	5	10	8	4	5	6	5	6.000
Pseudo data 1	7	3	8	5	6	5	3	5	4	3	5	4.909
Pseudo data 2	6	6	7	10	7	7	6	7	3	5	5	6.273
Pseudo data 3	3	6	5	7	4	7	4	7	4	7	7	5.545
...	.	.	.	.	.	.	.	.	.	.	.	.
...												
Pseudo data 10k	3	10	10	10	6	10	6	4	6	8	3	6.909

If the variance is taken of the mean of all 10,000 bootstrap samples, then for this implementation (which was done in MS Excel) this equals 0.306, giving an estimation error of $0.306^{1/2} = 0.553$, which is very close to the analytic equivalent of 0.560. To add process error, for each bootstrap sample a random value is generated from a Poisson distribution, using the mean for each bootstrap sample, which in this implementation gives the data shown in Table 4.36.

Table 4.36 *Simple bootstrap example: With process error added*

	Poisson mean	Simulated Poisson
Pseudo data 1	4.909	3
Pseudo data 2	6.273	4
Pseudo data 3	5.545	7
...	.	.
...		
Pseudo data 10k	6.909	5

The simulated Poisson values will then have both estimation and process error incorporated and hence so will their standard deviation. In this example, this equates to 2.508, which is very close to the analytic equivalent value of 2.513. In other words, the bootstrap procedure used in this way gives an estimate of 2.508 of the prediction error surrounding the mean of 6.0 for the estimated number of claims in 2015. The bootstrap samples could also be used to derive an estimate of the full distribution surrounding the mean (and hence confidence intervals for example).

For simple numerical examples such as this, an analytical approach might be able to be used, without needing to make use of bootstrapping. However, in a reserving context, the situation is considerably more complex and, depending on the statistical model being used to derive the reserves, an analytic approach may not be feasible, or may not produce the full distribution, and hence bootstrapping can be a practical alternative.

4.4.2 Bootstrapping in a Statistical Modelling Context

In general insurance reserving, the bootstrap procedure used is analogous to the way in which a GLM can be bootstrapped and then used in a forecasting context. However, it is helpful to start by considering how the more straightforward classic linear regression model can be bootstrapped.

The Classic Linear Regression Model

To understand this, it is first necessary to summarise the theory behind the classic linear regression model. For further details see Efron and Tibshirani (1993), which explains it in the context of bootstrapping.

- The dataset $\mathbf{x}$ consists of n points $\mathbf{x} = (x_1, x_2, \ldots, x_n)$, where each x_i is a pair of items, say, $x_i = (\mathbf{c}_i, y_i)$ with $y_i = \mathbf{c}_i \boldsymbol{\beta} + \epsilon_i$.
- In this case $\mathbf{c}_i$ is a $1 \times p$ vector $(c_{i1}, c_{i2}, \ldots, c_{ip})$, known as the *predictor* and $\boldsymbol{\beta}$ is the *parameter vector*, or *regression parameter*, $\boldsymbol{\beta} = (\beta_1, \beta_2, \ldots, \beta_p)^T$. y_i is known as the *response*.
- The error terms, ϵ_i, are each assumed to be from an unknown and identical error distribution F with mean zero, and which is independent of $\mathbf{c}_i$.
- The key assumption in the classic regression model is that

$$\mu_i = E(y_i | \mathbf{c}_i) = \mathbf{c}_i \boldsymbol{\beta} = \sum_{j=1}^{p} c_{i,j} \beta_j.$$

- So, in a very simple straight-line regression model, with the response variable denoted by y_i and the predictor variable as z_i (rather than x_i to avoid confusion with $\mathbf{x}_i$ above), $\mu_i = \beta_0 + \beta_1 z_i$ so that $\mathbf{c}_i = (1, z_i)$.
- The goal is to estimate the parameter vector $\boldsymbol{\beta}$ from the observed data.
- If $\boldsymbol{\Gamma}$ is defined as an $n \times p$ matrix with the ith row as $\boldsymbol{\gamma}_i$ (referred to as the *design matrix*), and $\mathbf{y}$ as a vector $(y_1, y_2, \ldots, y_n)^T$ then it can be shown that the least-squares estimate, $\hat{\boldsymbol{\beta}}$ is given by

$$\hat{\boldsymbol{\beta}} = (\boldsymbol{\Gamma}^T \boldsymbol{\Gamma})^{-1} \boldsymbol{\Gamma}^T \mathbf{y}. \tag{4.90}$$

This is equivalent to (4.54), which is the parameter estimation formula used in the ODP model.

- For the standard linear regression model, statistical theory provides information about how accurate $\hat{\boldsymbol{\beta}}$ is, through an analytical formula for the Standard Error. This is determined by first defining $\mathbf{G}$ as a $p \times p$ *inner product matrix*, $\mathbf{G} = \mathbf{\Gamma}^T \mathbf{\Gamma}$, which will have element $g_{hj} = \sum_{i=1}^{n} \gamma_{ih} \gamma_{ij}$ in row h, column j.

- Let σ_F^2 be the variance of the error terms, ϵ, in the model. Then $\sigma_F^2 = \text{var}_F(\epsilon)$. Hence, the Standard Error of the jth component of $\hat{\beta}$ is

$$\text{se}(\hat{\beta}_j) = \sigma_F \sqrt{G^{jj}}, \tag{4.91}$$

where G^{jj} is the jth diagonal component of the inverse matrix $\mathbf{G}^{-1}$.
- σ_F is estimated, using either the following formula:

$$\hat{\sigma}_F = \sqrt{\sum_{i=1}^{n} (y_i - \gamma_i \hat{\beta})^2 / n}, \tag{4.92}$$

or, a bias-corrected[31] version as follows:

$$\bar{\sigma}_F = \sqrt{\sum_{i=1}^{n} (y_i - \gamma_i \hat{\beta})^2 / (n - p)}. \tag{4.93}$$

This will mean that:

$$\bar{\sigma}_F = \hat{\sigma}_F \sqrt{\frac{n}{n - p}}. \tag{4.94}$$

Bootstrapping the Classic Linear Regression Model

A bootstrapping procedure is now implemented, which in this case is regarded as a type of non-parametric resampling bootstrap,[32] in which the residuals, rather than the data itself are resampled. A summary of the procedure is as follows:

- First, the $\hat{\beta}$ are estimated using (4.90). These parameters are entered into the model formula to determine the fitted values for each observation, as $c_i \hat{\beta}$.
- The residuals can then be determined as actual less fitted, that is

$$\hat{\epsilon}_i = y_i - c_i \hat{\beta}.$$

- To generate the bootstrap samples, these residuals are sampled randomly, with replacement, to generate

$$\epsilon^* = (\epsilon_1^*, \epsilon_2^*, \dots, \epsilon_n^*).$$

- The bootstrap responses are then derived:

$$y_i^* = c_i \hat{\beta} + \epsilon_i^*.$$

---

[31] This is the so-called *bias-corrected* formula for the estimate of σ_F, because it allows for the degrees of freedom, which is a function of the number of parameters, p and the number of data points n.

[32] Note that in some sources, such as Taylor and McGuire (2016), this type of bootstrap is referred to as semi-parametric, to distinguish it from a purely non-parametric bootstrap where the data itself is simply resampled with replacement.

- Next, these bootstrap responses are used, along with the original data, to produce the bootstrap regression parameters using (4.90) as

$$\hat{\boldsymbol{\beta}}^* = (\mathbf{C}^T \mathbf{C})^{-1} \mathbf{C}^T \mathbf{y}^*.$$

- If, say, 1,000 bootstrap samples were derived using the above procedure, there would then be 1,000 estimates for the parameters, which could then be used to draw conclusions about the reliability of the estimated parameters, for example by calculating the bootstrap Standard Error, and various quantile / percentile values etc. as per the simplified example given in Section 4.4.1.

- However, if the goal is just to derive the Standard Error for the components of $\hat{\boldsymbol{\beta}}^*$, then it can be shown[33] using (4.91) that, since, $\mathrm{SE}(\mathbf{y}^*) = \hat{\sigma}_F$:

$$\mathrm{SE}_{\hat{F}}(\hat{\boldsymbol{\beta}}_j^*) = \sigma_F \sqrt{G^{jj}}.$$

- In other words, the bootstrap estimate of the Standard Error for $\hat{\beta}_j$ is the same as the analytical one derived without bootstrapping, $\mathrm{SE}(\hat{\beta}_j)$ as in (4.91).

- Note that if it is deemed necessary to bias-correct the bootstrap standard errors, then (4.94) can be used, so that they are multiplied by

$$\sqrt{\frac{n}{n-p}}. \tag{4.95}$$

Bootstrapping a GLM in a Forecasting Context

The concept of a GLM was introduced in Section 4.3 in the context of the ODP model, so is not repeated here. When bootstrapping a GLM in a forecasting context, a similar procedure is applied as for the more simple classic linear regression model, but with two important differences.

First, the residual definition used must be consistent with the model, which will usually mean that the residuals are not simply equal to the actual less the fitted, as with the classic linear regression model. There are several different types of residual definition including "Pearson", "Deviance" and "Anscombe". The choice depends in particular on the form of the GLM error structure. The residual definition used for bootstrapping in a general insurance reserving context is discussed further in Section 4.4.3.

Second, if the GLM is to be used for forecasting or prediction type problems (as is the case for general insurance reserving) then an estimate of the total uncertainty surrounding the prediction error will need to incorporate process error as well, as referred to in the introduction to this chapter, in Section 4.1.3. The above bootstrap procedure for the classic linear regression model (and by extension a GLM model) will only produce an estimate of the *parameter* or *estimation* uncertainty, and so an additional step is required to incorporate process error. An example of how this additional step can be incorporated in a very simple situation was given in the numerical example in Section 4.4.1. How this can be achieved in

---

[33] See Efron and Tibshirani (1993) p. 112, Equation (9.29) for details.

a general insurance reserving context will depend on the model that is being bootstrapped; this topic is discussed further in the next section.

4.4.3 Bootstrapping in a General Insurance Reserving Context

The detail of the procedure in a general insurance reserving context will depend for example on the model that is being bootstrapped. However, it is helpful to outline a general procedure in the context of a non-parametric residual resampling bootstrap first, which is as follows:

(a) Define a statistical model that is appropriate for modelling the claims development process. This model will produce estimates of the future claims payments (i.e. the reserves) or the IBNR, depending on whether claims paid or claims incurred respectively are being modelled.
(b) Fit this model to an observed data sample (e.g. a data triangle).
(c) Determine appropriately defined residuals between the fitted statistical model and the observed data.
(d) Use Monte Carlo simulation to produce random selections of the residuals (with replacement).
(e) Use these randomly generated residuals to generate new "pseudo data", analogous to the observed data sample (but not the same).
(f) Re-fit the statistical model to each version of the pseudo data and produce forecasts of the future claims payments, ensuring that process error is incorporated in a suitable way.
(g) Finally, examine the distribution of the forecasts to produce estimates of the prediction error – i.e. related to uncertainty caused by both parameter and process error.

The above procedure is generic, in the sense that the statistical model referred to in step (a) can be any suitable statistical model. Two obvious choices are the ODP model and Mack's method, which, as explained in Section 4.1.5, can both be used to produce estimates that are equivalent to the standard column sum CL method. Bootstrapping each of these methods is explained in the remainder of this subsection. The choice between bootstrapping these two methods is discussed in the context of bootstrapping incurred claims data triangles, in Section 4.4.5.

Bootstrapping the Over-Dispersed Poisson Model

For ease of reference, the ODP model definition is repeated here, as per Section 4.3. With $X_{i,j}$ denoting the incremental claims for cohort i, in development period j, the ODP model (with a constant scale parameter) is defined as follows:

$$E[X_{i,j}] = m_{i,j} \text{ and } \text{Var}[X_{i,j}] = \phi E[X_{i,j}] = \phi m_{i,j},$$

$$\log m_{i,j} = \eta_{i,j},$$

$$\eta_{i,j} = c + \alpha_i + \beta_j, \text{ where } \alpha_0 = \beta_0 = 0.$$

Using standard GLM terminology, this defines a GLM with a logarithmic link function, with the variance being proportional to the mean. The parameter ϕ is an unknown scale parameter that is estimated as part of the fitting procedure.

As explained in Section 4.3, when the parameters of the ODP model are fitted to a data triangle, the results will equal those produced by the CL method (subject to certain constraints) – i.e. when the column-sum weighted chain ladder factors are applied to the leading diagonal to form the "fitted" data in the historical part of the triangle, the results are identical to the fitted results from the ODP model. In addition, the estimated claims reserves derived by using the fitted ODP parameters in the future part of the data triangle are the same as the estimated reserves produced using the CL method.

Because of this equivalence, when bootstrapping the ODP model, the deterministic chain ladder procedure can be used as the "model", instead of fitting a GLM model. However, the overall procedure will still be underpinned by the theoretical basis for bootstrapping a statistical model, as explained in Sections 4.4.1 and 4.4.2.

This theoretical basis is in some practical situations somewhat weakened, such as where an ODP model cannot be fitted to a particular data triangle (e.g. if the incremental value in the top right-hand cell of the triangle is negative), but where the deterministic CL method can still be applied. Similarly, some of the adjustments made to the standard deterministic CL method in practice can also be applied as part of the bootstrap procedure, without necessarily finding an equivalent GLM, which will also therefore somewhat weaken the underlying theoretical basis. This has led some authors (e.g. as in Pinheiro et al., 2003) to label the bootstrap procedure applied to the ODP model as a "mixed" model, since the residuals are still based on the definition used in the underlying ODP model, whereas the reserves are based on the CL method.

Although there may be some advantages to using a GLM framework to fit the model, such as flexibility in model structure, the ease with which the deterministic CL method can be re-applied to each iteration within the bootstrap procedure (including possibly making bespoke adjustments, such as curve fitting) has meant that this approach is typically used rather than re-fitting the GLM at each iteration.

The ODP version of bootstrapping the chain ladder method described here is a form of non-parametric bootstrapping applied to a non-recursive model.[34] Parametric bootstrapping is described in the final subsection of Section 4.4.5.

The Mack version of bootstrapping the chain ladder method is also a form of non-parametric procedure, but applied to a recursive model, and hence requires a slightly different approach compared to the ODP Bootstrap, as explained further below. These two approaches are referred to here as the ODP Bootstrap and Mack Bootstrap. Section 4.4.8 discusses bootstrapping other underlying reserving models.

---

[34] It is non-parametric in that the empirical residuals are resampled, rather than those implied by any underlying theoretical parametric distribution. Note that in some sources, such as Taylor and McGuire (2016), this type of bootstrap is referred to as semi-parametric, to distinguish it from a purely non-parametric bootstrap where no distributional assumptions are made in the model itself and the data is simply resampled with replacement.

As explained in Section 4.4.1, the statistical theory underlying the use of the bootstrap approach requires the "observations" that are being used for bootstrapping to be independent and identically distributed. Although it is perhaps not unreasonable in certain circumstances to assume that the cells in an incremental data triangle are independently distributed, it is not appropriate to assume that they are identically distributed. However, if defined in the correct way, the residuals (between the fitted model and the actual data) can be assumed to be at least approximately independently and identically distributed, or can be made so.

Thus, if the residuals are bootstrapped, instead of the observations themselves, this gives a reasonably robust bootstrapping procedure that can be used to estimate the parameter (i.e. estimation) errors of the underlying statistical model. The results, however, will only be valid if the assumptions of the underlying model (e.g. the ODP) that is subject to the bootstrap procedure are also valid.

Hence, if it can reasonably be assumed that the residuals are independently and identically distributed, then a residual can be taken from any cell in the historical triangle, and a suitable inversion formula applied using the fitted value in another cell in the triangle. The inversion formula will depend on the residual definition used.[35] This will create different versions of the pseudo data, to which the chain ladder procedure (as a proxy for fitting a GLM model such as the ODP) can be applied repeatedly.

The most commonly used residual definition in a general insurance reserving context for bootstrapping purposes is not simply the actual less fitted values, but rather the "Pearson" residuals. For the generic formulation of a GLM model, as per (4.48), the scaled Pearson residuals are defined as

$$r_{i,j}^s = \frac{X_{i,j} - \hat{m}_{i,j}}{\sqrt{\frac{\phi_j V(\hat{m}_{i,j})}{w_{i,j}}}}. \tag{4.96}$$

For the ODP model with constant scale parameters, $\phi_j = \phi$, $V(\hat{m}_{i,j}) = \hat{m}_{i,j}$ and $w_{i,j} = 1$, so that (4.96) becomes

$$r_{i,j}^s = \frac{X_{i,j} - \hat{m}_{i,j}}{\sqrt{\phi \hat{m}_{i,j}}}. \tag{4.97}$$

When simulation is used to sample from the residuals, to give $r_{i,j}^{s*}$, they are reconverted back to generate the pseudo data by rearranging (4.97) as

$$X_{i,j}^{s*} = r_{i,j}^{s*}\sqrt{\phi \hat{m}_{i,j}} + \hat{m}_{i,j}. \tag{4.98}$$

Therefore the scale parameter is effectively unwound when generating the pseudo data within the bootstrap procedure, and it does not need to be used when calculating the residuals during the bootstrap process, unless non-constant scale parameters are used (as discussed further in the relevant subsection of Section 4.4.3). The scale parameter still needs to be calculated, however, as it is needed at a later stage to add in process error, as

---

[35] For example, if the residuals were just defined as Actual less Fitted, the inversion formula would be the sampled residual plus fitted. Other residual definitions, such as Pearson's, will require a different inversion formula.

discussed below. Hence, when used in the bootstrap procedure, the unscaled residuals are calculated using

$$r_{i,j} = \frac{X_{i,j} - \hat{m}_{i,j}}{\sqrt{\hat{m}_{i,j}}}. \tag{4.99}$$

These residuals are "unscaled" because the scale parameter is removed from their formulation, so it is not strictly correct to refer to them as "scaled" residuals. If the sampled residuals in this form are denoted by $r_{i,j}^*$, then the equivalent re-conversion formula is

$$X_{i,j}^* = r_{i,j}^* \sqrt{m_{i,j}} + \hat{m}_{i,j}. \tag{4.100}$$

The scale parameter, consistent with the Pearson residuals definition of (4.70) in the ODP model, is estimated using

$$\hat{\phi} = \frac{\sum (r_{i,j})^2}{n - p}, \tag{4.101}$$

where n is the number of data points in the sample, p is the number of parameters, and the summation is over all the residuals in the triangle.

Since the ODP model is being bootstrapped, it makes logical sense to use the same approach as for the analytical version of the ODP model to estimate the uncertainty around the reserves. Section 4.3.5 explains how the MSEP is used for this, which is defined in mathematical terms, for a single value in the triangle, $X_{i,j}$ as

$$\text{MSEP}[\hat{X}_{i,j}] = \text{E}[(X_{i,j} - \hat{X}_{i,j})^2]. \tag{4.102}$$

Section 4.3.5 also shows that this can be broken down into the following components:

$$\underbrace{\text{MSEP}[\hat{X}_{i,j}]}_{\text{Prediction variance}} = \underbrace{\text{Var}[X_{i,j}]}_{\text{Process variance}} + \underbrace{\text{Var}[\hat{X}_{i,j}]}_{\text{Estimation variance}}. \tag{4.103}$$

Hence, as mentioned previously, the bootstrap procedure will need to incorporate an allowance for process variance so that it provides an estimate of the total prediction variance, as per (4.103). This is discussed further under "Incorporating Process Error" later in this section. As long as this is done, a form of the generalised bootstrap procedure outlined at the start of this section can be used to derive the results. For example, the variances for individual and combined cohorts can be calculated directly, and will be equivalent to those derived using the analytical formulae given in (4.72) and (4.73). The analytical formulae, however, do not produce an estimate of the full distribution of future claims outgo, whereas the bootstrap procedure does and can be used to derive percentiles of the predictive error distribution, for example. Before explaining the ODP Bootstrap procedure in more detail, the next subsection addresses an important aspect of aligning this with the analytical equivalent.

ODP Bootstrap – Degrees of Freedom Adjustment

If the Pearson residuals are used in the ODP Bootstrap, then, as explained in England and Verrall (1999), in order to properly align the bootstrap procedure with the analytic procedure, it is necessary to allow for the number of parameters used in fitting the model (i.e. to allow for the degrees of freedom in the procedure).

In the analytic GLM procedure, this will be allowed for automatically, but the bootstrap procedure will not make any such allowance; the fitted values are generated, regardless of how many parameters are being used. In other words, it is as if the bootstrap procedure is dependent only on n rather than adjusting for the number of parameters. So, in order to properly re-introduce this feature, an adjustment factor is needed, which is applied to the variance derived from the bootstrap procedure. This will be $n/(n-p)$, where p is the number of parameters. It can be thought of as a "degrees of freedom adjustment factor". This factor is multiplied by the variance which is produced by the bootstrap simulations, to give the estimation variance component (i.e. the $\text{Var}[\hat{X}_{i,j}]$ component of (4.103)) as

$$\frac{n}{n-p} \text{Var}_{\text{bs}}(R). \tag{4.104}$$

$\text{Var}_{\text{bs}}(R)$ is the Variance of the reserve produced by the bootstrap procedure, and R is the reserve. It can be seen that, as the number of parameters increases, the factor will also increase, and so the estimation variance produced by the bootstrap will increase. This adjustment factor is comparable to the bias-correction factor used when the classic linear regression model is bootstrapped, as given in (4.94) and (4.95) in Section 4.4.2.

Since one of the main benefits of using the bootstrap procedure is to analyse the full predictive distribution (incorporating process variance), the above approach needs enhancing so that when statistics other than the variance (e.g. percentiles) are derived from this distribution they will also be adjusted for the degrees of freedom. England (2001) suggests doing so by multiplying the residuals by the square root of the relevant factor, i.e. by $\sqrt{n/(n-p)}$, as an alternative to multiplying the variance by the factor, as per (4.104). Then, as long as process variance is incorporated into the procedure, by adjusting the residuals in this way, the full predictive distribution and the associated cash flows derived from the bootstrap procedure will also automatically include an adjustment for degrees of freedom (i.e. they will be "bias-adjusted") and can be analysed to produce any of the required statistics of interest. So, the unscaled residual formula (4.99) becomes

$$\tilde{r}_{i,j} = \sqrt{\frac{n}{n-p}} \left(\frac{X_{i,j} - \hat{m}_{i,j}}{\sqrt{\hat{m}_{i,j}}} \right). \tag{4.105}$$

These are then sampled with replacement, instead of those before the bias adjustment, to produce $\tilde{r}^*_{i,j}$ and reconverted back using a similar formula to (4.100) to create the pseudo data,[36] that is

$$X^*_{i,j} = \tilde{r}^*_{i,j} \sqrt{m_{i,j}} + \hat{m}_{i,j}. \tag{4.106}$$

---

[36] Equation (4.106) is not derived by rearranging (4.105), but instead a similar formula to (4.100) is used. This is so that the DOF adjustment is not unwound when deriving the pseudo data.

ODP Bootstrap Procedure

When applied in a general insurance reserving context to a single claims paid data triangle, the following steps are involved in the application of bootstrapping when using the ODP version of the chain ladder method as the underlying reserving model, as described above. Note that this description assumes a claims paid triangle is being used; other triangles can also be bootstrapped. The application to an incurred claims triangle is discussed in Section 4.4.5.

(a) Apply the standard deterministic CL method to the claims paid data triangle, to calculate development factors, taking a full column sum average for each period to determine the selected development factors for each development period. This is used instead of fitting an ODP model, due to the equivalence of the ODP model to the standard CL method, as discussed earlier in this section.

(b) Use the selected average development factors applied to the last diagonal, to derive the historical fitted values (the leading diagonal of this fitted triangle will therefore be equal to the actual values). The average development factors are effectively the model of the claims process that is fitted to the data.

(c) Convert the cumulative fitted data triangle to an incremental fitted data triangle. In this triangle, the fitted incremental value for the latest development point of both the first and last cohort will be equal to the actual incremental value for this period. This feature is the same as for the ODP model, as explained in Section 4.3.2.

(d) Calculate the bias-adjusted Pearson residuals using formula (4.105).[37] For an $n \times n$ triangle, there will be $N = \frac{1}{2}n(n + 1)$ values.

(e) Derive the bootstrap samples, by first selecting N residuals from these N residual values at random (with replacement, so there can be duplicates). This is equivalent to the "resampling" bootstrap procedure referred to at the start of Section 4.4.1.

(f) Derive the incremental pseudo data triangle using a rearranged version of the residual definition, that is use (4.106) with the fitted data from Step (c) above.

(g) This defines one bootstrap sample or simulation triangle of the "pseudo history". In effect, if the residuals associated with the incremental payments across all cohorts and development periods are deemed to be independently and identically distributed, this pseudo-history is assumed to be an example of one realisation that could have occurred with equal probability to that of the actual data triangle that was observed.

(h) Cumulate this triangle and calculate development factors from this pseudo triangle, as in Step (a) above.

(i) Apply these development factors to determine the estimated incremental future cash flows.

(j) Add process error to the incremental cash flows from the previous step, as explained further in the relevant subsection below, to produce revised incremental cash flows.

---

[37] Note that this is the form of residuals that are used for an ODP model. If a different model is being used, then a different definition of residuals will be needed.

These revised incremental cash flows represent one possible realisation arising from the bootstrap process.

(k) Repeat Steps (e) to (j) according to how many simulations are being used in the process (e.g 10,000).

(l) The set of simulated results from Step (k) will then represent an estimate of the full predictive distribution of future cash flows, with allowance for both estimation and process error. This can be used to determine the statistics of interest, such as the mean, variance and full distribution of future discounted and undiscounted cash flows (i.e. reserves) by cohort or in total. For example, the standard deviation of the bootstrap results (e.g. by cohort) then represents the bootstrap estimate of the prediction error (perhaps also referred to as the Standard Error).

The subsections below cover two important aspects of the implementation of bootstrapping the ODP version of the chain ladder model – incorporating process error and using non-constant scale parameters. This is followed by a description of how Mack's method can be bootstrapped. The following section then includes a worked example for both the ODP and Mack Bootstrap, to help explain how bootstrapping can be applied in practice. Further additional practical details and other considerations that apply when using bootstrapping in a reserving context are summarised after the examples – in Sections 4.4.5 and 4.4.6.

ODP Bootstrap – Incorporating Process Error

This section describes an approach for introducing process error into the bootstrap procedure for a non-recursive model such as the ODP model, as specified at Step (j) in the general procedure for this type of model given above. The approach, suggested in England (2001), is a form of parametric simulation resampling method, which mirrors that used in the simplified example given in Section 4.4.1. Other approaches are possible for this type of model, as discussed further in Section 4.4.6. Furthermore, a different approach is needed if the underlying model being bootstrapped is a recursive-type model, such as Mack's method, as explained further in the Mack Bootstrap section below.

In summary, the approach involves making an assumption regarding the distribution of each future cell of the triangle in each bootstrap sample, and then taking a random sample from this distribution. In mathematical terms, this involves, for each bootstrap simulation of a future triangle, proceeding as follows:

(a) For each cell (i, j) in the future triangle, simulate again from a suitable distribution, using the simulated incremental value in that cell as the mean, which is denoted by $\tilde{m}_{i,j}$, and the variance, which is denoted by $\hat{\phi}\tilde{m}_{i,j}$, with the scale parameter $\hat{\phi}$ estimated using (4.101). This additional simulation stage is designed to introduce the process uncertainty.

(b) Sum the simulated amounts in the future triangle, to give values by cohort and in total.

This is repeated for each bootstrap simulation. The resulting distribution of simulated amounts can then be used to produce the full predictive distribution, incorporating both estimation and process error.

The distribution that should be used in the second phase of the simulation (i.e. in Step (a) above), to be consistent with the overall ODP framework for the model, would be from an ODP with mean $\tilde{m}_{i,j}$, variance $\phi\tilde{m}_{i,j}$. An alternative, which England (2001) suggests can be easier to implement in practice, is to replace the ODP with another distribution (such as the Gamma), which is parametrised such that the mean and variance are correct.

Further details of this approach are contained in England (2001). It is used in the "BootChainLadder" function in the "ChainLadder" package in R, which is the software used to produce the results for the ODP Bootstrap worked example in Section 4.4.4.

ODP Bootstrap – Using Non-constant Scale Parameters

The description of bootstrapping the ODP model given here assumes that a single scale parameter is being used. However, it may be more appropriate to allow the scale parameter to vary, for example across development periods. England and Verrall (2006) recommend using non-constant scale parameters, or at least reviewing whether a constant scale parameter is appropriate. For example, if a review of the residual plots suggests that the variance of the residuals is significantly different at some development periods (i.e. the data is exhibiting heteroscedasticity), then allowing the residuals from all development periods to be capable of selection in the random bootstrap sampling process may not be appropriate, since the assumption underlying the use of bootstrapping is that all residuals are independent and identically distributed.

Instead, if the scale parameter is varied across development periods, this difference in variance can be allowed for. Further support for allowing the scale parameters to vary by development period is given in Carrato et al. (2015), where it is suggested that it is usually more appropriate to do so. In practice, for each particular application of the ODP Bootstrap, a decision will need to be made by the practitioner as to whether constant or non-constant scale parameters are used, based in part on an examination of the residuals. A comparison of results using constant and non-constant scale parameters can also be helpful to assess the importance of the decision.

Using the reformatted version of the scale parameter from (4.71), England and Verrall (2006) suggest that the scale parameter by development period can be written as

$$\hat{\phi}_j = \frac{n}{n-p} \times \frac{\sum_{i=0}^{n_j} \left(\frac{X_{i,j}-\hat{m}_{i,j}}{\sqrt{\hat{m}_{i,j}}}\right)^2}{n_j}$$

$$= \frac{\sum_{i=0}^{n_j} \left[\left(\frac{n}{n-p}\right)^{\frac{1}{2}} \left(\frac{X_{i,j}-\hat{m}_{i,j}}{\sqrt{\hat{m}_{i,j}}}\right)\right]^2}{n_j}, \tag{4.107}$$

where n_j is the number of residuals in development period j. This can be thought of as an average of the residuals across development periods, where the residuals are each multiplied by the bias-correction factor referred to in the "Degrees of Freedom Adjustment" paragraphs earlier in this section.

Then, when calculating the bias-adjusted residuals, the scale parameter by development period is used, so that (4.97) becomes

$$r_{i,j}^s = \sqrt{\frac{n}{n-p}} \left(\frac{X_{i,j} - \hat{m}_{i,j}}{\sqrt{\phi_j \hat{m}_{i,j}}} \right). \tag{4.108}$$

When simulation is used to sample from the residuals, the sampled residuals, $r_{i,j}^{s*}$, are reconverted back to generate the pseudo data using

$$X_{i,j}^* = r_{i,j}^{s*} \sqrt{\phi_j \hat{m}_{i,j}} + \hat{m}_{i,j}. \tag{4.109}$$

Further details are given in England and Verrall (2006), where it is explained (in the Appendix) that the above approach is effectively an approximation to the more formal "joint" modelling of mean and variance, whereby the scale parameters are estimated using an iterative maximum likelihood procedure. The approximation is used as a practical alternative that can be implemented in a bootstrap procedure. It is also noted in England and Verrall (2006) that the Mack method is formulated as a joint model of mean and variance, using the average of the squared bias-adjusted (weighted) residuals at each development period as the variance estimators. In that sense, bootstrapping the Mack method is comparable to the ODP Bootstrap using non-constant scale parameters.

An explanation of how non-constant scale parameters can be used in practice in the ODP Bootstrap is given in the second part of the worked example in Section 4.4.4.

Bootstrapping Mack's Method

An alternative to bootstrapping the ODP model, but still using an approach that is equivalent to the Chain Ladder (CL) method, is to bootstrap Mack's method. This section explains the procedure, including relevant mathematical details.

The description here follows England and Verrall (2006) and involves reformulating Mack's method as a GLM model, with the link ratios as the "response" variables, rather than the data itself, as for some other GLM reserving models, such as the ODP model.

Because Mack's method is a recursive model, the bootstrap procedure is slightly different to that for the ODP model, as each diagonal of the future part of the triangle needs to be built up successively using the underlying recursive structure. The procedure allows the process error to be included, to ensure that the resulting predictive distribution allows for both parameter and process error. It is generally preferable to fitting distributions (e.g. Lognormal) to the mean and variance produced by Mack's method (as referred to in the introduction to Mack's method given in Section 4.2).

To reformulate Mack's model with the development ratios as the response variables, (4.3) and (4.5) from Section 4.2.2 are each divided by the known cumulative claims, $C_{i,j-1}$, to become

$$E[F_{i,j-1}|C_{i,j-1}] = f_{j-1},$$

and

$$\text{Var}[F_{i,j-1}|C_{i,j-1}] = \frac{\sigma_{j-1}^2}{C_{i,j-1}}.$$

Or, equivalently

$$E[F_{i,j}|C_{i,j}] = f_j,$$

and

$$\text{Var}[F_{i,j}|C_{i,j}] = \frac{\sigma_j^2}{C_{i,j}},$$

where $F_{i,j}$, σ_j^2 and f_j are as defined in Section 4.2.2. This is the same as the formulation of Mack's model given in Mack (1999). The mean and variance can be seen to have the same form as in the GLM formulation in (4.48), with the weights, $w_{i,j}$ being equal to $C_{i,j}$ and non-constant scale parameters $\phi_j = \sigma_j^2$. Hence, this expresses Mack's model as a GLM that is fitted to the individual development ratios, $F_{i,j}$. The error terms in the GLM are assumed to follow a Normal distribution, which ensures that the estimator for σ_j^2 has the same form as in Mack's model, so that

$$F_{i,j} \sim N\left(f_j, \frac{\sigma_j^2}{w_{i,j}}\right).$$

Furthermore, a log-link function is used, as with the ODP model in (4.50), to ensure that the fitted development factors are greater than zero. When the model is fitted as a GLM to the ratios in this way, the fitted development factors, $\hat{f}_j$ will be the same as the column sum weighted average development factors, as in (4.7). The estimate of σ_j^2 is produced as for the original formulation of Mack's method, that is, as in (4.8):

$$\hat{\sigma}_j^2 = \frac{1}{I-j-1} \sum_{i=0}^{I-j-1} w_{i,j}(F_{i,j} - \hat{f}_j)^2, \tag{4.110}$$

where $w_{i,j} = C_{i,j}$.

As for the ODP Bootstrap, the equivalence of the Mack method to the CL method can be used to avoid the need to fit a GLM to each random sample within the bootstrap procedure. This means that Mack's method can, if required, be bootstrapped without the need for specialist statistical software. The bootstrap procedure for Mack's method is summarised below.

1. Create a triangle of individual development factors, $F_{i,j}$, and from these derive the CL development factors, $\hat{f}_j$, as per (4.7):

$$\hat{f}_j = \frac{\sum\limits_{i=0}^{I-j-1} C_{i,j} F_{i,j}}{\sum\limits_{i=0}^{I-j-1} C_{i,j}}. \tag{4.111}$$

If required, these could be smoothed by fitting a curve, as described in Section 3.2.6, which can be extrapolated into the tail if required.

2. Calculate $\hat{\sigma}_j^2$ using (4.110).

3. Determine the Pearson residuals, which for the GLM formulation of Mack's method, following (4.96), are

$$r_{i,j} = \frac{\sqrt{w_{i,j}}(F_{i,j} - \hat{f}_j)}{\hat{\sigma}_j}.$$

These can be bias-adjusted, which, as per Appendix A1.3 of England and Verrall (2006), should involve using different factors to those used with non-constant scale parameter when bootstrapping the ODP model, so as to be consistent with the original formulation of Mack's method. The factors based on Appendix A1.3 of England and Verrall (2006) are defined as $\sqrt{(n_j/(n_j - 1))}$, where $n_j = I - j$ (i.e. the number of residuals in each column associated with $\hat{f}_j$).

4. As per the ODP Bootstrap procedure, these residuals are randomly sampled, with replacement, and written as $r_{i,j}^*$, to generate the pseudo individual development factors, using

$$F_{i,j}^* = \frac{r_{i,j}^* \hat{\sigma}_j}{\sqrt{w_{i,j}}} + \hat{f}_j.$$

5. This will create a triangle of pseudo development factors, from which the pseudo column sum CL development factors, $\hat{f}_j^*$, can be calculated, as per (4.111), using the original weights, $C_{i,j}$:

$$\hat{f}_j^* = \frac{\sum\limits_{i=0}^{I-j-1} C_{i,j} F_{i,j}^*}{\sum\limits_{i=0}^{I-j-1} C_{i,j}}. \tag{4.112}$$

If the factors have been smoothed or extrapolated into the tail in Step 1, then this is replicated for these pseudo development factors.

6. Starting with the known cumulative claims in the leading diagonal of the data triangle, multiply the value in each row by the appropriate "bootstrapped" development factor from the previous step, to produce the bootstrapped values for the first future diagonal in the triangle. Denote the values in this diagonal by $C_{i,j}^*$.

7. Then, to introduce process error, a value is simulated from a Normal distribution[38] (as per the formulation of Mack's method as a GLM, but translated as cumulative claims) with mean equal to $C^*_{i,j}$ and variance as $\hat{\sigma}^2_{j-1} C_{i,j-1}$. This produces the bootstrapped cumulative claims in the first future diagonal of the triangle, incorporating process error.

8. Derive the bootstrapped values in the next diagonal, by taking the values from the previous step (i.e. after incorporating process error) and multiplying by the appropriate bootstrapped development factor. Introduce process error in the same way as for the previous diagonal (except that the variance formula will refer back to $C^*_{i,j}$ and so on).

9. Continue until all the future part of the triangle has been derived.

10. Calculate the bootstrapped ultimate and reserve for this bootstrap simulation for each cohort and in total.

11. Repeat the previous seven steps according to the number of required bootstrap simulations.

This procedure will then produce the full distribution of future cash flows, incorporating both estimation and process error, from which the mean and variance, along with any percentile values or other statistics can be derived.

4.4.4 Numerical Examples of Bootstrap Method

This section shows the application of the ODP Bootstrap method and the Mack Bootstrap method to the example data triangle used in the worked example of the ODP model given in Section 4.3.8, which was taken from Taylor and Ashe (1983). Both of these approaches give the same estimated reserves as the standard column sum CL method, but potentially different estimates of the prediction error and full distribution of future claims and the associated cash flows.

Numerical Example of ODP Bootstrap – Constant Scale Parameter

For this example, a constant scale parameter is used initially, and the impact of using non-constant scale parameters is considered in the next subsection. The example starts with the incremental claims paid triangle in Table 4.37.

Since the ODP model is being bootstrapped, the fitted incremental claims will be those shown in the ODP worked example, as given in Table 4.31. They will also be the same as those derived using the column sum average CL method, given the equivalence of the ODP model and the CL method. Using the fitted values and the actual incremental data shown in Table 4.37, the unscaled Pearson residuals can be derived for each cell in the triangle using (4.99), that is

$$\text{Residual} = \frac{\text{Actual} - \text{Fitted}}{\sqrt{\text{Fitted}}}. \tag{4.113}$$

---

[38] Note that the use of Normal distribution here might produce negative cumulative claims for some simulations. If this is deemed inappropriate, then it could be avoided by using a Gamma distribution, for example.

Table 4.37 *ODP Bootstrap example: Incremental Paid Claims*

Cohort	Development year									
	0	1	2	3	4	5	6	7	8	9
0	358	767	611	483	527	574	146	140	227	68
1	352	884	934	1183	446	321	528	266	425	
2	291	1002	926	1017	751	147	496	280		
3	311	1108	776	1562	272	352	206			
4	443	693	992	769	505	471				
5	396	937	847	805	706					
6	441	848	1131	1063						
7	359	1062	1443							
8	377	987								
9	344									

These are shown in Table 4.38.

Table 4.38 *ODP Bootstrap example: Unscaled Pearson Residuals*

Cohort	Development year									
	0	1	2	3	4	5	6	7	8	9
0	169	115	−112	−312	170	521	−236	−99	−87	0
1	−39	−55	−48	131	−178	−135	252	25	74	
2	−134	77	−46	−22	231	−404	207	59		
3	−93	204	−185	533	−391	−72	−262			
4	184	−158	122	−174	−21	176				
5	71	60	−79	−183	215					
6	78	−130	108	−29						
7	−161	−100	197							
8	−22	14								
9	0									

Each cell is then each multiplied by the degrees of freedom adjustment factor $\sqrt{n/(n-p)}$, as explained in the Degrees of Freedom subsection in Section 4.4.3. In this case, n = number of cells in triangle = 55 and p = number of parameters = $(2 * \text{cohort years} - 1) = 19$, so that the factor is 1.24. The resulting adjusted Pearson residuals are shown in Table 4.39. The significance of the shaded cell will become apparent in the next stage of the worked example.

Although the scale parameter is not needed at this stage (as a constant scale parameter is being used for this part of the example) it is needed in a subsequent stage when process error is added. In any case it is instructive to note that the parameter can be derived from Table 4.39, using the version of the formula given in the description of the ODP model at (4.71). This is the sum of the squares of the values in Table 4.39, divided by the number of observations, 55, which can be verified as 52,601 – the same value as shown in the ODP fitted parameter results given in Table 4.30.

For reference purposes, Table 4.38 is the same (ignoring rounding) as Triangle 4 on P291 of England and Verrall (1999) and Table 4.39 is the same as Triangle 4 on P9 of

Table 4.39 *ODP Bootstrap example: Unscaled Pearson Residuals with DOF adjustment*

Cohort	Development year									
	0	1	2	3	4	5	6	7	8	9
0	209	142	−138	−385	210	644	−291	−122	−107	0
1	−48	−67	−59	162	−220	−167	312	31	91	
2	−166	96	−56	−27	286	−499	256	73		
3	−115	252	−228	659	−483	−89	−324			
4	228	−195	151	−215	−25	218				
5	88	74	−97	−226	266					
6	97	−161	134	−35						
7	−199	−123	244							
8	−27	17								
9	0									

England (2001). The same results can also be produced using the spreadsheet "Bootstrap Models.xls" that accompanies Shapland and Leong (2010) and Shapland (2016). Finally, the "BootChainLadder" function in the "ChainLadder" package in R can be used to produce the same results. The relevant R code is in Appendix B.3.3, which also includes code for displaying selected types of diagnostic charts of the results, as well as for fitting a statistical distribution to the bootstrap simulations, as explained further in Section 4.10.4.

As expected, the residuals triangle given in Table 4.39 has a zero residual in the bottom left and top right cells of the triangle. As discussed after the examples, in Section 4.4.5, these two cells could be removed from the residual sampling process – they have been left in here to enable comparison of results with the other sources mentioned here.

The next stage is to repeatedly sample randomly from the residuals (including the DOF adjustment factor) given in Table 4.39, with replacement. This will mean that for each random sample, each residual has a 1 in 55 chance of appearing in a particular cell within that sample (assuming that none are excluded). For the purposes of this example, assume that the triangle in Table 4.40 is one such sample.

Table 4.40 *ODP Bootstrap: Example random selection of residuals after DOF adjustment*

Cohort	Development year									
	0	1	2	3	4	5	6	7	8	9
0	−195	−48	73	−107	−67	−161	151	−228	659	286
1	162	0	−199	−195	−167	−107	266	17	312	
2	31	0	−25	−483	286	−107	−27	88		
3	91	312	−122	151	96	−385	−483			
4	−499	209	−115	−56	−97	17				
5	−385	−195	252	256	−215					
6	88	659	151	96						
7	142	97	−228							
8	266	−97								
9	218									

It can be seen that in this particular sample, the residual of −48 for Cohort 0, delay 1, for example, has been taken from Cohort 1, delay 0 in the original residual triangle. The randomly selected residuals are then used to create one version of the pseudo data, as shown in Table 4.41.

The pseudo data is calculated using (4.106) as follows (in which the residual after applying the DOF adjustment is denoted by "AdjResidual"):

$$\text{Actual} = \text{AdjResidual}\sqrt{\text{Fitted}} + \text{Fitted}. \tag{4.114}$$

Continuing with the example calculations for Cohort 0 delay 1, in this equation, "AdjResidual" will be −48.384. The "Fitted" value is the incremental fitted claims for Cohort 0, delay 1 of 672,617 (i.e. 673 shown in thousands in Table 4.31). Hence, the pseudo data is equal to

Table 4.41 *ODP Bootstrap: Example pseudo data – incremental*

Cohort	\multicolumn{10}{c}{Development year}
	0	1	2	3	4	5	6	7	8	9
0	432	633	765	660	374	206	347	85	617	142
1	435	937	784	850	454	339	536	262	572	
2	428	927	946	546	792	335	353	295		
3	230	1,211	837	1,176	639	155	73			
4	321	1,029	770	883	450	375				
5	333	698	1,165	1,242	388					
6	375	1,627	1,175	1,193						
7	452	1,274	973							
8	348	877								
9	387									

$$-48.384 * \sqrt{672,617} + 672,617 = 632,935.$$

This is 633 in thousands, as shown in the shaded cell in Table 4.41.

The incremental pseudo triangle above is then cumulated and the CL method is applied to the resulting triangle to produce the pseudo development factors for that simulation. These can then be used to forecast the future incremental cash flows, excluding process error. The details are not shown here, as they are the same as for the deterministic CL method described in Chapter 3. To add process error, as per the approach described in the relevant subsection of Section 4.4.3, each incremental future claim amount produced by each bootstrap simulation is assumed to be the mean of an ODP distribution (or Gamma perhaps), with variance equal to the scale parameter multiplied by the mean, as per the formulation of the ODP distribution. Then, simulation is used to derive a random sample from this distribution, for each future cell of the triangle.

The overall process is repeated, according to the selected number of bootstrap simulations, to produce the full bootstrap sample, from which summary statistics and the estimated full distribution of future claims can be derived. Using this approach for this example produces the results shown in Table 4.42. The "Boot SE" column is the standard deviation of the bootstrap simulation results, and represents the bootstrap estimate of the prediction error. The results were produced using the "BootChainLadder" function within the "ChainLadder" package in R, with 5,000 simulations. The relevant code is in Appendix B.3.3.

Table 4.42 *ODP Bootstrap results summary (with DOF adjustment and ODP process error). Constant scale parameter*

| Cohort | Claims Paid | Reserve | | | |
		CL	Boot mean	Boot SE	As % reserve
0	3,901	0	–	–	
1	5,339	95	96	112	117%
2	4,909	470	479	225	47%
3	4,588	710	717	260	36%
4	3,873	985	1,000	308	31%
5	3,692	1,419	1,430	381	27%
6	3,483	2,178	2,190	501	23%
7	2,864	3,920	3,934	805	20%
8	1,363	4,279	4,332	1,067	25%
9	344	4,626	4,729	2,047	43%
Total	34,358	18,681	18,906	3,025	16%

Table 4.43 *Bootstrap example: Percentile results summary (with DOF adjustment and ODP process error). Constant scale parameter*

Item	Value	Ratio to mean
Mean	18,906	100%
50th percentile	18,705	99%
75th percentile	20,787	110%
90th percentile	22,822	121%
95th percentile	24,272	128%
99.5th percentile	27,571	146%

The column labelled "Boot SE" in Table 4.42 can be regarded as an estimate of the full prediction error, as it incorporates process error. The figures in the final column are very similar to the "Bootstrap/Simulation" results shown in Table 2 of England (2001) and to results that can be produced using the spreadsheet "Bootstrap Models.xls" that accompanies Shapland and Leong (2010) or Shapland (2016).

The results at various percentile levels are shown in Table 4.43. The uplifts to the mean, shown in the final column, are almost identical to the uplifts that can be calculated from Table 3 in England (2001). Hence, in this simple example, if reserves were booked at, say, 10% above the mean, or best estimate level, this would be equivalent to selecting a value at the 75th percentile from the bootstrap distribution.

As noted in the process error subsection of Section 4.4.3, a Gamma distribution can be used instead of an ODP distribution for the process error component. For this dataset, the R code in Appendix B.3.3 can be used to implement a Gamma distribution in this way. This shows that, for this dataset, the results are very similar to using an ODP distribution, with the overall Standard Error staying at approximately 16% and the ratio of the 99.5th percentile value to the mean reducing marginally to 145%. Other examples may show a greater difference, of course. In addition, alternative approaches are possible for the process error. For example, Section 4.4.6 describes a non-parametric resampling method.

Numerical Example of ODP Bootstrap – Non-constant Scale Parameters

To simplify the explanation of the ODP Bootstrap method, the previous results in this worked example assumed that the ODP model is being used with a constant scale parameter. However, as noted in the relevant paragraphs of Section 4.4.3, it may be appropriate to use scale parameters that vary, for example, by development period. To illustrate the use of non-constant scale parameters by development period for this worked example, the first step is to review the residuals by development period, using a constant scale parameter. These are shown in Table 4.44 and Figure 4.2. They are derived by taking the residuals in Table 4.39 and dividing each value by 229.35, which is the square root of the scale parameter of 52,601 referred to at the start of this example.[39]

Examination of Figure 4.2 suggests that there is less volatility in the residuals at the early and late development periods than in the other periods, indicating that a constant scale parameter may not be appropriate for this particular example. Ideally, there should be no discernible pattern by development year (or any other dimension of the triangle, for that matter), so to explore whether this improves if non-constant scale parameters are used, these first need to be calculated. Using the second version of the non-constant scale parameter formula at (4.107), the values for each development period are derived as the average of the squared values in the relevant column in Table 4.39. The square root of each of the resulting

Table 4.44 *ODP Bootstrap example: Scaled Residuals with DOF adjustment. Constant scale parameter*

	Development year									
Cohort	0	1	2	3	4	5	6	7	8	9
0	0.910	0.620	−0.603	−1.679	0.917	2.808	−1.269	−0.532	−0.468	0.000
1	−0.211	−0.294	−0.257	0.705	−0.958	−0.730	1.358	0.135	0.397	
2	−0.723	0.417	−0.246	−0.117	1.246	−2.176	1.117	0.317		
3	−0.499	1.099	−0.994	2.873	−2.106	−0.387	−1.412			
4	0.993	−0.850	0.660	−0.939	−0.111	0.949				
5	0.384	0.321	−0.423	−0.987	1.160					
6	0.422	−0.700	0.582	−0.154						
7	−0.866	−0.538	1.063							
8	−0.120	0.076								
9	0.000									

---

[39] The square root of the scale parameter is used, as per (4.97).

Table 4.45 *ODP Bootstrap example. Non-constant scale parameters*

	Development year									
	0	1	2	3	4	5	6	7	8	9
$\sqrt{\text{scale parm'}}$	139.9	142.3	153.0	318.1	282.6	386.6	296.7	83.9	99.6	0.0

scale parameters is given in Table 4.45.[40] As an example, for development period 7, the value of 83.9 is calculated as follows:

$$\sqrt{\underbrace{-122^2 + 31^2 + 73^2}_{\text{Sum of squares of col 7 of Table 4.39}} \Big/ \underbrace{3}_{\text{No. values in col 3}}} = 83.9.$$

The scaled residuals (with DOF adjustment) using non-constant scale parameters are shown in Table 4.46 and Figure 4.3.

These are calculated in a similar way as with constant scale parameters, by dividing the residuals in Table 4.39, by the square root of the non-constant scale parameters for the corresponding development periods. So, for example, the value of 1.666 in development period 5 for cohort 0 is calculated as 644 (being the residual in the corresponding cell in Table 4.39) divided by 386.6, which is the square root of the scale parameter for development period 5, shown in Table 4.45.

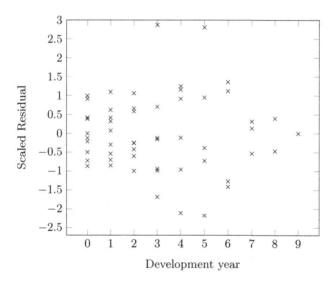

Figure 4.2 ODP Bootstrap example: Scaled residuals with DOF adjustment. Constant scale parameter

[40] The value of zero in development period 9, where there is only one value in the triangle, could be replaced if required by an alternative value – for example, in England and Verrall (2006), Appendix A1.3, the approach described involves using the minimum of the two previous values, giving a value of 83.9, which is the same as that given for the last development period in the Taylor and Ashe (1983) example in England (2010).

Table 4.46 *ODP Bootstrap example: Scaled Residuals with DOF adjustment.*
Non-constant scale parameters

Cohort	Development year									
	0	1	2	3	4	5	6	7	8	9
0	1.492	0.999	−0.904	−1.211	0.745	1.666	−0.981	−1.454	−1.079	0.000
1	−0.346	−0.474	−0.386	0.508	−0.777	−0.433	1.050	0.370	0.914	
2	−1.185	0.672	−0.369	−0.084	1.012	−1.291	0.863	0.866		
3	−0.819	1.772	−1.490	2.072	−1.710	−0.229	−1.091			
4	1.628	−1.370	0.989	−0.677	−0.090	0.563				
5	0.629	0.517	−0.634	−0.712	0.942					
6	0.691	−1.128	0.872	−0.111						
7	−1.420	−0.868	1.592							
8	−0.196	0.122								
9	0.000									

Figure 4.3 is then compared to the corresponding figure using a constant scale parameter, and there is some indication that using non-constant scale parameters produces a somewhat more uniform volatility of residuals across the development periods. In other words, there appears to be no discernible pattern. On this basis, it might be concluded that for this example, it is more appropriate to use non-constant scale parameters in the ODP Bootstrap procedure.

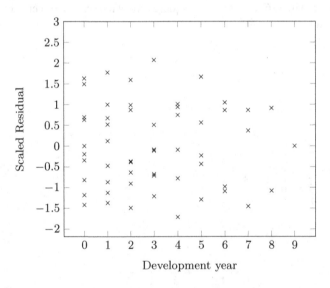

Figure 4.3 ODP Bootstrap example: Scaled residuals with DOF adjustment. Non-constant scale parameters

To produce bootstrap results using non-constant scale parameters, the residuals in Table 4.46 are used instead of those derived using a constant scale parameter. The non-constant scale parameters need to be used when deriving the pseudo data (as per (4.109)), but other than that the bootstrap procedure is the same as with constant scale parameters. Results can be produced using suitable specialist reserving software, where this feature is included, or alternatively it is also possible to produce results independently using, for example, the R software platform.

Examples of two published sources that produce these results for the same Taylor and Ashe (1983) data as used here are England and Verrall (2006) and Conort (2011). With thanks to Peter England, the first of these shows the results given in Table 4.47 (taken from Table 6 of that paper). The "SE" column here is equivalent to the prediction error.

Both sources show a CV across all cohorts combined of approximately 12%, compared to 16% when a constant scale parameter is used; such a difference could lead to materially different estimates of the reserves at higher percentiles, emphasising the potentially significant impact in practice of the choice between constant and non-constant scale parameters. The value of 12% is much closer to the CV of 13% derived using the Mack model (as given in Section 4.2.3). This is perhaps not surprising given the observation made in the relevant subsection of Section 4.4.3 regarding the comparability of the Mack model to the ODP Bootstrap using non-constant scale parameters.

Table 4.47 *ODP Bootstrap results summary (with DOF adjustment and ODP process error). Non-constant scale parameters*

Cohort	Mean reserve	SE	As % reserve
0	–	–	
1	96	43	45%
2	474	110	23%
3	716	142	20%
4	990	257	26%
5	1,428	389	27%
6	2,189	522	24%
7	3,940	732	19%
8	4,297	813	19%
9	4,660	1,296	28%
Total	18,790	2,203	12%

In practice, such comparisons will vary depending on the features of the particular dataset, and a judgement will need to be made on whether to use constant or non-constant scale parameters. If the latter are used, some form of smoothing of the parameters may be necessary in certain cases.

Numerical Example of the Mack Bootstrap

An example of how the Mack Bootstrap can be implemented in a spreadsheet is available at Verrall (2010a). With thanks to Richard Verrall, using this spreadsheet (with 1,000 simulations) applied to the same Taylor and Ashe (1983) dataset as used for the ODP Bootstrap example above and for the worked example of Mack's method in Section 4.2.3, gives the results shown in Table 4.48. These use residuals with the degrees of freedom adjustment factor applied to them and a Normal distribution for the process error, as per Section 4.4.3.

The results given in the columns labelled "SE" are the Standard Error, that is the prediction error, incorporating both parameter and process error. The results shown in the columns labelled "Mack Theory" are the results produced by applying the Mack formulae given in Section 4.2 and are taken from the results in Table 4.4. The CV% columns are just the SE divided by the bootstrap or Mack theory mean IBNR value. As expected, the Bootstrap IBNR is slightly different to the Chain Ladder IBNR, as a result of the fact that simulation is used to produce the results. Similarly, the SE and CV% for the Mack Bootstrap and Mack Theory are slightly different, but sufficiently close to suggest that the bootstrap is producing results that match the results from the analytic version of Mack's method. As with the ODP model, bootstrapping the Mack method has the advantage compared to its purely analytical equivalent that the full distribution of cash flows (and hence reserves) are produced as well as the mean and variance.

Table 4.48 *Mack Bootstrap example: Taylor and Ashe data – Summary results*

	IBNR		Mack Bootstrap		Mack Theory	
Cohort	Chain Ldr	Bootstrap	SE	CV%	SE	CV%
1	95	96	74	77%	76	80%
2	470	472	115	24%	122	26%
3	710	710	135	19%	134	19%
4	985	992	264	27%	261	27%
5	1,419	1,450	436	30%	411	29%
6	2,178	2,209	576	26%	558	26%
7	3,920	3,951	868	22%	875	22%
8	4,279	4,315	993	23%	971	23%
9	4,626	4,779	1,381	29%	1,363	30%
Total	18,681	18,974	2,530	13%	2,447	13%

Given the similarity of the CV% for the Mack Theory and Mack Bootstrap, the comment made in the previous subsection regarding the comparability of the analytical version of the Mack method and the ODP Bootstrap with non-constant scale parameters also applies to the Mack Bootstrap.

Some additional developments related to bootstrapping Mack's method can be found in Lo (2011), which includes suggestions, for example, of how to adapt the procedure to allow for the situation where the underlying independence assumptions of Mack's method may not hold for particular datasets. These suggestions are also of relevance when bootstrapping other models such as the ODP model.

4.4.5 Additional Details Related to Using Bootstrapping in Practice

The above description and numerical examples of the ODP and Mack Bootstrap were designed to explain the overall approach. When applying bootstrapping in practice there are numerous practical aspects that need consideration. Some examples of these are given below.

Scaling of the Bootstrap Mean

As noted in the introduction to this chapter, currently in a financial reporting context in particular, stochastic methods are not usually used to derive the best estimate or the booked figure for the reserves. In some cases, including when using the stochastic bootstrap method, it can be appropriate to adjust the mean that is produced by the method so that it reconciles with the selected best estimate for the reserves. This is usually done by scaling all the bootstrap results for each reserving category using either a multiplicative or additive (positive or negative) factor, such that the bootstrap mean equals the selected best estimate. This scaling can be done by cohort or in total across all cohorts combined. An additive scale factor will keep the standard deviation the same, and hence the CV will change, whereas a multiplicative factor will keep the CV unchanged. Similarly, an additive factor will change the ratio of a particular percentile value (e.g. 99.5th) to the mean, whereas a multiplicative factor will maintain the same ratio. The choice will depend on the circumstances.

Whenever the scaling causes a significant change to the bootstrap mean, it may be appropriate to investigate the reason for the difference, before adjusting the bootstrap results in this way. In some cases, it might be concluded that it is necessary to adjust the bootstrap procedure so that it produces mean results that are closer to the selected best estimates, rather than apply a large scaling factor. In other cases, it might be concluded that the bootstrap procedure is not likely to produce reasonable results and so other stochastic methods will need to be investigated.

After the bootstrap results have been scaled, they can, for example be used to estimate the implied percentile of the proposed booked reserves (by cohort or in total), thus providing an indication of the strength of the booked reserves. If the bootstrap results are used in this way without consideration of scaling the best estimate first, then this could produce a misleading indication of the strength of the booked reserves.

Selection of Residuals and Data in the Sampling Process

This subsection covers certain adjustments that are sometimes made to the residual selection process – namely partitioning (or "stratifying"), exclusion of outliers or zero residuals, and zero-averaging. In general, it is advisable to test the impact of any special adjustments that are made to the standard residual selection process.

In the ODP Bootstrap procedure described earlier in this section, the sampling of the residuals is carried out across the whole of the data triangle – i.e. to create the pseudo data for a particular cell in the triangle, any of the residuals can be sampled. In cases where the residuals appear to exhibit a non-random pattern, this might suggest that the underlying model has not captured the specific features of the data. If this is thought to be the case, then the technically correct approach to addressing this is to reconsider the underlying model that is being used, and then examine the residuals pattern for one or more alternative models. For example, using non-constant scale parameters by development period, as described in Section 4.4.3, may be helpful in some cases, but other model specifications may be required in other situations (e.g. to allow for non-random residuals by calendar period). A considerably less technically robust approach is to partition the triangle, so that the residual sampling is done within each partition. This will obviously change the underlying structure of the model and could have a significant impact on the resulting distribution of future claims. Partitioning the residuals may therefore be difficult to justify in practice as a better alternative to respecifying the underlying model.

Zero residuals may need excluding from the sampling process, in part depending on their underlying cause. If a standard CL model is being bootstrapped then the top right and bottom left cells in the triangle will be exact fits to the data, and hence the residuals will be zero. Some authors have argued that this justifies removing them from the residual sampling process (e.g. Shapland and Leong, 2010; Shapland, 2016), although a possible counter-argument is that they are a feature of the model that is being used, so it is that which should be changed if the existence of zero residuals in these cells is deemed to justify their exclusion.

In some situations, zero residuals may just be isolated instances at apparently random points in the triangle, without any specific underlying cause. In others, they may occur in parts of the triangle where the actual incremental claims are zero. If this is caused by a systematic pattern in the data (e.g. due to them being in a part of the triangle where the claims development is fully run-off) then this is likely to suggest they should be removed from the sampling process. In any case, investigating the cause, where possible, of any zero residuals may be appropriate. Where a decision is made to exclude one or more of them (either automatically or via separate identification) from the resampling process, the impact on the estimated summary statistics and claims distribution of excluding them compared to retaining them can then be investigated further before a final decision is reached.

In some cases, there can also be one or more residuals that, whilst not being zero, appear particularly high or low. If these are grouped in a particular part of the triangle, then this might suggest that the model is not fitting well to that part of the triangle, and perhaps an alternative model specification is required. In others, the data points corresponding to the unusually high or low residuals may be isolated outliers that perhaps should be excluded from the data, after which the model can be re-fitted to the remaining data.

A further adjustment to the standard sampling process that is sometimes applied is to "zero-average" the residuals – that is, make an adjustment to them so that the average residual is zero (e.g. deduct the mean value prior to adjustment from each residual when using them in the resampling process). The rationale for this is that, unless the average of the residuals is zero, then repeatedly resampling from them in the bootstrap process, and adding them to the fitted values, will mean that the bootstrap pseudo datasets will on average be higher or lower than the original fitted data. In other words, it could introduce an inherent bias into the resulting distribution of future claims outgo.

Whilst non-zero average residuals may be quite common in practice, the average may still be relatively small in absolute terms and they are not necessarily incompatible with the underlying distribution of residuals having a mean of zero. This is because the observed residual dataset is just one realisation from the underlying theoretical residual distribution. As for the other changes discussed above, it may be appropriate to measure the impact of any zero-averaging on the overall results before concluding that it should be implemented.

Application of Bootstrapping to Incurred Claims Data

Whilst the bootstrap procedure can in theory be applied to a data triangle consisting of any type of data used in reserving, with any cohort and development definitions, interpretation and use of the results will need to take into account the nature of the data being used.

The ODP and Mack Bootstrap procedures described above related to a claims paid triangle, and so the results will give an estimate of the distribution of the future claims and the associated cash flows in each future period. Because the claims paid to date are, by definition, fixed, they will not be a source of variability and so the distribution of the ultimate claims can be produced by adding the claims paid to date to the distribution of reserves. So, for example, the 99.5th percentile value for the ultimate claims will be the 99.5th percentile value for the reserves plus the claims paid to date (by cohort or in total).

When the bootstrap procedure is applied to an incurred claims triangle, the results will give an estimate of the distribution of the IBNR (including IBNER) claims – i.e. the difference between the estimated ultimate claims and the claims incurred to date. Assuming that there is sufficient incurred claims data to be satisfied that all the variability to ultimate has been captured adequately in the bootstrap procedure, or can be allowed for via a tail factor, the distribution of IBNR should allow implicitly for the variability in both the case reserves (i.e. effectively IBNER) and the pure IBNR. Hence, the distribution of the ultimate claims can be produced by adding the claims incurred to date to the distribution of IBNR. So, for example, the 99.5th percentile value for the ultimate claims will be the 99.5th percentile value for the IBNR plus the claims incurred to date (by cohort or in total). Furthermore, the distribution of unpaid claims (i.e. reserves) can be obtained by deducting the actual paid claims from the distribution of ultimate claims, obtained through an incurred bootstrap procedure.

Whilst using incurred claims in this way might in some circumstances produce less variability in the results than when paid claims are used in the bootstrap procedure, this seems logical, given that the case reserves provide some information regarding the unpaid claims, beyond that provided by the claims paid data alone. In addition, if the deterministic best estimate of the reserves is based partially or entirely on a projection of claims incurred data, then that provides some support for similarly placing some weight on the results of applying a stochastic method to the incurred claims data. In practice, application of a bootstrap procedure to both the paid and incurred claims data may be appropriate in some circumstances, and further investigation may be warranted if the results differ significantly.

When bootstrapping claims paid data, the results will automatically allow the cash flow variability to be analysed. However, when bootstrapping incurred claims data, an additional step will be required to achieve this. One approach is to model stochastically the ratio of paid to incurred claims over development time.

Shapland and Leong (2010) and Shapland (2016) suggest an alternative approach when bootstrapping the ODP model, involving running parallel paid and incurred bootstrap models, and then applying the paid development pattern to the results from the incurred model. Further details are available in Shapland and Leong (2010) and Shapland (2016).

One issue that becomes more relevant when bootstrapping incurred claims data is the possibility of needing to allow for negative development, since that is more common in incurred triangles than paid triangles. Bootstrapping the ODP model is usually possible when there are only isolated instances of actual negative incremental development, but bootstrapping Mack's model will usually cope better when there are repeated examples of these, and hence may be preferable with incurred claims data (or any other type of data, such as paid claims) that has this feature. However, where the future incremental values are required to always be positive, bootstrapping the ODP model will be consistent with this requirement, whereas the Mack model may produce future negative incremental values. Hence, the choice between bootstrapping these two models (or others) with any type of data will depend on the circumstances. In some cases, the negative incremental values can cause the model to produce negative fitted incremental values, which creates an issue in calculating the residuals, as the denominator in (4.113) will not be valid. One approach to dealing with this is to use a form of parametric bootstrapping, as explained in the subsection later in this section entitled "Parametric bootstrapping". Shapland and Leong (2010) suggest an alternative approach for dealing with this, which involves using absolute values of the negative incremental fitted amounts.

Dealing with Outwards Reinsurance

In practice, the distribution of future claims, net of reinsurance, will often be of interest. If the bootstrap procedure has been applied to the gross of reinsurance data, then the resulting distribution may or may not be appropriate to apply (e.g. in proportionate terms, by scaling) on a net of reinsurance basis. Some possible approaches for producing the distribution of reserves on a net of reinsurance basis using a bootstrapping approach are as follows:

1. Bootstrap the net of reinsurance data triangles directly. This might produce broadly reasonable results if the reinsurance is mainly proportional in nature, but it may not work well where there is non-proportional reinsurance or where the reinsurance programme has changed significantly over time.
2. Bootstrap the gross of reinsurance data and derive the net of reinsurance distribution by scaling the gross distribution by the ratio of net to gross best estimate reserves (i.e. use a multiplicative scaling approach, as referred to above in relation to scaling the mean). The bootstrap results will need to be split according to the reinsurance periods of cover so that the relevant contract terms can be applied correctly. This approach may be suitable where the reinsurance is proportional in nature, but where it is non-proportional other approaches may be better.
3. Bootstrap the gross of reinsurance data, and apply the reinsurance program explicitly to the simulated results. This approach might be possible, for example with Aggregate

Excess of Loss type reinsurance, but only where the programme applies at the level of granularity used in the bootstrap model (e.g by relevant reserve category and cohort).

4. As Step 2 above, but manually adjust the results to allow for any perceived reduction in variability caused by reinsurance. This adjustment might vary for different points on the distribution (e.g. the reinsurance may only impact the distribution at higher percentiles when the claims are large and affect the outwards reinsurance programme). It might also vary between classes of business, or might apply only where large claims/events are present.

5. Where there is some element of non-proportional reinsurance that applies on a per-claim basis, categorise the gross data for each reserving category according to size of claim – split between large claims that are expected to recover from the reinsurance and small claims that are not. Then, model the small claims using a bootstrap approach applied to the aggregate triangles, and model the larger claims using some form of frequency/-severity model. The latter will then allow the reinsurance recoveries to be determined directly for each modelled claim, by applying the known reinsurance programme terms. A bespoke model is normally needed for the large claim component of this approach, as none of the regularly used stochastic methods described in this chapter operate at an individual claim level.

All of these approaches have their limitations, and in practice the approach used will depend on a number of factors including in particular the data available and the nature / materiality of the reinsurance programme. Inevitably there will an element of judgement involved. If it is considered necessary to make a more accurate allowance for outwards reinsurance than is possible within an aggregated triangle type approach such as the bootstrap method, then a more sophisticated approach may need to be used, based perhaps on analysis that at least in part models claims at an individual level, as referred to above.

Some other suggested approaches are discussed in Chan and Ramyar (2014), including a "Conditional Ceded" approach, which it is suggested can capture the true dynamics of the reinsurance programme and hence model its impact on the distribution of reserves on a net of reinsurance basis.

Allowing for a Tail Factor

In some practical implementations of bootstrapping, the claims development will extend beyond the domain of the triangle. It will be necessary therefore to ensure that the underlying model that is being bootstrapped makes appropriate allowance for this. This can be achieved, for example, by using a curve fitting procedure akin to that used in deterministic claims reserving when deriving a tail factor. This is applied to each bootstrap sample, to produce both estimation and process variability in the "tail" component of the model, so that the cash flows projected by the model in this part of the data make appropriate allowance for variability.

Some commercially available reserving software packages will include this as a feature of the bootstrapping functionality, making this and other practical aspects of bootstrapping relatively straightforward for the reserving specialist to apply in practice.

A simple and very approximate approach to allowing for a tail factor, that was referred to in relation to Mack's method in Section 4.2.6, is to assume that the CV derived without the tail factor (including both process and estimation error) also applies to the reserve including the tail factor, either by cohort or in total. This simplistic approach ignores any additional volatility that might be present in the tail of the triangle, and so unless an appropriate adjustment can be made to allow for this, other approaches may be preferable.

Modifying the Standard Chain Ladder Approach

In order for the ODP Bootstrap procedure to be equivalent to the ODP model specified in Section 4.3, the standard Chain Ladder column sum weighted average development factors must be used in the implementation. However, alternative GLM formulations to the ODP which are not equivalent to this form of chain ladder model could also be fitted, and the same bootstrap procedure used to derive the prediction error. Section 4.4.8 discusses bootstrapping other claims reserving models.

Staying within the framework of a chain ladder model, it is quite possible to bootstrap such a model using different development factors, for example those averaged over a shorter period, or to use curve fitting to smooth the development factors (as described in Section 3.2) and to extrapolate beyond the domain of the data, as explained in the previous subsection. Whatever approach is used, it is reapplied at each stage of the bootstrap simulation, by modifying the first two steps in the standard procedure outlined in Section 4.4.3. However, the bootstrap procedure will not be as theoretically robust as it would be if a GLM-equivalent model was used.

Shapland and Leong (2010) and Shapland (2016) explain some implications of using approaches other than the full column sum in a GLM framework for the chain ladder bootstrap. They also discuss some other practical aspects of bootstrapping, such as dealing with missing values and outliers, adjusting for exposure and applying the approach to data that has partial first development or last calendar period data.

In some applications of the ODP Bootstrap, the level of variability in the more recent cohorts can be very significant, particularly for longer-tailed classes of business. This can lead to a potentially unrealistically high level of variability in the overall distribution of reserves. In the same way as with the deterministic chain ladder, this can be overcome by using an alternative method for the later cohorts, such as the Cape Cod or Bornhuetter–Ferguson ("BF") method. In a bootstrap context, the reserving "model" in the first step of the general procedure given in Section 4.4.3 will involve using a chain ladder for the earlier cohorts and, for example, an automated BF method for the more recent cohorts. The automation could involve taking an average loss ratio of, say, the previous five ULRs produced in each bootstrap simulation as the prior loss ratio assumption in the BF method, or simulating the prior loss ratios from a pre-selected distribution. In effect, this will "bootstrap" the BF method for these later cohorts. Such an approach may have limited theoretical justification, as it involves moving away from the underlying basis for the ODP Bootstrap, which by definition derives from its equivalence to the ODP model. It could also dampen the uncertainty too much in some cases, leading to an under-estimate of the level of the reserves at higher percentiles. The impact of these potential limitations will vary, depending on the circumstances.

If modifications are made to the basic bootstrapping approach involving a full column sum average CL approach, the impact on the results of any deviation from the fundamental requirement that the bootstrapping process is applied to independent and identically distributed values may need consideration.

Parametric Bootstrapping and Other Procedures

The ODP and Mack Bootstrap procedures described earlier in this section are both types of residual resampling bootstrap. The results of applying any such procedure to a single data triangle will depend critically on the size and variability of the data in that triangle (and the exact nature of the bootstrap procedure). So, in a 10×10 triangle, where there will be 55 observations, and it will not be known whether the extreme results that are simulated within the bootstrap procedure represent a 1 in 200 event, 1 in 1,000 event or indeed a 1 in 55 event. This uncertainty around the results will exist regardless of whether 100, 1,000 or 10 million simulations are used in the bootstrap procedure. The variability in the results will be constrained by the variability in the data itself, although if the process error is added using a parametric approach (as used in the worked examples of ODP and Mack Bootstrap given earlier in this section), that element of the overall prediction error will be less constrained.

Where there is a concern that this constraint materially affects the distribution of future claims produced by the bootstrap procedure (e.g. particularly at the higher percentiles if they are being used), then aside from making some form of approximate adjustment to the results to ensure that they are reasonable (e.g. perhaps based on calibrating them against a benchmark source), two examples as to how this constraint might be overcome in practice are as follows:

1. **Parametric Bootstrapping:** Rather than fitting a model to the incremental data triangle, and then sampling from the residuals between the actual and fitted data to produce the pseudo data, this type of bootstrapping involves making some form of parametric distributional assumptions regarding the incremental data, and then generating pseudo data directly from that distribution, without the need for any residual inversion formulae. Hence, if this approach is used, although the residuals might still be calculated, they are not sampled from. This can be an alternative approach where a non-parametric residual resampling procedure does not appear to work well – for example, as may be the case with some small datasets (which might have a limited range of residuals to sample from) or where the procedure produces one or more anomalous negative cumulative values. It should be noted that this description of parametric bootstrapping is consistent with that used in the statistical literature (e.g. Section 6.5 of Efron and Tibshirani, 1993). In some

other sources a different description may be used, which involves making distributional
assumptions about the residuals themselves, rather than the data.

2. **Distribution fitting:** Fit a distribution to the bootstrap simulation results (for example,
   using maximum likelihood) and look for any areas in the distribution which might
   suggest that the bootstrap results are producing misleading results. This can be done
   by comparing both the Value at Risk ("VaR") and the Tail Value at Risk ("TVaR") at
   a range of percentile levels for the simulated results against those derived from fitted
   distributions such as Normal, Gamma and Lognormal. The code in Appendix B.3.3 gives
   an example of using maximum likelihood to fit a Lognormal distribution to the results of
   bootstrapping the Taylor and Ashe (1983) dataset that was used in the worked example
   in Section 4.4.4. This approach is easier to apply to the reserves across all cohorts
   combined; if applied to individual cohorts, then any dependency between cohorts may
   also need to be allowed for in the approach used. Furthermore, where a distribution
   of cash flows is required (e.g. for discounting purposes) then such an approach can
   be difficult to apply. It may, however, be useful in some contexts, for example where
   bootstrapping is being used to assist in analysing an aggregate reinsurance contract that
   provides adverse development cover for the reserves across all cohorts combined for a
   particular category of business.

   Judgement will be involved in parametrising the distributions in these two approaches,
   and the results produced will need to be assessed for reasonableness, based, for example,
   on the practitioner's wider experience.

   For further discussion of parametric and non-parametric bootstrapping, see, for
   example, Björkwall et al. (2009) and Taylor and McGuire (2016).

4.4.6 Other Considerations Concerning the Application of Bootstrapping

Although bootstrapping is thought to be in reasonably widespread usage in a claims reserving context at present, the technique has a number of variations regarding how it can be used
in practice. It is quite possible that it will be subject to further debate and enhancement in
future, and hence the practitioner may need to consider modifying the bootstrap procedure
used in practice, based on these developments.

This section considers two particular aspects of bootstrapping in a reserving context that
have been discussed in the literature – namely, the residual definition used in the resampling
bootstrap and the allowance for process error.

Alternative Residual Definition in the ODP Bootstrap

When implementing the ODP Bootstrap procedure, an alternative formulation for the residuals to that given in Section 4.4.3, suggested in Pinheiro et al. (2003), involves using the
so-called *standardised* Pearson residuals. For the ODP model, these are defined as follows:

$$r_{i,j}^{**} = \frac{r_{i,j}}{\sqrt{1 - h_{i,j}}},$$ (4.115)

where $h_{i,j}$ is the diagonal of the *hat* matrix, as defined in (4.59) for the ODP model in Section 4.3.

If the standardised Pearson residuals are used, as per (4.115), then Pinheiro et al. (2003) and Shapland and Leong (2010) explain how this affects other aspects of the bootstrap procedure, such as the use of a degrees of freedom (or bias) adjustment factor. A more recent source that also discusses this topic, as well as other aspects of applying the ODP Bootstrap in practice, is Shapland (2016).

Alternative Way of Allowing for Process Error: Non-parametric Resampling Method

The description of the ODP Bootstrap procedure given in Section 4.4.3 referred to a parametric approach for allowing for process error, using, for example, an ODP or Gamma distribution. This section describes an alternative approach that does not rely upon selecting a distribution to introduce process error, but instead re-samples from the residuals a second time – i.e. it is non-parametric.

This approach is described in Pinheiro et al. (2003), where it is referred to as the "PPE" method and where the source of the procedure in a general statistical bootstrap context is stated as being Davison and Hinkley (1997). It involves using the second resampling of the residuals to enable calculation of the prediction error directly. It can be further explained by considering how the ODP Bootstrap procedure described in Section 4.4.3 can be modified to accommodate the residual resampling. To facilitate the modifications, it is first necessary to use the CL method in Step (a) of the procedure to derive the estimated future incremental claims derived from the original dataset (i.e. before any bootstrapping is done). Steps (b) to (i) are exactly the same, but are repeated here in abbreviated form for ease of reference. The whole procedure for the ODP Bootstrap, using non-parametric resampling for the process error, is then as follows:

(a) Apply the standard deterministic CL method to the claims paid data triangle to derive the column sum development factors, and use these to derive the estimated future incremental claims in each period for each cohort.

(b) Use the selected average development factors applied to the last diagonal, to derive the historical fitted cumulative values.

(c) Convert the cumulative fitted data triangle to an incremental fitted data triangle.

(d) Calculate the bias-adjusted Pearson residuals using formula (4.105). For an $n \times n$ triangle, there will be $N = \frac{1}{2}n(n + 1)$ values.

(e) Derive the bootstrap samples, by first selecting N residuals from these N residual values at random (with replacement, so there can be duplicates).

(f) Derive the incremental pseudo data triangle using a rearranged version of the residual definition, that is use (4.106) with the fitted data from Step (c) above.

(g) This defines one bootstrap sample or simulation triangle of the "pseudo history".

(h) Cumulate this triangle and calculate development factors from this pseudo-triangle, as in Step (a) above.

(i) Use the development factors from the previous step to derive the estimated future incremental claims for this bootstrap simulation.

(j) Now implement the second residual resampling to allow for process error. This begins by randomly selecting a residual from Step (d), that can be combined with the estimated future incremental claims in an individual cell, as derived from the previous step.

(k) Use this residual to derive an alternative prediction of the future incremental claims in each cell, by using the same rearranged residual formula as in Step (f) above, but with the fitted data being the future incremental claims according to the original chain ladder method derived from Step (a).

(l) Calculate the difference between the future incremental claims in Steps (i) and (k), which will represent the prediction error for that cell in the future triangle, for this bootstrap simulation.

(m) Implement the second residual resampling for each future incremental cell in the triangle – i.e. repeat Steps (j) to (l).

(n) Repeat steps (e) to (m) according to how many bootstrap simulations are being used in the process.

(o) Derive the distribution of the prediction error using the bootstrap simulations. This can be done for both the cohort totals, and the total across all cohorts (or indeed for individual future incremental cells if needed).

(p) To derive the kth percentile of the claims reserves themselves, add the kth percentile of the prediction error derived from the previous step to the original chain ladder estimate derived at Step (a). The percentiles and chain ladder estimate will be the individual cohort totals, or the total across all cohorts, according to which percentile is being calculated.

Pinheiro et al. (2003) describes a procedure that is consistent with that above, expressed in mathematical terms, which is applicable for GLMs with other definitions of the residuals.

Pinheiro et al. (2003) argues that the non-parametric resampling approach is more robust than the parametric procedure, but England and Verrall (2006) suggests that the results using the resampling approach will have extreme values that are only as large as the extreme values of the residuals. The approaches are discussed further in Björkwall (2011) and Björkwall et al. (2009), and in the latter both parametric and non-parametric approaches are outlined, including an approach referred to as "double bootstrap", which, although very computer-intensive, is suggested in Björkwall et al. (2009) as having some advantages.

Ultimately, the procedure used to add process error will be influenced by practical constraints, such as what approach has been used in the software available to the practitioner. The relative size of the process and parameter errors may also influence whether alternative approaches are investigated.

4.4.7 Assessment of Bootstrap Models and Results

Any individual application of bootstrapping with the ODP, Mack or any other model will include taking into account the general considerations that are applicable to all reserving exercises, such as understanding the business, as introduced in Section 1.2 and expanded upon in Chapter 5. For stochastic models, the testing of the appropriateness of the underlying model and the validation of the results are critical stages of the overall process, is discussed in Section 4.10, which includes examples of the possible approaches for the testing and validation of stochastic models in general, as well as specifically where bootstrapping has been used. For bootstrapping, this will involve assessing whether the underlying reserving model itself is appropriate, before implementing the bootstrapping procedure.

In addition to these specific tests for use in individual reserving assignments, some research has been done by a number of authors to review whether the results arising from chain ladder bootstrapping procedures are "accurate". Accuracy in this context has been assessed, for example, in two ways:

- First, hypothetical data can be simulated which follows a known model. A bootstrap procedure is then applied to a selected subset of this data, and the results compared to the known future data. An example of this approach is described in ROC (2007) and is commented upon in England and Cairns (2009).
- Second, the bootstrap procedure can be applied to subsets of historical publicly available data for individual insurance companies and the results compared with the actual claims outgo. Three research papers that have used this are Leong et al. (2012), Meyers and Shi (2011a) and Meyers (2015).

One of the key pieces of analysis in this research is to consider how well the stochastic methods predict the upper tails when the claims generating process is designed such that it conforms to a defined underlying model. This is done by calculating the proportion of "true" values from the known claim generating process that lie above, say, the 99th percentile generated by the stochastic model; if this is significantly below or above 1%, then the proponents of this approach suggest that this implies that the stochastic model is not estimating the tail accurately.

The overall conclusion cited by this research is that the ODP Bootstrap method (with constant scale parameter), particularly when applied without judgement or intervention from an experienced practitioner, may understate the tail of the distribution of reserves. Leong et al. (2012) suggests that this could be caused by the bootstrap method failing to capture so-called *systemic* risk, as defined in Section 4.1.3, and provide some possible solutions to including this type of risk in the results.

England and Cairns (2009) reviewed the work described in ROC (2007) and challenged some of the details. In particular, they pointed out that the version of Mack's method used in ROC (2007) used a fitted Lognormal distribution (as does that in Meyers, 2015), whereas other approaches can be used to provide a closer match to bootstrapping Mack's model (e.g. using the formulation described in the Mack Bootstrap subsection in Section 4.4.3); the conclusions in ROC (2007) and Meyers (2015) related to Mack's model therefore only apply to this "Lognormal" version of Mack's method.

England and Cairns (2009) also recommend that, when bootstrapping the ODP model, the reliability of the results at extreme percentiles can be improved by using non-constant scale parameters, and also suggest that parametric bootstrapping should be investigated in some circumstances, as discussed in the relevant subsection in Section 4.4.5. Finally, they also suggest that the inference drawn from differences between the proportion of "true" values from the underlying model that are above, say, the 99th percentile and the theoretical value of 1% may be invalid anyway. The use of this type of percentile test and other related model tests is discussed in more general terms in Jarvis et al. (2016).

Regardless of the merits or otherwise of this research, there can be no doubt that bootstrapping procedures need to be implemented with a good degree of judgement and careful assessment of the results, particularly when considering the tail of the distribution.

4.4.8 Bootstrapping Other Claims Reserving Models

There may be some situations where an analytic stochastic reserving method has been chosen as the preferred approach for a particular dataset, and that method only produces estimates of the mean and prediction error of the future claims, but a full predictive distribution of the claims is required.

Two possible ways of deriving this distribution are to assume a particular statistical distribution or to bootstrap the analytic reserving method. The first of these was mentioned in Section 4.1.4 and involves parametrising the chosen distribution using the derived mean and prediction error. However, the resulting distribution may not capture accurately the variability in the data that exists at different points on the distribution. Allowing for dependency across the cohorts can be difficult as well.

Bootstrapping the analytic method is an alternative approach that will enable the full predictive distribution to be estimated from the data itself. As noted at the start of the section, a bootstrapping procedure can, at least in theory, be applied to any appropriately specified claims reserving model. For example, any GLM which is deemed to be suitable for modelling a claims data triangle can be bootstrapped. Once the bootstrap procedure has been applied, the values at selected points on the distribution can be compared with those produced by a fitted statistical distribution, and a judgement made as to which results, or a combination thereof, should be used.

Bootstrapping two analytic methods – ODP and Mack – has already been discussed in Section 4.4.3. The remainder of this section provides brief details of how bootstrapping can be applied to a selection of other analytic reserving models. Further details are available from the sources that are cited here.

Lognormal model

This model is described in Section 4.7.7. It involves fitting a model to the logarithm of incremental claims, which are denoted by $Y_{i,j} = \log(X_{i,j})$. The model is formulated such that $Y_{i,j} \sim \text{Normal}(m_{i,j}, \sigma^2)$. The residuals and the associated pseudo data used for bootstrapping are defined in a similar way as for other models, that is

$$r_{i,j} = \frac{Y_{i,j} - \hat{m}_{i,j}}{\hat{\sigma}},$$

and

$$Y_{i,j}^* = r_{i,j}^* \hat{\sigma} + \hat{m}_{i,j}.$$

A similar bootstrap procedure is used as for other models, with the model being fitted to each randomly generated pseudo data triangle (on the log-scale) and then the fitted parameters being used to derive the estimated reserve for that bootstrap sample. The model would need to be fitted to each pseudo data triangle using appropriate statistical software, or in a spreadsheet or other bespoke software, as referred to in Section 4.7.7. As for other bootstrap procedures, process error needs to be incorporated. For each future cell in each bootstrap sample, this can be achieved via a similar approach as for the Mack Bootstrap procedure (Step 7) described in the final subsection of Section 4.4.3, but without the need for the recursive steps, by sampling from a Normal distribution with a mean equal to the generated pseudo incremental claims from the bootstrap procedure, and a variance of σ. The output from this final step is then translated from the log-scale by exponentiating, so that the reserves can be calculated for each bootstrap sample and the usual statistics produced from these results. Some additional details are given in the discussion section of England and Verrall (2006).

Other GLM-type models

In the same way that the ODP model can be bootstrapped, so can other GLM-type stochastic models. Some examples of these models, in their analytic form, are described in Section 4.7.11. The same bootstrap procedure as for the ODP model can be used with these models, except with a different GLM model fitted to each randomly generated bootstrap sample. However, the equivalence of the ODP model to the Chain Ladder model cannot be used with the other GLM models. Despite this, software does exist which enables the other models to be bootstrapped. For example, the "glm.Reserve" function within the "ChainLadder" package in R allows this.

Schnieper's model

This model is described briefly in Section 4.7.13, and a step-by-step bootstrap algorithm is given in Liu and Verrall (2009a). The procedure involves generating random bootstrap samples for each of the two triangles on which Schnieper's model is based – i.e. the development of IBNER claims and the development of IBNR claims. Since the model is recursive in nature, ratios are used rather than the claims data itself. As for bootstrapping other stochastic models, a bias correction is used for the residuals derived from the ratios, and process error is added (using a Normal distribution[41]) so that the full prediction error can be estimated.

---

[41] However, as noted in Liu and Verrall (2009a), other distributions can also be used.

Munich Chain Ladder ("MCL")

This method, described in Section 3.8.1, models both the paid and incurred data triangles within a single framework, taking into account their dependence. An approach to bootstrapping the MCL method is described in Liu and Verrall (2010). The MCL model is recursive in nature, and hence, as for Schnieper's model, the residuals of the paid and incurred development ratios are used in the bootstrap procedure, rather than the claims data. Liu and Verrall (2010) describe a bootstrap algorithm that is designed to allow for the dependence between the paid and incurred data triangles. It involves generating random bootstrap samples for four separate triangles – the paid and incurred development ratios, the ratio of the paid to incurred claims data, and the ratio of incurred to paid claims data. As for Schnieper's model, process variance is added using a Normal distribution.

Separation method

This is a deterministic method, described in Section 3.7, which uses both claim amounts and claim counts, the latter of which are assumed to be known. A key feature of the method is that it incorporates specific consideration of the impact of inflation on average claims costs. Björkwall et al. (2010) describes a procedure for bootstrapping this method, which is based on a parametric bootstrapping procedure described in Björkwall et al. (2009). In this paper, the procedure is applied to the Taylor and Ashe (1983) dataset and results are produced for different inflation assumptions, which are then compared to ODP Bootstrap results.

4.5 Bayesian Methods

In the context of stochastic claims reserving, Bayesian methods represent a distinct type of approach that have as their core feature the ability to incorporate information that is external to the data source for which a reserve estimate is required. The external data can be from a specific source (such as industry data for benchmark loss ratios) or can be derived based on the expert opinion of the reserving practitioner. The overall approach can be analytical in nature, although it is more commonly applied using simulation techniques.

Although there has been some support in the literature for the use of Bayesian methods to produce predictive distributions of claims reserves (e.g. see England and Verrall, 2006), at the time of writing they are not believed to be widely used amongst reserving practitioners. This does not mean that they should be ignored, since the fact that they allow external information to be incorporated into the analysis in a formal way should be an advantage, as many reserving exercises will involve the practitioner incorporating some element of external information or judgement into the process. Furthermore, Bayesian methods can be used to produce a full distribution of the future claims and their associated cash flows, rather than just an estimate of the prediction error. Hence, they provide an alternative to the bootstrap method outlined in Section 4.4. One feature of Bayesian methodology that is relevant in a reserving context is the use of "vague" or "non-informative" priors. This allows prior information (i.e. external to the data in a reserving context) to be factored in, even where there is significant uncertainty surrounding that prior information. In many ways, the deterministic BF method is Bayesian in nature,[42] but the derived ultimate is independent of how much uncertainty there is around the prior assumption; a stochastic Bayesian version of the BF method allows the degree of confidence there is in the prior to be factored into the reserving process.

An example of the general procedure in which Bayesian methodology is used in a stochastic reserving context is as follows:

1. Define a statistical model for the data triangle, with parameters that need to be estimated. For example, this could be the ODP model described in Section 4.3.
2. Define prior statistical distributions for one or more of the parameters of the statistical model in Step 1.
3. Select parameters for the prior distributions in Step 2. These are independent of the observed data triangle.
4. Update the parameters using the observed data triangle, to produce the posterior distribution of the parameters.
5. Using the posterior distribution of the parameters, together with a second simulation stage to incorporate process uncertainty, produce predictions of the future claims – i.e. produce a predictive distribution.
6. Analyse the predictive distribution to produce statistics of interest – such as mean reserve, prediction error of reserves and percentiles at selected values.

In theory the results can be produced analytically, but in a practical stochastic reserving context this is rarely feasible and so a simulation approach is often used, such as "Markov Chain Monte Carlo", or MCMC. The MCMC simulation approach produces an empirical posterior distribution of the model parameters and these are then used in the statistical model in Step 1 above to produce the required distribution of future claims. This requires specialist software, examples of which are freely available on the internet. For example, the "WinBUGS" software, described in Lunn et al. (2000), can be downloaded from www.mrc-bsu.cam.ac.uk/software/bugs/the-bugs-project-winbugs/.

Furthermore, some authors have provided example WinBUGS code for implementing certain types of stochastic Bayesian reserving methods – see, for example, Verrall (2004), Verrall (2007) and Beens et al. (2010). Further background on using WinBUGS to build Bayesian models can be found, for example in Scollnik (2001) and Scollnik (2004).

4.5.1 An Example of a Bayesian Stochastic Reserving Method

The term "Bayesian methods" is used here as a generic term for the family of stochastic reserving methods that are Bayesian in nature. A number of such methods have been

---

[42] Even where the prior for a particular cohort is derived from an average of estimated ultimates from earlier cohorts, as described in the second worked example in Section 3.4.3, this prior can still be regarded as external to the data for that cohort, since it is independent of the claims experience for that cohort.

proposed, but to demonstrate the principles behind the Bayesian approach and to contrast it with the other non-Bayesian stochastic methods discussed in this chapter, the following model specification is used, which is based on that in Verrall (2004) and Verrall (2007). Only a very high-level description, including a worked example, is given here, with further details being available in these two sources if required.

- The assumed statistical model is defined by initially considering the ODP model, which as per Section 4.3, can be defined in a cross-classified multiplicative form with the incremental claims $X_{i,j}$ being assumed to follow an ODP distribution, with mean of $x_i y_j$ and variance of $\phi x_i y_j$. The x_i and y_j parameters relate to the rows and columns respectively of the data triangle.
- To use the ODP model in the context of a Bayesian method which enables a Bornhuetter–Ferguson type approach to be used, it is reformulated as an Over-dispersed Negative Binomial (ODNB) model. The ODNB model is effectively a recursive reformulation of the ODP model.
- With this reformulation, the incremental claims $X_{i,j}$ have a mean of

$$\mathrm{E}[X_{i,j}] = (\gamma_i - 1) \sum_{m=0}^{i-1} X_{m,j}.$$

The γ_i parameters are related to the original x_i parameters, via a formula that is specified in Verrall (2004) and Verrall (2007).

- This formulation of the ODNB is similar to the more usual CL version of the ODNB model, defined in Section 4.7.10, which has a mean of

$$\mathrm{E}[X_{i,j}] = (\lambda_j - 1)C_{i,j-1} = (\lambda_j - 1) \sum_{m=0}^{j-1} X_{i,m}.$$

This is therefore recursive across the columns (i.e. the j parameters), compared to the reformulation above which is recursive across the rows (i.e. in the i parameters).

- This particular ODNB formulation is chosen, since like the ODP model it produces reserve estimates that are the same as the deterministic CL method, but it is also more intuitively related to the BF method than the ODP formulation, because the prior distribution of the row parameters does not affect the estimation of the column parameters. This is explained further in Verrall (2004) and Verrall (2007).
- As for the classical ODP or ODNB model, in practice the usual tests as to whether this is a suitable model for the data would need to be carried out, but are omitted here to help focus on the overall procedure.
- The prior information is introduced by assuming that the row parameters, x_i follow independent Gamma distributions with parameters α_i and β_i. The mean and variance of the Gamma distribution are α_i/β_i and α_i/β_i^2 respectively. This can rewritten by defining the mean as M_i, so that the variance is M_i/β_i. Thus for a given mean, the variance can be changed by using different β_i, such that a higher value will produce a lower variance and vice-versa. In other words, the higher the value of β_i the more confidence there is

deemed to be in the prior row parameter assumptions. These are transformed into the γ_i parameters, for use in the model, as referred to above.

Bayesian methodology is used for the row and column model parameters, rather than maximum likelihood, as per the classical ODNB model described in Section 4.7.10. However, the scale parameter, ϕ, is the same as that derived from the GLM approach. A full Bayesian approach could include assuming a distribution for this parameter and then estimating it along with the other parameters, but for simplicity, it is kept the same as for the GLM approach. The numerical example follows that given in Verrall (2004) and Verrall (2007), and is implemented in the WinBUGS software. For illustration purposes, the same Taylor and Ashe (1983) data is used as for the other worked examples of the stochastic methods in this chapter.

Varying the row parameters can be thought of as a means of implementing a Bornhuetter–Ferguson (BF) method, as per Section 3.4, but within the context of a stochastic model. By changing the variability around the prior assumptions for the row parameters, the example demonstrates how the results change between a Chain Ladder method and a BF method, according to whether a high or low variance respectively is associated with the row parameters.

Three versions of the model are implemented, with specifications summarised in Table 4.49.

Table 4.49 *Bayesian MCMC example: Model specifications*

Model no.	Description	Variance of priors
1	Bayesian Chain Ladder	Very high
2	Bayesian BF All cohorts	Very low
3	Bayesian BF All cohorts	Medium

The assumptions and estimated reserves using the deterministic CL and BF methods are given in Table 4.50. The Bayesian model results are summarised in Table 4.51, and in Table 4.52 the values of the Total reserves at selected percentiles are shown, together with the ratio to the mean Total reserve. All figures are in thousands unless stated otherwise.

The results were produced using the R software, from which it is possible to call Win-BUGS directly; this approach makes it relatively straightforward to produce MCMC results using different model structures and assumptions. The WinBUGS model code follows that given in the Appendix of Verrall (2007). The R code (incorporating the underlying Win-BUGS model code) is too long to include in Appendix B, but is available from the author on request. Ten thousand simulations were used (half of which were discarded as part of the "burn-in" procedure in WinBUGS). In practice, more simulations may be necessary, but this number is sufficient for this purely illustrative example.

An explanation of the columns in Tables 4.50 and 4.51 follows:

- The "Latest" column in Table 4.50 is the claims paid to date in the Taylor and Ashe (1983) data triangle.

Table 4.50 *Bayesian MCMC example: ODNB assumptions*

Cohort	Latest	Assumptions		Reserve	
		% dev	BF prior	CL	BF
1	5,339	98	5,500	95	96
2	4,909	91	5,500	470	480
3	4,588	87	5,500	710	737
4	3,873	80	5,500	985	1,115
5	3,692	72	5,500	1,419	1,527
6	3,483	62	6,000	2,178	2,308
7	2,864	42	6,000	3,920	3,467
8	1,363	24	6,000	4,279	4,550
9	344	7	6,000	4,626	5,585
Total	30,457	N/A	N/A	18,681	19,865

Table 4.51 *Bayesian MCMC example: ODNB results summary*

Cohort	Model 1		Model 2		Model 3	
	Reserve	CV%	Reserve	CV%	Reserve	CV%
1	96	116	94	115	96	120
2	472	46	480	44	468	47
3	708	37	735	34	717	37
4	990	31	1,115	27	993	31
5	1,427	26	1,519	22	1,436	27
6	2,192	23	2,310	18	2,208	22
7	3,932	20	3,466	14	3,832	19
8	4,302	24	4,538	12	4,413	20
9	4,762	41	5,581	11	5,398	20
Total	18,881	16	19,837	9	19,561	12

- The "% dev" column is the standard chain ladder percentage developed values for this data triangle, as shown in the worked example of the Mack method in Table 4.6.
- The BF priors are selected example values, which are the same as in the example in Verrall (2007), so that the results can be compared.
- The deterministic reserve columns show the reserve derived using the standard chain ladder and BF methods. For example, for cohort 4, the CL reserve is $3,873,311 * (1/79.73\% - 1) = 985k$ and the BF reserve is $5,500,000 * (1 - 0.7973) = 1,115k$. As expected, the CL results are the same as those in the "IBNR" column of Table 4.4 for Mack's method example, as the IBNR is the same as the reserve in this example.
- In this example, the BF method produces a higher reserve than the CL method for all cohorts except cohort 7. This is a function of the fact that for this cohort, the BF prior (of 6m) is lower than the CL ultimate (of 6.8m, derived from 2864k/42.22%), whereas for all other cohorts the BF prior is higher. This is as per the discussion of the deterministic BF method in Section 3.4.1.

- For each model, two columns are shown in Table 4.51 – the mean reserve derived from the MCMC results and the Coefficient of Variation (CV) of the reserve. This will incorporate both estimation and process error.

Table 4.52 *Bayesian MCMC example: ODNB percentile results summary*

	Model 1	Ratio	Model 2	Ratio	Model 3	Ratio
Mean	18,881		19,837		19,561	
50%	18,670	99%	19,760	100%	19,385	99%
75%	20,670	109%	21,030	106%	21,040	108%
90%	22,750	120%	22,230	112%	22,641	116%
95%	24,181	128%	22,950	116%	23,590	121%
99.50%	28,210	149%	25,221	127%	26,340	135%

One of the advantages of Bayesian MCMC methods over some other analytical stochastic methods is that, like the bootstrap method (e.g. applied to the ODP and Mack models), they can be used to estimate the full distribution of future claims. Table 4.52 shows a range of percentile results for the three different models in this numerical example.

The results for each model are now discussed briefly.

Model 1 – Bayesian ODNB Chain Ladder

For Model 1, high variances (with a standard deviation of value 10m) are assigned to the priors, which is designed to replicate the chain ladder method, where no weight is assigned to the priors. The reserve for Model 1 (of 18,881k) is very close to the CL reserve, suggesting that this approach is indeed replicating the chain ladder method. Any difference is a function of the simulation used within the MCMC process.

The CVs by cohort and in total for Model 1 are very similar to those for the ODP Model, as shown in Table 4.29 and for the ODP Bootstrap model (both with a constant scale parameter), as shown in Table 4.42. This is as expected, since, as explained in Section 4.7.10, the ODNB model will produce the same prediction error as the ODP model. The overall shape of the distribution of the Total reserves is also very similar to the ODP Bootstrap results (with a constant scale parameter), as can be seen by comparing the percentile ratios for Model 1 in Table 4.52 with those in Table 4.43, which are virtually identical. Thus, Model 1 is replicating the ODP Bootstrap stochastic chain ladder results. It is included to allow comparison of the results with the other Bayesian models. The results are very similar to the chain ladder MCMC results in Table 2 of Verrall (2007).

Model 2 – Bayesian ODNB BF Low Variance

For Model 2, very low variances (with a standard deviation of value 1) are assigned to the priors, which is designed to replicate the BF method, but within the context of a stochastic model. This is borne out by the fact that the reserve for Model 2 is close to the deterministic BF reserve. The CV is, as expected, markedly lower than for the CL method, where higher variances are assigned to the priors. The results are very similar to the Negative Binomial model results in Table 7 of Verrall (2007). In practice, it would be unusual to have such confidence in the priors, so an alternative variance assumption is tested in the next model.

Model 3 – Bayesian ODNB BF Medium Variance

Model 3 is the same as Model 2, except that the variance is set at a potentially more realistic level, with a standard deviation value of 1m (i.e. about 20%) of the priors. The reserve for Model 3 is lower than the deterministic BF reserve and the Model 2 reserve, but it is still closer to the BF method than the CL method. The CV is higher than Model 2, which is expected given the higher variance assigned to the priors, but still reasonably far below that for the Chain Ladder results of Model 1. The results are very similar to the Negative Binomial model results in Table 8 of Verrall (2007).

This example demonstrates the material impact that changing the assumptions around the priors and their associated variance can have on the results for this particular Bayesian model. It might be quite reasonable for the overall variability around the estimated reserves to be lower for a model that incorporates prior assumptions, compared to a Chain Ladder method, particularly for more recent cohorts where the reserves under a CL approach can sometimes be very unreliable. However, if such a model were being used in practice, then judgement would be used in relation to the prior (and other) assumptions, for example to avoid assigning undue confidence (i.e. low variance) to the priors, which could lead to potentially unrealistically low variability around the reserves.

This particular Bayesian model does have a number of limitations, such as having no tail factor functionality and similar positivity constraints as the ODP model. However, the model was chosen to illustrate how Bayesian methodology can be used to produce similar types of results to the more commonly used bootstrap approaches. Other model specifications and types of Bayesian models have been developed (as referred to in the next section) and others may be developed in future.

4.5.2 Further Details of Bayesian Stochastic Reserving Methods

This section has outlined one example of a Bayesian stochastic reserving model. There have been a number of other Bayesian models proposed in the literature. For example:

- Gisler and Wüthrich (2008) propose a Bayesian CL model, which they show is equivalent to the standard CL model, when non-informative priors are used. Bühlmann et al. (2009) uses a recursive representation of this model to develop results on a one-year basis (for example, as per the MWCL model, which is described in Section 4.6).

- Merz and Wüthrich (2015a) describe a non-informative prior gamma–gamma Bayesian CL model that they use to analyse the total run-off prediction uncertainty (as referred to in Section 4.6.5). This model is also explained in Merz and Wüthrich (2015b), where a worked example demonstrates that very similar results can be obtained for this Bayesian model (using non-informative priors) and Mack's method.

- England et al. (2012b) describes a Bayesian ODP model using uniform and gamma prior distributions for the row parameters. Links to the chain ladder and Bornhuetter–Ferguson method are explored. The models in the paper all use a constant scale parameter and assume no tail factor is required.

- Taylor (2013) describes Bayesian reserving models as a type of "random effects model" where some or all of the parameters are subject to a prior distribution. The paper suggests that most Bayesian random effects models in the literature have tended to implement such effects only for the row parameters. This paper outlines chain ladder type models that also allow for random effects on the column parameters. The relationship with Bayesian models in other papers, such as England et al. (2012b), is also discussed.

- Verrall et al. (2012) describes the use of a different form of simulation methodology known as "Reversible Jump Markov Chain Monte Carlo" (RJMCMC) methods. The methodology was applied to a model selection problem, where the model allows tail factors to be fitted as part of the process. The authors suggest that the RJMCMC approach has advantages over other approaches, in particular that it relies less on ad hoc procedures as for some other approaches. WinBUGS code is included as an appendix to the paper.

- Meyers (2015) first assesses the performance of two non-Bayesian methods, namely the Mack method and the ODP Bootstrap method, using the data referred to in Section 2.5.4, available in Meyers and Shi (2011b). These are found to have limitations, which the author seeks to address by suggesting a number of alternative Bayesian MCMC-type models. For use on Incurred data, these include a model referred to as the "Correlated Chain Ladder" (CCL) model, which allows for a particular form of dependency between cohorts to be incorporated. For Paid data, two models are suggested that allow for payment year trends – the Leveled Incremental Trend (LIT) model and the Correlated Incremental Trend (CIT) model. Finally, a third model for use on Paid data is suggested – the Changing Settlement Rate (CSR) model – which as its name implies is designed to allow for the possibility of a speeding up (or slowing down) of claims settlement rates. Meyers (2015) provides access to the underlying software and data that are used in all the examples given there. The Bayesian methods are developed in the R software, as used for the worked examples earlier in this section, but with JAGS (see Plummer (2003) or Plummer (2013)), rather than WinBUGS being used.

4.6 Merz–Wüthrich Method

4.6.1 Introduction

The subject of measuring reserve risk over a one-year time horizon was covered in the introduction to this chapter, in Section 4.1.6. It is explained there that this risk can be measured by estimating the uncertainty surrounding the so-called *Claims Development Result* or "CDR", which is the difference between an estimate of the undiscounted ultimate claims made now, and an estimate made in a year's time, taking into account the claims development and emergence of new information during the year. The CDR can therefore be thought of as the profit (or loss) in the reserves over a one-year horizon.

This section describes one approach to estimating reserve risk over a one-year time horizon, attributed to Michael Merz and Mario Wüthrich. The method is an analytic approach, and so does not rely upon simulation. Essentially, it uses the same assumptions as Mack's method, except that it considers uncertainty over a one-year period, whereas Mack's does so over the lifetime of the liabilities; in others words it is effectively the one-year equivalent of Mack's method.

The approach (which for convenience is referred to here as the Merz–Wüthrich Chain Ladder or "MWCL" method) is documented in Merz and Wüthrich (2007b)[43] and Merz and Wüthrich (2008a), and was initially motivated by the requirements of the Swiss Solvency Test ("SST") which has a one-year basis for reserving risk.

The MWCL method does not include the functionality required to include a tail factor and only produces an estimate of the uncertainty surrounding the CDR, but not a full distribution of the CDR. However, it is also possible to apply a bootstrap procedure to produce the full distribution, allowing for a tail factor if required, as discussed further in Section 4.8.2.

Despite the fact that the method does not produce a distribution around the CDR, and does not include tail factor functionality, it is a key method for estimating the one-year reserve risk, partly because it features as part of the Solvency II regulations. In particular, it was one of the methods used by a working group to evaluate the reserve risk parameters used in the Solvency II standard formula (as per EIOPA, 2011) and is also a component of some of the standardised methods cited in relation to determining "Undertaking Specific Parameters", which, under the regulations, can be used to adjust the standard formula (as explained in CEIOPS (2010)).Other approaches to estimating one-year reserve risk are summarised in Section 4.8, including those which allow a tail factor to be used and which produce a full distribution of the CDR.

The mathematical formulation of the MWCL method is given next in this section. For readers who do not wish to read the mathematical formulation, the worked example in Section 4.6.4 will help explain the method and how it can be implemented in practice. Although the mathematical formulation might appear complex to some readers, the worked example demonstrates that the method is in fact relatively straightforward to implement in practice (in a spreadsheet for instance). Its close relationship to the Mack method also means that it can be implemented within the same spreadsheet or programming framework as Mack's method.

Since the original formulation of the MWCL method, it has been extended to a multi-year time horizon, rather than just a single year. This work is discussed briefly at the end of this section, in Section 4.6.5.

---

[43] The authors have advised that the work underlying this paper is incomplete and was superseded by Merz and Wüthrich (2008a), which should therefore be regarded as the original reference source for this approach.

4.6.2 Mathematical Formulation of the MWCL Method

The MWCL method is based around modelling a claims paid data triangle where no tail factor is required, and is defined mathematically as follows. This section only provides a high-level summary of the key formulae in the MWCL method. Further details are given in Merz and Wüthrich (2008a) and Wüthrich et al. (2009).

- As for other methods, such as Mack, $C_{i,j}$ represents the cumulative claims payments for cohort i at development year j. i is valid over 0 to I and j over 0 to J. Thus, an $I \times J$ data triangle is being used, where $C_{i,J}$ represents the ultimate claims for cohort i.

- The reserve at time $t = I$ is denoted by

$$R_i^I = C_{i,J} - C_{i,I-i},$$

and at time $t = I + 1$ is denoted by

$$R_i^{I+1} = C_{i,J} - C_{i,I-i+1}.$$

- The claims data at time $t = I$ is denoted by

$$D_I = \{C_{i,j}; i + j \le I \text{ and } i \le I\},$$

and at time $t = I + 1$ by

$$D_{I+1} = \{C_{i,j}; i + j \le I + 1 \text{ and } i \le I\} = D_I \cup \{C_{i,I-i+1}; i \le I\}.$$

- The model assumptions are:[44]

  - Cumulative claims payments $C_{i,j}$ are independent across cohorts.
  - $(C_{i,j})_{j \ge 0}$ are Markov processes and there exist constants f_j, σ_j such that, for all $1 \le j \le J$ and $0 \le i \le I$,

$$E[C_{i,j}|C_{i,j-1}] = f_{j-1}C_{i,j-1},$$

$$\mathrm{Var}(C_{i,j}|C_{i,j-1}) = \sigma_{j-1}^2 C_{i,j-1}.$$

In Merz and Wüthrich (2008a) it is noted that these model assumptions are weaker than those used in Mack (1993a), specifically that the Markov process assumption is replaced in Mack's method by an assumption related only to the first two moments.

- As per Mack's method, the MWCL model uses chain ladder factors, defined, at time I, as follows:

$$\hat{f}_j^I = \frac{\displaystyle\sum_{i=0}^{I-j-1} C_{i,j+1}}{S_j^I}, \text{ where } S_j^I = \sum_{i=0}^{I-j-1} C_{i,j},$$

---

[44] Slightly different notation is used here compared to that in Merz and Wüthrich (2008a), for consistency with the notation used for other methods in this book, such as Mack's method.

and, at time $I + 1$ by

$$\hat{f}_j^{I+1} = \frac{\sum\limits_{i=0}^{I-j} C_{i,j+1}}{S_j^{I+1}}, \text{ where } S_j^{I+1} = \sum\limits_{i=0}^{I-j} C_{i,j}.$$

The only difference to the definition used for Mack's method is the use of the I and $I + 1$ superscript notation, since the model considers the variability in the $[I, I + 1]$ time interval.

- The variance parameters are estimated using the same formula as in Mack's method, so that

$$\hat{\sigma}_j^2 = \frac{1}{I - j - 1} \sum\limits_{i=0}^{I-j-1} C_{i,j}(F_{i,j} - \hat{f}_j)^2. \tag{4.116}$$

- Also as per the standard chain ladder model and Mack's method, $E[C_{i,j}|D_I]$ is estimated, given $C_{i,I-i}$, by

$$\hat{C}_{i,j}^I = C_{i,I-i}\hat{f}_{I-i}^I, \ldots, \hat{f}_{j-2}^I \hat{f}_{j-1}^I,$$

and $E[C_{i,j}|D_{I+1}]$ is estimated, given $C_{i,I-i+1}$, by

$$\hat{C}_{i,j}^{I+1} = C_{i,I-i+1}\hat{f}_{I-i+1}^{I+1}, \ldots, \hat{f}_{j-2}^{I+1} \hat{f}_{j-1}^{I+1}.$$

- Similarly,
  $E[R_i^I|D_I]$ is estimated, given $C_{i,I-i}$, by

$$\hat{R}_i^{D_I} = \hat{C}_{i,J}^I - C_{i,I-i},$$

and
$E[R_i^{I+1}|D_{I+1}]$ is estimated, given $C_{i,I-i+1}$, by

$$\hat{R}_i^{D_{I+1}} = \hat{C}_{i,J}^{I+1} - C_{i,I-i+1}.$$

- The CDR is defined as the difference between the claims reserve estimated at time I and that estimated at time $I + 1$, adjusted for the claims paid between time I and $I + 1$. Merz and Wüthrich (2008a) first define what is referred to as the "true" CDR for cohort i in accounting year I to $I + 1$ as

$$\mathrm{CDR}_i(I + 1) = \mathrm{E}\left[R_i^I|D_I\right] - \left(X_{i,I-i+1} + \mathrm{E}\left[R_i^{I+1}|D_{I+1}\right]\right)$$
$$= \mathrm{E}\left[C_{i,J}^I|D_I\right] - \mathrm{E}\left[C_{i,J}^I|D_{I+1}\right]. \tag{4.117}$$

$X_{i,I-i+1}$ denotes the incremental payments in accounting year I to $I + 1$, that is

$$X_{i,I-i+1} = C_{i,I-i+1} - C_{i,I-i}.$$

- However, the "true" CDR is unknown and needs to be estimated. This is achieved by replacing the expected ultimate claims $\mathrm{E}\left[C_{i,J}^I|D_I\right]$ and $\mathrm{E}\left[C_{i,J}^I|D_{I+1}\right]$ in (4.117) by their estimates $\hat{C}_{i,J}^{I}$ and $\hat{C}_{i,J}^{I+1}$, to produce an estimator for the true CDR, which Merz and Wüthrich (2008a) refer to as the "observable" CDR for cohort i, defined as follows:

$$\widehat{\text{CDR}}_i(I+1) = \hat{R}_i^{D_I} - \left(X_{i,I-i+1} + \hat{R}_i^{D_{I+1}}\right)$$
$$= \hat{C}_{i,J}^{I} - \hat{C}_{i,J}^{I+1}. \tag{4.118}$$

- The goal is to derive an estimate of the prediction error of the expected value of 0 for the observable CDR over the accounting period. In a similar way as when considering the prediction error over all future years (e.g. as in Mack's method), when doing so over a single year, the prediction error in the expected CDR of zero is measured using a Mean Squared Error of Prediction (MSEP), conditioned on the data triangle. Hence, in mathematical terms, for a single cohort, the goal becomes deriving an estimator for

$$\text{MSEP}_{\widehat{\text{CDR}}_i(I+1)|D_I}(0) = E\left[\left(\widehat{\text{CDR}}_i(I+1) - 0\right)^2 |D_I\right]. \tag{4.119}$$

- This estimator is denoted by

$$\widehat{\text{MSEP}}_{\widehat{\text{CDR}}_i(I+1)|D_I}(0). \tag{4.120}$$

- Merz and Wüthrich (2008a) show how this estimator can be broken down into process and estimation error as follows:

$$\widehat{\text{MSEP}}_{\widehat{\text{CDR}}_i(I+1)|D_I}(0)$$
$$= \underbrace{\widehat{\text{Var}}\left(\text{CDR}_i(I+1)|D_I\right)}_{\text{Process error}} + \underbrace{\widehat{\text{MSEP}}_{\widehat{\text{CDR}}_i(I+1)|D_I}\left(\widehat{\text{CDR}}_i(I+1)\right)}_{\text{Estimation error}}. \tag{4.121}$$

Note that this equality only holds because the underlying derivation[45] involves a linear approximation.

- The same source derives a formula for this:

$$\widehat{\text{MSEP}}_{\widehat{\text{CDR}}_i(I+1)|D_I}(0) = \underbrace{\hat{C}_{i,J}^2 \frac{\hat{\sigma}_{I-i}^2/(\hat{f}_{I-i}^I)^2}{C_{i,I-i}}}_{\text{Process error}}$$
$$+ \underbrace{\hat{C}_{i,J}^2\left[\frac{\hat{\sigma}_{I-i}^2/(\hat{f}_{I-i}^I)^2}{S_{I-i}^I} + \sum_{j=I-i+1}^{J-1} \frac{C_{I-j,j}}{S_j^{I+1}} \frac{\hat{\sigma}_j^2/(\hat{f}_j^I)^2}{S_j^I}\right]}_{\text{Estimation error}}. \tag{4.122}$$

---

[45] See Merz and Wüthrich (2008a) and Wüthrich et al. (2009) for further details.

- For the MSEP across aggregated cohorts, Merz and Wüthrich (2008a) show that this can be estimated using

$$\widehat{\text{MSEP}}_{\sum\limits_{i=1}^{I} \widehat{\text{CDR}}_i(I+1)|D_I}(0) = \sum_{i=1}^{I} \widehat{\text{MSEP}}_{\widehat{\text{CDR}}_i(I+1)|D_I}(0)$$

$$+ 2 \sum_{1 \leq i < k \leq I} \hat{C}_{i,J}^I \hat{C}_{k,J}^I \left[\frac{\hat{\sigma}_{I-i}^2/(\hat{f}_{I-i}^I)^2}{S_{I-i}^I} + \sum_{j=I-i+1}^{J-1} \frac{C_{I-j,j}}{S_j^{I+1}} \frac{\hat{\sigma}_j^2/(\hat{f}_j^I)^2}{S_j^I} \right]. \qquad (4.123)$$

4.6.3 Mathematical Relationship Between MWCL Method and Mack's Method

The MWCL method is relatively straightforward to implement in practice, particularly if the mathematical relationship between it and Mack's method is understood. Hence, before providing the worked example, the mathematical relationship between the two methods is explained, bearing in mind that Mack's method estimates the uncertainty in the total estimated future claims to ultimate (i.e. in the reserve), whereas the MWCL method estimates the uncertainty of the CDR over a one-year time horizon. This follows the explanation given in Merz and Wüthrich (2008a), in which it is noted that the relationship only holds where the MWCL formula uses a linear approximation, as referred to immediately after (4.121).

First, consider the MSEP for a single cohort. Table 4.53 shows a comparison of the relevant formula for the process and estimation error components of the MSEP.

Table 4.53 *Comparison of MSEP formula for Mack and MWCL methods: Single cohort*

Error	Mack(4.18)	MWCL(4.122)
Process	$\hat{C}_{i,J}^2 \sum\limits_{j=I-i}^{J-1} \frac{\hat{\sigma}_j^2/(\hat{f}_j^I)^2}{\hat{C}_{i,j}}$	$\hat{C}_{i,J}^2 \frac{\hat{\sigma}_{I-i}^2/(\hat{f}_{I-i}^I)^2}{\hat{C}_{i,I-i}}$
Estimation	$\hat{C}_{i,J}^2 \sum\limits_{j=I-i}^{J-1} \frac{\hat{\sigma}_j^2/(\hat{f}_j^I)^2}{S_j^I}$	$\hat{C}_{i,J}^2 \left[\frac{\hat{\sigma}_{I-i}^2/(\hat{f}_{I-i}^I)^2}{S_{I-i}^I} + \sum\limits_{j=I-i+1}^{J-1} \frac{C_{I-j,j}}{S_j^{I+1}} \frac{\hat{\sigma}_j^2/(\hat{f}_j^I)^2}{S_j^I} \right]$

Hence, it can be seen that for the process error, the MWCL method just has the first term of the corresponding Mack formula, that is at $j = I - i$. For the estimation error, the first term in the MWCL formula is the same as that in the Mack formula, but all subsequent terms are scaled by the factor $C_{I-j,j}/S_j^{I+1}$ (which will be less than 1).

For the aggregated cohorts, the process error will be the sum of the process error for the individual cohorts. For the estimation error, it is only necessary to compare the covariance terms, which form part of the estimation error, in (4.20) and (4.123), as shown in Table 4.54.

Table 4.54 *Comparison of MSEP formula for Mack and MWCL methods: Aggregated cohorts – Covariance term*

Method	Covariance term
Mack (4.20)	$2 \sum\limits_{1 \leq i < k \leq I} \hat{C}_{i,J} \hat{C}_{k,J} \sum\limits_{j=I-i}^{J-1} \frac{\hat{\sigma}_j^2}{\hat{f}_j^2} \frac{1}{S_j^I}$
MWCL (4.123)	$2 \sum\limits_{1 \leq i < k \leq I} \hat{C}_{i,J}^I \hat{C}_{k,J}^I \left[\frac{\hat{\sigma}_{I-i}^2 / (\hat{f}_{I-i}^I)^2}{S_{I-i}^I} + \sum\limits_{j=I-i+1}^{J-1} \frac{C_{I-j,j}}{S_j^{I+1}} \frac{\hat{\sigma}_j^2 / (\hat{f}_j^I)^2}{S_j^I} \right]$

Hence, as for the estimation error for a single cohort, the difference between the two formulae is that the first term in the MWCL formula is the same as that in the Mack formula, but all subsequent terms are scaled by the factor $C_{I-j,j}/S_j^{I+1}$. Using these relationships between the two methods, the worked example in the next section demonstrates how results using the MWCL method can be produced by adapting the calculations used to implement Mack's method.

4.6.4 Worked Example of MWCL Method

To enable implementation of the calculations using the relationship between Mack's method and the MWCL method, this worked example involves the application of the method to the Taylor and Ashe (1983) dataset. It should be noted that this example of the MWCL method involves the linear approximation version of the MWCL method, as referred to immediately after (4.121).

Starting with the process error, the calculations and results are shown in Tables 4.55 and 4.56.

As per the first comparison table in Section 4.6.3, the MWCL process error includes the first diagonal (at index value $I - i$) from the Mack process error. Hence, the diagonal in Table 4.55 is the same as the first diagonal (below the stepped line) of the corresponding Mack process error results given in Table 4.10. The calculations shown in Table 4.56 are set out in the same way as those for Mack's method, with the "Sum" column in the MWCL table being the diagonal only. This produces a process error of 1,335,912 (or 1,336k as shown in the table) across all cohorts combined.

For the estimation error, it is first necessary to calculate the scaling factor, $C_{I-j,j}/S_j^{I+1}$, as referred to in Section 4.6.3. This is shown in Table 4.57. The items in the first row, labelled "Col Sum" are the denominators of this factor (i.e. S_j^{I+1}) and are equal to the sum of the items in the relevant column in the example triangle. The items in the second row, labelled "Diag" are the numerators (i.e. $C_{I-j,j}$) and are equal to the relevant diagonal entry in the triangle.

Table 4.55 *Merz–Wüthrich method worked example: Process error results (×1,000) – Part 1*

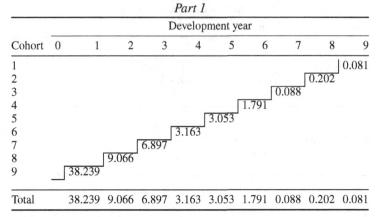

Cohort	0	1	2	3	4	5	6	7	8	9
1										0.081
2									0.202	
3								0.088		
4							1.791			
5						3.053				
6					3.163					
7				6.897						
8			9.066							
9		38.239								
Total		38.239	9.066	6.897	3.163	3.053	1.791	0.088	0.202	0.081

Table 4.56 *Merz–Wüthrich method worked example: Process error results – Part 2*

Cohort	Sum	Ult($\hat{C}_{i,J}$)[a]	Product	ProcError[a]
1	8.076E–05	5,434	2.385E+09	49
2	2.017E–04	5,379	5.834E+09	76
3	8.764E–05	5,298	2.460E+09	50
4	1.791E–03	4,858	4.227E+10	206
5	3.053E–03	5,111	7.975E+10	282
6	3.163E–03	5,661	1.014E+11	318
7	6.897E–03	6,785	3.175E+11	563
8	9.066E–03	5,642	2.886E+11	537
9	3.824E–02	4,970	9.445E+11	972
Total	6.258E-02	N/A	1.785E+12	1,336

[a] Ultimate and Process Error values are in thousands. All other values are in units.

Table 4.57 *Merz–Wüthrich method worked example: Calculation of scaling factors*

Item	Development year								
	0	1	2	3	4	5	6	7	8
Col Sum	3,671	11,615	17,912	21,931	21,655	19,828	17,331	13,430	9,173
Diag	344	1,363	2,864	3,483	3,692	3,873	4,588	4,909	5,339
Scaling factor (%)	9.37	11.74	15.99	15.88	17.05	19.53	26.47	36.56	58.21

Then, the estimation error can be calculated, as per the first Mack/MWCL comparison table in Section 4.6.3. The calculations and results are given in Tables 4.58 and 4.59. So, the first diagonal in Table 4.58 is the same as the first diagonal in the corresponding estimation error table for Mack's method, as given in Table 4.12. All other diagonals are scaled by the factors in Table 4.57, remembering that it is necessary to refer to the previous column, as per the construction of the formulae.

For example, for cell (7, 5) it was shown in Section 4.2.3 that the value in Mack's method example is 0.000627. For the MWCL method, this is multiplied by the scaling factor in

Table 4.58 *Merz–Wüthrich method worked example: Calculation of estimation error for single cohorts (×1,000) – Part 1*

Cohort	0	1	2	3	4	5	6	7	8	9
					Development year					
1										0.112
2									0.116	0.065
3								0.116	0.042	0.065
4							0.435	0.008	0.042	0.065
5						0.627	0.085	0.008	0.042	0.065
6					0.597	0.107	0.085	0.008	0.042	0.065
7				1.313	0.095	0.107	0.085	0.008	0.042	0.065
8			1.206	0.210	0.095	0.107	0.085	0.008	0.042	0.065
9		3.953	0.142	0.210	0.095	0.107	0.085	0.008	0.042	0.065

Table 4.59 *Merz–Wüthrich method worked example: Calculation of estimation error for single cohorts ('000) – Part 2*

Cohort	Sum[a]	Ult($\hat{C}_{i,J}$)	Product	EstError[a]
1	1.125E–04	5,434	3.321E+09	58
2	1.817E–04	5,379	5.256E+09	72
3	1.395E–04	5,298	3.916E+09	63
4	5.511E–04	4,858	1.301E+10	114
5	8.286E–04	5,111	2.165E+10	147
6	9.055E–04	5,661	2.902E+10	170
7	1.716E–03	6,785	7.899E+10	281
8	1.819E–03	5,642	5.790E+10	241
9	4.708E–03	4,970	1.163E+11	341

[a] Ultimate and Estimation Error values are in thousands. All other values are in units.

column 4 of Table 4.57, i.e. by 17%. This equals 0.00010695, or 0.107 when multiplied by 1,000, as shown in cell $(7, 5)$ of Table 4.58. The columns are then summed to produce the "Sum" column in Table 4.59, multiplied by the square of the ultimate (i.e. $\hat{C}_{i,J}^2$) and finally square roots are taken to get the estimation component of the prediction error[46] for each cohort.

In order to calculate the estimation error across all cohorts combined, the covariance term needs to be calculated. The calculations are set out in Table 4.60.

They are set out in the same way as for Mack's method, as in Table 4.14. Because the scaling factors, $C_{I-j,j}/S_j^{I+1}$, are inside the second (nested) summation of (4.123), rather than scale the Mack covariance values, it is easier just to use the sum of the scaled items

---

[46] Which is synonymous with the term "Standard Error" used in the context of the Mack method in Section 4.2.

in the calculation of the MWCL estimation error for the single cohorts, as given in Table 4.59. So, for example, the value of 23,896m in cell (5, 8) in Table 4.60 is calculated as

$$\hat{C}_{5,J}\hat{C}_{8,J}\left[\frac{\hat{\sigma}_{I-5}^2/(\hat{f}_{I-5}^I)^2}{S_{I-5}^I} + \sum_{j=I-5+1}^{J-1}\frac{C_{I-j,j}}{S_j^{I+1}}\frac{\hat{\sigma}_j^2/(\hat{f}_j^I)^2}{S_j^I}\right]$$
$$= 5,111\text{k} \times 5,642\text{k} \times 0.0008286.$$

The value of 0.0008286 derived from the formula in square brackets is taken from Row 5 of the "Sum" column in Table 4.59.

The covariance term in (4.123) is then calculated by summing the values in the triangle part of Table 4.60 and multiplying by two. Combining the square root of this (which is 1,025k) with the "ProcError" and "EstError" values in the final columns of Tables 4.56 and 4.59 respectively gives the results summary shown in Table 4.61.

The same results can be produced if the Taylor and Ashe (1983) data triangle is entered into the spreadsheet that accompanied Bühlmann et al. (2009). As explained in the next section, they can also be produced via the "CDR" function in the "ChainLadder" package. One further source which gives the same results as those shown in Table 4.61 is the R code produced by Arthur Charpentier in Charpentier (2012).

The values in the Prediction Error columns represent an estimate of the full prediction error of the Claims Development Result, split between process and estimation error. To compare the one-year CDR prediction error with the multiple year Mack prediction error, variances need to be used. The ratio of the relevant variances for each of the Process, Estimation and Total items to the equivalent Mack values are shown in the final three columns of Table 4.61. This ratio is one way of measuring how much of the reserve risk is expected to run-off over the first calendar year after the valuation date, which in this case is

Table 4.60 *Merz–Wüthrich example: Calculation of covariance term*

Cohort	0	1	2	3	4	5	6	7	8	9	Ult[a]
			Development year								
1		3,287	3,238	2,969	3,124	3,460	4,147	3,448	3,037	5,434	
2			5,177	4,747	4,994	5,531	6,630	5,513	4,856	5,379	
3				3,591	3,778	4,184	5,014	4,170	3,673	5,298	
4					13,684	15,156	18,165	15,106	13,306	4,858	
5						23,974	28,735	23,896	21,048	5,111	
6							34,777	28,921	25,474	5,661	
7								65,692	57,863	6,785	
8									50,999	5,642	
9										4,970	
Ult[a]		5,434	5,379	5,298	4,858	5,111	5,661	6,785	5,642	4,970	

[a] Ultimate values are in thousands. All other values are in millions.

Table 4.61 *Merz–Wüthrich example: Results summary ('000)*

Cohort	Prediction Error			As % of Mack Variance		
	Process	Estimation	Total	Process	Estimation	Total
1	49	58	76	100%	100%	100%
2	76	72	105	71%	79%	75%
3	50	63	80	23%	54%	36%
4	206	114	235	81%	79%	81%
5	282	147	318	59%	63%	60%
6	318	170	361	41%	47%	42%
7	563	281	630	51%	53%	52%
8	537	241	589	36%	41%	37%
9	972	341	1,030	57%	56%	57%
Cov		1,025			57%	
Total	1,336	1,175	1,779	51%	56%	53%

a little over 50%. The related topic of how the risk margin for insurance liabilities runs-off over time is discussed further in the next section.

4.6.5 Extension of MWCL Method to the Multi-year Time Horizon

Since it was originally formulated, the MWCL method has been extended to produce an estimate of the variability in the CDR for each future year, rather than just over the first future year. This allows the uncertainty over an ultimate time horizon to be broken down into its emergence over time. The details can be found in Merz and Wüthrich (2015a), which includes formulae to explain how the prediction error at ultimate, according to Mack's method, can be broken down into the prediction error in each future period, according to the MWCL method. As for the MWCL method, the results only apply where there is no tail factor and only in the context of the standard CL model.

The extended MWCL method has been implemented in the R software, via the "CDR" function in the "ChainLadder" package, and Merz and Wüthrich (2015a) gives an example of the relevant R code that can be used to produce results for the data shown in that source. For consistency with the numerical examples given elsewhere in this book, including that for the MWCL method over a single year given earlier in this section, Tables 4.62 and 4.63 show the results of applying the extended MWCL method to the Taylor and Ashe (1983) dataset. The relevant R code is given in Appendix B.3.4.

Key points to note regarding these results are as follows:

- The first CDR column shows the estimated Prediction Error of the CDR for the first future accounting period, and as expected is the same as the MWCL results shown in the Total column of Table 4.61 – i.e. a total of 1,779k across all cohorts combined.
- The second and subsequent CDR columns show the estimated Prediction Error CDR for the corresponding accounting periods derived using the extended MWCL method.

Table 4.62 *Extended MWCL method: Example for Taylor and Ashe (1983) dataset.*
All figures in '000

Cohort	Reserve	CDR Prediction Error for future accounting period no.									$\sqrt{\text{Sum}^2}$
		1	2	3	4	5	6	7	8	9	
1	95	76	0	0	0	0	0	0	0	0	76
2	470	105	61	0	0	0	0	0	0	0	122
3	710	80	91	56	0	0	0	0	0	0	134
4	985	235	61	82	51	0	0	0	0	0	261
5	1,419	318	234	58	82	52	0	0	0	0	411
6	2,178	361	329	243	59	86	54	0	0	0	558
7	3,920	630	391	359	266	64	94	60	0	0	875
8	4,279	589	555	345	318	237	57	84	53	0	971
9	4,626	1,030	539	511	317	294	219	52	77	49	1,363
Total	18,681	1,779	1,178	885	608	429	268	129	97	49	2,447
As % Mack		53%	23%	13%	6%	3%	1%	0%	0%	0%	

Table 4.63 *Extended MWCL method example. Risk margin run-off. All figures in millions*

	Risk measure									
	Stdev CDR			Var CDR			Reserves			
Cohort	Period	All	R/O	Period	All	R/O	Clm Period	Res start	All	R/O
1	1.78	5.42	100%	3,165	5,988	100%	5.23	18.68	55.92	100%
2	1.18	3.64	67%	1,387	2,824	47%	4.18	13.45	37.24	67%
3	0.89	2.46	45%	784	1,437	24%	3.13	9.27	23.78	43%
4	0.61	1.58	29%	369	653	11%	2.13	6.14	14.51	26%
5	0.43	0.97	18%	184	284	5%	1.56	4.02	8.37	15%
6	0.27	0.54	10%	72	100	2%	1.18	2.45	4.35	8%
7	0.13	0.27	5%	17	28	0%	0.74	1.28	1.89	3%
8	0.10	0.15	3%	9	12	0%	0.45	0.53	0.62	1%
9	0.05	0.05	1%	2	2	0%	0.09	0.09	0.09	0%

- The final column shows the square root of the sum of squares of the individual year CDRs. This is the same, by cohort and in total, as the results for the Mack worked example given in Section 4.2.3 – i.e. a total of 2,447k. Hence, as an example, for Cohort no. 3:[47]

$$\underbrace{79.846^2}_{\text{PE CDR(1)}} + \underbrace{91.093^2}_{\text{PE CDR(2)}} + \underbrace{56.232^2}_{\text{PE CDR(3)}} = 17,835$$

$$= \underbrace{133.549^2}_{\text{(Mack SE)}}.$$

[47] Values are shown in thousands to 3 d.p. here to demonstrate equality of the relevant values, but they are shown without d.p. in the table for ease of presentation.

In other words, the sum of the variances of the CDR for each future accounting year is equal to the variance of the reserve across all future accounting years. The final row of Table 4.62, labelled "As % Mack", shows the ratio of the square of the total CDR Prediction Error for each future accounting year, as a percentage of the Mack Variance. This shows how the reserve risk runs-off over time. Hence, in the first year, the value is 53%, as per the results for the MWCL worked example summarised in Table 4.61, with the percentage gradually decreasing thereafter.

Risk Margin Run-off

Merz and Wüthrich (2015a) also explains how the results from applying the extended MWCL approach can be used to derive the risk margin for the run-off of insurance liabilities, using a range of different approaches. In particular, three approaches are described, based on the cost of capital loading method using a standard deviation, variance and reserve-based risk measure. The results of applying these three different measures can then be used to show how a cost of capital risk margin runs-off over time. The risk measure is effectively equivalent to the shareholder capital on which dividends, representing the cost of capital, are payable. The method requires a provision (the risk margin) to be made now for these dividends in respect of all future time periods. The ratio of the provision at the second and subsequent future accounting periods to the provision at the start of the first accounting period can then be used to analyse the run-off of the provision over time, using different risk measures. Full details are given in Merz and Wüthrich (2015a).

As an illustration, continuing with the same Taylor and Ashe (1983) example described here, the results for all cohorts combined, as shown in Table 4.63 and Figure 4.4 can be derived. Points to note regarding these results are as follows:

- For the run-off pattern using the first risk measure, the standard deviation of the CDR (shown in the columns labelled "Stdev CDR" in the table), the CDR in the period is taken from the Total row in Table 4.62. This is then summed over all future periods to produce the next column in the table (labelled "All"). Finally, the run-off pattern ("R/O" in the table) is derived by dividing these values at each accounting period by the value at the start of all accounting periods. For example, at period 4, the value of 29% is 1.58/5.42.

- For the variance risk measure, shown in the next three columns of the table labelled "Var CDR", the standard deviations are squared and the run-off pattern derived in the same way as for first risk measure.

- For the reserve risk measure, shown in the final four columns of the table, the run-off of the claims over time according to the CL method is used to derive the reserves at the end of each future accounting period. The starting position is the total reserve of 18.68k, as shown in Table 4.62. Subsequent values are derived using the projected incremental claims, as derived for the deterministic CL method example and summarised in Table 3.21 in Chapter 3. Hence, from that table, the projected incremental claims in the first future period are 5,227 (shown as 5.23k in Table 4.63), which is subtracted from the

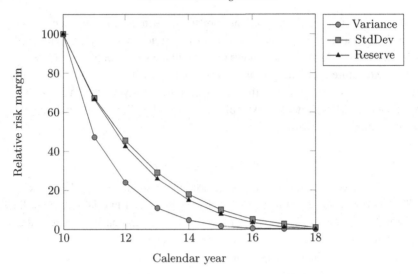

Figure 4.4 Extended MW method example: Risk margin run-off measures

starting reserve of 18.68k, to produce the reserves at the start of the next period of 13.45k, as shown in Table 4.63.

- These reserve values can then be summed across all future periods to give the values in the "All" column in the table, and the run-off pattern derived in a similar way as for the other two risk measures. This type of volume-based scaling is consistent with certain regulatory approximations, such as those described in a Solvency II context (e.g. see European Commission, 2010, paragraphs TP.5.32 and TP.5.41).
- Figure 4.4 summarises the three run-off patterns in graphical form. This suggests that, for this dataset, the regulatory approximation using a reserve risk measure produces marginally lower run-off risk margins compared to a standard deviation-based approach, whereas the variance measure is significantly lower. Although this comparison will vary with different types of business, Merz and Wüthrich (2015a) suggest that, since typically used risk measures such as value-at-risk and expected shortfall are often closer to the standard deviation-based risk measure than to the variance-based risk measure approach, that approach should be preferred.

This section provides only a brief summary of the MWCL method over a multi-year time horizon. Further details, including the relevant mathematical formulation, are given in Merz and Wüthrich (2015a).

It is also possible to use a simulation-type approach (e.g. bootstrapping) on a multi-year time horizon basis, as discussed in Section 4.8.2. Such an approach has the advantage that it can be used to produce a full distribution of the CDR for each future year.

Röhr (2016) describes an alternative approach to deriving the MSEP formulae in Mack's model and the MSEP formulae on a multi-year time horizon basis, using an error propagation formulation. An example is also given in Röhr (2016) using the same Taylor and Ashe (1983) dataset as in the worked example here; the results agree with those given in Table 4.62.

4.7 Other Stochastic Methods

This section provides a brief summary of a selection of other stochastic methods not described in detail elsewhere in this chapter. The first five sections include reference to some more recently developed methods, and the remaining sections cover some older methods. Some of the methods are not distinct in themselves, but are variations on others described elsewhere in this chapter. References are provided for sources giving further details if required. In the final subsection, references are also given for a number of further methods.

4.7.1 Individual Claim Models

This group of models, sometimes referred to as "micro-level" loss reserving, "triangle-free" reserving or "granular" models, build upon the criticism that can be levelled at the significant majority of other stochastic and deterministic reserving models, which use aggregated claims data – i.e. that they are ignoring potentially powerful information that might exist within individual claims records.

It was several years ago when the question was raised in England and Verrall (2002) regarding whether, with the increasing use of computer power, it would be better to examine this individual claims data, rather than aggregated data. However, for some reason, this type of model has not yet become popular, although there has been continued effort to develop such models. Perhaps the combination of even more computer power and increasingly reliable data records might promote a greater use of these models in future, even if it has not yet done so. Some examples of their use for third party motor bodily injury are given in Taylor and McGuire (2007) and McGuire (2007). Antonio and Plat (2014) provides several further references for individual claim models, including Taylor et al. (2008), Zhao et al. (2009) and Zhao and Zhou (2010). They also propose a further model, based on an approach referred to as "Position Dependent Marked Poisson Process", which is described in a number of other papers referenced in Antonio and Plat (2014). They show the results of applying their model to a large database of individual claims for european general liability business, and conclude that, for this dataset, their model outperforms a number of more traditional aggregate models, based on a comparison of results from an out-of-sample test applied to both types of model.

Further reference to "triangle-free" reserving using an individual claim type model can be found in Parodi (2013) and Parodi (2012). The approach described in these sources involves a separate model for frequency and severity using individual loss information, along with a simulation approach for the IBNR distribution. It is argued that this type of triangle-free reserving can lead to a more accurate estimate of the uncertainty surrounding a projection of ultimate losses. Other sources that discuss granular reserving include Marcuson et al. (2014), Lee (2010) and Pezzulli and Margetts (2008).

4.7.2 Stochastic Bornhuetter–Ferguson Methods

In Section 4.5, an example of a Bayesian method was described, which can be regarded as a stochastic version of the deterministic Bornhuetter–Ferguson (BF) method. It is also possible to construct a different type of stochastic BF model, using so-called "frequentist" statistical methodology, rather than Bayesian methodology. This involves formulating the model in a GLM-type framework and then using Maximum-Likelihood Estimation (MLE) techniques to derive the results, in a similar way as for other GLM-based stochastic reserving methods such as the ODP model.

The advantage of this approach over a Bayesian approach is that there is no need for specialist software (e.g. WinBUGS), and in fact, the results can be produced relatively easily in a spreadsheet. However, unlike the Bayesian BF approach, the Stochastic BF approach does not produce an estimate of the full distribution of the reserves and the associated cash flows.[48] Rather, it produces an estimate of the MSEP, split between three component parts – Process, Parameter and Prior Standard Error.

A number of sources have described a Stochastic BF approach, including Mack (2008), Merz and Wüthrich (2008c), Alai et al. (2009, 2011) and England et al. (2012b). The description here follows Alai et al. (2011), in which the implementation of the approach is simplified from that given in the previous paper by the same authors (i.e. Alai et al., 2009). The approach is abbreviated here by "ASBF".

The key features of the ASBF method are described briefly below. Mathematical details are available in Alai et al. (2011). The method is considerably more complex than its deterministic counterpart, but despite this, it is possible to implement the method in practice using a spreadsheet. For example, the approach used in the spreadsheet which accompanied Merz and Wüthrich (2008c), in relation to Example 6.26 of that reference, can be used for this purpose. Specialist reserving software that incorporates an ASBF-type method will of course make the practical application more straightforward.

- The incremental claims are assumed to be independent of each other and to follow an ODP distribution with a constant scale parameter. The parameters can be estimated, as per the ODP model, using the standard column sum CL method.
- It is assumed that there are prior estimates of the ultimate claims available for each cohort where the ASBF is to be applied and that these are independent of the incremental claims.
- The reserve is calculated in the same way as for the deterministic BF method.
- It is also assumed that there is an estimate of the variability of the prior estimates.
- The prior estimates do not have to be independent of each other, and if they are not, then assumptions related to their correlation are required as inputs.
- It is assumed that no tail factor is required, although it might be possible to adjust the approach to allow for this.

---

[48] Although a bootstrap procedure could perhaps be used to produce the full distribution of reserves. See, for example, Carrato et al. (2015) for some preliminary thoughts on this.

- Analytical formulae exist that allow an estimate of the variability around the BF reserve to be derived, both for individual cohorts and for cohorts in aggregate. The variability is measured by the MSEP, as for some other stochastic methods in this chapter.

- The formulae may appear somewhat complex at first sight, but can be implemented in a spreadsheet, as mentioned above. Since this is an analytical method, no simulation is required.

- The MSEP is made up of process error and parameter estimation error, split between that related to the ODP parameters and the prior parameters. If the prior parameters are assumed to be dependent across the cohorts, then the MSEP under the ASBF approach incorporates an additional term to allow for this.

This approach and the Bayesian BF-type approach described in Section 4.5 represent two alternatives for implementing the BF method in a stochastic model context. The third way is to introduce the BF method in a bootstrapping procedure. This is described in Section 4.4.5, in the subsection that deals with modifying the standard chain ladder approach in a bootstrapping procedure.

4.7.3 Double Chain Ladder

This relatively recently developed model, described in Miranda et al. (2012) and referred to as DCL here, is based on an extension of a model specified in Verrall et al. (2010), which was developed further in Miranda et al. (2011). This section provides a brief summary of the model. Further details can be found in the papers referred to here.

The model uses both paid claim amount and claim number/count data triangles, and hence is only applicable where both of these data types are available. The claim count data will need to be reasonably well defined, and the data considerations mentioned in the context of deterministic Frequency–Severity methods mentioned in Section 3.5.2 will also need to be borne in mind. In many cases, particularly for direct insurance, claim count data should be available, so this model could therefore have potentially wide application for this type of business.

One of the key features of this model is that it divides the reserves into two elements – that related to claims that have been reported but not settled (which, in the context of the DCL model, are referred to as RBNS) and that related to claims incurred but not reported, i.e. the IBNR. In addition, both reporting and settlement delay are explicitly included within the model structure. This means that any tail beyond the triangle is allowed for automatically, and is related to these delays, rather than being some form of extrapolation of the observed development factors, as is often the case with other methods.

Another key feature of the model is that it is based on consideration of the contribution that individual claims and their development make to the aggregate claim amount and count data triangles, but does not require actual individual claims data. Although methods that make use of individual claims data might have some appeal, as discussed in Section 4.7.13, in practice, such data may not always be available. Hence, the DCL model represents a possible alternative to using only claim amount data triangles, by first making use of claim count data triangles and also formulating the model so that it is built up from consideration of individual claims development. If parameterised in the appropriate way, it can be shown to produce results that are identical to the standard chain ladder, and in that sense it could be seen as another stochastic version of the deterministic chain ladder method.

The standard DCL model will produce both an estimate of reserves, and a Standard Error, but it is also possible to use bootstrapping to produce the full distribution of reserves.

As with all methods, the DCL model is based on certain assumptions. These include assuming that each individual claim has only one payment associated with it. Although this will often not be true in practice, the authors of Miranda et al. (2012) suggest that many claims will be dominated by one payment, and so believe this assumption is unlikely to be materially wrong.

As a relatively new method, it may take time for practitioners to experiment with the method on real data, but this is made relatively straightforward as it has been implemented in a package in the R software – see Miranda et al. (2013b) for further details and Miranda et al. (2013e) for a description of the package.

Further details, including the mathematics that define the model, are available in Miranda et al. (2012). The model is also discussed in two other papers – Miranda et al. (2013d) and Miranda et al. (2013c). The first of these considers how to incorporate prior knowledge related to both the number of zero-claims and the claims development inflation, and the second considers a modification to the DCL, using inflation derived from the incurred triangle in a way comparable with the deterministic Bornhuetter–Ferguson method. A further enhancement to the model, referred to as "RBNS preserving Double Chain Ladder", is described in Hiabu et al. (2016), which also discusses various other extensions to the DCL model.

4.7.4 Smoothing-type Models

As mentioned in Section 3.2, when using the deterministic Chain Ladder method, it can sometimes be helpful to fit some form of mathematical curve to the development ratios to smooth the pattern and/or extrapolate them to estimate a tail factor. When the goal is to derive a point estimate of the reserves, this smoothing process does not necessarily need to be done within the framework of a stochastic model. However, if a more formalised smoothing procedure is required, and where estimation of reserve variability is needed, then the practitioner can make use of one or more stochastic models that incorporate some element of smoothing to the parameter model fitting process. These models are usefully summarised in Björkwall et al. (2011), which refers to a range of existing stochastic smoothing-type models, and also proposes a further GLM-type smoothing model. The existing smoothing models referred to in this paper are as follows:

- Hoerl-curve and Lognormal type models (including Wright's model), as described in Sections 4.7.12 and 4.7.7 respectively, including use of the Kalman filter for smoothing the parameter estimates.

- The PTF-type models, as described in Section 4.7.8.
- Generalised additive models ("GAMs"). This family of models are similar to GLMs, except that the predictor (i.e. η in (4.129)) can include non-parametric smoothing functions for some covariates, rather than just linear parametric functions. Typical smoothing functions include cubic smoothing splines, locally weighted regression smoothers and kernel smoothers. The degree of smoothing in these models can be controlled via a parameter, with no smoothing at one extreme (i.e. the model fits exactly to the data-points) and full smoothing at the other extreme. Smoothing models can be constructed such that when there is no smoothing, the model is equivalent to the ODP GLM (and hence the Chain Ladder method) and when there is full smoothing for the development time parameters, the model is equivalent to the Hoerl curve. A number of papers consider these models in the context of claims reserving including Verrall (1996) and England and Verrall (2001). The latter describes a GAM framework that incorporates a number of well-known stochastic reserving models as special cases, as well as features that allow smoothing and extrapolation of factors.
- Generalised linear mixed models ("GLMMs"). Antonio and Beirlant (2008) define a semi-parametric regression model that has an interpretation as a GLMM. This paper uses a Bayesian approach to derive a full predictive distribution, and explains how a smoothing-type model can be particularly useful for large datasets where the development frequency is, for example, quarterly. An introduction to GLMMs in an actuarial context is given in Antonio and Beirlant (2007).

Although some of these have been described elsewhere in this chapter, it is useful to include them here so that they can be seen in the context of the family of smoothing-type models.

Björkwall et al. (2011) comment that two of the above models – GAMs and GLMMs – might be considered too sophisticated to become popular in reserving practice, and so suggest an alternative smoothing model, which is a form of log-linear GLM, incorporating a calendar period parameter. The specification of this model begins with a similar formulation to the GLMs described in Section 4.7.11, summarised mathematically in (4.130), except that there is an additional calendar-year term, so that

$$\eta_{i,j} = c + \alpha_i + \beta_j + \gamma_k.$$

These parameters are smoothed by introducing a matrix formulation of the above equation, together with new parameters that are related to the original parameters above, via smoothing curves. These curves incorporate parameters that can be used to control the degree of smoothing in the cohort, development and calendar time direction. Once the model is fitted to the data, bootstrapping can be used used to provide an estimate of the full distribution. Björkwall et al. (2011) show numerical results of the model when applied to the same Taylor and Ashe (1983) dataset used for other stochastic methods in this chapter. They also show how their model framework can incorporate a "normal" GLM and a Hoerl-curve model, which gives the flexibility to use these types of models if preferred.

4.7.5 Multivariate Stochastic Reserving Methods

This section provides a brief summary of a group of analytical stochastic methods that are designed to be used with multiple reserving categories simultaneously. These are usually referred to as multivariate methods and can be used to derive an estimate of the uncertainty of the aggregated reserves, across all categories combined. This is not achieved by applying a method to the aggregated data, but to the individual categories in a multivariate framework. They therefore represent an alternative to applying one or more stochastic methods independently to each reserving category and then combining them using the approaches outlined in Section 4.9.

The main advantage of using multivariate approaches is that they automatically allow for the correlations/dependency that may exist between the data triangles, whereas the methods for combining reserves across reserving categories described in Section 4.9 require assumptions to be selected, which can be very difficult in practice to determine, as discussed in Section 4.9.3.

Whether the implied correlation/dependency within a multivariate approach is appropriate for all pairs of reserving categories will need careful consideration before the results are used in practice, however, and in some cases the practitioner might prefer the flexibility of being able to directly influence the correlation/dependency between categories by using methods that require them to be input assumptions.

One limiting factor of multivariate methods is that the same reserving method is used for each reserving category, whereas in practice it may be appropriate to use more than one stochastic method, to suit the characteristics of each category, and then combine the results for each category. Most multivariate methods also assume that the data triangles for all reserving categories are of the same size, which may not be the case in some practical situations.

Where other approaches are used to produce the aggregate distribution of reserves (such as those described in Section 4.9), multivariate methods can be used as an alternative approach to provide comparative results. Used in this way, and even where they may have some limitations, such methods may still be preferable to deriving comparative results simply by applying a stochastic method to the aggregate data itself.

As well as being used to produce results for combinations of reserving categories, some multivariate stochastic methods can be used to produce results for triangles where the relationship between them is "structural", as is the case for paid and incurred triangles for a single reserving category. When used in this way, they are effectively stochastic alternatives to certain deterministic methods, such as the Munich Chain Ladder method, described in Section 3.8.1.

A number of different multivariate stochastic methods have been developed. This section describes briefly three methods, Multivariate Mack, Multivariate time-series model and a General Multivariate Chain Ladder model, which are all alternative types of analytic stochastic multivariate versions of the chain ladder method. They produce the mean and prediction error, rather than the full distribution, although this could be estimated as for the single triangle versions of the models. Simulation-type approaches can also be used to produce a full distribution of the aggregate reserves. These are described in Section 4.9, which considers the broader subject of combining results across different reserving categories.

Other multivariate approaches not discussed here include a multivariate additive loss reserving model and a combined multivariate chain ladder and additive model, both of which are described in Chapter 8 of Merz and Wüthrich (2008c). Multivariate methods are also discussed in Chapter 5 of Merz and Wüthrich (2015b).

Multivariate Mack Method

One of the first multivariate stochastic methods was that described in Braun (2004), which is a multivariate version of Mack's method. The approach is founded on the assumption that the correlation between two reserving categories can be captured through the correlation between the development factors at corresponding development points in each category. The correlation assumption can vary by development period, but not by cohort, as per the chain ladder approach, which Mack's method is based upon. This approach to correlation may not capture appropriately other correlation effects, such as those that are inflation-related and which therefore have a calendar period impact.

Braun (2004) develops recursive formulae for the prediction error, split between process and estimation error, for the sum of estimated ultimate claims for two data triangles, for individual cohorts and for all cohorts combined. These are closed-form analytical formulae comparable with the Mack method formulae for a single triangle, given in Section 4.2.

Additional details of the approach are given in Braun (2004), where a worked example is given using Reinsurance Association of America (RAA) data, which demonstrates that the approach is relatively straightforward to implement in practice, at least for the two triangle example shown in the paper.

Multivariate Time-series Model for Chain Ladder

This approach, described in Merz and Wüthrich (2007a), is a multivariate extension of the BBMW chain ladder time-series model referred to in Section 4.2.8. As mentioned in that section, the BBMW model is closely related to the Mack method, in that the estimation error formulae in the latter represents a linear approximation of the corresponding BBMW formulae. The BBMW formulae contain the additional cross-product term referred to in Section 4.2.8.

Merz and Wüthrich (2007a) provides closed-form formulae for the prediction error, split between process and estimation error, for the sum of the estimated ultimate claims for N correlated run-off triangles, for individual cohorts and for all cohorts combined. These formulae are similar to those in Braun (2004), except that additional cross-product terms arise, as per the univariate comparison of the Mack method with the BBMW method. There is one additional difference between the formulae in Braun (2004) and in Merz and Wüthrich (2007a), related to the definition of correlation. The multivariate time-series model will give a larger estimation error component than the Multivariate Mack method referred to in Braun (2004), although in Merz and Wüthrich (2007a) it is noted that for the practical examples examined by the authors, the differences were small, or even negligible. This is demonstrated for the worked example in Merz and Wüthrich (2007a), which uses the same data as in Braun (2004), and where the prediction Standard Error between the two approaches is virtually identical.

General Multivariate Chain Ladder Model

This method, abbreviated here by GMCL and described in Zhang (2010), is founded on an underlying multivariate structure for the chain ladder estimators, which is in contrast to the two previous methods which use univariate estimators. This structure was described in two earlier papers (Pröhl and Schmidt, 2005; Schmidt, 2006), which define a multivariate chain ladder model using matrix notation. It has also been used in another paper, Merz and Wüthrich (2008b), which specify formulae for the prediction error of the multivariate chain ladder model.

The GMCL model is, as the name implies, a more generalised form of the multivariate chain ladder model described in the earlier papers mentioned above. For example, it includes an intercept term, as with the ELTF models described in Barnett and Zehnwirth (2000) and referred to in Section 4.7.8. Hence, in the same way that the use of an intercept term can in some cases overcome the limitations of the standard chain ladder model, so can it in a multivariate context.

As for the deterministic Munich Chain Ladder method described in Section 3.8.1, the GMCL model also allows structural relationships between triangles (e.g. paid and incurred) to be modelled. The model uses a technique known as "Seemingly Unrelated Regressions" to estimate parameters, which facilitates flexible model structures (such as the use of an intercept term) as well as modelling of correlations between reserving categories. Full details are available in Zhang (2010), which includes a discussion of the relationship between the GMCL model and other models such as those described in Pröhl and Schmidt (2005) and Merz and Wüthrich (2008b).

For practitioners who wish to experiment with the GMCL model, it is worth noting that it has been implemented in the R software environment under the "ChainLadder" package, using the "MultiChainLadder" function. See Gesmann et al. (2014) for further details.

4.7.6 Poisson Model

This model is believed to have been first specified in Hachemeister and Stanard (1975) and was originally formulated for use with claim counts. It is defined in relatively straightforward mathematical terms, with the incremental claims, $X_{i,j}$ being assumed to be independent of each other and follow a Poisson distribution with mean $\alpha_i \beta_j$. The α_i parameters represent the row effects, and the β_j parameters represents the column effects and are specified as summing to 1. Hence, the expected ultimate claims for cohort i are represented by the α_i parameters, with β_j being the expected pattern of claims emergence across development time. The interpretation of the claims emergence pattern will depend upon how the data triangle is constructed, so that, for example, if the triangle represents the reported claim numbers by development period, then β_j represents the estimated number of claims reported in development period j.

The Poisson model has a cross-classified multiplicative structure, which results in the parameters having an intuitive meaning, as described above. However, as noted in Verrall (1991), if the model is re-expressed in additive form, by taking logarithms, then statistical analysis of the resulting additive model is more straightforward. The additive form of the model is a type of log-linear regression model, as described in Section 4.7.7.

Because the underlying distribution is Poissonian, the incremental claims must be positive, which makes it suitable for use with claim number triangles. Without modification, this restriction makes the model difficult to apply to claim amounts, where there is potential for negative incremental claim movements (e.g. as is the case with some incurred claims amount data). However, it is relatively straightforward to extend the model so that it can be used where some of the incremental values are negative. The technique used is called "quasi-likelihood" and is described in Renshaw and Verrall (1998).

Various sources have proved what is described in Merz and Wüthrich (2008c) as "the remarkable fact" that if the standard statistical fitting procedure of Maximum Likelihood Estimation (MLE) is used to fit the α_i and β_j parameters in the Poisson model, then the reserve derived using these parameters is identical to the standard column sum average Chain Ladder reserve estimates. Turning this the other way round, it means that using a standard column sum Chain Ladder method can be thought of as equivalent to fitting a cross-classified Poisson statistical model using the standard MLE fitting procedure. An example source which shows this equivalence is Verrall (2000a), which discusses two alternative MLE approaches – conditional and unconditional. The former assumes that the cumulative claims to date in the triangle are fixed, so that no other values could possibly have been observed. In contrast, the unconditional approach assumes that these values are treated as one possible realisation of the values that could have been obtained. The unconditional approach is argued in Verrall (2000b) as being more appropriate, and it will produce higher prediction errors than the conditional approach. Mack (1991) also shows the equivalence of the Poisson model to the Chain Ladder method, using an alternative approach referred to as the "method of marginal totals".

The Poisson model is not believed to be used for stochastic claims reserving without the use of an "over-dispersion" parameter, in which case it becomes the more often referenced ODP model, with a single scale parameter, as explained in detail in Section 4.3. By including this parameter, the additional uncertainty that tends to exist in claim amount data, compared to claim number data, will usually make the ODP model (and particularly bootstrapping thereof) more suitable for use with such data than the Poisson model.

4.7.7 Lognormal or "Log-linear Regression" Model

A version of this model, for incremental claims, was originally documented in Kremer (1982) and was one of the earliest examples of a stochastic model suggested for use in a

general insurance claims reserving context. As with many other analytic forms of stochastic
reserving method, it produces an estimate of the mean reserve and the variance, rather
than the full distribution. The Lognormal distribution is often used in general insurance
applications, due its skewness, and so in many ways it has an intuitive rationale for use
in stochastic claims reserving. This model is sometimes also referred to as "Log-linear
regression", as is the case in Christofides (1990).

The model involves taking logarithms of the incremental claims and assuming that these
follow a Normal distribution. This means that the model will not work when one or more of
the incremental claims are negative, although the negative values could be excluded from the
fitting process, which may not distort the results too much if the number of negative values
is small. The formulation of the model assumes that all cohorts have the same development
pattern, as per the deterministic chain ladder method. This model will produce estimated
reserves that are similar, but, not necessarily the same as the Chain Ladder method, and
sometimes materially different. It is difficult to explain the model further without reference
to the mathematical formulation of the model, which follows.

Mathematical Formulation of the Log-linear Regression Model

As per Section 2.5.6, the incremental claims for cohort i in development period j are
denoted by $X_{i,j}$. The model can have various forms, but an example of one particular
relatively simple form (as per England and Verrall, 1999) is as follows:

$$\log[X_{i,j}] = m_{i,j} + \epsilon_{i,j}$$
$$m_{i,j} = \eta_{i,j}$$
$$\eta_{i,j} = c + \alpha_i + \beta_j \text{ with } \alpha_0 = \beta_0 = 0, \tag{4.124}$$

where $\epsilon_{i,j}$ is assumed to be Normally distributed $N(0,\sigma^2)$, and so $\log[X_{i,j}]$ will be Normally
distributed $N(m_{i,j},\sigma^2)$. It is therefore similar in structure to the ODP model, except that a
Normal distribution is used rather than a Poisson distribution, and the logarithms of the
incremental claims are used as the "response" variables, whereas in the ODP model the
response variables are the incremental claims themselves. In addition, the link function in
the GLM is just an equality, rather than logarithmic. The parameters α_i and β_j represent
cohort and development period effects and, along with the c parameter can be estimated
using maximum likelihood, which because the error terms, $\epsilon_{i,j}$ are Normally distributed, is
equivalent to a least-squares approach – i.e. minimising the residual sum of squares. The
variance, σ^2, is estimated by dividing the residual sum of squares by the degrees of freedom
(i.e. the number of data points in the triangle less the number of parameters in the model).

Hence, the model assumes that the logarithm of incremental claims are Normally distributed with mean $m_{i,j}$ and variance σ^2. The model is formulated slightly differently in
Christofides (1990), but is essentially the same as that defined above. The model can be
fitted using the linear regression facility in standard statistical software, or, as Christofides
(1990) shows, in a spreadsheet. Once the model is fitted, it can be used to estimate the
variance of the individual future claim amounts and the sum of these amounts (i.e. the

reserve). Christofides (1990) shows how this can be done in a spreadsheet. It can also be done, for example, in the R software, as explained in Gesmann (2013a). The model will fail to work if there are negative incremental claims, although, as explained in Christofides (1990), the data could in theory be modified (e.g. by adding a constant to all the values) to remove the negatives.

Other forms of this type of model to that described above are possible, by selecting a different form for $\eta_{i,j}$, so that, for example, calendar period effects are included. England and Verrall (2002) explain how the Lognormal model can also be adapted to allow smoothing of the row parameters using the Kalman filter. Smoothing models are described further in Section 4.7.4.

4.7.8 Extended Link Ratio Family and Probabilistic Trend Family of Models

These two families of models, abbreviated to ELRF and PTF are described in Barnett and Zehnwirth (2000). The ELRF models are regression models which include factors to allow for the row and column effects of the data triangle. They build on the observation made in Section 3.2.6 that the deterministic chain ladder can be shown to be equivalent to a form of linear regression on each successive pair of columns in the data triangle.

The PTF models are quite different to the ELRF models. In particular, they are fitted to the logarithm of the incremental claims, and in that sense they may be regarded as broadly similar to the log-linear regression model outlined in Section 4.7.7, but with some important differences including the fact that they may include parameters to allow for calendar-period effects.

The two families of models are described in the same section here because in Barnett and Zehnwirth (2000) the ELRF models are described as a bridge between the standard chain ladder models and the PTF models. Accordingly, the ELRF models are first explained in more detail.

ELRF Models

The notation used here is consistent, as far as possible, with the rest of this book, which will mean that it is slightly different to that used in Barnett and Zehnwirth (2000). Under these models, the cumulative claims are modelled as

$$C_{i,j} = a_0 + a_1 i + bC_{i,j-1} + \epsilon_i. \tag{4.125}$$

This can be rewritten as

$$X_{i,j} + C_{i,j-1} = a_0 + a_1 i + bC_{i,j-1} + \epsilon_i, \tag{4.126}$$

where $X_{i,j}$ denotes the incremental claims. Then (4.126) can be rearranged to express the model in terms of incremental claims:

$$X_{i,j} = a_0 + a_1 i + (b-1)C_{i,j-1} + \epsilon_i, \tag{4.127}$$

where a_0 can be regarded as the intercept, a_1 as the trend and b as the slope or ratio between successive cumulative claims; ϵ_i is the error term, which has variance of the form $\sigma^2 C_{i,j-1}^{\delta}$. When $a_0 = a_1 = 0$ and $\delta = 1$ the model is a linear regression model with zero intercept, and hence is equivalent to the column sum CL model, as pointed out in Section 3.6.2. Barnett and Zehnwirth (2000) also identify other types of model in the ELRF family by using different constraints for the parameters.

In Section 3.2.6, it was noted that an advantage of expressing the chain ladder model in a regression framework is that it enables all the usual statistical diagnostics to be used to check the validity of the underlying assumptions, for example using residual plots. This advantage applies equally well to the ELRF models, with the additional benefit that the models represent a more flexible framework than the standard CL model. In Barnett and Zehnwirth (2000) it is suggested that, when applied in practice, it is often the case that the statistical diagnostics, including residual plots, imply that ELRF models (and hence the standard CL models) are not appropriate models, from a theoretical perspective, to use to estimate reserves or the variability around those reserves, and that if they are used then they could therefore produce misleading results. Whether this is true in any individual case will depend on the particular dataset being considered. Furthermore, if the practitioner determines that a CL or ELRF-type model is potentially inappropriate, they might consider whether or not adjustments can be made to the standard procedure or whether judgement can be used to adjust the results to overcome at least some of the limitations of the model. The general topic of testing whether a particular deterministic or stochastic model is appropriate is discussed further in Sections 5.5 and 4.10.3 respectively, and applies equally well to the testing of ELRF models. Where the practitioner determines that a CL or ELRF-type model is inappropriate, then an alternative approach is needed, and Barnett and Zehnwirth (2000) suggest using a PTF model.

PTF Models

This family of models, originally described in Zehnwirth (1994), are fitted to the logarithm of the incremental claims data and include parameters to allow for cohort, development and calendar period effects. Mathematically, the PTF models are expressed as follows:

$$\log[X_{i,j}] = \alpha_i + \sum_{k=1}^{j} \gamma_k + \sum_{t=1}^{i+j} \iota_t + \epsilon_{i,j}. \tag{4.128}$$

The α parameters are related to cohort trends, the γ parameters to development period trends and the ι parameters to calendar period trends. The ϵ parameters are the error terms from a Normal distribution with mean 0 and variance $\sigma_{i,j}^2$, which may vary by cohort, development and calendar period.

In summary, Lognormal distributions are fitted to the incremental claims in each cell of the historical triangle. In statistical modelling terminology, as for the log-linear regression model, in the PTF models the logarithms of the incremental claims are used as the "response" variables. Lognormal distributions and their correlations (arising through the estimated parameters) are also used for each future cell of the triangle. Predictive distributions for the reserves by cohort, calendar period and in total can then be produced using simulations from these correlated Lognormal distributions.

Since the models are fitted to the logarithm of incremental claims, they will not be suitable where there are negative values, without some adjustment.

Different versions of the model can be fitted if required for example, with parameters for one or more of the cohort, development and calendar period dimensions. Barnett and Zehnwirth (2000) and Zehnwirth (1994) discuss other variations on the standard model including using error terms with different variances and dealing with the potentially common situation where the cohort, development and/or calendar period model parameters are correlated with each other (termed "multi-collinearity"). Diagnostic statistical analysis, including residual plots, can be used to decide which model is most suitable in any particular case.

The PTF and ELRF models (and variations or related models) are implemented in at least one commercially available specialist reserving software product. A version of the PTF family of models can be implemented in the R software, as shown in Gesmann (2013b). This produces results very similar to those shown in Tables 3.1, 3,2 and 3.3 of Barnett and Zehnwirth (2000).

4.7.9 Clark's LDF Model

This model, originally described in Clark (2003), involves defining a specific functional form for the loss emergence, combining this with an estimate of ultimate loss and selecting parameters using maximum-likelihood. It enables estimates of the reserve and the variability around that reserve to be estimated and can be implemented in a spreadsheet relatively easily. It is included within the Casualty Actuarial Society's Exam 7 syllabus (May 2017). The key components of Clark's model are described below, in general rather than mathematical terms. For the mathematical details, see Clark (2003).

- First, an expected cumulative loss emergence pattern is defined, ranging from 0% to 100%. Clark (2003) suggests using either a loglogistic (i.e. an Inverse Power) or Weibull curve, but noting that other forms can also be used. This represents the proportion of claims that are paid or incurred up to a particular development point, with development being measured to the average accident date. The loss development factors (i.e. LDF) referred to in the title of this subsection are the inverse of this emergence pattern. An exposure adjustment factor is used for time points which are less than the full exposure period.

- The same pattern is assumed for all cohorts, and the model is designed to work when it is strictly increasing over time. If negative claims development is expected, then a different model will need to be used. The use of the emergence pattern in this way means that it is straightforward to implement the model where there are unevenly spaced evaluation dates.

- In order to select the best emergence pattern, it is first necessary to define how it can be used to estimate the incremental claims in each development period. The model uses one of two approaches – LDF or Cape Cod.

- The LDF approach introduces an additional parameter for each cohort – the estimated ultimate claims – and multiplies this by the expected emergence pattern in the relevant period to determine the estimated incremental claims. For an emergence pattern with two parameters, the number of parameters to be fitted to the data under Clark's LDF approach will be the number of cohorts plus two.

- The Cape Cod approach is similar to the standard deterministic Cape Cod method, as described in Section 3.6. An appropriate exposure base first needs to be determined, which should be the best available measure which is thought to be proportional to the estimated ultimate losses for each cohort. For example, the premiums adjusted for rates and/or claims inflation/trend could be used. The exposure base selected is not a parameter of the model, but a set of selected values that are input to the model. As per the deterministic Cape Cod method, a single Expected Loss Ratio is selected, which when multiplied by the selected exposure measure for each cohort produces the estimated ultimate losses for that cohort. This is then multiplied by the expected emergence pattern in the relevant period to derive the estimated incremental claims in each period. For an emergence pattern with two parameters, the number of parameters to be fitted to the data under Clark's Cape Cod approach will be three.

- The next step is to select the "best" parameters for each of the LDF and Cape Cod approaches. The fitting procedure suggested by Clark (2003) is the Maximum Likelihood approach.

- Clark (2003) implements this approach by assuming that the incremental losses in each development period follow a particular statistical distribution (the Over-Dispersed Poisson distribution) and that there is a constant ratio of variance to mean (represented by a scale parameter). Clark (2003) gives the MLE formulae for determining the fitted parameters using the loglogistic and Weibull emergence patterns.

- Once the fitted parameters have been determined, formulae given in Clark (2003) can be used to derive estimates of the variance of the reserves, broken down into process and parameter elements.

- The model assumes that the incremental claims in each development period for each cohort are independent and identically distributed. As for other stochastic methods, such an assumption can be tested, for example using residual plots. This and the assumption of a constant scale parameter, along with the mathematical approach used to estimate the variance of the reserves, are noted in Clark (2003) as all leading to the possibility that the model will understate the variability in the reserves. This obviously needs to be borne in mind when using the results from the model.

This model has been implemented in the "ChainLadder" package in the software environment, R, using the "ClarkLDF" and "ClarkCapeCod" functions. This enables Clark's LDF and Cape Cod results to be produced very quickly, along with associated residual graphs etc. Clark (2003) shows an example of the model, applied to the Taylor and Ashe (1983) data that has been used for the examples of many of the other stochastic methods described elsewhere in this book. Using the "ChainLadder" package in R, the results of applying a range of Clark models to this dataset are shown in Table 4.64. The relevant R code is in Appendix B.3.5.

Table 4.64 *Clark and other model results for Taylor and Ashe (1983) data*

Model	Reserve	SE	CV
Clark LDF – Loglogistic uncapped	35,580	6,777	19.0%
Clark LDF – Loglogistic capped	28,952	4,862	16.8%
Clark LDF – Weibull	21,170	3,865	18.3%
Clark – Cape Cod – Loglogistic capped	29,656	3,403	11.5%
Clark – Cape Cod – Weibull	22,188	2,704	12.2%
ODP no tail factor	18,681	2,946	15.8%
Mack with tail factor	20,219	2,584	12.8%

The table also shows the equivalent results for the ODP and Mack stochastic methods. The various Clark model results in this table are very similar to the corresponding results given in Clark (2003). A number of points in relation to the application of the model to this particular dataset can be noted:

- The loglogistic curve used with both the LDF and Cape Cod versions of the model appears to give very high reserve estimates, due to its "heavy tail". Even capping the curve at, say, 240 months, as mentioned in Clark (2003) still appears to produce much higher estimates, compared to the other stochastic methods.

- In contrast, if a Weibull curve is used, the reserves are much lower, reflecting the lighter tail of this curve. They are also much closer to the results for the two other stochastic methods.

- For both curve types, the CV is much higher for the LDF method than for the Cape Cod method. This is noted as a general observation in Clark (2003) and is a result of using the extra information within the Cape Cod method – i.e. the exposure measure.

- In this case, the Clark Cape Cod with a Weibull curve produces reserve and CV results that are reasonably close to one of the other Stochastic methods – the Mack method with a tail factor.

- In practice, as for any stochastic model, in deciding whether the Clark model is appropriate to use, the types of diagnostic tests referred to in Section 4.10.3 can be applied. For this dataset, the code in Appendix B.3.5 includes a function to produce the residual and QQ plots for each version of the Clark model.

Clark (2003) contains further details of the model, including mathematical formulae for deriving the fitted parameters.

4.7.10 Over-dispersed Negative Binomial Model

This is a form of GLM which, as for the ODP model, produces reserve estimates that are the same as the deterministic chain ladder method. It is believed to have been originally suggested in Verrall (2000a). It is abbreviated here by ODNB. Mathematically, it is expressed recursively, with the incremental claims $X_{i,j}$ having a mean of $(\lambda_j - 1)C_{i,j-1}$ and a variance of $\phi\lambda_j(\lambda_j - 1)C_{i,j-1}$, where λ_j can be regarded as equivalent to the chain ladder development factor (which is denoted by f_j in the description of the Mack method in Section 4.2, for example). It can also be expressed as a model of the cumulative claims, $C_{i,j}$ with mean $\lambda_j C_{i,j-1}$ and the same variance as for the incremental claims. When the mean is expressed recursively in this way, it may be thought of as a more intuitively logical stochastic representation of the chain ladder method than the ODP model.

If one or more of the λ_j values is less than one, the variance will be negative and hence the model does not work, but a Normal approximation can be used to overcome this if required, as explained in England and Verrall (2002), although they also question the appropriateness of this, given the usual preference for using skewed distributions in a general insurance context. As for the ODP model, over-dispersion is allowed for via the ϕ parameter. England and Verrall (2002) explain how the ODNB can be specified as a GLM and so fitted using suitable GLM software, and also how the process, estimation and hence prediction variance can be derived. The same source also explains that, in practice, although the split between process and estimation variance may be different to the ODP model, the overall prediction variance will be very similar[49] and hence in practice they suggest that it does not matter which model is fitted. At present, the ODNB is not thought to be widely used in practice. However, it does have a particular relevance in the context of Bayesian methods, as explained in Section 4.5.1, and it can be shown that the Normal approximation to the ODNB model is closely related to Mack's model, and will produce the same results.

4.7.11 Gamma and Other Related GLM Models

A generalised family of GLMs for use in stochastic reserving can be defined as follows:

$$E[X_{i,j}] = m_{i,j} \text{ and } \mathrm{Var}[X_{i,j}] = \frac{\phi_{i,j}V[m_{i,j}]}{w_{i,j}} \tag{4.129}$$

$$m_{i,j} = g^{-1}(\eta_{i,j}),$$

where $V[m_{i,j}]$ is the so-called variance function and $g()$ is the link-function, both of which can take various forms, depending on the chosen model structure.

One particular sub-group within this generalised family of GLMs is defined as follows:

$$E[X_{i,j}] = m_{i,j} \text{ and } \mathrm{Var}[X_{i,j}] = \phi E[X_{i,j}]^z = \phi m_{i,j}^z, \tag{4.130}$$

$$\log m_{i,j} = \eta_{i,j},$$

$$\eta_{i,j} = c + \alpha_i + \beta_j, \text{ where } \alpha_0 = \beta_0 = 0.$$

---

[49] In fact, theoretically the prediction variance will be identical, but the practical implementation may involve approximations, which may mean that they are slightly different. The theoretical equivalence arises because the ODNB model can be regarded as a recursive version of the ODP model.

Table 4.65 *Gamma and other GLM models example results for Taylor and Ashe (1983)*
data ('000)

Cohort	ODP Reserve	ODP CV%	Gamma Reserve	Gamma CV%	Gaussian Reserve	Gaussian CV%	Inverse Gaussian Reserve	Inverse Gaussian CV%
1	95	116.3	93	48.4	101	309.9	102	24.7
2	470	46.0	447	36.0	497	78.4	456	30.9
3	710	36.8	611	29.1	806	57.1	518	23.3
4	985	30.8	992	25.7	973	47.8	961	24.0
5	1,419	26.4	1,453	24.2	1,370	38.1	1,466	25.4
6	2,178	22.7	2,186	24.1	2,139	28.3	2,155	28.1
7	3,920	20.2	3,665	25.7	4,089	19.4	3,338	33.1
8	4,279	24.5	4,122	28.5	4,404	22.8	3,955	33.3
9	4,626	42.8	4,516	36.9	4,793	54.8	4,428	35.0
Total	18,681	15.8	18,086	14.9	19,173	21.9	17,378	15.9

As noted in Section 4.3.2, the ODP model is from this family of models, with $z = 1$. When $z = 0$ the model is a Gaussian model, when $z = 2$ it is a Gamma model and when $z = 3$ it is an Inverse Gaussian model. It is also possible to let z be fitted between a value of 1 and 2, in which case it is referred to as a Compound Poisson model. All of these models are sometimes referred to as being from the Tweedie family of GLMs. These models can be fitted to the data in a similar way as for the ODP model. For example, they can be fitted using the "glm.Reserve" function within the "ChainLadder" package in R, by setting the "var.power" parameter equal to the equivalent z value. If this is done for the Taylor and Ashe (1983) dataset, then the results for the ODP, Gamma, Gaussian and Inverse Gaussian models are as shown in Table 4.65. The relevant R code is in Appendix B.3.6.

Some of these models have been discussed in various papers. For example, England and Verrall (1999) provide the results of applying several stochastic methods to the Taylor and Ashe (1983) dataset, which allows a comparison to be made with the results here. This comparison shows that the results in Table 4.65 for the ODP and Gamma models are the same as for the "Poisson GLM" and "Gamma GLM" model results respectively in Tables 1 and 2 of England and Verrall (1999).

4.7.12 Wright's Model

This model, which was originally defined in Wright (1990), is based upon consideration of the claims payment process, using an underlying risk-theoretic basis that considers both the systematic and random components. This leads to a model that includes the following mathematical formulation, using notation consistent with the rest of this book, rather than with Wright (1990):

- The incremental paid claims, $X_{i,j}$ are assumed to be the sum of $N_{i,j}$ claims, each denoted by $X_{i,j}^{(k)}$, so that

$$X_{i,j} = \sum_{k=0}^{N_{i,j}} X_{i,j}^{(k)}.$$

- The claim numbers, $N_{i,j}$, are assumed to be independent Poisson distributed, with

$$E[N_{i,j}] = e_j a_j \kappa_i j^{A_i} \exp(-b_i j),$$

where κ, A and b are unknown constants that need to be estimated, and e_i is a known measure of exposure that is an input to the model.

- The individual claims are assumed to be independent random variables, with

$$E[X_{i,j}^{(k)}] = m_{i,j} = \exp(\delta(i+j))k j^\lambda,$$

and

$$\mathrm{Var}[X_{i,j}^{(k)}] = v(E[X_{i,j}^{(k)}])^2,$$

where k and λ need to be estimated and $\exp(\delta(i+j))$ is intended to represent claims inflation, with δ being the constant force of inflation and $(i+j)$ being calendar time. In the variance above, v is the constant of proportionality – in other words, the variance of the individual claims is assumed to be proportional to the mean squared.

- Combining the model for claim numbers and claim amounts gives

$$E[X_{i,j}] = e_i a_j \kappa_i j^{A_i} \exp(\delta(i+j))k j^\lambda,$$

and

$$\mathrm{Var}[X_{i,j}] = (1+v)\exp(\delta(i+j))k j^\lambda E[X_{i,j}].$$

- This model can be put into a GLM framework. Wright (1990) shows how the Kalman filter approach can be used to derive smoothed parameter estimates across the cohorts, as a means of avoiding over-parameterisation.
- Renshaw (1994a) showed an alternative formulation of Wright's model as a GLM, which is closer to that of other stochastic GLM-type models, such as the ODP model. This has

$$E[X_{i,j}] = \exp(u_{i,j} + c + \alpha_i + \beta_i \log(j) + \gamma_i j + \delta t),$$

and

$$\mathrm{Var}[X_{i,j}] = \phi_{i,j} E[X_{i,j}] = \phi_{i,j} m_{i,j}.$$

- This can be rewritten by defining

$$\log(E[X_{i,j}]) = \eta_{i,j},$$

where

$$\eta_{i,j} = u_{i,j} + c + \alpha_i + \beta_i \log(j) + \gamma_i j + \delta t.$$

This represents a GLM with a logarithm link function and variance which is proportional to the mean.

- England and Verrall (2002) note that, by ignoring the $u_{i,j}$ and δt terms, the above formulation of Wright's model is similar to another type of model, known as the Hoerl curve.

Wright has defined a further stochastic model, described in Wright (1992), which uses the concept of Operational time. It is explained briefly in Section 4.7.13.

4.7.13 Further Stochastic Methods

This section provides a brief reference to some other stochastic reserving methods, not described elsewhere in this chapter. Some of these are reasonably well developed, with documented description and analysis of the approach; others are more tentative or at an early stage of development. Where known, one or more suggested reference sources are provided, which can be used to obtain further details of these methods, if required. The first three methods share a common feature in that they all seek to model both paid and incurred claims data in a single model framework.

There are other existing stochastic methods not described here, and enhancements to existing methods or entirely new approaches are likely to be developed in future. The practitioner can use various sources to explore these other methods and to keep up to date, including those available online which may be updated when new approaches are developed. One such source is Merz and Wüthrich (2015b), referred to in the introduction to this chapter.

Munich Chain Ladder (MCL) method

A deterministic version of this method is described in Section 3.8.1, which seeks to take into account the correlation between paid and incurred claims development data. However, the original paper in which the method was suggested (Quarg and Mack, 2004) also included discussion of a stochastic formulation, based on a generalisation of Mack's model, although it did not go as far as deriving an estimate of the prediction error of the MCL method reserve estimates. Since then, further efforts have been made to formulate the MCL method in a stochastic model framework, including Merz and Wüthrich (2006) and Merz and Wüthrich (2014), so that prediction errors can be estimated. The latter describes the "Modified Munich Chain Ladder Method", which is designed to be fully consistent with the stochastic model assumptions set out in the paper. There is also reference in this paper to a further stochastic model, which, like the MCL model seeks to allow for the relationship between paid and incurred claims, but which is described as providing more stable results than the Modified MCL method (see the "PIC" model description below).

Paid-Incurred Chain ("PIC") method

This is a Bayesian-type method which was originally described in Merz and Wüthrich (2010) and was later enhanced in Happ and Wüthrich (2013) so that it respected dependence properties within the data. The PIC method is similar to the Modified MCL method in that it seeks to model both paid and incurred claims data simultaneously. The PIC model produces a single estimate of the claims reserve derived from this data, and also has the desirable feature that as well as allowing the prediction error to be derived analytically, the full predictive distribution can be derived using simulation. Developments of the PIC model include allowing for tail factors (see Merz and Wüthrich, 2012), and formulation within a one-year time horizon (see Happ et al., 2012). A version of the PIC method has been implemented in the "ChainLadder" package in R, using the "PaidIncurredChain" function.

Extended Complementary Loss Ratio ("ECLR") method

This analytical stochastic method is described in Dahms (2007) and, like the three previous approaches above, makes use of both paid and incurred claims data. It relies on the assumption that the reported claims (i.e. the case reserves, being the difference between incurred and paid claims) represent a good measure of the remaining exposure. It combines projections of both the reported and paid claims to produce a single reserve estimate, as well as an estimate of the prediction error. A further feature of this approach is that it can be used where the data triangles are in some way incomplete – for example, where the full development history is not available.

Chain Ladder Link Ratio Family ("CLFM") model

This model is described in Bardis et al. (2012) and is an extension to Mack's method and to the methods described in Murphy (1994). It is intended to provide a model which allows estimates of the MSEP to be derived when the Chain Ladder method is used with development factors other than just the column sum average or simple average. The CLFM model uses recursive formulae as for Mack's method. It can be implemented using the "ChainLadder" package in R, by first using the "CLFMdelta" function and then the "MackChainLadder" function.

Wright's operational time model

This model uses both claim amount and claim number triangles, and utilises the concept of Operational time. This concept was introduced in the context of deterministic Frequency-Severity methods in Section 3.5.4. An application of the concept in a stochastic model context is given in Wright (1992). In summary, the stochastic model postulates a relationship between the mean operational time and the mean claim size. Hence, the time dimension is no longer development time, as with most other stochastic methods, but operational time – i.e. the cumulative proportion of claims settled at a particular development time. The functional form of the model involves six parameters that need to be estimated using quasi-Maximum Likelihood (e.g. in GLM software). Inflation is introduced by using an exponential relationship between development time and operational time. As for many other stochastic methods, the model allows an estimate of the prediction error to be derived. Full details are given in Wright (1992), and a practical implementation is explained in Bain (2003).

Schnieper's model

This model, described in Schnieper (1991) was originally proposed for use in pricing casualty excess of loss reinsurance business, although that paper also made reference to its use in a reserving context. Later, two papers (Liu and Verrall 2009a, 2009b) built on the original work to put it in the context of a stochastic model, in both analytic and bootstrap form. Schnieper's model is formulated around dividing the total incremental incurred claims development in each period into two components – one relating to movements in claims reported prior to the period (i.e. IBNER) and the another to new claims reported in that period (i.e. IBNR). Merz and Wüthrich (2008c) state (in Section 10.2) that Schnieper's model was the basis for the popular Mack method, originally suggested in Mack (1993a). Formulae are given in Liu and Verrall (2009b) for the MSEP of the reserves under the Schnieper model, and a bootstrapping algorithm is given in Liu and Verrall (2009a) to enable the full predictive distribution to be derived.

Continuous Chain Ladder

This refers to the approaches outlined in Miranda et al. (2013a) that recognise the underlying triangular structure associated with aggregated claims data, but do so using the more granular individual claims data. Cohort and development periods are considered within this structure, arranged in a two-dimensional density; calendar periods are not considered. Building on the observation that a Chain Ladder model is an (inefficient) form of histogram estimator of this density, the Continuous Chain Ladder approach uses a kernel smoother as the estimator of the continuous bivariate density, which is assumed to have a multiplicative structure. Once the kernel density has been fitted to the data, it can then be used to estimate the future claims.

Age-period-cohort models

This refers to a family of models that are based on a type of model used in epidemiology and demography. They involve parameters in three directions in the triangle – for the cohort (α), development (β) and calendar period (γ) effects, and have a formulation written as

$$\mu_{i,j} = c + \alpha_i + \beta_j + \gamma_{i+j-1} + \delta,$$

where δ is an overall level parameter. They are broadly similar therefore to the PTF-type model described in Section 4.7.8 and to the generalised form of GLM described in Section 4.7.11, with the specific inclusion of calendar period, as well as cohort and development period effects in the model structure.

Kuang et al. (2008a, 2008b, 2011) consider these models, using a form where the logarithm of the incremental claims are assumed to be Normally distributed. These papers show results of applying the model to a dataset used in Barnett and Zehnwirth (2000), which describes another type of model that includes parameters in three directions – namely the PTF-type models, as summarised in Section 4.7.8. Brydon and Verrall (2009) also discuss this type of model, in order to study the allowance made for inflation in the standard Chain Ladder method.

4.8 Methods Using a One-year or Multi-period Time Horizon

4.8.1 Introduction

This section outlines approaches which have been developed to estimate reserve risk over a one-year time horizon. One of these approaches, the MWCL method, has already been discussed in detail in Section 4.6, and so this section focuses on other approaches. Alternatives to the extended version of the MWCL over a multi-period time horizon (as explained in Section 4.6.5) are also discussed in this section.

The subject of measuring reserve risk over a one-year time horizon was introduced at the start of this chapter, in Section 4.1.6, so that background is also not repeated here, other than to note for ease of reference the definition of the "Claims Development Result", or CDR. For an individual or fixed group of cohorts, the CDR over one year is the difference between the estimated undiscounted ultimate now and what it is one year from now. It can also be expressed as the difference between the reserve now (i.e. the "opening" reserve) and the reserve in a year's time (i.e. the "closing" reserve), less the estimated claims paid in the year.[50] The CDR is a prospective measure, in the sense that, like a reserve, it represents future values that are unknown and hence need to be estimated. It might also be termed the run-off result or the Prior-Year-Development (PYD), although these terms are effectively retrospective measures, made in relation to a prior period. Some authors (e.g. Robbin, 2012) have described the variablity in the CDR by reference to the variablity in the "retro" estimate of the reserve at the end of the year, where this is defined as the reserve at the end of year plus the estimated paid in the previous year. Whichever way it is defined, estimating the uncertainty in the CDR over the next year is the goal of applying stochastic methods on a one-year basis – ideally determining the full distribution of the CDR.

Implementation of these methods is a key input for determining the reserve risk component of capital requirements at a specified level of confidence, on a one-year basis. This section focuses on a description of the methods themselves, rather than a detailed discussion of their application in the context of determining capital requirements. In the description that follows, it is assumed that the reserves are established on a best estimate basis, so that the uncertainty in the CDR arises solely from the claims process, rather than as a result of any margin or deficit in the reserves. In reality, many companies do hold a margin above a best estimate, but this is effectively a buffer that can, depending on the regulatory provisions, be regarded as an offsetting item against any capital provision that is required to be held in respect of the uncertainty in the CDR (i.e. the reserve risk component).

Robbin (2012) identifies three drivers of uncertainty in the CDR, which are helpful in considering how this might be estimated:

- Inherent volatility of the ultimate claims. A best estimate can be derived of the ultimate at the start of the year, but the claims experience during the year may be inconsistent with this, due for example to random process error, systemic changes in the underlying claims process and estimation error in this estimate;

---

[50] Reserve now minus Reserve in a year's time minus claims paid in the year.

- The emergence of new information during the year, which can relate to a combination of specific claims experience for the relevant reserving category being analysed (paid, incurred, numbers etc.) and external information such as court rulings or emergence of potential new sources of claim, that could influence future claims payable; and
- The methodology and data used to derive an updated best estimate one year later. Although this might be based on the approach used a year ago, in practice, updates might be made, due at least in part to the emergence of information during the year.

The goal is to estimate the uncertainty in the prospective CDR, recognising that this uncertainty can arise from one or more of these sources:

The methods used to do this can be divided into the following groups.

1. Simulation-based methods.
2. Recognition pattern methods.
3. Analytical methods.
4. Other methods.

As with stochastic methods on an ultimate time-horizon basis, these can all be applied to paid and/or incurred data triangles, but the interpretation of the results will vary, depending on which data has been used. Furthermore, as with ultimate time-horizon stochastic methods, it is sometimes difficult to incorporate all the detailed adjustments and combinations of methods that are used in the best estimate reserving process, and some scaling of the "pure" one-year results produced from these methods may be necessary in practice for some cohorts and reserving categories. This scaling can potentially invalidate the assumptions underlying the methods, and hence the results need careful interpretation when it is used.

Before summarising these methods, it is important to recognise that in practice it may be difficult to establish whether all sources of uncertainty in the CDR have been allowed for appropriately. This is partly because the reserve at the end of the year will be influenced by the emergence of unpredictable shocks or new external information during the year, all of which may influence the selection of the reserve at the end of the year and which may be difficult to allow for within the method used. The methods described here may therefore not capture some of this variability, and hence in practice it may be necessary to overlay an element of judgement in selecting the final results on a one-year basis.

4.8.2 Simulation-based Methods

These take a variety of forms, but the general procedure for a single reserving category, where claims paid data is used, following Ohlsson and Lauzeningks (2008), is set out below:

1. Begin with a best estimate of the reserves at the start of the year, and define an algorithm which has been used to derive those reserves and which can be repeated at a future time point. The algorithm might be a standard chain ladder procedure, or may involve other methods, such as Bornhuetter–Ferguson, and may include other features such as

smoothing of development factors and estimation of tail factors. The requirement for it to be repeatable means that it cannot involve any element of subjective judgement.

2. Simulate the next year's claims paid development. The methods used to do this can include bootstrapping or parametric simulation from a statistical distribution, and can also include an allowance for process error.

3. Reapply the algorithm from Step 1 to the data triangle including the next year's claims development simulated in Step 2, to produce the best estimate reserve at the end of the year for that simulation.

4. Calculate the run-off result (i.e. the CDR) as the Reserve at the start of the year from Step 1, less the simulated claims paid during the year from Step 2, less the Reserve at the end of the year from Step 3. Alternatively, the ultimate claims estimate from Step 3 can be subtracted from the ultimate claims estimate from Step 1. With this definition, if the CDR is positive, then a run-off surplus (i.e. profit) has emerged, and if it is negative, then a deficit (i.e. loss) has emerged.

5. Repeat steps 2 to 4 for a suitable number of simulations, so that the full distribution of the CDR can be estimated.

6. Derive the required statistics of the CDR, such as the standard deviation and the chosen percentile values. This can be done for individual cohorts or across all cohorts combined.

This procedure is sometimes referred to as the "actuary in the box" or "re-reserving" approach, which is a reference to the algorithm used to derive the reserves.

An enhancement to the above procedure, mentioned in Ohlsson and Lauzeningks (2008), is to make an explicit adjustment for inflation. For example, if a chain ladder method has been used in the algorithm, then this could follow an inflation-adjusted approach, as described in Section 3.2.9, perhaps with future inflation (including during the first year of projection) being stochastic.

The procedure above can be extended to simulate claims for more than one future year, so that the uncertainty in the CDR for each year between now and ultimate can be estimated. This is comparable with the extension of the analytical MWCL method to a multi-period time horizon, explained in Section 4.6.5. However, a simulation-based approach has the advantage that it can be used to derive the full distribution of the CDR in each future year, and a tail factor can be incorporated into the process.

Where more than one reserving category is involved, once the procedure has been applied to each category, the aggregate distribution of the CDR can then be derived using one or more of the approaches outlined in Section 4.9.

Different practical implementations of the above general procedure have been suggested. An example, based on using the ODP Bootstrap is summarised as follows:

1. Apply the Chain Ladder method to the data triangle and then calculate the reserve. This reserve is then the reserve at the start of the year in the CDR calculation.

2. Generate the residuals (using a suitable definition) implied by the Chain Ladder method.

3. Randomly sample from these residuals and derive a simulated past data triangle.

4. Apply the Chain Ladder method to the simulated past data triangle and calculate the estimated future claims in the next year.

5. Add process error to the estimated future claims from the previous step, using a suitable statistical distribution, to generate the simulated claims in the next year, incorporating process error.

6. Derive a new data triangle using the actual past data and the simulated future claims in the next year, from the previous step.

7. Apply the Chain Ladder method to the data triangle in the previous step.

8. Calculate the reserve from the triangle in the previous step. This is then assumed to be the expected reserve at the end of the year in the CDR calculation, conditioned on the claims that have emerged in the year.

9. Derive the CDR as the reserve from Step 1 less the simulated claims in the next year from Step 5 and the reserve at the end of the year from Step 8.

10. Repeat Steps 3 to 9 according to how many simulations are required, and derive the required statistics and distribution of the CDR.

As described in Section 4.4.3 in relation to the Mack Bootstrap procedure, it is also possible to bootstrap the development factors. If this is done, then the procedure above becomes a bootstrap version of the analytic one-year MWCL method described in Section 4.6. The obvious advantage over the analytic MWCL method is that a full distribution of the CDR can be estimated. Further details are given in De Felice and Moriconi (2006) and Boumezoued et al. (2011). England et al. (2012a) also show how applying the "actuary-in-the-box" simulation approach to the Taylor and Ashe (1983) data gives very similar prediction error results to the results derived from the analytical version of the MWCL method, as shown in Table 4.61 in Section 4.6.4. The same source also shows how this type of simulation approach can be used to derive results on a multi-period basis, comparable to those produced using the extended MWCL method described in Section 4.6.5.

The points made in Section 4.4.3 in relation to the practical application of bootstrapping when it is used to estimate ultimate reserve risk, such as allowing for degrees of freedom and selecting the residuals, are equally relevant here. In addition, variations in the above procedure (using either an ODP or Mack Bootstrap) beyond the pure chain ladder approach are possible in practice. For example, adjustments may include curve fitting to the raw development factors and estimation of a tail factor. In all cases, it is important that the expected CDR is still zero, so some scaling might be required if the adjustments remove this feature. It is also important to recognise that many "re-reserving" approaches will be relatively automated, and so may fail to capture fully all the sources in variability of the CDR, as discussed in the introduction to this section, and so may understate the variability. Use of more than one type of method can help address this, with the usual application of judgement often being required.

Depending on the exact formulation of the method and the number of reserving categories being modelled, this approach can involve lengthy simulation times. If these become unmanageable, then other methods can be used, such as those outlined in the paragraphs that follow.

4.8.3 Recognition Pattern Methods

This type of approach, sometimes also referred to as "Emergence Pattern" methods, can be useful when the ultimate reserve risk has either been estimated or derived from external sources, but where there is insufficient data to allow simulation or analytical methods to be used to derive the one-year reserve risk. It is based on the observation that the ultimate reserve risk must "emerge" over time between now and ultimate. So, in a one-year reserve risk context, the method relies upon being able to estimate what proportion of the ultimate reserve risk emerges during the next year. Typically, for short-tail classes the ultimate reserve risk emerges quickly, whereas for longer-tail classes the emergence is slower. The ultimate reserve risk can, for example, be expressed as the difference between, say, the reserves at the 99.5th percentile and the best estimate of the reserves. This can be by cohort or in total. Other approaches are possible, such as using the Coefficient of Variation (CV) of the reserve on an ultimate time horizon basis, and finding a way to adjust this so that it is on a one-year basis. Options for estimating an emergence pattern for the ultimate reserve risk include:

- Use an estimated claims payment pattern and apply this to the ultimate reserve risk at the required percentile, or to the CV. For example, if for a particular cohort it is estimated that, say, 50% of the reserve is expected to be paid next year, then assume also that 50% of the ultimate reserve risk for that cohort "emerges" over the year. The claims payment pattern can either be expressed as the proportion of the total estimated future claims (by cohort or in total) that are paid in each year, or using a "disposal rate" approach (as described in Section 3.5.3 in relation to Frequency Severity Method no. 2), where the claims paid in each period are expressed as a proportion of the reserve at the start of that period. In the former case, the pattern sums to 100%, whereas in the latter it does not. In either case, this represents a very simplistic approach that effectively involves selecting a scaling factor that is multiplied by the reserve risk at ultimate or by the CV, to produce the comparable one-year horizon measure.
- Use the payment pattern, but add an element of stochastic variation by assuming the proportion paid in each year follows a chosen statistical distribution (e.g. Normal). This can help overcome an obvious limitation of the pure payment pattern approach which may not capture the variability in the CDR caused by emergence of shocks or new information in the year.
- Apply one or more other one-year reserve risk methods to the data and then, in conjunction with an estimate of the ultimate reserve risk, derive an emergence pattern, with appropriate smoothing of the pattern if necessary. This can be a useful exercise, since the reasonableness of the emergence pattern can sometimes be easier to assess than the standalone one-year reserve risk results. However, if the data is sparse, this may be

very difficult, and considerable judgement may be needed to adjust the results in order to produce even a broadly reasonable pattern. England et al. (2012a) explores using a bootstrapping approach to derive an emergence pattern, fitted to either the distribution of ultimate claims or reserves.

- Use suitable benchmark industry or other data to derive emergence patterns by reserving category. This can be done, for example, by applying analytical- or simulation-type methods to the benchmark data on both a one-year and ultimate time horizon basis, and then translating the results into emergence patterns, which can then be applied to estimates of ultimate reserve risk for the data being analysed. Benchmark one-year reserve risk metrics could also be applied directly to the best estimate reserves for the data being analysed.

Some of these approaches and others are discussed further in White and Margetts (2010) and England et al. (2012a).

4.8.4 Analytical Methods

The MWCL method, which is explained in Section 4.6, is an analytical method that can be used to estimate one-year reserve risk. There are very few other one-year analytical approaches that have been suggested to date. One such approach is based on adapting Bayesian–MCMC-type methods. This approach was used by some participants in the AISAM-ACME (2007) study and involves using the Bayesian methodology described in Section 4.5, but adapted to produce simulations of the one-year run-off result. Further details are available in AISAM-ACME (2007).

One other simplistic method that might be labelled as analytic, which uses only the historical data, and so does not rely upon a reserving model, is specified in the Solvency II Undertaking Specific Parameters regulations as "Method 1". In summary, this involves analysing the data in prior years to compare the reserve selected at the start of the year (on a best estimate basis) with the sum of the paid claims in the year and the reserve established at the end of the year. Further details can be found in CEIOPS (2010).

4.8.5 Other Methods

A number of other approaches can be used to estimate one-year reserve risk, some of which should only be regarded as broad approximations, and others which rely upon benchmarking from other sources, and hence which are not distinct methods themselves. Even where more detailed methods appear to work, it can still be useful to use benchmarking or approximate methods to provide comparative results. Some examples of these approaches are given below:

- **Time-scaling or VaR equivalence**: This involves using the ultimate reserve risk distribution and selecting an appropriate point on that distribution as an approximation for the risk over a one-year time horizon. For example, White and Margetts (2010) suggest

Table 4.66 *Solvency II Standard Formula one-year reserve risk parameters*

No.	Class of business	Parameter
1	Motor vehicle liability insurance and proportional reinsurance	9%
2	Other motor insurance and proportional reinsurance	8%
3	Marine, aviation and transport insurance and proportional reinsurance	11%
4	Fire and other damage to property insurance and proportional reinsurance	10%
5	General liability insurance and proportional reinsurance	11%
6	Credit and suretyship insurance and proportional reinsurance	19%
7	Legal expenses insurance and proportional reinsurance	12%
8	Assistance and its proportional reinsurance	20%
9	Miscellaneous insurance and proportional reinsurance	20%
10	Non-proportional casualty reinsurance	20%
11	Non-proportional marine, aviation and transport reinsurance	20%
12	Non-proportional property reinsurance	20%

that, if the reserves can be assumed to run-off over n years, then the 99.5th percentile on a one-year basis can be approximated by using the $(99.5\text{th})^n$ percentile on an ultimate basis. The AISAM-ACME Solvency II study provides some data on VaR equivalence, which suggests a similar relationship, in that the longer the term of the liabilities the lower the equivalent percentile on an ultimate basis. See Section 4.4 of AISAM-ACME (2007) for further details. These approaches would typically only be used where all other methods fail to work, or to provide approximate results to compare with those from other methods.

- **Benchmarking**: Depending on the context, there may be industry or other company data available that can be used to derive suitable one-year reserve risk benchmarks. Some smoothing of these benchmarks may be necessary in practice, and the general comments made in Section 6.9.1 regarding benchmarking apply equally well here. An example of a benchmark is the Solvency II Standard Formula one-year reserve risk parameters. These are summarised in Table 4.66 and represent the ratio of the Standard Deviation of the CDR to the reserve at the start of the year. The source for these is paragraph SCR.9.23 in EIOPA (2014d). They are designed to be used on net of reinsurance data, and do not include any allowance for catastrophe risk.

4.8.6 Further Reading

Examples of additional sources that discuss estimation of one-year reserve risk are listed below:

De Felice and Moriconi (2006) is one of the earlier papers on one-year reserve risk, where the concept of "Year-end-expectation" or YEE was introduced to contrast with the usual ultimate view, which in this paper is termed "Liability at Maturity".

ISVAP (2006) uses the YEE approach with the ODP chain ladder model with a constant scale parameter to study the solvency capital requirements on a sample of Italian motor insurance companies.

Bühlmann et al. (2009) studies one-year reserve risk within the context of a Bayesian chain ladder model.

Diers and Linde (2013) extends the one-year CDR concept by including premium risk as well in a multi-year context.

Robbin (2012) outlines a method that is designed to address a potential limitation in the approach adopted in the Solvency II Standard Formula for reserve risk, whereby the parameters do not vary by cohort and hence are independent of the stage of development of the run-off of reserves.

Saluz and Gisler (2014) argues that repeated successive observations of CDRs with the same sign are not necessarily inconsistent with an underlying best estimate reserving process. This is achieved by introducing a Bayesian claims reserving model, which allows evolving external information to be incorporated into the model.

Dal Moro and Lo (2014) sets out a challenge to produce methods on a one-year basis that provide closed-form solutions that are applicable to BF, Cape Cod and mixed chain ladder/BF/Cape Cod methods. Although the paper does not provide any methods itself, it does summarise briefly the range of approaches for one-year risk that have been developed so far, and gives the authors' views of where further research in this field is needed.

Merz and Wüthrich (2015b) considers derivation of the CDR and its associated uncertainty for a number of different stochastic models, including Bayesian and non-Bayesian Chain Ladder, BF, and multivariate Gaussian.

4.9 Combining Results Across Categories

The description of the stochastic methods included in this chapter has focused on the application to a single data triangle. In practice, in most reserving work there is likely to be more than one data triangle, representing different reserving categories, such as class of business, claim type or claim size. There may also be categories where the data is not in a triangle format, such as relating to large event losses. Although the variability of each of these will be of interest, the key focus in most situations will be the overall uncertainty and the distribution of the reserves across all categories combined.

Whilst the stochastic results at or around the mean level of reserves can be added together for this purpose, in order to produce combined results at other points on the distribution (e.g. at higher percentiles), the dependency between the categories will first need to be considered. The dependency relationship will relate specifically to the factors such as inflation that influence the future claims outgo for each category, which might be different to the relationship that might exist between, for example, expected loss ratios for different classes before the business has been written. Whether the relationship is the same at different points of the distribution of future claims will also need to be considered. The degree of volatility in the distribution of future claims for each reserving category will also be relevant, since if this is generally quite low, the level of correlation between the categories will have less impact when the results are aggregated at different points on the distribution. The converse is also true of course.

At this stage, it is worth noting the distinction between dependency and correlation. Correlation between reserving categories in the context of determining an aggregate distribution of the future claims outgo refers to the tendency of the future claims for each category to move in the same (or perhaps opposite) direction – i.e. whether the underlying factors that cause the future claims for one category to increase also cause the future claims for another category to increase, and if so the strength of this connection. Dependency refers to the broader relationship that might exist between the future claims for each category – for example, it may be that there is only a tendency for both sets of future claims to move in the same direction in the tail of the distribution of future claims. Using a single correlation factor to describe the relationship between reserving categories is effectively one type of dependency; it may not always capture the nature of the relationship between the two categories, and a more complex dependency structure may need to be allowed for in order to construct a suitable aggregate distribution.

If the nature of the dependency is relatively straightforward, such that the correlation is thought to be "perfect" or extremely strong between all categories, and the same at all points of the distribution, then the results at each percentile for each category can be added together.[51] In practice, this will not usually be the case, and hence adding the results together in this way will produce unreasonably high estimates for the aggregate reserves at higher percentiles, for example. In most situations, there will be some level of independence between the categories, which will give rise to a diversification benefit when the results are combined across categories, so that, for example, the value at the 99.5th percentile for the total reserves will be less than the sum of the values at the 99.5th percentile for each category. Hence, some allowance needs to be made for this diversification benefit. The approach used to allow for this benefit will depend on several factors, including whether the stochastic method(s) used have produced the full distribution for each reserving category, or just the mean and prediction error. It will also depend on the nature of the dependency between the different categories, which should preferably be considered first before deciding on the approach used to combine the results across the reserving categories. This is discussed further in Section 4.9.3.

The simplest form of relationship is that all pairs of reserving categories have the same type of dependency with each other, but the level of dependency is potentially different for each pair. The dependency can take various forms, with a simple, but commonly used example being through the use of correlation coefficients for each pair of categories. A more

---

[51] Note that if, for example, bootstrapping has been used and the number of simulations used is too small, then this approach may produce inaccurate results for both the individual reserving categories and for the combined portfolio. The obvious solution is to increase the number of simulations, and / or perhaps use the same random number set for each reserving category within the bootstrap process.

complex form of dependency can, for example, be characterised as one which varies across different points of the distribution of future claims outgo for each category. In a reserving context, for example, two classes of business might only show positive correlation at extreme values for the future claims outgo, which is difficult to capture with a single correlation coefficient. Alternatively, there could be correlation across all points of the distribution, but the degree of correlation could be greater in the tail of the distribution (so-called "tail dependency"). A further point to consider is whether the nature of the relationship between all pairs of reserving categories is the same. It is quite possible, for example, for some reserving categories to only have a tail dependency between them, but others to have a more straightforward relationship.

In some situations, in order to derive an approximate variance or distribution of the aggregate reserves, it might be possible to apply one or more stochastic methods to an aggregate data triangle, rather than apply a method that combines the results from the individual reserving categories. Considerable care will be needed, however, in interpreting the results of such an approach, as any significant lack of homogeneity between the reserving categories will be masked and it will not be possible to model the impact of different correlation or dependency assumptions between the reserving categories using this aggregated approach. In some cases, such an approach can produce misleading results.

Hence, in most situations, a technique will need to be employed to aggregate the results derived by applying the stochastic methods to the individual reserving categories. It is not uncommon, when doing so, however, to use a more aggregated grouping of categories than is used for the main best estimate reserving exercise. In some cases, a hierarchical approach might be used, whereby the full set of reserving groups are initially combined into a number of subgroups, for which the results are combined (using the chosen aggregation technique) and then the aggregated results for each subgroup are used in a further stage of aggregation. The grouping used for this purpose, even if it does just mirror that used for the best estimate reserving exercise, can have a potentially material impact on the derived diversification benefit, and hence needs to be considered carefully. For example, testing the sensitivity of the results to the grouping used can sometimes be helpful.

The remainder of this section outlines some of the techniques that have been developed for aggregating stochastic results across reserving categories. It is divided into subsections that deal with the situation where a full distribution has been estimated (e.g. via bootstrapping), and where only the mean and prediction error have been estimated. There is then a section that deals with the topic of determining the nature and extent of the dependency/correlation between reserving categories and consideration of the associated impact on the aggregate results. Since this section only represents a preliminary introduction to the topic, the final subsection gives some suggested further reading.

An alternative to deriving results independently for each reserving category, and then using one of the procedures described in this section to aggregate the results, is to use methods which allow multiple reserving categories to be modelled in a multivariate framework which already has the relationship between categories embedded within it. These approaches are discussed in Section 4.7.5.

4.9.1 Aggregation Techniques Where the Full Distribution has Been Estimated Using Bootstrapping

For techniques such as bootstrapping which produce simulated values for each reserving category, three types of approaches that can be used to allow for the correlation/dependency between categories are "Location Mapping", "Re-sorting" and "Distribution Aggregation". These are described below. Location Mapping and Re-sorting are discussed further in Shapland and Leong (2010), Shapland (2016) and Kirschner et al. (2002, 2008).

Location Mapping

In this approach, which can be considered as a type of "synchronous bootstrapping" technique, the correlation is allowed for within the bootstrapping routine itself, rather than through a specified correlation matrix as with the Re-sorting approach. The basic idea is that when the bootstrap procedure is applied to the first category, the "location" in the triangle (i.e the row and column) of each simulated residual is noted and that location is then used when selecting residuals for the other categories. This is designed to retain the underlying correlation structure that is implicit within the original data. One key advantage of this approach is that it does not require selection of correlation factors, which in practice can be highly judgemental, although this also means that the sensitivity of the combined results to different correlation assumptions cannot therefore be tested. Furthermore, depending on the nature of the data, the synchronised bootstrapping procedure may not provide an accurate representation of the true underlying dependency between the categories; for example if the model being bootstrapped has similar deficiencies for two particular reserving categories, then Location Mapping might infer an inappropriate dependency that only exists because of the common model deficiencies. Also, this approach is more difficult to implement if the triangles are not of the same size for all categories, or if different data points have been excluded. Other synchronous bootstrapping procedures which allow different correlation structures to be allowed for are possible, using the principles of a technique known as "Seemingly Unrelated Regressions". Further details are given, for example, in Taylor and McGuire (2007).

Re-sorting

This approach involves sorting (i.e. ordering or ranking) the individual simulation results produced by the bootstrap models for each category and then defining a procedure that re-sorts them, such that when they are added together, the combined results have a chosen target dependency between the categories. A decision needs to be made as to the specific results from the bootstrap simulations for each reserving category that will be used in the sorting procedure – for example, individual cash flows, reserves by cohort or total reserves across all cohorts combined. One approach is to use the total reserves and then, if aggregated values for other items (e.g. reserves for a particular cohort, or even CDR when a one-year time horizon is being used) are required, the bootstrap simulation results are sorted in the same order as for the total reserves, thus implying the same dependency as for the total reserves. In other words, although the sort order for the other items might be different to that for the total reserves, the resulting aggregate distribution for the other items is assumed to have the same level of dependency implied by that for the total reserves. When there is a strong degree of positive correlation between two categories, the procedure should be designed so that simulated values of comparable ranks have a tendency to be combined in the aggregated results. Different procedures can be used to combine the results in this way – two of which are based on the Iman–Conover and copula techniques. A brief introduction to copulas is given later in this section. The Iman–Conover procedure can be used to induce a target rank correlation between reserving categories. However, it appears to have less coverage in the literature (at least in an insurance context) than copulas, and no further reference is made to it in this book. Details are available in Iman and Conover (1982), Mildenhall (2005), and also in Shapland (2016), where it is used for a worked example in an ODP Bootstrap context. The procedure used can, where necessary, be designed to introduce more complex associations between the categories – for example, some types of copula will produce more dependency in the tail of the distribution than in other parts of the distribution, which can be appropriate in some circumstances. In general terms, the Re-sorting approach can be used where the triangles for each category are of different sizes. One disadvantage of this approach compared to Location Mapping is that the dependency relationship (e.g. correlation and type of copula) needs to be estimated or selected, which in practice can be highly judgemental, but at least it allows the impact of alternative correlation factors and dependency structures on the combined results to be tested.

Distribution Aggregation

This is similar to Re-sorting, but instead of combining the actual simulation results, a cumulative distribution function (CDF) is first created from the results for each reserving category as required, and then these are aggregated. The CDF can either be an empirical one, derived directly from the simulation results, or it can be derived by fitting an analytical distribution (e.g. Lognormal) to the simulation results, and using that to define the CDF for each reserving category. Once these so-called "marginal" distributions have been defined, and the correlation matrix has also been determined, they are combined into an aggregate distribution using a similar method as with the Re-sorting approach.

Some examples of the approaches that can be used to combine bootstrap results in either the Re-sorting or Distribution Aggregation methods are described in more detail in the following subsections. Before these are discussed, a brief introduction to copulas is given.

Brief Introduction to Copulas

In general terms, copulas can be thought of as describing dependence structures. In statistical terms, a copula is a multivariate probability distribution for which the marginal probability distributions are all uniform. In essence, they can be regarded as mathematical functions that allow individual (or "marginal") statistical distributions (of any type) to be combined to form a multivariate distribution. Thus, embedded within any multivariate distribution there will be underlying marginal distributions; a copula defines the dependence structure and so "joins" those distributions into the multivariate form. A wide range of copulas have been specified that allow different types of dependencies between the marginal distributions to be modelled. Examples include those referred to as Elliptical copulas (e.g. Gaussian and t-copula) and Archimedean copulas (e.g. Gumbel, Clayton, Cook–Johnson and Frank copula).

A number of copulas have the feature that they allow tail dependency to be modelled (e.g. the t-copula and the Gumbel copula), but the widely referenced Gaussian copula does not. The nature of the tail dependency will vary according to the copula. For example, Archimedean copulas can be parametrised to produce more dependency in the upper tail of the distribution than in the lower tail, so that the dependency is strongest for larger values, which may be a feature in certain general insurance reserving contexts. This is discussed further, for example, in Venter (2002).

The procedures outlined below include some mathematical details in order to explain how copulas can be implemented in practice when aggregating reserving results produced using stochastic methods. As an introduction to the topic, it is helpful to consider a simple example in this particular context. If there are two random variables, denoted by X_1 and X_2, and their simulated realisations are both sorted in ascending order, then the rank correlation will be $+1$ (i.e. perfect positive rank correlation). In contrast if one of the random variables is sorted in ascending order and the other in descending order, then their rank correlation will be -1 (i.e. perfect negative [i.e. inverse] rank correlation). Hence, the resulting aggregate distributions of X_1 and X_2 will have rank correlation between these two random variables of $+1$ and -1 respectively. In other words, the sort order gives embedded dependency in the aggregate distribution in the range $(+1, -1)$, with the dependency in this case being measured by the rank correlation. In practice, something in between the two extremes will usually be required. In the Re-sorting context used here, for example, the copula simulation procedures are effectively a tool to determine the order in which the realisations are sorted before they are summed, to give a required level of dependency.

There is a wide body of literature on copulas for those readers requiring additional details, which cover topics such as the properties of different copulas and the selection of the type of copula. For example, Wang (1998), Frees and Valdez (1998) and Embrechts et al. (2001) are three papers which discuss the use of copulas in an insurance context. Shaw et al. (2011) is also a useful reference source on the subject of dependency in economic capital modelling.

The next two subsections each describe a Monte Carlo simulation procedure that can be used to combine bootstrap results using a particular type of copula, known as the Gaussian or Normal copula. The first subsection deals with the situation where the bootstrap results themselves are aggregated using a Re-sorting approach. The second procedure is similar to the first, except that instead of using the bootstrap results, a distribution is fitted to the results for each reserving category, and the results are then aggregated using a distribution inversion approach. The choice between these two approaches will depend partly on the reasonableness of the distributions implied by the bootstrap simulation results for each reserving category, and partly on which one is favoured within the relevant software being used.

Although only Gaussian copula procedures are described here, in theory similar procedures can also be used for other copulas, such as the t-copula. For further details of such procedures see, for example, Piwcewicz (2005) and Tang and Valdez (2009), which describe a t-copula procedure.

As explained in Wang (1998), in many situations, the practitioner will only have a broad indication of the correlation parameters, without knowing the underlying multivariate distribution and/or dependency structure that exists. In these situations, the Gaussian copula procedures outlined here are a relatively straightforward way of simulating the required correlated results.

One other type of copula which can be useful is the Independence copula, which, as its name implies, assumes that each of the categories are independent, and hence no correlation assumptions are necessary. Alternatively, the Gaussian copula can be used with zero correlation, which produces the same results. Either of these approaches can be useful in a reserving context to estimate the maximum diversification benefit that could be assumed (absent any offsetting or negative correlation), as compared with the other extreme of full correlation that is implied by adding the results across the individual categories, after sorting into their rank order. Then, when non-zero correlation is assumed, together with a copula designed to model the chosen dependency structure, the resulting diversification benefit can be compared to the fully independent case.

Re-sorting Approach Using a Gaussian Copula

The Gaussian copula is a well-known type of copula which represents the dependence structure of the multivariate Normal (or Gaussian) distribution. Implementing the Gaussian copula in the context of the Re-sorting approach involves the following Monte Carlo simulation procedure. This is based on a version of the Gaussian copula Monte Carlo simulation algorithm from Section 8 of Wang (1998). Since that version is based on using distributions rather than bootstrap simulations, it is modified along the lines of the description of the bootstrap Re-sorting algorithm given in Kirschner et al. (2008).

The procedure is applied to a situation where there are, say, n reserving categories, for which the bootstrap procedure has been used to produce, say, m simulations. It is assumed here that the Re-sorting algorithm is being applied to the total reserves (i.e. future claims outgo) for each reserving category, across all cohorts combined.[52] Denote these simulations by the matrix, X, which has entries $x_{a,b}$, where a relates to the simulation number and b relates to the reserving category, so that a ranges from 1 to m and b from 1 to n.

1. Determine the rank order of the X entries within each reserving category (i.e. within each column of X). In other words, for each reserving category order the m simulations to determine a value between 1 and m for each simulation value.

---

[52] As noted in the initial description of the Re-sorting approach, if aggregated results are required for other items, such as the future claims outgo for individual cohorts, one possible approach is to use the same simulation sort order derived by applying the Gaussian copula procedure for the total reserves.

2. Re-order the X entries according to this ranking to produce a matrix that is denoted by X^r, with entries $x^r_{a,b}$. In effect, the resulting sorted values in each column in X^r are equivalent to an estimated cumulative distribution function of the future claims outgo for each reserving category. These marginal distributions (and hence the mean and all percentile values) for each reserving category are unchanged, as they have simply been re-ordered.

3. Select the target rank correlation factors, $r_{i,j}$ between each pair of reserving categories, i and j. These will each have a value between -1 and +1. Rank correlation is used, rather than linear correlation, as the procedure determines the sort order such that the aggregate results have the required target rank correlation factors. Determination of correlation factors is discussed further later in this section.

4. Convert the target rank correlation factors to linear correlation factors, $\rho_{i,j}$, for use in a multivariate Normal distribution using:[53]

$$\rho_{i,j} = 2\sin\left(\frac{\pi r_{i,j}}{6}\right).$$

5. Arrange these linear correlation factors in a correlation matrix, Σ and ensure that this matrix is positive semi-definite ("PSD"), which will be needed so that the factors are internally consistent. This is discussed further at the end of Section 4.9.3, which covers the determination of dependency between reserving categories and includes reference to sources that explain the techniques for checking if a matrix is PSD, and modifying it if that is not the case.

6. Apply a technique known as Cholesky decomposition[54] to construct the lower triangular matrix B, such that $\Sigma = BB'$.

7. Generate, using some form of simulation,[55] a vector $Y = (y_1, y_2, \ldots, y_n)'$ of standard Normal distribution values from $N(0, 1)$.

8. Define $Z = BY$. $Z = (z_1, z_2, \ldots, z_n)'$ will have the appropriate joint probability distribution function defined by the correlation matrix.

9. Set $u_i = \Phi(z_i)$, where Φ is the standard $N(0, 1)$ cumulative distribution function. $U = (u_1, u_2, \ldots, u_n)'$ is then a vector of correlated values between 0 and 1 from a Gaussian copula.

10. Repeat Steps 7 to 9 inclusive m times and then determine the rank values of the U values within each reserving category. Denote these rank Normal values by S, which has entries $s(a, b)$.

11. To produce simulated values from the aggregate distribution with the required target rank correlation and a Gaussian copula dependency structure, first use the S ranks to "look-up" values from the original simulated data, to produce a further matrix, X^r_c, which has entries defined as follows:

$$x^r_c(a, b) = x^r_{s(a,b),b}.$$

---

[53] This is the formula for converting Spearman's rank correlation factors, as per Section 8 of Wang (1998).

[54] Cholesky decomposition is explained in many widely available sources, including wikipedia. For a reference within a risk management context, see Sweeting (2011), Section 10.3.3.

[55] This is easily achieved in many software packages, for example using MS Excel with the function NORM.S.INV(RAND()).

12. Then, denote the simulated values from the aggregate distribution by V, which will have m entries as follows:

$$v(a) = \sum_{i=1}^{n} x^r_{s(a,i),i}.$$

This can then be used to determine the percentiles, for example, for the aggregate distribution.

A simplified example will help demonstrate the last three steps of this procedure. Assume that there are just 5 ($=m$) simulations for 3 ($=n$) reserving categories, with simulated values X and re-ordered values X^r, as shown in (4.131). X^r is effectively the sort order assuming perfect positive rank correlation, as explained in the brief introduction to copulas earlier in this section.

$$
X = \begin{matrix} 1 \\ 2 \\ 3 \\ 4 \\ 5 \end{matrix}
\begin{bmatrix}
57 & 136 & 233 \\
26 & 128 & 271 \\
72 & 188 & 211 \\
7 & 109 & 257 \\
32 & 161 & 205
\end{bmatrix}
\; ; \; X^r = \begin{matrix} 1 \\ 2 \\ 3 \\ 4 \\ 5 \end{matrix}
\begin{bmatrix}
7 & 109 & 205 \\
26 & 128 & 211 \\
32 & 136 & 233 \\
57 & 161 & 257 \\
72 & 188 & 271
\end{bmatrix} .
\tag{4.131}
$$

Then, assume rank correlations have been selected, converted to linear correlation factors and then organised in a correlation matrix. Assume also that the $N(0, 1)$ values have been generated, as per Steps 3 to 9 inclusive, to produce m values for U. These values are then ranked in order to produce an example of the matrix S, as shown in (4.132). The rank values in S are then used to look-up values in X^r, as per step 11 above, to produce X^r_c, as shown in (4.132). Then, because the rank values are the same as in S, the rank correlation is then also the same, as required.

$$
S = \begin{matrix} 1 \\ 2 \\ 3 \\ 4 \\ 5 \end{matrix}
\begin{bmatrix}
5 & 4 & 2 \\
3 & 2 & 5 \\
2 & 1 & 3 \\
4 & 5 & 4 \\
1 & 3 & 1
\end{bmatrix}
\; ; \; X^r_c = \begin{matrix} 1 \\ 2 \\ 3 \\ 4 \\ 5 \end{matrix}
\begin{bmatrix}
72 & 161 & 211 \\
32 & 128 & 271 \\
26 & 109 & 233 \\
57 & 188 & 257 \\
7 & 136 & 205
\end{bmatrix} .
\tag{4.132}
$$

In other words, when the rows of X^r_c are summed, as per Step 12, the resulting sum will represent the distribution of total claims across the three reserving categories, with values that are correlated according to the initially selected target rank correlation factors.

Distribution Aggregation Approach Using a Gaussian Copula

This is similar in concept to using the Gaussian copula in the Re-sorting approach, but instead of the simulation results being combined using a copula, distributions are used. It is based on that given in Wang (1998), but the same procedure is also given in other sources, such as Frees and Valdez (1998). It involves the following procedure, applied to a situation where there are, say, n reserving categories, for which the marginal distributions $(F_1, F_2, \ldots, F_n)$ have been derived. As noted above, these marginal distributions are derived from the simulation results and can be in either a user-defined empirical form or a statistical distribution form. The procedure below applies where these marginal distributions relate to the total reserves (i.e. estimated total future claims outgo) for each category, across all cohorts combined.

1. Determine the rank correlation factors, $r_{i,j}$ between each pair of reserving categories, i and j. Convert them to linear correlation factors for use in a multivariate Normal distribution context, as per the Re-sorting Gaussian copula procedure. Arrange the linear correlation factors in a correlation matrix, Σ.
2. Apply Cholesky decomposition to construct the lower triangular matrix B, such that $\Sigma = BB'$.
3. Generate, using some form of simulation, a vector $Y = (y_1, y_2, \ldots, y_n)'$ of standard Normal distribution values from $N(0, 1)$.
4. Define $Z = BY$. $Z = (z_1, z_2, \ldots, z_n)'$ will have the appropriate joint probability distribution function defined by the correlation matrix.
5. Set $u_i = \Phi(z_i)$, where Φ is the standard $N(0, 1)$ cumulative distribution function. $U = (u_1, u_2, \ldots, u_n)'$ is then a vector of correlated values between 0 and 1 from a Normal copula.
6. Apply the inverse transform $F_i^{-1}(u_i)$ to generate simulated values for each reserving category. In other words, find the value, x_i, of the distribution, for say, the ith reserving category, such that $F(x_i) = u_i$.
7. Repeat Steps 3 to 6, say, 10,000 times (or however many simulations are deemed appropriate) to produce 10,000 vectors of simulated values $(F_1^{-1}, F_2^{-1}, \ldots, F_n^{-1})$, or $(x_1, x_2, \ldots, x_n)$, which are correlated according to the selected rank correlation factors, with a Gaussian copula dependency structure.

The simulated x_i values can then be summed across the reserving categories to produce simulations of the aggregate distribution with the required rank correlations between the reserving categories, and a Gaussian copula dependency structure, from which, for example, percentile values can be derived.

4.9.2 Aggregation Techniques When Only the Mean and Prediction Error Have Been Derived

For a number of the stochastic reserving methods described in this chapter, the output will include only an estimate of the mean and prediction error of the reserve, rather than a full distribution. In this scenario, to estimate the mean reserve for the sum of the reserving categories, the means for the individual categories can be added together. However, to estimate the variance of the sum of the categories, and possibly also the distribution of this sum, it is necessary to consider how the categories are correlated or dependent on each other, and to allow for this when combining across classes. Two approaches are outlined here for creating the combined results – the Variance–Covariance approach and the Copula approach, using assumed distributions.

Variance–Covariance Approach

This is a straightforward approach that uses the statistical property that the variance of the sum of correlated random variables is the sum of the variances and covariances of all the variables. The linear correlation coefficient is used, which may not be appropriate if, for example, the distribution of reserves for individual categories is skewed or the dependency between the categories varies across the distribution. However, the results using this approach can be obtained very quickly and it can produce broadly reasonable results in some circumstances.

To apply this approach, it is first necessary to estimate the correlation between each pair of categories (e.g. in respect of the total reserves across all cohorts). Once this is done, the correlations and the variances of the individual categories can be combined to create the covariance between each pair of categories. These covariances are then summed, together with the individual variances, to derive the variance of the sum of the reserving categories. If a distributional assumption can be made regarding the total reserve across all categories (e.g. a Lognormal) then the total reserves at different percentiles can also then be estimated, for example using the approach described in Section 6.9.4. The sensitivity of the results to different correlation assumptions can be tested with this approach.

In mathematical terms, assuming there are, say, n reserving categories, this procedure is summarised as follows:

- Denote the $n \times n$ linear correlation matrix by Σ, with entries $\rho_{i,j}$ representing the estimated correlation between category i and j, with diagonal entries of value 1.
- Create an $n \times n$ diagonal matrix, denoted by S, which has the entries $(\sigma_1, \sigma_2, \ldots, \sigma_n)$ on the diagonal, representing the standard deviations of each reserving category, and 0 elsewhere.
- Determine the Variance–Covariance matrix V using

$$V = S\Sigma S.$$

V will have diagonal entries equal to σ_i^2, representing the variance of category i and other entries equal to $\sigma_{i,j}$ representing the covariance between category i and j, so that:

$$\sigma_{i,j} = \rho_{i,j} \sigma_i \sigma_j.$$

- If all the n^2 entries of V are summed this will give the variance of the sum of the reserving categories.
- An assumption can then be made regarding the distribution of the aggregate reserves, and parameters estimated using the derived mean and variance from the previous step. Percentile values can then be calculated using this distribution.

Copula Approach Using Assumed Distributions

An alternative to the Variance–Covariance approach, again when only the mean and prediction error have been derived for each reserving category, is to make an assumption regarding the statistical distribution of the future claims for each reserving category and then to combine the categories using the copula distribution aggregation approach, as outlined in Section 4.9.1.

This has the advantage that it can be used to test the effect of different dependency structures between the reserving categories, as compared with using the straightforward linear correlation of the Variance–Covariance approach. A disadvantage, however, is the need to select the copula type and the distributions for each reserving category, which will be very judgemental if the stochastic reserving methods have not produced an estimate of the full distribution of future claims. In summary, the procedure is as follows:

1. For each reserving category, select a suitable distribution for the future claims (typically in aggregate across all cohorts). This could be empirical or statistical (e.g. Lognormal).
2. Select suitable parameters for each of these distributions, using the mean and variance determined from applying the chosen stochastic reserving method to each class (for a statistical distribution example see Section 6.9.4).
3. Select an appropriate copula to reflect the assumed dependency between the reserving categories.
4. Where required by the chosen copula, select suitable correlation factors between the reserving categories.
5. Apply an algorithm to determine the joint distribution. The Distribution aggregation approach for the Gaussian copula outlined in Section 4.9.1 is an example of such an algorithm, with the marginal distributions F_i being the distributions defined in Step 1 above.
6. Calculate the variance, percentiles etc. for the joint distribution.

4.9.3 Determining Dependency Between Reserving Categories

In some relatively rare situations, the future claims for each reserving category might be believed to be highly or fully correlated with each other across all points of the distribution of future claims. In this case, the values at any point on this distribution for all categories combined can be derived by summing the corresponding values across individual categories. However, in almost all situations, it is not appropriate to assume perfect correlation. Where this is the case, the theoretically correct approach is to investigate the nature of the dependency between the future cash flows of the reserving categories and select an appropriate model of this dependency structure. In practice, in most general insurance reserving contexts an assumption of correlation between reserving categories is made, since more complex dependency structures are usually very difficult to model and parametrise. The copula approach described earlier, however, can be used to introduce, for example, tail dependency, if required.

Hence, in many situations, unless the underlying stochastic model already allows directly for dependency between the reserving categories (e.g synchronous bootstrapping or in multivariate model, as described in Section 4.7.5), it is necessary to determine or select appropriate correlation factors between the reserving categories. This is a notoriously challenging task, that inevitably involves a considerable degree of judgement.

It is important to emphasise that the context in which the correlation factors are being used may influence the nature of the correlation that needs to be estimated. For a reserving exercise (i.e. on an ultimate time horizon basis), the correlation that needs to be estimated relates to the relationship between future claims outgo of each reserving category, whereas when considering a one-year time horizon in a Solvency II-type context, it is the correlation between the Claims Development Results. These can both be different to the correlation that might exist between reserving categories before the business is written. In the latter situation, for example, two property reserving categories might both be exposed to future windstorm losses in a particular region, and hence there is strong correlation between their performance. However, in a reserving context, if the reserves are being estimated only for the earned component (i.e. in respect of past accidents/events only) then this future "frequency" correlation does not exist, although there could still be severity correlation arising from, for example, the impact of the settlement of a disputed claim that affects more than one category. In addition, the correlation that might be observed between historical ultimate loss ratios for particular reserving categories might be partly caused by there being a similar underwriting cycle affecting premium rate adequacy across the categories. However, these pricing-related correlations do not directly impact the correlations between future claims outgo which are those that are of interest when combining results in a stochastic reserving exercise. It is worth noting that these comments also apply if a more complex dependency structure is being used to define the relationship between the reserving categories.

The correlation can also be related to different aspects of the future claims development. For example, there could be correlation amongst future calendar years across categories (e.g. due to inflation or a change in case reserving philosophy that affects several classes).

Although the nature and level of correlation between reserving categories might vary according to the item that is being aggregated (e.g. total reserves, reserves for individual cohorts, one-year CDR etc.) it may only be feasible in practice to select correlation factors for one of these items and for these to be used, for example, in a Re-sorting algorithm (as explained in the introduction to Re-sorting). This does not, however, preclude testing the sensitivity of the aggregated results to different values for these correlation assumptions.

Hence, bearing these points in mind, examples of approaches that can be used to select correlation factors between each pair of reserving categories are as follows:

- **Expert Judgement**: Apply expert judgement and general reasoning to select a descriptive summary of the strength of the correlation, taking into account all relevant knowledge that is available from underwriters, claims staff, actuaries etc. Types of description could be "High", "Low", "Medium", "Low-Medium", "High-Medium" etc. These descriptions are then translated into numerical values using judgementally selected values. For example, Low might be 10%, High might be 60% and Medium 30%. If a more complex dependency structure were being used, such as using a copula, then it may also be possible to make use of expert judgement on the nature of the dependency between reserving categories. For example, there may be some categories for which expert judgement indicates that there is more likely to be tail dependency at the upper extreme of outcomes for future claims outgo.

- **Analysis of data**: In theory, various types of historical data for each reserving category could be analysed to attempt to discern correlation or dependency structures. However, in practice, this can be challenging. Some suggestions have been made in the literature for such analyses – for example, Piwcewicz (2005) outlines an approach using rank correlation applied to historical incremental claims for each reserving category. Even where a successful analysis can be carried out, it is likely that the final selected assumptions will need to take into account output from one or more of the other approaches described here.

- **Benchmarking**: Depending on the context, there may be external sources to the particular entity for which reserves are being estimated, that can be used to select correlation factors. There may also be sources available that enable the overall diversification benefit to be benchmarked at selected points on the distribution. An example a benchmark source is taken from Paragraph SCR.9.31 in EIOPA (2014d), see Table 4.67.

Table 4.67 *Correlation (%) matrix for Solvency II standard formula*

Class	1	2	3	4	5	6	7	8	9	10	11	12
1	100	-	-	-	-	-	-	-	-	-	-	-
2	50	100	-	-	-	-	-	-	-	-	-	-
3	50	25	100	-	-	-	-	-	-	-	-	-
4	25	25	25	100	-	-	-	-	-	-	-	-
5	50	25	25	25	100	-	-	-	-	-	-	-
6	25	25	25	25	50	100	-	-	-	-	-	-
7	50	50	25	25	50	50	100	-	-	-	-	-
8	25	50	50	50	25	25	25	100	-	-	-	-
9	50	50	50	50	50	50	50	50	100	-	-	-
10	25	25	25	25	50	50	50	25	25	100		
11	25	25	50	50	25	25	25	25	50	25	100	
12	25	25	25	50	25	25	25	50	25	25	25	100

Class	Description
1	Motor vehicle liability insurance and proportional reinsurance
2	Other motor insurance and proportional reinsurance
3	Marine, aviation and transport insurance and proportional reinsurance
4	Fire and other damage to property insurance and proportional reinsurance
5	General liability insurance and proportional reinsurance
6	Credit and suretyship insurance and proportional reinsurance
7	Legal expenses insurance and proportional reinsurance
8	Assistance and its proportional reinsurance
9	Miscellaneous insurance and proportional reinsurance
10	Non-proportional casualty reinsurance
11	Non-proportional marine, aviation and transport reinsurance
12	Non-proportional property reinsurance

Table 4.67 shows the correlation matrix between classes of business that is used (at the time of writing) in the standard formula under the Solvency II regulations to derive the combined standard deviations across classes. These are tail correlations and are intended to represent the relationship between lines of business (premium and reserves combined) at the one-year 99.5th VaR level. As with the use of benchmarking in all aspects of reserving, the relevance of the benchmark is critical. Hence the example given here is only strictly relevant when used in a Solvency II one-year CDR context. Their use in other contexts, such as those involving ultimate time-horizons, may not be appropriate.

In practice, a combination of the above approaches may prove to be the most useful. The selected correlation matrix must be symmetrical (i.e. upper right part is a mirror image of the lower left part). It must also satisfy certain other constraints, since, for example one reserving category cannot be strongly positively correlated with two other categories, if those two categories are strongly negatively correlated with each other. These constraints can be met by ensuring that the matrix is positive semi-definite (or positive definite for some algorithms). Various techniques are available for testing whether a matrix satisfies the required constraints, and if it does not, modifying it so that it does. For example see Budden et al. (2008). The subject is also discussed in Shaw et al. (2011) and in Joubert and Langdell (2013).

4.9.4 Further Reading

When aggregating the results from applying stochastic reserving methods to individual reserving categories, given the inherent difficulty in determining the nature of the dependency between categories and/or estimating correlation factors, it can be helpful to explore the impact on results of using alternative aggregation techniques and assumptions. The findings from carrying out such analysis will inevitably vary according to the circumstances, but it may be instructive to compare them in general terms against those documented in relevant sources in the general insurance (and other) literature. Two examples of such sources are Tang and Valdez (2009) and Piwcewicz (2005), both of which show results using a range of different copula assumptions.

4.10 Stochastic Reserving in Practice

Section 1.2 outlines the key elements of a generic reserving exercise, covering the key stages of Background, Data, Analysis and Reporting. These elements apply equally well to the application of stochastic or deterministic reserving methods. Hence, any application of stochastic reserving methods can benefit from being seen in the context of these stages, rather than in isolation. Chapter 5 summarises further details that apply in practice to each stage of a reserving exercise and Chapter 6 also covers a number of additional reserving topics. Whilst many of these apply to both deterministic and stochastic reserving methods, there are some additional aspects that apply in particular to stochastic reserving in practice. This section summarises these, broken down into the following subsections:

- Background – Initial considerations.
- Data – Initial considerations.
- Analysis – Selection and testing of methods.
- Analysis and Reporting – Validation and communication of results.

For any individual application, there may be other aspects beyond those outlined here, which will be related, for example, to the nature of the business being analysed.

The results from applying stochastic methods can be materially different depending on how the practical aspects summarised here (and others specific to the exercise) have been dealt with. Furthermore, because of their inherent complexity, there is perhaps a greater risk with stochastic reserving methods that the application can become an end in itself, with certain practical aspects being overlooked. An extreme example would be not allowing for the existence of an aggregate outwards reinsurance policy that (absent credit risk) caps the downside at some multiple of the current best estimate of ultimate claims for a particular group of cohorts. Although such an obvious oversight might appear unlikely, other practical aspects can exist which are less obvious without a careful consideration of the broader aspects of the reserving exercise. Experienced practitioners are likely to consider all the key elements as a matter of course, but even for these, taking stock of the situation in a broader context can be beneficial periodically, as can peer review of the overall process.

4.10.1 Background – Initial Considerations

Prior to the selection and application of appropriate stochastic reserving methods, there are a number of initial considerations, which are summarised in this subsection.

The practical application of stochastic methods in any individual exercise will, as for deterministic methods, depend critically on the context. The two main contexts – assessing the uncertainty surrounding a proposed reserve figure and as an input to capital modelling – as well as other possible contexts such as reinsurance programme design, were identified at the start of this chapter, and so are not repeated here. The context will impact upon aspects such as whether a full distribution of future claims is required, or whether an estimate of the mean and prediction error will suffice. These aspects will in turn influence method selection and the format of results that are required (e.g. whether distribution of full cash flows by future calendar period are required).

In situations where stochastic methods are being applied by the same team within an insurer, or by an external consultant, who are responsible for applying deterministic methods (for example to determine a point estimate for financial reporting purposes) then the relevant background topics, such as the nature of the business will usually already be known to them. However, where the application of stochastic methods is being done by a separate team, or as a standalone consulting project, the consideration of the various background matters that apply in a deterministic reserving context can be an important early task before the selection and application of these methods.

4.10.2 Data – Initial Considerations

As well as the general data-related matters that apply to any reserving exercise, such as what data is available within the required timescale, there are a number of specific points that apply to the application of stochastic reserving methods. These are summarised in this subsection.

Reserving Categories

For practical reasons, grouping of data for stochastic reserving sometimes needs to be different to that used for deterministic best estimate reserving and so choosing this grouping and deciding how to deal with individual distorting claims, events or groups of policies/-contracts usually needs to be done at an early stage of the process. In general terms, the number of subdivisions used in stochastic reserving is usually fewer than that used in deterministic reserving, and in some cases a "hierarchical" type grouping approach might be used when results are combined across categories, as referred to in Section 4.9. This reduces the number of dependency assumptions that need to be made, but this will need balancing against creating groups that are too heterogeneous in nature. For categories where triangle-based stochastic reserving methods cannot be used, an appropriate approach will need to be selected, based on the available data and nature of the category. Simulation using Frequency–Severity assumptions and suitable mathematical distributions is an approach that can be used for such categories, for example.

Paid or Incurred Data

As explained in the context of the bootstrap method (in Section 4.4.3), the interpretation of the results of applying a stochastic method to either paid or incurred claim amount data differs slightly, but both should provide information on the uncertainty attached to the estimated ultimate claims. The choice as to whether paid or incurred claims data or both should be used when applying a stochastic method to the data will be influenced partly by the same considerations that apply when using a deterministic method. Hence, issues such as the length of the development tail and the consistency of the case reserving strength over time will be important factors to consider, as will the overall stability of the data. Furthermore, bearing in mind that the point estimate of reserves is usually based on the application of deterministic, rather than stochastic methods, one approach to deciding on whether to use paid or incurred claims data when applying stochastic methods to a particular reserving category, is to mirror the balance between these two types of data that was used in selecting the best estimate of reserves. However, if stochastic cash flow projections are required, then paid data will need to be used or a paid claim pattern selected as an input and subsequently used to generate the cash flows.

Cohort and Development Period Frequency

The uncertainty of the future claims for any particular category of reserves is not related to the cohort and development period frequency selected as part of the reserving process.

However, depending on the stochastic method used, the results can be influenced by these selections. For example, as pointed out in Carrato et al. (2015), when applying the Mack method to a quarterly triangle, the prediction variance will be higher (other things being equal) than when it is applied to an annual triangle. Consideration therefore needs to be given to the selection of cohort and development period frequency when applying stochastic methods, and in some cases the sensitivity of results to alternative options may need investigating.

Determining Correlations / Dependency

Assuming that aggregate results across all categories combined are required, after applying the relevant stochastic methods to the individual reserving categories, it will usually be necessary to make assumptions regarding correlation/dependency, as discussed in Section 4.9. The reserving grouping (and hence the data to be collated) can be partly influenced by this requirement, since it may not always be possible to calibrate the correlation/dependency for some subdivisions of the business (and hence a hierarchical approach, as referred to above, may sometimes be used).

Outwards Reinsurance

Allowing for outwards reinsurance can be more complex in a stochastic reserving exercise than a deterministic one, and hence consideration of the data required can be an important early step in the reserving process. However, in some cases this complexity may be justified. For example, with some types of non-proportional reinsurance, a stochastic approach that involves modelling the distribution of claims by size, along with their impact on the reinsurance programme, can produce a more credible estimate of the expected recoveries than a purely deterministic approach. Some of the approaches that can be used are discussed in the context of the bootstrap method in the outwards reinsurance subsection of Section 4.4.5, and these are also applicable to a number of other stochastic methods.

4.10.3 Selection and Testing of Methods

Initial Selection of Stochastic Methods

The previous sections in this chapter have focused mainly on the details of the relevant stochastic reserving methods. Understanding the range of methods available is an important prerequisite for selecting a suitable method(s), since without that knowledge, method selection can be constrained. Practitioners will often have their preferred method(s), or will default to that used in a previous or similar exercise. Practical constraints such as which methods are readily available in the reserving software that is in use in the particular context, and what data is available can also be relevant. Periodic reassessment of the suitability of the range of methods that are available can be appropriate, however, particularly if there have been significant changes in business mix or claims profile over time.

For stochastic methods, factors such as whether a full distribution is required (with or without full cash flows) and whether an ultimate or one-year time horizon is required in the given context will influence the initial method selection. If bootstrapping is selected as an approach to use, then depending on the software being used, there may be a decision to be made as to which underlying stochastic method is bootstrapped. For example, if the ODP or Mack Bootstrap is being used, then an initial review of the claims development data may be required to determine the existence of negative values in the data, as this will influence which of these two models is bootstrapped. This is discussed further with reference to incurred claims data in Section 4.4.5.

Testing the Appropriateness of Methods

Once one or more stochastic methods have been selected as initial possible approaches to use, then unless a previous (but still relevant) assessment has been made of the appropriateness of those methods, this will be an important early part of the application in practice. Assuming that the required data is available, then the core component of this assessment is to consider whether the assumptions that underpin the method are valid for the particular dataset. General reasoning can form part of this – for example, if the method requires cohorts to be independent, and it is known that some claims or events affect more than one cohort simultaneously, then this assumption is potentially invalid. Review of diagnostic charts and application of statistical tests are other ways of assessing method suitability. These are discussed further in separate subsections below.

In practice, it is not always possible to test the validity of all assumptions, and even if it is, the results may be inconclusive or may apply to some reserving categories but not others. Where there is any doubt about the validity of the underlying assumptions, this creates additional corresponding uncertainty in the validity of the results themselves, and hence it may be appropriate to treat them with more caution than would otherwise be the case. This is particularly important if the potential lack of validity of one or more of the assumptions is thought to lead to an inherent bias (upwards or downwards) in the results, noting however that in some situations it might not be obvious whether this leads to higher or lower results. Applying more than one method and comparing the results can help mitigate the risk of relying too much on a potentially inappropriate method.

In certain situations, due for example to data limitations, it may be obvious that one or more of the assumptions for the available stochastic methods are either wholly or partially invalid. An example might be the independence of cohorts, which applies for example to Mack's method and to a number of other stochastic methods. This assumption may be invalidated in some circumstances, for example it might be argued that this is the case where underwriting year data is used and where an individual catastrophe event affects more than one cohort. Adjustments to the data to help address this issue, such as removal of such events, may be possible in some cases. Alternatively, a simulation approach could be used whereby an assumption is made regarding the mathematical distribution of future claims (e.g. Lognormal or Gamma) for each cohort along with an appropriate dependency assumption between the cohorts.

Where the chosen method requires an assumption of independence of cohorts to be made, then this and one or more other assumptions may at least be partly invalidated in a number of practical applications of stochastic methods. However, the reality will often be that estimation of reserve uncertainty is still required. Unless it is practical to use other methods where the assumptions are not invalid, the use of, for example, an alternative benchmarking type approach that is not reliant solely on the data itself and the careful use of judgement in interpreting the results are possible ways forward in such situations, along with clear communication of the associated limitations of the results.

As well as selecting and testing the validity of the stochastic methods to apply to individual reserving categories, unless an integrated approach is being used across multiple categories, it will usually be necessary to produce results for the sum of the reserves across all categories. The approaches that can be used for this are outlined in Section 4.9.

Once the results across the combined categories have been produced, an assessment of the overall reasonableness of these results and by category can be made. Testing the sensitivity of results to alternative data subdivisions can also assist in determining whether this subdivision is a key factor in the overall process design. Consideration of the overall reasonableness of results can form part of an iterative review of whether the stochastic model(s), and the detailed methodological and parameter selections that have been made in the model implementation are appropriate. Although this may appear somewhat circular, an assessment of the reasonableness of the results is an essential part of any reserving exercise. For stochastic methods, this is discussed further in Section 4.10.4.

Diagnostic Charts – Residual Plots

Most stochastic methods involve a model which is fitted to the data. Hence, the residuals, which represent a measure that summarises the difference between the actual and fitted data, suitably formulated depending on the chosen model, can be calculated. Probably the most commonly used diagnostic charts are plots of these residuals. For triangle-based methods, it is common to review such plots by cohort, development period, calendar period and by size of fitted value. An example of such plots is shown in Figure 4.5, which shows the results of applying Mack's method to the Taylor and Ashe (1983) dataset. As well as residual plots, two other charts are shown in Figure 4.5 – a stacked bar chart giving the latest and estimated future claims, the Mack Standard Error, and a chart showing the forecast development patterns. These charts were produced using the "plot" functionality from "MackChainLadder" within the "ChainLadder" package in R (see R Core Team, 2013 and Gesmann et al., 2014 for further details). The relevant R code is in Appendix B.3.1.

If the model is a "good" fit to the data, then there should be no obvious trends in the residuals, for each of the different dimensions (development, cohort, calendar periods or by size of fitted value) and the residuals should be randomly distributed around zero. For the Taylor and Ashe (1983) dataset, as pointed out in Barnett and Zehnwirth (n.d.), there is some evidence in Figure 4.5 of an upward trend in the residuals in the last three calendar years. This might suggest that a calendar year effect needs to included within the stochastic model. Examples of models which allow such an effect to be modelled are the PTF model, described in Section 4.7.8; the log-linear smoothing GLM described in Section 4.7.4;

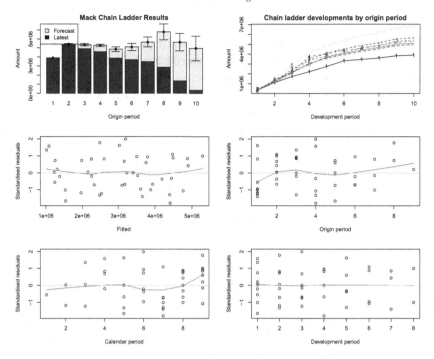

Figure 4.5 Examples of diagnostic charts – Mack's method for Taylor and Ashe (1983) data

Wright's model, described in Section 4.7.12; and the age-period-cohort model, described in Section 4.7.13. Shapland (2016) also describes a GLM Bootstrap approach that allows calendar year effects to be modelled.

Outliers in the residuals can sometimes have a large effect on the results of bootstrap stochastic models, and can be identified using Box–Whisker plots for example, as explained in Shapland and Leong (2010). The possible exclusion of the data related to such identified outliers may then follow, after considering whether they could be representative of the "true" variability in the data, in which case they should be retained.

As well as no obvious trends in the residuals, the variability for each dimension should also be broadly similar, otherwise this is evidence of the model not capturing the variance structure of the data – i.e. there is heteroscedasticity within the data.

Depending on the model being used, it may be possible to introduce additional parameters to attempt to remove the heteroscedasticity. For example, for the ODP Bootstrap method, this may be achieved by using non-constant scale parameters to adjust for development period heteroscedasticity, as discussed in the relevant subsection of Section 4.4.3. Mack's method also effectively does the same.

Shapland and Leong (2010) suggest that plots of the standard deviations of the residuals by development period, and the range relativities thereof, can be used to assist in identifying heteroscedasticity in the residuals. They show an example of these additional plots, using the Taylor and Ashe (1983) dataset.

Diagnostic Charts – Other Types

Although residual plots are probably the most commonly used diagnostic chart used when testing stochastic reserving methods, there are some other types which can be useful. This section briefly summarises a selection of these.

A type of chart that Barnett and Zehnwirth (n.d.) suggest can in some cases identify deficiencies in a model that are not shown clearly in the residual plots described above, is derived from the application of so-called *out of sample predictive testing*. The idea, based on an approach often used in time series applications, is to remove a diagonal from the data triangle, apply the stochastic method, and then compare the actual and fitted data on that diagonal, for example by drawing plots of this data or of the implied residuals. These *one-step-ahead prediction errors* or *validation residuals* can sometimes show patterns that are not evident in the other types of residual plots, and hence can be a useful part of overall method validation.

Carrato et al. (2015) suggest that one-way summary plots of actual and fitted values can be useful when building a model. They give examples showing actual and expected claim amounts by payment year, as a simple means of identifying whether a model is appropriately capturing payment year trends in the data. They suggest that one advantage of this type of plot, compared to residual plots, is that by showing the magnitude of the payments, they enable the practitioner to focus on the model fit where payment levels are high.

A further type of diagnostic approach suggested in Carrato et al. (2015) is "heat maps". These can take various forms, but Carrato et al. (2015) shows some examples that use a triangle format for the ratio of Actual versus Fitted values, where low values of this ratio, values around 100% and high values are each colour coded. A well-fitting model should in theory show a random pattern of colour, with no obvious clustering of particular colours in the triangle.

Statistical Tests

Examination of charts such as residual plots can be helpful in assessing the chosen stochastic model, but this relies upon a subjective visual inspection. This can be supplemented, if required, by using one or more statistical tests. For example, to review the distribution of the residuals, it can be useful to test whether they come from a Normal distribution. Even where the stochastic model does not require the residuals to be Normally distributed (such as the ODP Bootstrap) this can still be a useful test, as it should help highlight if, for example, there is any obvious skewness in the distribution of the residuals. One such statistical test of Normality is known as the "Shapiro–Wilk" test, which produces a p-value that can be used to test the null-hypothesis that the residuals are Normally distributed. Thus, if the p-value is less than, say, 5%, this would suggest that the residuals do not appear to follow a Normal distribution. Other tests that can be used to test the residuals are discussed in Shapland and Leong (2010). The statistical Normality tests can also be supplemented by drawing a Normality or "Q–Q plot", which shows the actual quantiles of the residuals against the theoretical quantiles from a Normal distribution. It should be

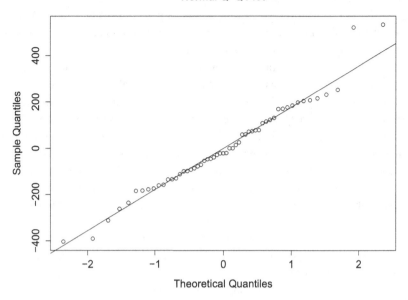

Figure 4.6 Normality test of residuals using Q–Q plot – ODP model for Taylor and Ashe (1983) data

emphasised that if the residuals pass a Normality test, this does not necessarily mean that the stochastic model is a "good" fit, but rather it assesses whether the residuals follow a Normal distribution.

A further type of probability plot is a "P–P" plot, which is similar to a Q–Q plot, but uses the cumulative distribution functions of the actual data and the theoretical distribution. This type of plot is referred to in Carrato et al. (2015) as being useful for checking the validity of the assumptions of the ODP model, for example.

An example Q–Q plot for the Pearson residuals derived from the ODP Bootstrap model applied to the Taylor and Ashe (1983) dataset is given in Figure 4.6. The residuals are those shown in Table 4.38 in the ODP Bootstrap worked example in Section 4.4.4. The Shapiro–Wilk test for Normality produces a p-value of 19.03%.

Examination of Figure 4.6 and the Shapiro–Wilk p-value of 19.03% both suggest that the residuals are reasonably close to a Normal distribution, although the plot suggests there are perhaps some large value outliers that in practice might warrant further investigation.

4.10.4 Validation and Communication of Results

As with deterministic reserving methods, an assessment of the reasonableness or validity of the results is an essential stage of the overall process of applying stochastic reserving methods, before they are communicated to the relevant personnel. The detail of this assessment will depend on the methods being used and the context of the exercise, but it will always involve the practitioner applying their judgement and experience. Where the results are being used to estimate reserve levels at higher percentile values (e.g. at the 99.5th percentile) then it is particularly important to validate the reasonableness of these results, as they are generally less reliable than at lower percentiles.

Most reserving software packages that include stochastic methods will have a range of numerical and graphical analyses that will assist with the validation of the results. Some possible examples are given below, but there will be others in use by practitioners. The graphical and numerical analyses described here also provide a useful summary of some of the ways in which results from stochastic models can be summarised and communicated, depending on the context and audience. Some of the approaches described below are only relevant for stochastic methods that produce a full distribution of the reserves (such as bootstrapping). Where this is not the case, these approaches may still be relevant if a statistical distribution has been fitted using the estimated mean and prediction error.

- **Reconciliation of stochastic results with deterministic results**. As noted in the intro-duction to this chapter, in many practical situations where stochastic methods are used and where a point estimate of the reserves is required, the latter will be determined using a combination of deterministic, as opposed to stochastic reserving methods, often over-laid with an element of judgement as well as various ad-hoc adjustments. The resulting selected best estimate of the reserves can then be compared with the best estimate of the reserves produced by the stochastic method(s) – for individual cohorts and reserving categories, as well as in total. This comparison provides both a sense check of the results of the stochastic method(s) and determines whether any scaling of the stochastic results is necessary. Where there are material differences, this can lead to further investigation to ensure that the stochastic method has captured any special features which have been allowed for in the deterministic analysis. In many cases, however, it will be impossible to replicate the bespoke adjustments that have been made in the deterministic analysis, par-ticularly if these are ad-hoc in nature. Where scaling of the stochastic results is deemed necessary, this can be by cohort, reserving category or in total, and can be multiplicative or additive. Any such scaling needs to be done with caution; this is discussed further in the description of the ODP and Mack Bootstrap in the relevant paragraphs of Section 4.4.5, but the points made are equally applicable to other stochastic methods.

- **Graphical review of results**, which can include graphs of:

  - Paid and Incurred claims development, for individual cohorts and groups of cohorts scaled to ultimate claims at one or more percentile levels (or at mean plus selected multiples of prediction error), as compared to a best estimate basis. These will be similar to those explained in Section 5.8, and allow a simple visual assessment to be made. They can be helpful in identifying any material anomalies in the results – for example, such as unreasonably high ultimate claims at a higher percentile for one or more cohorts, when the graph of incurred claims shows a reasonably long period of flat development, and low outstanding claims.

- A "fan-chart" showing the known historical claims development and the estimated distribution of cumulative claims for individual or groups of cohorts at each future development point. As well as being useful in stochastic reserving, this type of chart is used in a range of other contexts, particularly to show uncertainty in time series data and in forecasts. They typically show the uncertainty increasing as the development reaches further into the future. In a stochastic reserving context, it can be useful to compare the actual claims development up to the most recent valuation date with the uncertainty shown in the fan-chart produced at the previous valuation date(s).
- For bootstrap methods, a histogram of simulation results, perhaps with fitted "kernel density", by cohort and in total.
- Also for bootstrap methods, a curve of the cumulative distribution function of simulation results, by cohort and in total, perhaps with one or more fitted distributions also shown.
- Box-whisker plots of reserve or ultimate, by cohort and in total.
- Box-whisker plots of actual past data against simulated data, similar to that suggested in Barnett and Zehnwirth (n.d.), and referred to in the discussion of residual plots in Section 4.10.3.

Such graphs can also be useful in communicating the results to non-technical audiences, so that they can also form a view on the reasonableness of the results produced by stochastic methods, without necessarily having any detailed knowledge of those methods.

An example of the last four types of graph is shown in Figure 4.7. This example is taken from applying the ODP Bootstrap to the Taylor and Ashe (1983) dataset, as discussed in Section 4.4.4. "IBNR" in this figure is equivalent to the claims reserve. The figure was produced using the "BootChainLadder" function in the "ChainLadder" package in R. The relevant code is in Appendix B.3.3.

- **High-level reasonableness checks of numerical diagnostics**, including reviewing by reserving category, across cohorts and in total:

  - Prediction errors;
  - Coefficients of Variation ("CV");
  - Ultimate loss ratios ("ULR") at selected percentile levels;
  - Ratio of reserves at selected percentile levels to best estimate reserves; and
  - Minimum and maximum values of simulated reserves.

The last three only apply where the stochastic model produces a full distribution, such as with bootstrapping. In general, other things being equal (and there being similar business volumes across the cohorts), the absolute value of the prediction errors will be higher for the more recent cohorts. Also, the prediction errors in absolute terms for each reserving category should be higher for all cohorts combined than any individual cohort, and the total across all categories combined should be less than or equal to the sum of the prediction errors for each category. Certain factors can in practice change these generalisations in a particular situation, and some random noise will inevitably mean that the results will not always conform to these patterns.

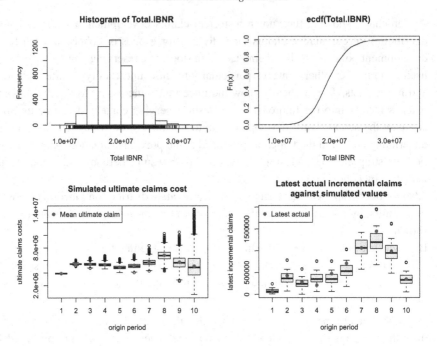

Figure 4.7 Examples of results graphs – ODP Bootstrap model for Taylor and Ashe (1983) data

- **Comparison of results against benchmarks**, related to both market/industry level data and to other entity-specific analyses that may be available to the practitioner. These benchmarks can be by reserving category and/or cohort or on a combined basis and can include the same numerical diagnostics referred to above, thus forming part of the overall reasonableness checks of the results. The points made in Section 6.9.1 regarding ensuring suitability of the benchmarks are equally relevant in the context of stochastic reserving. For some practical situations, such as a start-up company, the data available for the specific entity may be relatively sparse or non existent, and hence application of stochastic methods may either be impractical or may produce unreliable results. In this case, benchmarking will play a more central role in determining the estimates of reserve variability. Over time, when more data becomes available, more weight can be given to the results of stochastic methods applied to the entity's own data, along with a comparison to suitable benchmarks.

- **Backtesting of results**, for example by considering the reasonableness of the percentiles (derived from the stochastic reserving methods) that are implied by known historical movements in ultimate claims. Hence, if such movements always appear at the very high percentiles, then this suggests there is perhaps more variability in the data than is being captured by the stochastic methods. Distortions caused by any changes in reserve margin philosophy need to be allowed for when interpreting the output from such analysis. This type of analysis only works where there is sufficient history, but where there is it can allow those not involved in the detail of the reserving process to assess whether the results are consistent with the historical movements in reserves. Where appropriate, further backtesting can be done on industry or synthetic data, as opposed to data specific to a particular reserving exercise. This can help judge the reasonableness or otherwise of particular stochastic methods. Section 4.4.7 gives examples of some existing research to test a bootstrap procedure in this way.

- **Applying stress & scenario tests** to review, for example, the reasonableness of the percentiles (derived from the stochastic reserving methods) that are implied by the uncertainty in certain large claims or events, as advised by the claims team.

Other sources that include suggestions for stochastic reserving model selection and validation include Shapland and Leong (2010), Shapland (2016), Hayne et al. (2005), IAA (2010: Section IV.D) and Lloyd's of London (2014).

5

Reserving in Practice

5.1 Introduction

The main purpose of this chapter is to summarise a number of key generic aspects of reserving in practice. It is designed mainly for the practitioner who is new or relatively new to reserving, although some of the content may also be helpful to more experienced practitioners when they are reviewing their overall reserving process.

This chapter builds on the various other practical aspects of reserving covered in other chapters. For example, some of the practical details of applying specific reserving methods are covered in Chapters 3 and 4 for deterministic and stochastic methods respectively. Chapter 6 also covers a number of reserving topics that may be relevant to applying methods in practice, such as the use of Actual vs Expected methodology and discounting.

Inevitably, in any reserving exercise there will be points of detail that depend, for example, on the nature of the business being analysed and the purpose of the exercise. Chapter 7 summarises some of this detail for a selected range of specific contexts, such as UK Motor and US Workers' Compensation.

Hence, many of the chapters have material which is relevant to reserving in practice. Together, this content refers to the "art" of reserving, to supplement the other parts of the book, which focus more on the "science" of reserving; reserving work in practice will typically involve both of these aspects.

In Chapter 1, a reserving exercise was summarised as having four key stages – Background, Data, Analysis and Reporting–with brief details being provided on each of these stages. The current chapter provides some additional practical details of selected aspects of each of these stages. The first two stages – Background and Data – have already been covered in earlier chapters in sufficient detail, so this chapter only includes relevant additional points, to help put these stages in the context of the overall reserving process. The Analysis stage is covered in several sections, relating to various aspects of this stage of the process, which forms the core of most reserving exercises.

The focus in this chapter is on using deterministic methods to derive a point estimate of the reserves, for example, for financial reporting purposes. However, estimation and communication of uncertainty is often an important part of many reserving exercises (including those where the only objective is to derive a point estimate), and so, before the section on documentation and reporting, there is a section covering this topic.

As noted in Chapter 1, most reserving carried out in practice will involve using software in one form or another. Whilst this will influence various practical aspects of the process, the mechanics of applying the methods using the software are only one element; there are many others that can have a significant influence on the results of the exercise. One of the aims of this chapter is to identify some of these other elements.

5.2 Background

Before any data is processed in the reserving software, the first stage of a reserving analysis will usually involve a number of preliminary background activities, as identified in Chapter 1. These cover topics such as understanding the context and the business that is subject to the reserving exercise and determining whether a point estimate, range or full distribution of outcomes is required. All of these can influence various aspects of the process, such as the data requirements and method selection.

If the objective is to establish a point estimate, then agreement on the basis for the reserves is necessary (e.g. best estimate or other). For the purpose of this chapter it is assumed that the objective of the exercise is to determine a point estimate of the reserves on a best estimate basis, the intended meaning of which is given in Section 1.1. Hence, when applying the chosen reserving methods, the target reserve estimates are intended to represent the average across the range of values for the possible future claims outgo. In practice, unless a stochastic method is being used, it will not usually be possible to estimate this average mathematically – the practitioner will need to apply the methods in such a way that they do not contain any explicit or implicit margins for prudence or optimism. This is certainly a feature of the "art" of most reserving exercises, given the inexact nature of this definition. It partly involves seeking to ensure that each key judgement made as part of the reserving process is based on an approach that can, as far as practicable, be justified as being reasonable on the basis of the available internal and external evidence. An experienced practitioner will usually be able to form a view on such judgements, based on their knowledge and experience, and hence reach an opinion on whether a particular reserve selection represents a best estimate or not.

The uncertainty surrounding the best estimate of reserves and the determination of a possible margin in excess of this figure are considered in Section 5.9.

Once these basic details have been established, some level of planning the exercise may be necessary, depending on the scale of the project and whether it has been done before by the relevant team. If this is not the case, then where access to the output from the most recent reserving exercise is possible, this will often need examining, for example to understand the previous methodology and assumptions.

Most of the detail of this background stage has already been covered in Chapter 1 and so is not repeated here. Three topics for which some additional detail is helpful relate to understanding the business, reserving governance and compliance with professional and regulatory requirements. These are therefore considered in separate sections below.

It is inevitable that the details of the tasks and topics that will be covered in any preliminary background stage of a reserving exercise will depend on the context for the exercise – for example, who is doing the work, for what insurance entity and with what scope. This section focuses on the generic features that apply in one form or another to most reserving exercises. To supplement this, Chapter 7 provides some examples of the key features of reserving in a range of specific contexts, which will influence the matters to be addressed as part of the background and subsequent stages of the reserving exercise.

5.2.1 Understanding the Business

The quality of the output from a reserving exercise will almost certainly be improved if the practitioner has an appropriate level of understanding of the business being analysed. The extent of existing knowledge of the business will influence the amount of work required, and hence the relevance of the activities described here. At one extreme, if the project is being done for an entirely new insurance client of an external consulting firm, then the effort is likely to be greater than at the other extreme where it is being done by a practitioner employed by the insurance entity who has carried out several similar exercises in the recent past. However, even in the second of these, periodic consideration of the types of matters outlined here can be beneficial in ensuring that all relevant background information on the business is gathered prior to the detailed data and analysis stages beginning.

Understanding the business can be achieved in various ways – for example, through one or more of a combination of desktop examination of documents, meetings with management and analysis of relevant data. The detail will inevitably depend on the context, including for example whether the exercise is internal to the entity concerned or is part of an external consultancy engagement. The reserving categories being considered will also influence the detail, so that, for example, a large diverse portfolio of business which might have changed materially over the relevant period will likely require more in depth analysis than one consisting of a single class of business, which has remained fairly stable. This section refers mainly to generic information that might be gathered during a reserving exercise to assist in understanding the business; other examples are given for the specific contexts in Chapter 7.

Examples of the types of information that might be gathered either in meetings with management or by requesting relevant documents are summarised below, divided into various categories representing different aspects of the business. This is based partly on the subdivisions given in Appendix G of Friedland (2013). Whilst some of this information may be included in a list of the data and information to request for the purposes of carrying out a reserving exercise, it does not include the numerical data – this is covered in Chapter 2 and in the next main section, entitled "Data".

- **General company information:** Ownership and leadership of the business (including any recent changes thereof), business plans, regulatory correspondence, financial/statutory reports and strategy documents.
- **Information Technology:** Use of different IT systems for different parts of the business; Details of any recent changes in IT processes and their potential impact on the business,

including premium and claim processing; Reliance on third-party firms and/or market bureaux for such processing; Data sources for reserving and their consistency with financial reporting requirements.

- **Accounting and Audit:** Impact of accounting requirements on data used for reserving purposes, including any differences between financial reporting and reserving data segmentation; Internal or external auditor reviews or commentary related to reserving.

- **Claims:** Summary of case reserving philosophy and procedures by class of business, and any changes in these affecting the data used in the reserve study. The detail will depend on the classes of business but may include matters such as any changes in key claims personnel or external loss adjusters, reliance on third party claims bureaux or agents, any changes in automated reserve levels, information on recent claim frequency and severity trends, fraudulent claims tracking, impact of recent and pending legal and other external developments and the findings from any independent claims audits that have been conducted.

- **Marketing and Underwriting:** Summary of competitive landscape, including position in the underwriting cycle and business strategy by line of business and region/currency; key distribution channel trends, recent marketing initiatives or targeting/avoidance of any particular areas within each class of business; nature of coverage provided for each class of business and the associated drivers related to claims incidence, reporting and development. Where relevant, information might also be collated to help assess the position of each reserving category within the reserving cycle.

- **Pricing:** Rate change information over relevant historical period by class of business in recent years; any rate filing information where relevant; information from the pricing team on projected loss ratios by class of business.

- **Reinsurance:** Summary of outwards reinsurance programme protecting the relevant cohorts, including utilisation and any known reinsurance exhaustion, disputes or bad debt. The details needed will depend on the complexity and materiality of the relevant programmes.

Other categories of information may also be collated, partly dependent on the nature of the business and the scope of the exercise. For example, further details of known catastrophe or large loss exposures and a summary of the investment portfolio may be needed in some cases. Changes in business mix will also often be considered at this early stage, since that can have an impact on the reserving categories that are used.

Some authors have suggested the use of a questionnaire to gather the relevant information across the range of different topics, but the approach used in practice will depend on the circumstances. A list of questions for management, for example, is given in Berquist and Sherman (1977: Appendix B) and Friedland (2013: Appendix G). Other sources provide checklists for data and information that can be gathered in particular contexts – for example, see Jones et al. (2006), which does so separately for London Market, Personal Lines and Commercial Lines.

Once the basic details such as the purpose of the reserving exercise have been established, some practitioners prefer to gather the required numerical reserving data and carry out an initial graphical and/or numerical review before meeting with management, as this will often identify questions that only arise upon inspection of the data. There may therefore be an element of iteration in this part of the process. Examples of this type of preliminary graphical review of the data are given in the data chapter, in Section 2.5.5. In other cases, the practitioner will collate certain other types of data to enable various numerical and graphical analyses to be carried out in order to enhance or validate information gained through a review of documents or meetings with management. This may consist of databases or summary data covering aspects such as policy records, claims and exposure information. An example of the type of analyses that might be conducted in one particular context – London Market Casualty business – is given in Section 7.3.6.

The relevant information that is gathered so as to understand the business which is the subject of the reserving analysis can have an impact on many aspects of the overall process, including the data that is required, the methods to use and the assumptions to select when applying those methods.

5.2.2 Reserving Governance

Most reserving work will be carried out within some form of overall governance framework. This can either be relatively informal, but in many cases it will be specified in a more formal document. It can also be different for work carried out by internal practitioners compared to that done as part of an external professional services engagement. In general terms, however, the governance framework will impact on various aspects of the work, including:

- Reserving process and philosophy: This can impact the steps involved in each reserving exercise, as well as the target basis for the reserving levels that are produced by that process (e.g. best estimate, or best estimate plus margin to an agreed level of prudence). In some cases, there may be a distinction between the basis used for the core technical analysis (e.g. best estimate) and that used for the booked reserves (e.g. including some form of margin), for financial reporting purposes.
- Structure of teams: Which staff do the core reserving work, who reviews that work, including use of peer review, and who takes responsibility for the final results.
- Use of committees to review results: For internal reserving work, it is common for the governance framework to include one or more committees (e.g. audit or reserving) who are responsible for reviewing the output from the detailed reserving work, modifying the judgements if necessary and making a recommendation to the Board, for example.
- For internal work, the extent to which external professional services firms are used to review the results produced by the internal team.
- Agreed procedures for certain aspects of the work, such as documentation of key judgements and compliance with professional standards.

For any individual reserving exercise, consideration at an early stage of the governance framework that applies can be helpful in planning the programme of work. For example, this can help ensure that sufficient time is built into the plan to allow for actions following any feedback from relevant persons or committees. The governance framework itself may need updating periodically, in part depending on any changes in the relevant professional or regulatory requirements, which are covered in the next section.

5.2.3 Compliance with Professional and Regulatory Requirements

As for the governance framework, at an early stage of any reserving exercise it is appropriate for the practitioner to consider the professional and regulatory requirements which apply. In this context, "professional" refers to actuarial (or other) standards and guidance produced by local and international professional actuarial organisations, and "regulatory" refers to the relevant insurance or other regulator. These will vary depending on the professional designation of the practitioner (if any), the scope of work and the jurisdiction of both the insurance entity concerned and where the work is being carried out.

Some of the professional and regulatory requirements will be mandatory and others will be in the form of advisory or guidance material. They can range from high-level ethical-type conduct requirements to principles-based technical standards or more specific requirements or guidance in particular areas.Virtually all aspects of reserving work can be affected by professional requirements – including data selection and checking, application of judgement, assumption selection, use of models, consideration of reserve uncertainty, documentation of the work and communication of the findings. These requirements can arise from both local professional standards and from international standards (e.g. as produced by the International Actuarial Association). The standards will normally specify, for example, what disclosure is required in the communications regarding compliance with those standards.

Regulatory requirements will depend on the nature of the work and jurisdiction, but where relevant, they may have a direct impact on the scope of work itself, as well as on the other aspects of reserving work referred to above in relation to professional standards.

It will be the responsibility of the practitioner and the relevant insurance entity to determine the requirements that apply and hence design the reserving exercise accordingly. These requirements can change over time, and hence the practitioner and the relevant entity will need to ensure they take account of the latest requirements.

5.3 Data

As noted in Chapter 1, collation and review of data can sometimes take up a disproportionate amount of time in a reserving exercise, particularly if errors are found. However, this cannot always be avoided, and given the inherent uncertainty that surrounds estimation of reserves, time spent avoiding adding to this uncertainty by collating appropriate and accurate data is usually justified.

The process of collating the data and selecting the reserving methods to use can be somewhat circular. For example, it may not be possible to determine whether a development-based method can be used until the triangle data has been collated and reviewed. However, in many practical situations, it will be reasonably obvious from the start whether such methods are likely to be appropriate (e.g. they might have been used many times before in the relevant context) and so collation and checking of the required data can proceed immediately.

The stage of understanding the business will usually suggest possible subdivisions of the data for the purpose of estimating the reserves, and so this will impact on the required data. This stage may also help identify known or potential exposure to special items, such as large losses, catastrophe claims or large individual contracts, which will also determine whether separate data needs collating for these items. Similarly, other aspects of the process will influence the data to be collated, including, for example, the significance of the outwards reinsurance, and how that will be reflected in the data to be analysed (e.g. gross and reinsurance, or net of reinsurance, and possibly separate data by reinsurance category).

Chapter 2 provides a generic list of the range of different types of data and other related qualitative information that are used in reserving. Chapter 7 also provides some examples of this data and information in a range of different contexts, including possible subdivisions of data that might be used for reserving purposes in those contexts.

Other topics related to reserving data, such as identification of data issues are covered in Chapter 2 and so are not repeated here. However, one additional practical point that can be made before discussing the analysis stage concerns the reliance that might be placed on certain types of reserving data. For example, in cases where it is anticipated that one or more of the key reserving assumptions will be based on qualitative (or "soft") information that has been provided, the practitioner may wish to consider whether it is necessary to corroborate this information with additional quantitative data, where possible. Hence, this may result in additional data being collated before the detailed analysis stage can commence.

5.4 Analysis

This stage forms the core of a reserving process. The different components of the analysis stage and the associated practical issues that arise are covered in separate sections, which follow. It is assumed, unless stated otherwise, that the analysis is being applied to data gross of reinsurance, but most of the points will apply to other data types. Allowance for reinsurance is considered in Section 6.2.

One feature of any reserving exercise that has the potential to affect any stage of the process, but perhaps can have more impact on the analysis stage than the other stages, relates to behavioural factors. In a reserving context, this refers to a broad range of factors that might influence the choices and judgements made by the reserving practitioner when carrying out their role in the reserving exercise. An example is so-called "anchoring", which refers to the tendency to place too much reliance on one piece of information, leading to possibly biased results. Other biases might also feature in reserving, such as "status quo", which refers to the tendency to be unduly influenced by an existing value. In a reserving context, this could, for example, be the estimated ultimate claims as at the previous valuation date. Many other behavioural factors might be present in any reserving exercise, and the practitioner may benefit from periodic consideration of whether they exist in the particular environment and context in which they are operating. A suitable reserving governance framework may help to mitigate any material risks that might arise due to the influence such factors. Examples of sources that discuss this subject in more detail in a general insurance context, covering topics such as behavioural economics, prospect theory and framing include Rothwell (2016), Fulcher and Edwards (2013) and Jones et al. (2006).

5.5 Method Selection

Chapter 3 explains several deterministic reserving methods, including information on the circumstances in which they might be appropriate to use (or not to use). Deciding on which method(s) to use for any particular reserving exercise involves considering a range of factors. This consideration applies separately to each reserving category and will often involve initially determining whether triangle-based methods are appropriate or not, taking into account the underlying claims process. In many situations, such methods will be appropriate and the next step is then to consider which of these type(s) of method should be used, taking into account a range of factors such as the following (in no particular order):

- **Data availability:** The extent of the development history available will impact on whether methods such as the Chain Ladder approach can be used. The type of data that is available or appropriate to use will also influence the method selection; for example, claim number data will be available or appropriate in some circumstances, but not others. Even if a particular method is thought to be appropriate for one or more reserving categories, if the required data is not available within the time constraints of the reserving exercise, then this will make it impractical.
- **Cohorts:** It is often the case that different methods will be used for different cohorts. Typically, development-based methods can be used for the older cohorts, whereas loss-ratio-based methods (including the BF method) are usually more appropriate for the more recent cohorts. In addition, whilst the same reserving method might be used for a group of cohorts, the underlying assumptions might be different for subsets of cohorts within that group. For example, if a Chain Ladder approach is being used, and the claims development pattern appears to have changed across the cohorts, then this might require different development factor assumptions to be made.
- **Method assumptions:** Determining whether a particular method can be used will in part depend on whether the assumptions underlying that method are valid for the relevant reserving category. It may not be possible to determine this at the outset, before the data has been collated and analysed, so there may be a degree of iteration between the various stages of the analysis.
- **Software:** Whilst the type of methods that are available in the software being used to carry out the reserving should preferably not be a main driver of method selection, it is a factor that may influence this, from a practical perspective. In many cases, the software being used will already be known to incorporate methods that are expected to be used anyway.

- **Previous reserving exercises:** In many cases, except for situations such as a start-up entity, there will have been previous reserving exercises carried out by the same or different team. In either case, reviewing the methods used previously can assist in determining what might be appropriate for the current exercise, although reconsideration of this may be advisable from time to time, rather than always defaulting to what was done before.

Where triangle methods are deemed appropriate, preliminary analysis of the claims development data in numerical and/or graphical form can help both with method selection and with identification of particular features of the data that need further investigation (e.g. unusual claims development patterns for particular cohorts). This can also be helpful in deciding on the type of data to which the selected methods will be applied – for example, claim numbers and/or amounts, and if amounts whether to claims paid and/or incurred. Hence, the sequencing of data collation and method selection can vary depending on the circumstances, and can be iterative.

Where standard triangle-based methods are deemed inappropriate for certain reserving categories, other methods will need to be used, based on an understanding of the underlying claims process. An example of the methods that can be used for one such category – latent claims – is given in Section 6.6.

In many situations it will be appropriate to select more than one method for each reserving category, each of which might also be applied to more than one data type (e.g. claims paid and incurred amounts). This may be the case even if it is clear that there is one method/data combination that is obviously more appropriate than the others. The final selected results for each cohort can then be based on a suitable selection from the results of applying each method to the different data types, as explained further in Section 5.7.

One final consideration which may not directly influence method selection, but is nevertheless likely to be of interest to the practitioner, concerns knowledge regarding the reserving methods that are used by others in the same country and context. This can be based partly on the experience and market knowledge of the practitioner and their colleagues, but may also be supplemented by relevant surveys that may be conducted from time to time. For example, the 2016 survey described in ASTIN Committee (2016), covered 42 countries and looked at the use of deterministic and stochastic reserving methods, along with various other aspects of their application in practice. This showed that the Chain Ladder, Initial Expected Loss Ratio Method and the Bornhuetter–Ferguson method are the most commonly used deterministic methods (amongst survey respondents), with Frequency–Severity (i.e. Average Cost) methods also being quite widely used. Further details are available in ASTIN Committee (2016).

5.6 Applying the Methods

Having decided on the set of methods to use, assuming the required data has been collated and checked, the selected methods can then be applied to the data.

In some cases, particularly where there are a significant number of reserving categories, an automated approach might be used. This might involve an automatic download of the data in the required format and then the application of a pre-agreed standard approach for each selected method to the relevant data. For example, a CL method might be applied to each of paid and incurred claims for all reserving categories and cohorts, with pre-agreed tail-fitting procedures and possibly data exclusions, together with an automated BF method for, say, the most recent three cohorts, using user-defined prior loss ratio assumptions. The results from the automated calculations may then be reviewed by the practitioner and adjustments made where necessary.

In other cases, a more manual approach will be used, based on working through the various stages of each method within the chosen software environment.

In some contexts, the automation might involve the use of development patterns which have been rolled-forward from previous exercises, which are reapplied to the updated data triangles to produce some initial results. The use of roll-forward procedures such as this is discussed further in Section 6.3.

Various bespoke adjustments might be made within the application of each method to individual reserving categories, depending on the features of the business being analysed. For example, where there has been a significant change in business mix within an individual reserving category that is expected to impact upon the application of the chosen reserving methods, then this will need consideration, as explained in Section 6.9.2. Other examples of bespoke adjustments are given for some of the specific contexts covered in Chapter 7.

For the application of individual methods to particular types of data, results might be produced using alternative scenarios for the key assumptions, particularly those where there is most uncertainty. An example might be producing results using, say, "Low", "Medium" and "High" tail factors. These alternatives all contribute to the output from this part of the analysis.

It is helpful to document the methods used, key judgements and any bespoke adjustments made at each stage of the process. Various options exist for such documentation, ranging from using a "notes" facility in the relevant software to pre-defined templates by reserving category.For many reserving projects, there will be an agreed procedure for peer-review and sign-off of the work. This can relate both to the judgements and options selected within each method for each reserving category, and to the final selected results (as discussed below). This will define, for example, the particular individual reserving specialists responsible for each part of the process.

Whatever approach is used to derive the initial set of results for the chosen methods, the output for each reserving category will comprise various schedules, the content of which will depend on factors such as the purpose of the exercise, the software being used and the style and content preferences of the relevant insurer, consulting firm or individual practitioner. Two types of numerical schedule are one which summarises the results for an individual method (applied to a particular data type such as claims paid or incurred) and one which shows the results for all methods that have been used (applied to all relevant data types). The former will be used as a diagnostic tool for judging the reasonableness of the results for an individual method, and will therefore depend on the nature of that method. An example of this type of schedule for the CL method is shown in Table 5.1.

Table 5.1 *Example results table: Single method*

As at date:	31-Dec-14	Analysis by:	DJH
Reserving category:	Taylor & Ashe data	Peer review by:	FEG
Unit:	Thousands		
Data type:	Claims Paid		
Method:	Chain Ladder		
Version:	No tail factor		

	Actual data				Estimated			
Cohort	Prem	Paid	OS	Inc	UltPrm	UltClm	IBNR	Reserve
2005	5,333	3,901	–	3,901	5,333	3,901	–	–
2006	7,333	5,339	11	5,350	7,333	5,434	84	95
2007	6,875	4,909	25	4,934	6,875	5,379	445	470
2008	6,875	4,588	46	4,634	6,875	5,298	664	710
2009	6,111	3,873	39	3,912	6,111	4,858	946	985
2010	5,914	3,692	464	4,155	5,914	5,111	956	1,419
2011	6,316	3,483	952	4,435	6,316	5,661	1,226	2,178
2012	6,316	2,864	1,320	4,185	6,316	6,785	2,600	3,920
2013	6,000	1,363	2,049	3,412	6,000	5,642	2,230	4,279
2014	5,714	344	669	1,013	5,714	4,970	3,957	4,626
Total	62,788	34,358	5,574	39,932	62,788	53,039	13,107	18,681

	Loss Ratios				Dev Fac		
Cohort	Paid	Inc	IBNR	Ult	Paid	Inc	IBNR/OS
2005	73%	73%	0%	73%	1.00	1.00	N/A
2006	73%	73%	1%	74%	1.02	1.02	786%
2007	71%	72%	6%	78%	1.10	1.09	1813%
2008	67%	67%	10%	77%	1.15	1.14	1447%
2009	63%	64%	15%	79%	1.25	1.24	2425%
2010	62%	70%	16%	86%	1.38	1.23	206%
2011	55%	70%	19%	90%	1.63	1.28	129%
2012	45%	66%	41%	107%	2.37	1.62	197%
2013	23%	57%	37%	94%	4.14	1.65	109%
2014	6%	18%	69%	87%	14.45	4.91	592%

The second type of schedule, an example of which is given in Table 5.2, can be used to compare the results from several methods applied to one or more data types, and act as a summary of the selected results. Although based on the Taylor and Ashe (1983) data used in the worked examples in the earlier chapters, most of the figures shown in these tables are for purely illustrative purposes only. The items shown in both of these example schedules should be self explanatory.

The format and detailed content of these types of schedule will vary considerably between different reserving exercises, with many types of data and diagnostics being shown in them. In this case, for example, Table 5.2 includes the estimated ultimate claims from the previous analysis as well as a comparison of Actual vs Expected development. These can be helpful items when deciding on the final selected results, as explained further in Section 6.3.

Table 5.2 *Example results table: Results selection*

As at date:	31-Dec-14	Analysis by:	DJH
Reserving category:	Taylor & Ashe data	Peer review by:	FEG
Unit:	Thousands		
Results status:	Draft		

Cohort	Actual data					Claims ultimate using			
	Prem	Paid	OS	Inc	UltPrm	CL Pd	CL Inc	IELR	BF Pd
2005	5,333	3,901	–	3,901	5,333	3,901	3,940	4,000	3,901
2006	7,333	5,339	11	5,350	7,333	5,434	5,542	5,500	5,435
2007	6,875	4,909	25	4,934	6,875	5,379	5,379	5,500	5,389
2008	6,875	4,588	46	4,634	6,875	5,298	5,298	5,500	5,325
2009	6,111	3,873	39	3,912	6,111	4,858	5,053	5,500	4,988
2010	5,914	3,692	464	4,155	5,914	5,111	5,316	5,500	5,219
2011	6,316	3,483	952	4,435	6,316	5,661	5,944	6,000	5,791
2012	6,316	2,864	1,320	4,185	6,316	6,785	6,853	6,000	6,331
2013	6,000	1,363	2,049	3,412	6,000	5,642	5,868	6,000	5,914
2014	5,714	344	669	1,013	5,714	4,970	5,218	6,000	5,929
Total	62,788	34,358	5,574	39,932	62,788	53,039	54,410	55,500	54,223

Cohort	Selected				Prev's Ult (30/9)		A vs E (Paid)		
	Method	Ult	IBNR	Reserve	Value	Change	Act	Exp	A – E
2005	CL Inc	3,940	39	39	4,006	– 66	68	95	– 27
2006	CL Inc	5,542	193	203	5,617	– 75	425	490	– 65
2007	CL Paid	5,379	445	470	4,841	538	280	196	84
2008	CL Paid	5,298	664	710	4,768	530	206	188	18
2009	CL Paid	4,858	946	985	4,372	486	471	391	80
2010	CL Paid	5,111	956	1,419	4,753	358	706	590	116
2011	CL Paid	5,661	1,226	2,178	5,434	226	1,063	923	140
2012	CL Paid	6,785	2,600	3,920	6,988	– 204	1,443	1,561	– 118
2013	BF Paid	5,914	2,501	4,550	5,529	384	1,062	923	139
2014	BF Paid	5,929	4,916	5,585	5,861	68	377	351	26
Total	N/A	54,417	14,485	20,059	52,171	2,246	6,101	5,708	393

Cohort	Loss Ratios				Dev Fac		
	Paid	Inc	IBNR	Ult	Paid	Inc	IBNR/OS
2005	73%	73%	75%	74%	1.01	1.01	N/A
2006	73%	73%	75%	76%	1.04	1.04	1804%
2007	71%	72%	80%	78%	1.10	1.09	1813%
2008	67%	67%	80%	77%	1.15	1.14	1447%
2009	63%	64%	90%	79%	1.25	1.24	2425%
2010	62%	70%	93%	86%	1.38	1.23	206%
2011	55%	70%	95%	90%	1.63	1.28	129%
2012	45%	66%	95%	107%	2.37	1.62	197%
2013	23%	57%	100%	99%	4.34	1.73	122%
2014	6%	18%	105%	104%	17.23	5.85	735%

These types of results schedules can prompt a number of questions to be raised or new issues to be identified by those who are involved in their review, which may need addressing before the results can be finalised.

When preparing the second type of schedule, which summarises the results from the range of methods that have been applied to the data, it can be helpful if the practitioner identifies in any accompanying commentary which methods are deemed potentially reasonable for which cohorts, or whether some should be ignored for certain cohorts. If this is not done, then when others are reviewing the schedules they may wrongly infer that the results from all methods are equally plausible, and this is sometimes not the case.

5.7 Selection and Analysis of Results

The process of selecting the final results from amongst those derived by the various methods, or otherwise, is arguably the most critical stage, as it leads directly to the reserves produced by the overall process, rather than the results of applying a single method to the data. This is an example of where the experience of the reserving practitioner can have a significant impact on the selected results.

For the purposes of this section, the "final results" are assumed to be those that are selected as part of the core technical reserving analysis, rather than the possibly different values that are finally selected by the insurer, for financial reporting booking purposes, for example. For that purpose, the procedures that are used will depend on the insurer's own reserving philosophy and governance process. For example, the reserving governance process may involve the relevant committee making the final selection of reserves that will be booked, taking into account broader considerations. The information provided as part of the reporting stage of the reserving exercise, including that related to uncertainty, as covered later in this chapter in Section 5.9, would usually be expected to make an important contribution to the selection process for the final reserves to be used in the relevant context. Commercial, regulatory and strategic factors, for example, may also have differing levels of impact, depending on the context.

There are no golden rules or universally applicable procedures that define the factors to take into account when selecting the final results derived from the core technical reserving analysis, although some considerations are summarised below.

- There may be a default or standard "in-house" preferred approach, such as the CL method applied to incurred claims for all cohorts except the last two, for which the BF method on incurred claims is used, with priors based on a combination of business plan loss ratios and averages across earlier cohorts.
- The estimated ultimate claims derived in the previous reserving exercise might form a reference point for the latest selected results, although the risks associated with "anchoring" results may be relevant here, as referred to in the brief discussion of behavioural factors in Section 5.4. If there was such an exercise, then it is often helpful if it includes an estimate of the expected claims paid and/or incurred movement during the period

between valuations. If this exists, then the schedule as at the current valuation date can also incorporate Actual vs Expected comparisons, as shown in Table 5.2, which can then be factored into the results selection process, if required. Actual vs Expected analyses are discussed further in Section 6.3.

- A review of a range of numerical and/or graphical diagnostic tools can assist with the process. Some practitioners favour using graphs to select between results from various methods or to determine if an initial set of selected results is reasonable; this approach is discussed in Section 5.8. Others prefer to focus on numerical diagnostics such as ultimate loss ratio, ultimate to incurred ratio, ultimate to paid ratio, IBNR to case ratio and IBNR as % premium. The values of these items for individual cohorts and the overall trend across the cohorts can be reviewed. Experience and judgement are used to make results selections based on a review of these diagnostics. Some of these diagnostics are shown in the example results in Tables 5.1 and 5.2.

- It may be obvious that the results from certain methods are unreasonable, so these can be ignored – e.g. if the estimated ultimate is below incurred to date for a particular cohort, for which it is known that claims incurred are still developing upwards. Similarly, it may be clear that most or all the results of applying a particular method to one type of data are obviously unreasonable for one or more cohorts. In some cases, it will be known in advance what methods and data combinations are likely to produce unreasonable results, and so these combinations need not be used, or the results may be discarded at an early stage of the process.

- There may be specific features of the data such that it is anticipated that some methods will be likely to produce results that are invalid (although perhaps not obviously wrong) for particular cohorts – for example, an unusual large claim early in the development of a cohort might mean that a CL-type method will produce unreasonably high results for that cohort.

- The relative materiality of each reserving category, and the variability of the estimates across the different methods (applied to different data types) and cohorts can help decide where the selection effort should be focused. For example, for categories where the majority of methods produce results that are reasonably similar to each other for several cohorts, the final selection for that category is likely to be easier than where this is not the case. Similarly, if there are results from individual methods that are materially different to those produced by a number of other methods for one or more cohorts, then, once the reason for this has been established, a decision can be made as to whether to include them or not in the consideration of the selected ultimate. Finally, in some cases, where the application of the same reserving method to paid and incurred claims data produces different results for one or more cohorts, then rather than perhaps just taking averages to produce the selected results, further investigation of this is sometimes justified, in an attempt to understand what is causing the disparity in results. Treatment of divergent results such as this in the context of the Chain Ladder method is discussed in Section 3.2.5.

- In some cases, there may a number of methods that all produce equally reasonable results for one or more cohorts for a particular reserving category, with no obvious way to choose between them. In this case, some practitioners combine the results produced by the different methods using some form of simple or weighted averaging to produce the selected ultimate. The weights used might be based on a review of the relevant numerical and graphical diagnostics, and on the practitioner's judgement and experience. The use of graphs in this way is discussed in Section 5.8.

- Knowledge of both the general and specific aspects of each reserving category relating to factors such as market cycle, profitability, claims environment and other emerging trends might influence the selected results, and may not be obvious from the data.

The key objective here is to select results for each reserving category and cohort that are designed to be appropriate or reasonable given the context and basis of the reserving exercise. In some cases, this might mean selecting results for one or more cohorts that are outside the range of results produced by the various methods, perhaps due to allowance for special features such as the existence of large claims. Being able to explain the rationale for the selections made from amongst the results produced by the various methods (or not) can be very helpful in communicating and documenting the process. Such documentation can be a good discipline in its own right, but in any case might be a requirement of the relevant professional standards that will apply in some circumstances. As noted above, some practitioners use graphs as part of the final results selection process, and so this approach is discussed in further detail in Section 5.8.

The above process might only produce an initial draft set of results. During the process of selection and peer review of these initial results, various questions can arise. These might be questions for the reserving practitioner who carried out the review, such as requests for alternative analyses and assessment of the impact of other ad-hoc adjustments or further subdivision of reserving categories. They might also include questions that need to be discussed with the relevant insurer's staff (e.g. claims and underwriting). Furthermore, depending on the context, there may be governance steps in the process (e.g. reserving committees) or a comparison of the results with those produced by other reserving practitioners involved in the process, such as internal actuarial reserve estimates being compared with external actuarial estimates, or review by group reserving departments in larger insurance corporations. These questions and any further governance steps or comparisons of results from different sources can all result in there being a degree of iteration from the initially selected draft results to the final selected results.

A final point to note regarding this stage is that, although the overall objective is often to select a single estimated reserve value, the process of producing results using more than one method applied to different data types also produces useful information on the uncertainty surrounding the selected estimate, which can form an important part of the communication of the results. This is discussed further in Section 5.9.

5.8 Using Graphs to Review and Select Results

As explained in Chapter 2, graphs can be used at the data analysis stage, even before any reserving methods have been applied to the data, to provide insight into the underlying claims development. However, they become even more powerful when used as an integral part of applying the methods and when reviewing or communicating the results. Most of the graphs used in this way are of the claims development, in various forms. This subsection provides examples of some of the types of graph that can be used within various stages of the process, but there will be other examples used by practitioners, in part dependent on the functionality of the reserving software being used. Such software may have a range of other useful graphing functionality, such as magnification of particular areas of the graph and highlighting individual cohorts.

It is assumed in the examples that follow that development-based methods, such as the CL method, have been deemed appropriate to use for at least some of the cohorts (since otherwise, the use of this type of graphical review is less relevant). It is also assumed that graphs are being used for claim amount data only, although very similar graphs can also be used for other types of data such as premiums or claim numbers.

Types of Graph and Their Uses in Reserving

There are two main types of graph that are typically used in practice:

- **Graph type 1 – Individual cohorts**: Individual cohort graphs showing claims paid, incurred and possibly also outstanding. These are usually shown on a cumulative basis, but incremental graphs can also reveal features that are sometimes otherwise not evident on the cumulative graphs. They can either be unscaled or scaled to the estimated ultimate values, as derived from one or more methods. When in an unscaled format, they can also either be raw amounts or loss ratios, in which case they might also include the estimated ultimate as a horizontal line. The ultimate values included on these graphs can either be the latest available ultimate or the estimated ultimate as it has changed over time. Alternative ultimates on different bases can also be shown, represented as different horizontal lines. When used as part of the process of applying a CL-type method, these graphs can also sometimes include the "fitted" development pattern, which may vary by cohort, depending on the approach used.

- **Graph type 2 – Grouped cohorts:** All cohorts (or a subset thereof) are shown on the same graph, usually with each cohort scaled to its own chosen ultimate value, so that the vertical axis scale is expressed as a percentage, with the ultimate value therefore being represented by a horizontal line at 100%. These can be shown separately for claims paid and claims incurred or with both data types shown on the same graph. The graphs are usually shown in scaled cumulative format, but scaled incremental graphs can also sometimes be useful. Unscaled graphs of this type can also be used, which are not then dependent on the ultimates, but are helpful in reviewing the relative volumes of claims in each cohort and the overall claims development profile, before scaling to ultimate. As with individual cohort graphs, in some contexts, this type of graph can also include the fitted development pattern.

Some examples of both of these types of graph (all on a cumulative basis) are given in the description of the CL method, in Section 3.2.7. An example of an incremental graph (on an unscaled basis) is given in Section 2.5.5.

Variations on the main two types of graph may be used, and other types of graph are also used by some practitioners to assist with the reserving process, such as those showing development factors (e.g. for curve fitting purposes, as per Section 3.2.8) or the change in estimated ultimates over time.

These graphs can be used in three distinct ways – first, to assist when making judgements and assessing the reasonableness of results within the application of a single method; second to help choose between methods and make the final selection of estimated ultimates; and third, to help communicate to others the reasonableness of this selection. For the first of these, when using the CL method for example, the graphs can be used to assist in identifying individual cohorts or development ratios which are outliers and which might therefore need to be excluded from the estimation of average development factors. They can also be used to identify any changes in the overall claims paid and/or development pattern over time that may need to be allowed for in the reserving process. Further examples of how graphs can be used in this way in the context of the CL method are given in Section 3.2.7.

For the second and third types of usage – determining and communicating the final selected ultimate claims and hence reserves – the two types of graph described above can be created using the estimated ultimate values for each of the various methods that have been applied to the data, to help choose between the methods. Alternatively, the graphs can be created using an initial draft selection for the ultimates (e.g. based on a weighting across methods that might vary by cohort or on an automated roll-forward of the previous estimated ultimates). The objective in all cases is to help judge the reasonableness of potential ultimate selections, based on the observed claims development for individual and groups of cohorts.

For cohorts where a development factor based approach is deemed appropriate, examination of the relative position of each cohort on the grouped cohorts graphs can be a useful way of judging the reasonableness of the ultimate. This is usually considered mainly at the latest development point for each cohort on a paid and incurred basis on the scaled graphs, although examining the relative position at earlier points can also be useful. If the relative position is deemed inappropriate for a particular cohort, then use of the scaled graph in this way can be used to select an alternative position. This is then implemented numerically using the "% developed" value that is implied by the alternative selection. Where this alternative position is chosen, for example, to be at approximately the midpoint of the position on the graph of the other cohorts (or a suitable subset thereof) at the same development point, it is effectively equivalent to selecting the median cumulative development factor to ultimate implied by the other cohorts. Other alternative positions will be equivalent to other functions of the development factors implied by the other cohorts.

If this approach is used, then, by selecting a revised % developed from which the estimated ultimate is calculated for a particular cohort, a judgement is being made that this is an appropriate reserving basis to use. This judgement can be justified or explained in a number of ways, such as:

- By reference to benchmark % developed factors derived from suitable sources.
- By showing that a similar ultimate value can be derived through the full application of, for example, a CL method to the the relevant data triangle, or through a suitable weighting across the results from applying a number of such methods, with different underlying assumptions.
- By comparing the selection with the range of results already produced by the different methods that have been applied to one or more data types.

The approach of selecting the % developed, based on a visual inspection of the graph may involve departing from the principles or assumptions underlying one or more of the relevant reserving methods. However, it can be appropriate where, for example, the application of such methods results in reasonable estimates for some cohorts, but not others, and so some form of manual adjustment using an approach such as this may be justified for one or more cohorts. The reserving practitioner's experience and judgement will play an important role in ensuring that the final selected ultimates are reasonable, based on the available data and other relevant information.

These graphs can also be used by reserving practitioners to present the results as part of a technical review process, including peer review. Finally, they are also a powerful way of communicating the results of a reserving exercise to individuals who may not be reserving specialists, and, in so doing, help them to fulfil their role in the reserving governance process.

Limitations of Using Graphs

Although graphs represent a powerful diagnostic tool, their use in this way does have its limitations. In particular, as previously noted, they are typically only relevant for reserving categories where there is both sufficient historical data and where development-based methods such as CL are appropriate. It is also difficult to make much use of graphs for cohorts that have very limited claims development. Where the business mix, underlying case reserving or speed of claims payments have changed across the cohorts, then the graphs can be potentially misleading if an allowance for such changes is not taken into account. Similarly, the inclusion of large losses or catastrophes in the data can also lead to invalid inferences being made from the graphs. Finally, in practice the results selection process will rely on both graphical and numerical analysis, so that even where graphs play a key role in that process, it will normally be appropriate to review the numerical diagnostics (such as Ultimate Loss Ratios and IBNR to premium ratios) implied by any ultimates selected at least partly using graphical analysis, before finalising the results. Indeed, some practitioners make little or no use of graphs, preferring instead to focus on numerical diagnostics to assist, for example, in the selection of assumptions within the application of individual reserving methods and in the selection of final results from the range of different methods that have been used.

Hence, just as with numerical reserving data and results, care needs to be taken in using and interpreting the types of graph described here for reserving purposes. If used appropriately, however, graphs can be a very useful part of the reserving practitioner's toolkit.

5.9 Estimating and Communicating Uncertainty

In the previous sections of this chapter it was assumed that the reserving exercise involved deriving a single point estimate of reserves, on a best estimate basis. In many contexts, consideration of the uncertainty surrounding this estimate and provision of information regarding this uncertainty to the recipients of the results may also be appropriate. Such consideration can arise for various reasons. For example, professional standards can require communication of uncertainty. In addition, information on this uncertainty can be useful as an input to the process of determining a margin in excess of a best estimate, for financial reporting purposes. In other cases, the determination of a range for the reserves may be part of the required scope of work. Finally, even where these reasons are not a feature, the recipients of the output of a reserving exercise may benefit from being provided with information related to reserve uncertainty, as this may not always be apparent when a single point estimate is the focus of the exercise. In other words, it will help to avoid the situation where the recipients may gain a false impression as to the reliability of the point estimate, which represents a predicted value of the uncertain future claims outgo.

It is worth noting that, strictly speaking, once a single value for the reserves (i.e. the point estimate) has been selected it is not itself "uncertain", since it is a fixed and known figure. The uncertainty arises because the reserve value is an estimate of an unknown item – the future claims outgo (or more simply, the "outcome"). So, the uncertainty relates to the fact that the actual, but currently unknown, future claims outgo may turn out to be different to the known current reserve value, which is simply an estimate of that outgo. This is usually what is meant by the term "reserve uncertainty", and that is its intended meaning when used here. The different types of uncertainty involved in estimating the future claims outgo are considered further in the introduction to stochastic methods in Chapter 4.

There is a considerable volume of published material on this topic. Some of the relevant sources are cited in this section and elsewhere in this book. One additional source that provides a useful summary of the work carried out between 2004 and 2009 in relation to reserving and uncertainty by the UK actuarial profession's General Insurance Reserving Taskforce (GRIT) and its successor body, the Reserving Oversight Committee (ROC), is Gibson et al. (2009).

The remainder of this section is divided into three subsections – the first deals with approaches that can be used to assess reserve uncertainty, followed by a section that considers the use of reserve ranges and finally a section that concludes with a discussion of the terminology used to communicate reserve uncertainty.

5.9.1 Approaches for Assessing Reserve Uncertainty

A number of approaches exist for assessing reserve uncertainty. These range from complex numerical or analytical techniques to broader high-level qualitative consideration of the impact of different sources of uncertainty. In practice, a combination of the approaches described here might be used, depending on the context, as they may provide complementary perspectives on the uncertainty. As with selecting a point estimate for the reserves, the practitioner's judgement and experience will need to be used to select the approaches that are appropriate for each reserving exercise, for example taking into account the complexity and benefits of each approach.

Three different approaches are considered here for assessing reserve uncertainty, as summarised below:

- **Use a stochastic method:** Chapter 4 describes a number of stochastic methods that can be used to provide either an estimate of the variance around a point estimate of reserves or the full distribution of future claims outgo (possibly by future calendar period). Depending on the data, assumptions and methods used, only certain aspects of the uncertainty may be captured, as discussed further in Section 4.1.3. In some situations, the mean value from the stochastic methods is scaled to the point estimate of reserves derived from deterministic methods, as explained further in Chapter 4. The use of stochastic methods may first involve a detailed application to individual reserving categories, with the results then being aggregated (allowing for dependency) to produce an aggregate distribution. Where a full distribution has been estimated, the equivalent percentile value on that (possibly scaled) distribution of any single value for the reserves can then be determined. Some insurers use such an approach as part of the process for establishing the margin in excess of the best estimate, so that the resulting booked reserves (i.e. best estimate plus margin) used for financial reporting purposes are set in accordance with their agreed reserving philosophy. In other cases, where only very high-level indicative information regarding reserve uncertainty is required, stochastic methods may be used in a much more approximate way, for example by applying them to an aggregate data triangle for all reserving categories combined and then showing a single overall "fan chart" for each cohort or group of cohorts; such simplified analysis will have its limitations, as referred to in Section 4.9.
- **Use a range of deterministic reserving methods or stress/scenario tests:** In many deterministic reserving exercises, more than one method will be used, applied to different data types (e.g. paid and incurred claims) and using a range of alternative assumptions (e.g. varying tail factors). Some of the results may also be based on the use of benchmarks, derived from other sources. Others may be based on suitable stress or scenario tests, related to factors that will influence the future claims outgo. Hence, the output from these different approaches can provide information on how the reserve estimate varies according to the methods and assumptions that are used. This in turn might be used to make broad inferences regarding reserve uncertainty, although the extent to which the various methods have captured the full range of potential outcomes or include results

which are deemed unreasonable needs to be taken into account before drawing any conclusions and communicating the results.

- **Review sources of uncertainty:** This approach involves identifying the specific sources of uncertainty associated with the future claims outgo for the relevant reserving categories, often also with some assessment of the potential quantitative impact of those sources. Collaboration between different personnel at the insurer, including reserving, claims and underwriting staff and the committees involved in the reserving governance process (e.g Reserving, Audit and Board) can be helpful as part of this process. The sources will vary according to the context, but can be external, such as those relating to the claims or legal environment, and/or internal, such as those relating to the insurer's processes and procedures, or to specific large claims or events.

  The quantification of the impact of the sources represents an estimate of the amount by which the best estimate of the reserves might change due to each source. This may be derived using detailed analysis, based on the particular circumstances of individual claims for example, or from more approximate approaches. In some cases, information on the potential likelihood associated with a particular source might be provided, and in others only information on the severity associated with the source will be given, perhaps including sensitivity analyses. Where more than one source is identified, consideration will need to be given to whether they are independent of each other, or whether there is any dependency between them. The overall impact can be either an increase or reduction in reserves, or both, and it is common for it to be expressed as a range, which may not necessarily be symmetric. Ideally, if a range is quoted, then some indication of the meaning of the range is helpful, as discussed further below. The results of this type of approach can usefully be arranged in a simple table, which lists the various sources of uncertainty, along with the estimated range of their potential numerical impact on the reserves, where available, and possibly some indication of the likelihood associated with each source. This then enables the practitioner as well as the relevant persons or committees involved in the reserving governance process to review the sources and their impact, and to consider whether any proposed margin within the booked reserves is appropriate, given the magnitude of the impact of the key sources of uncertainty. Chapter 7 refers to some of the particular sources of uncertainty associated with a number of reserving contexts.

Other approaches to assessing reserve uncertainty may also be used in practice, allowing for the many qualitative as well as quantitative aspects that are relevant. For consideration of a general framework for measuring reserve uncertainty, see MacDonnell and MUQ Working Party (2016).

5.9.2 Use of Reserve Ranges

Where the output from a consideration of reserve uncertainty using one or more of the approaches described in the previous section is some form of range for the reserves, it is helpful if the associated commentary makes clear the meaning of this range. Similarly, where an individual reserve value is assigned some form of qualitative label or attribute (such as "high"/"low" or "unlikely to be exceeded") to describe its perceived level of prudence or optimism, it may be appropriate to clarify the meaning of the attribute. In some cases, the scope of work will require a particular type of range or individual reserve value to be determined. In others, the scope may be less clear. In both cases, ensuring that the requirement is fully understood at the start of the exercise can avoid misunderstandings later in the process.

Some examples of the types of ranges that might be referred to in practice include a range of:

- best estimates/reasonable best estimates/central estimates;
- reasonable/plausible/probable outcomes or estimates; and
- possible outcomes.

Where multiple terms are given above for the same type of range, it is assumed that these are synonymous with each other. For example, "a range of reasonable best estimates" is assumed to be synonymous with "a range of best estimates". The meaning of such terms can vary according to the context and market practice in the relevant territory, but some possible definitions that might form the basis of their meaning are summarised below. These definitions will need adapting to the particular circumstances, and are not intended as firm definitions, merely potential example definitions.

A Range of Best Estimates

This will typically represent the range of reserves, within which all values could be described as being a best estimate. Such a range arises due to the inherent uncertainty in the reserve estimation process – there is usually no single known "best estimate" of the reserves in any given situation – even though an individual practitioner might have firm views on what their proposed best estimate should be. In effect, if several reserving practitioners were given identical data and relevant information (other than each other's reserve estimates), they would inevitably all derive different, although potentially quite similar, best estimate reserve values. The range of values that they would produce effectively defines a range of best estimate values. However, since the true boundaries of this range are in practice unknown, individual practitioners will need to derive an estimated range of best estimates. For example, they can do so by varying the assumptions used as part of the application of the relevant reserving methods to derive estimates that they consider to be towards the top and bottom of the range, but always ensuring that the resulting reserves can all still be reasonably described as best estimates. Alternatively (or in addition) they can use their judgement and experience to apply approximate positive/negative percentage loadings to their best estimate of reserves.

A Range of Reasonable Outcomes/Estimates

This will typically represent a range that is wider than a range of best estimates, but not as wide as the range of possible outcomes, described below. Hence, it may exclude any outcomes which are considered unlikely or remote, but include outcomes that are perhaps reasonably optimistic and pessimistic. Defining what a "reasonable outcome" or "reasonable estimate" is can be difficult in practice without using additional explanatory language or perhaps quantitative values, such as percentiles. This is discussed further in the communicating uncertainty paragraphs below.

A Range of Possible Outcomes

This will usually represent a much wider range than the previous two types of range and effectively is intended to encompass all possible outcomes for the future claims, regardless of how extreme those outcomes might be perceived. It has limited use in practice in a financial reporting context, due to the relatively wide set of values that it usually encompasses, although the overall distribution of the range of outcomes may be relevant in other contexts. The lower value will often not be as low as zero or below, although such extremes are certainly possible in some situations, particularly if there are recoveries due from other parties, or if the reserve definition includes a deduction for specific items such as future premiums. In many situations, however, the lower value might be bounded by the case reserves, or a proportion (less than or greater than 100%) thereof. It might be determined by the use of very optimistic (but still technically possible) assumptions, or by selecting a value at a suitably low percentile if the full distribution of future claims has been estimated. The higher value, on the other hand, may have no obvious theoretical limit, although in some cases it may be bounded by a suitable aggregation of policy limits. It can be determined in a similar way as for the lower value, using very pessimistic assumptions, or by selecting reserves at a suitably high percentile level.

The interpretation of a range of reasonable estimates and the range of possible outcomes is discussed further, for example in Gibson et al. (2009).

5.9.3 Terminology Used to Communicate Uncertainty

The nature of the communication of uncertainty will depend on the particular circumstances, including the requirements or needs of the recipients of the output from the reserving analysis and the detail of the work that has been carried out to assess the uncertainty. Part of the communication will involve describing the analysis that has been done to assess the uncertainty. At one extreme, high-level qualitative statements might be provided, and at the other a description of the results of applying one or more stochastic methods to the data might be described in a detailed report.

Whatever approach is used, careful and consistent use of terminology to describe the uncertainty is vital in ensuring the communication is fully understood. This may involve using some form of label or attribute to describe the estimated reserve value or range, which can help in understanding the uncertainty attached to the estimate. Examples of the labels that might be used when referring to individual reserve amounts or to the upper or lower points of a range (and hence to the overall range) include:

- best estimate/reasonable best estimate/central estimate;
- low/high;
- optimistic/pessimistic; and
- prudent/cautious.

Qualitative statements that refer to the chance or probability that the actual future claims will be above or below a single reserve figure or within or outside a particular range may also be used in some contexts. These statements might include terms such as "likely", "very unlikely", "remote" or "extremely unlikely".

Providing a clear definition or additional explanation for these labels and terms can assist in clarifying their meaning. The use of the term "best estimate" and by implication its synonyms given above has already been discussed earlier in this chapter, in Section 5.2, and in Chapter 1, so is not discussed further here. For the others, the definition may involve providing some form of approximate probability or chance to help clarify the intended meaning. Possible example definitions/clarifications (in words and expressed as probability statements) are summarised in Table 5.3.

Table 5.3 *Example definitions of terms used for level of prudence or optimism in given reserve values*

Label/term[a]	Example meaning[b]	Example probability
Fairly high	Fairly unlikely that the outcome will exceed the value	75% chance of outcome being below the value
Fairly low	Fairly likely that the outcome will exceed the value	75% chance of outcome being above the value
Low	Likely that the outcome will exceed the value	90% chance of outcome being above the value
High	Unlikely that the outcome will exceed the value	90% chance of outcome being below the value
Very low	Very likely that the outcome will exceed the value	95% chance of outcome being above the value
Very high	Very unlikely that the outcome will exceed the value	95% chance of outcome being below the value
Extremely low	Extremely likely that the outcome will exceed the value	99% chance of outcome being above the value
Extremely high	Extremely unlikely that the outcome will exceed the value	99% chance of outcome being below the value

[a] In any of these, an equivalent term for "Low" that might be used is "Optimistic", and equivalent terms for "High" might be "Cautious" or "Pessimistic".

[b] Variations on these are possible. For example, in Jones and ROC Working Party (2007), the meaning corresponding to "Very high" above is "Possible, but very unlikely that the outcome will lie above this estimate".

These will need adapting to the particular circumstances and are not intended as firm definitions, merely potential examples; many variations may be used in practice. For example, the probability statements may all be preceded by suitable phraseology to make it clear that they are only approximate indications, rather than exact and known measures of chance. Some other examples are given in the vocabulary table in Jones and ROC Working Party (2007), which also provides further commentary on quantifying and communicating reserve uncertainty.

Where "remote" is used as part of the attribute assigned to a range or single reserve figure, then the definitions shown in Table 5.3 for "very unlikely" or "very likely" could be used. For example, the phrase "it is estimated that there is a remote chance that the actual future claims will be above the upper point of the range" could be interpreted as meaning that there is estimated to be approximately a 5% (or lower) chance that the actual future claims will be above the relevant value.

Where percentile values or probabilities are used to communicate uncertainty, this may imply that the uncertainty has been quantified to a high degree of accuracy. However, in practice this accuracy may not be justified; for example, where the uncertainty has been based on significant elements of judgement, the practitioner will need to find a way to ensure that the information on uncertainty is interpreted appropriately. In some cases, use of only qualitative statements may be preferable, or careful use of suitable caveats that convey the inherently approximate nature of any quantitative measures that are provided.

5.10 Documentation and Reporting

The majority of reserving work will be done in a context where it needs to be documented and reported in some form or other. Documentation here is taken to mean the creation of documents that are not typically provided outside of the team doing the reserving work (i.e. they are "internal" documents), whereas reporting is taken to mean the creation of more formal documents that are used to communicate the results to relevant stakeholders (e.g. persons or parties beyond the reserving team). These written documents may be supplemented by an oral presentation of results, as part of the overall process for communicating results to relevant stakeholders.

The content and design of reserving documentation and reporting will be influenced by a range of factors such as the complexity of the reserving exercise and any relevant professional or regulatory requirements. In addition, the type of reporting will be influenced by its intended purpose and the knowledge and experience of the intended users of the documents. The overall objective will usually be to ensure that the key messages and findings from the work are fully understood by those users.

Common items included as part of the reporting for general insurance reserving work include:

- Executive Summary
- Background: Introduction, Purpose, Scope, Authors, Professional standards compliance statement

- Limitations and Assumptions
- Data
- Methods
- Commentary on methods used and the key judgements for each reserving category, large loss and other relevant components of the reserves
- Commentary on Analysis of Change
- Commentary on Uncertainty
- Numerical results – Summary and detailed
- Graphs of claims development
- Appendices: Content depends on context.

Some of these will only be provided in certain circumstances, and in others only a high-level description of the key points will be required. Documentation will tend to focus on the more technical aspects of the reserving exercise, sometimes with the overall objective that, when combined with the external reporting that has been produced, these two sources will allow another suitably experienced reserving practitioner to understand the rationale for the key judgements that have been made, and reproduce the results if necessary.

6

Selected Additional Reserving Topics

6.1 Introduction

This chapter contains a range of additional reserving topics that are not covered elsewhere in the other chapters, including:

1. A section covering the estimation of outwards reinsurance.
2. A section describing the use of roll-forward and Actual vs Expected procedures.
3. A number of sections covering the approaches used to estimate reserves for certain specific types or category of reserve, specifically Unallocated Loss Adjustment Expenses, reinsurance bad debt and latent claims.
4. A section that covers discounting of reserves for the time value of money.
5. A section that summarises reserving under Solvency II.
6. A final section covering a number of miscellaneous topics (e.g. using external data sources and deriving Unearned Premium reserves).

6.2 Outwards Reinsurance

This section considers the approaches that can be used for estimation of outwards reinsurance in a reserving exercise, where the purpose of the exercise is assumed to be the derivation of a point estimate of the reserves, rather than a distribution of future claims. Outwards reinsurance may also be referred to as "ceded" reinsurance in some contexts.

Allowing for outwards reinsurance across a full distribution is considered in the stochastic methods chapter. In particular, some of the approaches that can be used are discussed in the context of the bootstrap method in the outwards reinsurance subsection of Section 4.4.5, and these are also applicable to a number of other stochastic methods. In some cases, even where the principal aim of the exercise is to establish the allowance for outwards reinsurance in respect of a single point estimate, stochastic methods can be helpful, because they can be used to make allowance across the full range of possible future gross claims outgo. Depending on the nature of the outwards reinsurance, this may produce a more accurate estimate of the outwards reinsurance reserve than, for example, netting down a single point estimate of the reserve.

The reserving of inwards reinsurance business (i.e. reinsurance business written by a company) is a different topic. However, the methods that can be used are generally similar to those for insurance business, with some specific approaches being used, where appropriate. Some of the specific reserving contexts covered in Chapter 7 make reference to such approaches.

The impact of outwards reinsurance on a particular reserving exercise can vary between being negligible (or none at all) to being very material. Within a particular reserving exercise, the materiality can also vary between reserving categories and cohorts. In addition, the nature of the outwards reinsurance can vary, ranging from relatively straightforward and stable over time, to highly complex and variable.

The purpose of considering outwards reinsurance in a reserving exercise will depend on the context. Although it will often be to enable the reserves net of reinsurance to be estimated, other purposes include pricing or commutation of outwards reinsurance. This section assumes the purpose is to estimate the reserves net of reinsurance.

The approach that is used to determine the credit that is given for the outwards reinsurance programme will be partly influenced by its complexity and materiality. The approach used might also vary according to the type of reinsurance, as explained below.

In general terms, the two main approaches that can be used to derive reserves on a net of reinsurance basis are to derive them directly by applying suitable reserving methods to the data net of reinsurance, or to apply those methods to the data gross of reinsurance and then deduct an allowance for outwards reinsurance, based on a separate estimation of that allowance. Both of these approaches use mainly aggregate claims data. A third approach, based on deterministic or stochastic modelling of individual claims, can also be used in some circumstances. These three approaches are discussed further below, followed by a section that discusses selecting the final results for outwards reinsurance.

6.2.1 Applying Reserving Methods to Net of Reinsurance Data

In theory, all of the deterministic and stochastic methods described in Chapters 3 and 4 can be applied to data on a gross or net of reinsurance basis, and also to the reinsurance data. By applying them to data on a net of reinsurance basis, the results will obviously also be on that basis.

Such an approach has the advantage that it is simple to apply and for certain types of reinsurance it can reduce the volatility in the data triangles caused by infrequent large claims that can exist in gross of reinsurance data. However, its main disadvantage is that, depending on the reserving method used and the type of reinsurance in place, the results may make an inappropriate allowance for reinsurance, particularly where the reinsurance programme has changed significantly over time. For example, where the reinsurance programme is non-proportional in nature and where the excess point has increased materially beyond claims inflation across the cohorts, then (other things being equal) the expected development pattern for the older cohorts may be shorter than that expected for the later cohorts, where the excess point is higher. This will mean that if a CL method is applied to the data, without adjustment, the implied reserves for the later cohorts may be understated. Other mis-estimations may apply for different types of reinsurance and/or where the claims experience has impacted the reinsurance programme in different ways across the cohorts.

Hence, this approach tends to be of most use where the reinsurance programme has been reasonably stable over time, is proportional in nature or where it is relatively immaterial.

In most situations it is necessary to also derive reserves on a gross of reinsurance basis, so that the reinsurance reserves can be derived by deducting the net from the gross reserves. This is so that the appropriate credit for reinsurance can be shown in the financial statements, since a common accounting practice is to show expected future recoveries from reinsurance as an asset in the balance sheet, and gross reserves as a liability (so that net reserves are not shown explicitly). Hence, where the allowance for reinsurance is being made by applying reserving methods to the data net of reinsurance, a separate application is usually also required for the gross of reinsurance reserves. Similar reserving methods to those used for the net of reinsurance data can be used, as can comparable assumptions (such as claims development patterns), with the objective being to derive reserves on a consistent basis. This consistency may be difficult to achieve in some cases, however, particularly where the reinsurance programme is complex, and in these situations an alternative is to use the other approach of deriving a separate allowance for outwards reinsurance, which is then deducted from the gross reserves, as described in the next section.

6.2.2 Deducting an Allowance for Outwards Reinsurance

Two broad methods can be used to derive the credit for reinsurance to be deducted from the gross reserves within this approach. First, the reinsurance data can be projected to ultimate, using similar reserving methods as for the gross or net data. Second, some form of ratio approach can be used, combined with specific reinsurance allowance in respect of individual large claims or events.

Projecting the reinsurance data to ultimate can involve applying one or more reserving methods to the data, but the same point that arises when projecting net of reinsurance data regarding consistency with the gross results also applies. In some cases (e.g. perhaps for the less developed cohorts), a simple IELR approach can be used. This will involve selecting a suitable ultimate loss ratio (by cohort) to be applied to the estimated ultimate reinsurance premiums, perhaps related only to the relevant subset of the reinsurance programme. As with using the IELR method for gross reserving, there will be judgement involved in selecting that loss ratio. It may also be necessary in some contexts to consider whether any special adjustment is needed to allow for the impact of features of the reinsurance programme on the reinsurance premiums used for this purpose, such as reinstatement premiums and profit commissions. Where relevant, this is referenced in the individual sections in Chapter 7.

Assuming that appropriate data is available to determine the amount of reinsurance recoveries received in respect of the claims paid and of the amount due on gross outstanding claims (which should normally be the case), then the key judgement to be made in the reserving exercise in respect of outwards reinsurance is the estimation of the reinsurance IBNR, or "RI IBNR". Once this has been estimated, the net IBNR can be calculated and added to the net outstanding claims to derive the net reserves. This observation is partly what prompts the use of a ratio-type method to determine the reinsurance credit. This involves determining, for example, a number of so-called "reinsurance to gross" (or "R/G") ratios for each cohort, calculated as the ratio of reinsurance premiums, reinsurance claims paid, outstanding and incurred to their gross counterparts. These R/G ratios are then used to infer a suitable R/I IBNR ratio, which is then applied to the gross IBNR, to estimate the R/I IBNR. In some cases, for some cohorts it may be necessary to infer R/G ratios observed for groups of other cohorts (allowing for any changes in business mix and/or reinsurance programme across the relevant cohorts). The use of a ratio approach such as this may work reasonably well for some types of reinsurance (e.g. proportional, such as Quota Share) but less well for others (e.g. non-proportional, such as Excess of Loss).

In practice, either of the two methods might be done separately for each type of reinsurance, and possibly also for different sizes of claim within each reserving category. In addition, various bespoke adjustments may be necessary, depending on the context. For example, where there are large losses present in the data that have given rise to excess of loss reinsurance recoveries, it is important to consider whether the methodology used makes a valid extrapolation for the purposes of determining the reinsurance credit. A ratio method might, for example, need to allow for this when making inferences from the potentially relatively high observed R/G ratios to avoid assuming that this ratio will also apply to the gross IBNR, when there may be limited scope for that IBNR to give rise to reinsurance recoveries.

6.2.3 Modelling Individual Claims

The previous two approaches rely mainly on aggregated claims data, perhaps supplemented by specific estimation of reinsurance recoveries for large claims and/or catastrophes. In some cases it can be helpful to estimate the reinsurance recoveries using a deterministic or stochastic model of individual claims. This might be done, for example, where the reinsurance programme is non-proportional in nature and where there is limited past reinsurance claims history from which estimates of future recoveries can be derived.

The approach involves first understanding the business that has been written, to determine the possible range of individual claim sizes that might be possible (e.g. for London Market business, the line size written will influence this). Examination of the distribution of the incurred gross claim amounts, including those that are not currently large enough to generate reinsurance recoveries, can then be used to identify either deterministic scenarios for the claim size distribution, or the parameters for a statistical distribution so that the recoveries can be modelled stochastically. Input from claims, reinsurance and underwriting staff can then be sought to help validate the selected assumptions. Once the assumptions have been made, the deterministic scenarios or results from applying the stochastic model (e.g. using Monte Carlo simulation) are then applied to the relevant reinsurance programmes, to estimate the reinsurance recoveries. In practice, sensitivity testing of these results using alternative assumptions may be appropriate before the results are finalised.

6.2.4 Selecting the Final Results for Outwards Reinsurance

As noted above, the approach used may vary between reserving categories and cohorts. In some cases, the final selected results for particular categories and cohorts might be based on a combination of the approaches described above.

Whatever method is used, a range of numerical and graphical diagnostics can be used to help assess whether the selected reinsurance credit is reasonable. These can include review of claims development graphs on a net and reinsurance basis, review of the implied R/G IBNR and R/G ultimate ratios, and of the implied reinsurance ultimate loss ratios. The latter will be impacted by both the estimated ultimate reinsurance claims and the corresponding ultimate reinsurance premiums. The methods described earlier in this section have focused on estimating the claims part of this, but in some cases it may also be necessary to estimate the ultimate reinsurance premiums (and hence the future reinsurance premiums payable, which are effectively a liability for the insurer), part of which may be dependent on the claims recoverable from the reinsurance. Where relevant, this is discussed further in the individual contexts in Chapter 7.

An additional consideration that may influence the final selection is the extent to which there is a potential for non-recovery of reinsurance. This can be due, for example, to the possibility of there being a default in respect of the amount payable by the reinsurers, caused by financial difficulties or disputes. Any adjustment for such matters is normally considered as a separate adjustment to the reserves, rather than within the main calculations of the net of reinsurance reserves. Further consideration of this topic is provided in Section 6.5.

The precise methodology used may vary between reserving categories and will depend on many factors, including the context for the reserving exercise. Some additional considerations are considered in the context-specific sections in Chapter 7.

There appear to be relatively few sources in the literature that focus on the topic of reserving for outwards reinsurance. Some examples include Czapiewski et al. (1993) and Chrin and Fell (2010). Chan and Ramyar (2014) also discusses the topic in the context of netting down a gross distribution, as referred to in Section 4.4.5.

6.3 Roll-forward, Actual vs Expected and Fast Close Procedures

6.3.1 Introduction

For reserving work carried out as at a particular valuation date, in many situations encountered in practice there will have been a similar exercise carried out as at one or more previous valuation dates (e.g. the previous quarter or year-end). Where this is the case, one approach for deriving the reserve estimates as at the valuation date that can be used in addition to or even instead of the application of reserving methods as at that date is to derive reserves using some form of Actual vs Expected ("A vs E") or "roll-forward" procedure.

This is usually done in a partially or fully automated way, which may enable the time taken to produce reserve estimates after the current valuation date to be reduced. However, more time will be required prior to that in order to derive the necessary assumptions for the A vs E or roll-forward procedure. It is relevant only for cohorts where there is an estimated ultimate value already in existence at a prior valuation date, and so with annual cohort and development frequency data, the procedure is not relevant for the latest cohort. These procedures can in theory be applied to gross, net and reinsurance data, although they are probably more commonly used on a gross basis, as there can be practical difficulties when doing so on a net or reinsurance basis.

The timescale for production of results after a valuation date can be reduced further by using a "fast close" type procedure, which involves applying the reserving methods as at a date, say, one month prior to the valuation date, and then using a roll-forward procedure once the claims data is available at the valuation date.

A brief description of the following procedures is provided in the subsequent sections:

1. Static roll-forward
2. Methodology roll-forward
3. Actual vs Expected
4. Fast close.

The grouping of different approaches in this way is not uniform or well defined within or between different market sectors and countries, and these terms may be used in different ways in other contexts.

It is assumed in all cases that the reserves that are subject to the relevant procedure have been established on a best estimate basis. Other bases, such as that used for the booked reserves (which may include a margin above a best estimate), can also be used, and the principles are broadly the same, except that interpretation of the results will vary accordingly.

6.3.2 Static Roll-forward

This represents the simplest type of roll-forward procedure and is likely to be of use mainly for relatively immaterial reserving categories, or where the claims development is reasonably stable or predictable.

It can take various forms, but in its simplest form the reserves at the previous valuation date (i.e. those based on the previous ultimates) are reduced by the claims paid in the period between valuation dates. Although it might seem reasonable to describe the reserves derived in this way as having been "rolled-forward" from the previous reserves, it involves nothing more than retaining the previous estimated ultimate claims; when the claims experience in the relevant period justifies a change to those estimates this type of roll-forward approach will not reflect that. Hence, this particular approach might also be referred to as "static ultimates". Other types of static roll-forward, as identified in Bruce and ROC Working Party (2015), include maintaining the Ultimate Loss Ratios, IBNR or reserves established at the previous reserving exercise. These approaches may result in a change in ultimate between valuation dates, but they do not make any direct allowance to reflect the observed experience in the relevant period. All of these approaches are relatively straightforward to implement in an automated way. Hence, the results can be produced very quickly once the data is available at the latest valuation date, perhaps using different forms of static roll-forward for different reserving categories and cohorts if required.

In practice, any form of static roll-forward type approach can be supplemented by examining the claims development in the relevant period and assessing whether ultimates produced by the selected roll-forward approach are reasonable, or whether they need amending. The types of graph described in Section 5.8 can be used to help make this judgement, by visually reviewing the observed paid and/or claims development since the previous valuation date, to assess whether the ultimate claims produced by the automated roll-forward approach appear reasonable, given the claims development since then. Numerical diagnostics can also be used to assess this, such as the proportion of estimated reserve or IBNR established at the previous valuation date that has been eroded by the actual paid and incurred claims development since then. This might be seen as a type of A vs E approach, because some consideration is being given to the expected development in the period, but without any explicit estimation of that movement as part of the process (which is what distinguishes the more detailed A vs E approach, discussed in Section 6.3.4).

Automated static roll-forward approaches that are supplemented by some form of manual review of the claims development in the relevant period, as described above, are highly judgemental in nature. However, they can sometimes help quickly identify whether any of the previously estimated ultimate claims for particular reserving categories need changing at the latest valuation date. In some cases, a conclusion might be reached for some cohorts or reserving categories that none of the static roll-forward approaches produces reasonable ultimates. If this is the case, an alternative approach might be needed, such as one or more of those described in the subsequent sections.

6.3.3 Methodology Roll-forward

Depending on the claims reserving method used, it may be possible to reapply the methods and assumptions used at the previous valuation date to the updated data at the current valuation date. In other words, the methodology (and assumptions) established at one valuation date are "rolled-forward" to the subsequent valuation date. Depending on the reserving software being used, this approach may take some time to design and implement initially, but once that is done it can usually be applied more quickly than carrying out a full reserving exercise, including selecting new assumptions, at the latest valuation date. The detail of the implementation will vary according to the reserving methods being used. For example, where a single CL-type method has been used at the previous valuation date, the selected development pattern can be reapplied to the data as at the latest valuation date. Where other reserving methods, or a blended approach, has been used to select the ultimates, it requires more careful implementation.

For CL-type methods, where the actual development is similar to the expected development implied by the previously selected claims development pattern, then this type of roll-forward approach will mean that the estimated ultimate claims will be similar between valuation dates.

In practice, even where it is felt that the same methods and assumptions selected at the previous valuation date are still valid at the latest valuation date, an automated re-application may not produce reasonable estimates for all reserving categories and cohorts. Hence, as with the static roll-forward approaches, numerical diagnostics and graphs can be used to determine where the results of the methodology roll-forward need amending. These will need careful scrutiny and there can be many factors to take into account to ensure reasonable results are obtained. These factors are very similar to those that apply in the next type of approach – A vs E – so they are considered in the relevant paragraphs below.

6.3.4 Actual vs Expected Approach

An alternative to both the static and methodology roll-forward approaches involves deriving an estimate, as at the previous valuation date, of the expected future claims development up to the next valuation date.[1] This can then be used in two ways. First, it can be compared with the actual claims development in the same period, to help determine whether the estimated ultimate claims at the previous valuation date need revising and, if so, in what direction and by what order of magnitude. Second, it can be used as part of a numerical procedure to provide automated "A vs E" ultimate claims. In the latter, the procedure can vary according to the underlying reserving method(s) being used. For example, if a standard CL method is being applied to the incurred claims data, then one possible procedure is for the "A vs E"-based ultimate to be designed to be the same as if the development pattern were applied to the incurred claims at the latest valuation date. This is then equivalent to the methodology roll-forward approach and involves the following procedure (for a single cohort, applied to incurred claims):

1. Define the estimated ultimate claims as at the previous valuation date as "ULT1". For this simple example, this will have been derived using a development factor derived from a CL method, applied to the incurred claims as at that date.
2. As at the previous valuation date, using the incurred claims development pattern derived from the CL method applied to that data, estimate the expected movement in incurred claims up to the next (i.e. current) valuation date. Label this "E".
3. Then, as at the current valuation date, observe the actual movement in incurred claims since the previous valuation date. Label this "A".

---

[1] Some practitioners define A vs E so that it relates to any observable quantity, rather than just claims development, but since the focus here is claims reserving, a more restricted definition is used. See Bruce and ROC Working Party (2013) for an example of a broader definition of A vs E.

4. Calculate the Actual less Expected value – i.e. A – E.
5. The updated A vs E ultimate at the current valuation date is then the previous ultimate, ULT1, plus (A – E) multiplied by the incurred claims CL development factor at the current valuation date. The CL development pattern used here is the same as that used to derive ULT1 and E.

Hence, in summary, this procedure takes the estimated ultimate claims at the previous valuation date and adds a multiple of the difference between the actual and expected movement in the period, with the multiple being the incurred claims development factor to ultimate. If the actual movement is close to the expected movement according to the incurred CL development pattern, then the updated ultimate at the current valuation date will be similar to the estimated ultimate at the previous valuation date, which is logical within the context of this particular procedure. It is equivalent to the methodology roll-forward approach using the CL pattern, as the same ultimate at the current valuation date can be derived by multiplying the incurred claims at that date by the corresponding CL development factor (which is effectively the rolled-forward factor from the previous valuation date).

Other procedures can be designed that take different multiples of the A – E difference, depending on the underlying claims reserving methodology being used. For example, if reserving methods such as IELR or BF are being used, then the prior loss ratio assumptions will need to be allowed for in the calculation of the expected claims movement. Whatever procedure is used, there are several factors to take into account when implementing them in practice. Many of these points also apply to the methodology roll-forward type approach. They include the following:

- It can be appropriate to set tolerances for the observed A – E by cohort, whereby if the difference is within a specified range, no adjustment is made to the previously estimated ultimates.
- Similarly, if the observed A – E for particular cohorts, or groups of cohorts, is materially above the specified tolerance levels (in absolute terms – i.e. positive or negative) then further investigation may be warranted, as the automated A vs E ultimates may not be appropriate, and a reselection and application of suitable reserving methods at the current valuation date may be necessary.
- In some cases, even where the actual and expected movements are close to each other for a number of cohorts, there may be other cohorts for the same reserving category where this is not the case, which may need taking into account. For example, if, say, the oldest two cohorts have shown actual movements in excess of expected that has necessitated an increase in ultimate claims for those cohorts, then that might mean the tail factor implied by the new ultimates has also increased. This increase might need taking into account for the later cohorts, as they will be potentially impacted by the increased tail factor, and hence their ultimates may need to increase, even if the A vs E analysis for those cohorts suggests otherwise.
- A vs E can be applied to more than one type of data, such as paid and incurred claim amounts. Different conclusions might be drawn from the A vs E analysis for these data

types, which will need consideration in selecting the estimated ultimate claims at the current valuation date.

- In some instances, the A vs E analysis might suggest that the reserving model (i.e. the methods and assumptions) selected at the previous valuation date needs revising. This will involve examining the data at the current valuation date and selecting a new model at that date. It will then replace any ultimates based on an automated A vs E approach and provide a revised model that can potentially be used in future A vs E analysis.

- Various bespoke adjustments may often be needed in practice to allow for particular features of the data, such as large losses and catastrophes. This might include making special allowance for both new losses or catastrophes that have occurred since the previous valuation date and for material movements in existing losses or catastrophes.

- The example procedure described above assumes that a single incurred CL method has been used to derive the selected ultimate claims for all cohorts, which is then used to estimate the expected claims development, for use in the automated A vs E analysis. In practice, the reserving methods used may vary by cohort, and for an individual cohort the selected ultimate may be based on a weighted combination of estimates derived from different methods, possibly applied to different types of claims data, as explained in Section 5.7. In this case, there may not be a single reserving method that can be used to derive the expected claims development between valuation dates for all cohorts. Hence, different A vs E formulae may be needed for different cohorts, and where a weighted approach has been used for individual cohorts an alternative approach to deriving the expected claims movement will be needed (e.g. by selecting a suitable paid or incurred claims development pattern).

Regardless of the detail of the A vs E approach used, it is important to recognise that the expected claims movement will itself be an estimate. Hence, any automated A vs E procedure will need overlaying with a degree of judgement, particularly when deciding how to reflect any observed differences between actual and expected movements in the updated ultimate claims estimates. Examination of numerical and graphical representations of the data can be helpful here. Where there is significant doubt regarding how the actual claims movement should be reflected in the updated estimated ultimate claims for particular reserving categories/cohorts, then where justified on materiality grounds and if time permits, there can be no substitute for a re-application of appropriate reserving methods using updated assumptions, applied to data as at the current valuation date.

Various practical aspects of A vs E approaches are discussed further in Bruce and ROC Working Party (2013). Staudt (2012) presents a more mathematical approach to the subject.

6.3.5 Fast Close Procedures

In some practical reserving contexts, the timetable for finalising results after the relevant valuation date is such that there is insufficient time to perform a full reserving analysis including all the necessary reporting requirements and governance steps using data as at that date. One way of dealing with this is to implement a "fast close" procedure. In summary, this is broadly similar to the roll-forward or A vs E approaches described in the previous sections, except that the as-at date where the initial analysis is carried out is usually, say, only one month prior to the target valuation date (although it can be longer). Another way of describing this is to refer to there being an "early close", which enables timely production of the required reserving information and results as at the required valuation date.

Bruce and ROC Working Party (2015) discuss the fast close process in detail, and suggest that it can involve the following five stages:

1. **Early close:** The reserving data is collated as at a date prior to the valuation date and a full reserving exercise is carried out, or a previously derived set of ultimate claims are rolled-forward. Given that the date of this exercise will most likely be an intermediate point within the usual periods of development frequency in the data triangles (e.g. as at the end of month two within a quarter for triangles that use a quarterly development frequency) special allowance will need to be made for this if development factor methods are being used. For example, before applying the development factor methods to the data, the incremental claims in the leading diagonal can be grossed-up to produce estimated quarter-end positions, or the triangle can be re-cast so that the development period is the same in each column. Specific consideration may be needed in relation to any new large claims/events that have occurred in the period immediately before the early close as-at date, and hence for which there may be limited information on which to base estimated reserves. For example, in some cases, a decision might be made to select a "holding position" for these losses, and then update the estimates in subsequent stages of the analysis. If a methodology roll-forward or an A vs E approach is used as part of the overall Fast Close process, then the necessary methodology and assumptions are also determined at this stage.

2. **Roll-forward analysis:** Using data as at the valuation date, carry out the mechanical/automated aspect of the roll-forward analysis, using the relevant methodology and assumptions determined as part of the early close stage. This can involve one or more of the static roll-forward type approaches, a methodology roll-forward, or an A vs E type approach. The detailed approach might vary by reserving category and cohort. Any "holding position" reserves established for individual claims/events at the early-close as-at date are also updated at this stage if relevant new information is available.

3. **Select roll-forward ultimates and reserves:** Apply judgement to the results of the previous stage (e.g. using graphs and numerical diagnostics) to select final ultimates and hence reserves as at the valuation date.

4. **Collate results:** Collate the results from the previous stage into the relevant exhibits and other management information schedules. Bruce and ROC Working Party (2015) label this stage as the fast close part of the overall process, but acknowledge that this is not a widely held definition.

5. **Reporting:** This moves beyond the previous stage to prepare all the various reports required, including those that are needed to meet the relevant governance and internal/external regulatory requirements.

Whilst some of these stages could be grouped together, the above breakdown serves to identify the complexity that can be involved in implementing a workable fast close process. Further detail is given in Bruce and ROC Working Party (2015).

6.3.6 Additional Remarks Regarding These Approaches

Wherever the previously estimated ultimate claims have not simply been retained at the latest valuation date, it can very useful to show (by reserving category and cohort) the difference between these updated ultimates and those based on the simple roll-forward of the previous reserves (i.e. based on retaining the previous ultimates). This will apply regardless of whether a roll-forward, fast close or complete re-application of reserving methods has been used to derive the updated ultimates. It will help to demonstrate whether the previously estimated ultimate claims needed changing, and if so by how much, based on the claims experience and/or emergence of new information since the previous valuation date. Commentary can then be provided as part of the reporting of results, where appropriate, to explain the cause of these changes. The cause may relate to a wide range of factors, with a common one being "unexpected claims development" – i.e. the actual movement was not in line with the expected movement at the previous valuation date. If this is the case, then additional analysis and explanation of the unexpected claims development may be helpful.

It is important that any such comparisons are done only for cohorts that have the same exposure definition at the two timepoints (e.g. they both relate to the same period of earned exposure), otherwise the change in ultimates will be in part due to the change in exposure period, which makes it difficult to interpret.

The overall objective of any procedure of this type is, to varying degrees, to speed up or make more efficient the overall reserving process. This will allow more time to be spent on the key areas of judgement rather than on other less value-added aspects. However, a further consideration when designing the procedure is the extent to which it is deemed desirable for it to be expected to produce a set of ultimates as at the relevant valuation date that are consistent (within a chosen degree of materiality) with those which would be produced if a full reserving exercise were carried out at that date. If this is deemed desirable, then for certain reserving categories or cohorts, if there is sufficient doubt that the procedure will produce results that are expected to be consistent in this way, then a reselection and application of the methods and assumptions as at the valuation date may be necessary. This may not be practical in some contexts (e.g. a fast close type application) in which case the practitioner might wish to consider whether the procedure can be designed so as to mitigate any material risk that it could produce significantly different reserve estimates compared with those that might be produced by a fully updated reserving analysis at the valuation date.

Finally, when using the terms "roll-forward", "A vs E", "fast close" or "early close" in any communication of reserving analysis, it is helpful if the reserving practitioner explains what approach has been used to derive the relevant estimated ultimate and reserve values. This is because in some cases it could involve one of the more straightforward approaches described above (e.g. a static roll-forward of ultimates), and in others a more sophisticated fast close type approach may have been used; the interpretation of the results in each case might be different.

6.4 Unallocated Loss Adjustment Expenses

This category of reserve refers to claims handling expenses that are not allocated to individual claims. It usually relates to internal costs associated with handling claims, but can also include other related administrative costs and external claims handling costs not allocated to individual claims. It is commonly referred to as Unallocated Loss Adjustment Expenses or ULAE (e.g. in the USA in particular), but can also be referred to as just internal adjustment expenses (IAE) or as indirect expenses. It contrasts with Allocated Loss Adjustment Expenses (ALAE), which typically relates to external costs associated with individual claims. ALAE are commonly added to the indemnity element of individual claims and included within the claims data used for reserving purposes. Hence, when applying reserving methods to such data, the implied reserve will also include an allowance for future ALAE costs. However, ULAE will not be included, and where the scope of the reserving exercise includes estimation of ULAE, a separate calculation will therefore be necessary.

This section discusses some examples of the methods that can be used to estimate ULAE. Compared with the techniques used to estimate indemnity and ALAE costs, those used for ULAE are generally much less sophisticated, and there has been less research into suitable methods. This is most likely due to the fact that reserves for ULAE are in most cases relatively immaterial compared to the other reserve categories; typically the ULAE reserve will be less than 10% of the total reserve and often considerably less.[2] The proportion will depend to some extent on the type of business, volume and complexity of claims. In addition, if it is assumed that the insurer will continue to write new business in future, then it may be appropriate (depending on the relevant accounting basis) to also assume that the contribution to ULAE (and other) expenses from this new business will continue in future at the same rate. Where this does apply, the ULAE reserve is likely to be a lower proportion of the total reserve than in cases where it does not apply, such as for an insurer in run-off (other things being equal).

The ULAE reserve is sometimes just referenced as a loading to the other reserves, without necessarily much justification, given their relative immateriality. In some cases, although the ULAE reserve might be small relative to the size of the total reserves, if the profitability of the business is relatively low, then the impact of, for example, a change in the approach used to derive the ULAE reserve can be significant relative to the declared profits.

Before considering the different methods that can be used to estimate ULAE, it is worth noting that there are a range of different types of expenses that can be included in ULAE, and consideration of this is advisable before attempting to estimate a ULAE reserve. The precise types of expense that are appropriate to include in the ULAE reserve will depend

---

[2] For example, data from Lloyd's of London in 2012 showed that ULAE as a proportion of gross reserves, across the market as a whole was 1.2% – see Slide 29 in Lloyd's of London (2012) for further details

partly on the jurisdiction of the relevant entity and any associated regulations or market practice. Some of the expenses are proportionate to the volume of claims activity and some are fixed. The core component of most ULAE reserves will be the salaries and other employment-related costs of the claims handling staff, and will cover their time spent at all stages of claim development from notification, through to handling of outstanding claims and finally to settlement. Other types of expense that may be included in ULAE are IT, management, accommodation, utility and other administrative costs associated with the claims department.

The jurisdiction and any associated regulations will also influence the exposure basis to be used for the ULAE reserve. For example, ULAE could relate to just the future expenses in handling incurred claims (other than ALAE) as at the valuation date, whether reported or not. Alternatively, it could relate to the future expenses in handling claims related to business that has been written up to the valuation date, which will cover both incurred claims and claims in respect of unearned exposures.

ULAE reserves are usually established so that they are the same on a net and gross of reinsurance basis. However, some of the ULAE costs incurred could be related to the outwards reinsurance programme, such as time spent by the insurer's staff estimating the impact of reinsurance on individual claims. It might be argued that these costs should be excluded when establishing ULAE reserves on a gross of reinsurance basis, but in practice such a distinction may not be made. In addition, although most outwards reinsurance contracts cover ALAE, but not ULAE, if any special contracts do exist that allow partial or full recovery of ULAE costs, then this will need to be factored into the calculation of the overall reserves, including ULAE, on a net of reinsurance basis.

In the description of ULAE methods below, it is assumed that an appropriate definition of costs and exposure basis has been agreed and that reserves for gross and net ULAE are the same.

6.4.1 Methods for Estimating ULAE Reserves

When estimating reserves for indemnity and ALAE costs, the standard reserving methods described elsewhere in this book will typically involve estimating ultimate claims for each cohort, with the reserves being derived by deducting the claims paid to date. However, for ULAE, the approaches used tend to estimate the reserve directly through estimation of the future or unpaid ULAE costs at the relevant valuation date, rather than deriving an estimate of the ultimate ULAE costs. In addition, although some organisations may track ULAE by reserving category, ULAE reserves are usually established as a single item relating all reserving categories combined.

Three types of method for estimating ULAE reserve are described separately below – ratio, count and projection. A further section discusses some other possible methods. As with reserving methods for indemnity and ALAE costs, the method chosen in any particular reserving context will depend on a range of factors including the available data and the materiality of the ULAE reserve. Applying more than one method can be beneficial, as can comparing the estimated ULAE reserve with that established at previous valuation dates and with that implied using suitable benchmarks (e.g. the ULAE reserve as a proportion of total gross reserves for other insurers, where available).

Ratio Methods

These methods all involve using data related to known ULAE costs in a prior calendar period, and expressing those costs as a ratio of a claim-related volume measure (typically paid claims) in the same period. This ratio is then multiplied by an estimated future volume measure (typically the claims reserve) to derive the ULAE reserve. Where more than one past period is used, an average ratio can be derived, or the trend of the ratio over time can be taken into account when selecting the ratio that is used to estimate future ULAE costs. If the business is written in currencies other than the currency in which the ULAE costs are incurred, then this will need to be taken into account when using ratio methods.

Different types of ratio method have been suggested in various sources. Table 6.1 summarises four of these. In each case, the ULAE reserve is derived by calculating the ratio for one or more relevant past calendar periods according to the specification in the table, selecting a suitable average or trended value from these, and then multiplying it by the relevant future volume measure as defined in the last column of the table. For example, for a single calendar year, the first three types of ratio method use the following:

$$\text{ULAE reserve} = \frac{\text{Paid ULAE}}{\text{Claims paid}} \times \text{Volume measure.}$$

Because the volume measures are all related to the estimated future claims (i.e the reserve), each type of method produces an estimate of future ULAE costs. In all cases, the data used for the calculations relate to gross of reinsurance values (although as noted above, the past ULAE costs themselves may include internal claims handling costs related to calculating or making reinsurance recoveries). All the methods implicitly assume that the past ULAE can be used in some way as an input to the estimation of future ULAE.

The simplest type of ratio method is the first shown in Table 6.1, Type 1, labelled "Simple", which should be self-explanatory, so that the Claims Paid represent the claim payments in the same past calendar period as for the Paid ULAE. This and Types 2 and 3 in the table are a type of ULAE ratio method that is sometimes referred to as a "Paid-to-Paid" method, for obvious reasons. They all assume that payments for ULAE are proportional to

Table 6.1 *ULAE reserving methods: Types of ratio formula*

#	Name	Numerator	Denominator	Volume measure
1	Simple	Paid ULAE	Claims Paid	Claims reserve
2	Simple with IBNR total	Paid ULAE	Claims Paid	IBNR + multiplier × case estimates
3	Simple with IBNR split	Paid ULAE	Claims Paid	Pure IBNR + multiplier × (case estimates + IBNER)
4	Expected Paid	Paid ULAE	Expected Paid	Any of above

payments for claims and that the timing of ULAE payments follows the timing of payments for claims. They work best when the relationship between ULAE and claims is stable over time, and hence for portfolios that are in run-off, or which are contracting or growing significantly they may under- or over-state the future ULAE costs.

Ratio method Type 2 is a slightly more complicated approach. It is designed to recognise that future claims handling costs would normally be expected, other things being equal, to be higher for claims that are not yet open, compared with those that are already open. This is because for the latter there will be no future costs associated with opening the claim file, for example.

As shown in the table, this is allowed for in the volume measure by separating the IBNR and case reserves element. The latter then includes a multiplier value, which is less than one and which should represent an estimate of the proportion of total ULAE costs for individual claims that are expended at the time a claim is opened, with the rest relating to those incurred between then and the final settlement of the claim.

A typical value for the multiplier might be 50%, but it will depend on the nature of the business and the associated claims. The use of a 50% factor is referred to in some sources (e.g. Rietdorf and Jessen, 2011) as the "50/50 rule". Method Type 2 will always produce a lower ULAE reserve than method Type 1, assuming the multiplier is less than one. Hence, it may help address the criticism sometimes made regarding method Type 1 that it overstates the required ULAE reserve (e.g. see Mango and Allen, 2009). However, selecting a suitable value for the multiplier is not necessarily straightforward, and an element of judgement will almost certainly be required. Ratio method Type 2 is of the same form as the simplification cited in the european Solvency II regulations. For example, see Technical Annex 2 and Guideline 71 in EIOPA (2014b). In some sources, such as Friedland (2013), this version of the ratio method is referred to as the "Classical Paid-to-Paid" method. In others (e.g. see Buchwalder et al., 2005 and Rietdorf and Jessen, 2011) when a multiplier of 50% is used it is referred to as the "New York Method". Ratio method Type 3 seeks to refine the calculations by recognising that IBNR normally includes two elements – "Pure IBNR", which is in respect of newly reported claims, and IBNER, which is in respect of movement for already reported claims. Hence, the multiplier is applied to both the case reserves and the IBNER element.

Finally, ratio method Type 4 is similar to Types 1 to 3, except that in the denominator of the ratio instead of using the actual past paid claims, an estimate of the corresponding expected paid claims in each past calendar period is used in the calculations. In theory, any of the future volume measures could be used in combination with the ratio, although in the paper where this approach was originally suggested (Mango and Allen (2009)) it is described as being used with the claims reserve – i.e. as per the simplest of ratio methods described here – Type 1. The rationale for using expected values is that in some cases the actual claims data may be very limited, or it may be distorted by unusually large claims for example.

Other versions of these ratios are possible. For example, in Kittel (1981) it is suggested that instead of using just calendar period claims paid in the denominator of the ratio, the average of calendar period claims paid and claims incurred could be used, in an attempt to better reflect the volume of claims activity corresponding to the calendar period paid ULAE in the numerator of the ratio. Ohlsson (2016) also suggests an alternative formula based around calibrating the "New York Method", as referred to above.

All ratio methods for estimating ULAE have limitations of varying degrees, principally connected with the fact that they try to use past ULAE costs as a guide to future ULAE costs, without explicit consideration of factors such as future inflation, changing claims staffing levels, claim profiles, etc., all of which may affect the actual future ULAE costs. Hence, in practice they may produce ULAE reserves that are too high or too low, depending on the circumstances. Where materiality justifies further analysis, alternative methods can be considered.

Tables 6.2 to 6.4 show a worked example for ratio methods Types 1 to 3, applied to purely illustrative data. All figures are in £000 unless stated otherwise.

Table 6.2 *ULAE worked example for ratio methods: Calculation of past ratios*

Calendar Yr	Paid ULAE	Claims Paid	Ratio
2013	3,673	163,611	2.24%
2014	3,562	171,737	2.07%
2015	2,984	159,778	1.87%
Total	10,219	495,127	2.06%
		Selected	2.00%

Table 6.3 *ULAE worked example for ratio methods: Current claims reserves*

Item	Value
Case	190,567
Pure IBNR	14,902
IBNER	140,995
Total IBNR	155,897
Reserve	346,464

Table 6.4 *ULAE worked example for ratio methods: ULAE reserve*

Method Type	Multiplier	
	50%	75%
1	6,929	6,929
2	5,024	5,976
3	3,614	5,271

Table 6.2 shows the observed ULAE ratios for three previous calendar years, along with the implied ratio across all years combined and the selected ratio that will be used in the ULAE reserve calculation. As the table shows, there seems to be a falling trend in the ratio

over time; in practice this would be investigated further before the selected ratio is finalised. In this case, a ratio of 2% is selected.

Table 6.3 shows the current gross claims reserves across all cohorts combined, which will have been estimated using suitable methods. In this case, the analysis has enabled the Pure IBNR to be identified separately, which allows ratio method Type 3 to be applied; in many cases this may not be possible.

Table 6.4 then shows the calculation of the ULAE reserve for each type of ratio method, with two alternatives for the multiplier in method Types 2 and 3 being used – 50% and 75%. The calculations follow the earlier description of the different types of ratio method, so that, for example, using method Type 2 with a 50% multiplier, the ULAE reserve is

$$\text{ULAE reserve} = \underbrace{2\%}_{\text{Selected ratio}} + [\underbrace{155{,}897}_{\text{Total IBNR}} + \underbrace{50\%}_{\text{Multiplier}} \times \underbrace{190{,}567}_{\text{Case reserves}}]$$

$$= 5{,}024.$$

As expected, method Type 1 produces higher ULAE reserves than the other two types of method, and similarly method Type 3 produces the lowest reserve, as both the IBNER and case reserves are reduced by the multiplier, before applying the ULAE ratio. The choice of which type of ratio method to use in practice, and if relevant, the value of the multiplier to use, will depend on knowledge of the costs involved in claims handling and the extent to which there is sufficient information to justify using a lower ULAE ratio for case reserves (possibly including an IBNER element), which is effectively what using a multiplier less than 1 implies.

Worked examples of various types of ULAE ratio method are also given, for example, in Chapter 22 of Friedland (2013).

Projection Methods

For this type of method, rather than using ratios derived from past data, an explicit projection of future ULAE costs is made, allowing for the expected run-off of the claims reserves to which they relate. In theory, a projection-based method could involve detailed projections at a transaction level, based on the ULAE costs of each relevant claim-department transaction such as closing, negotiating, settling and reopening claims, together with separate projection of the different types of overhead costs associated with the claims handling function. However, this will be very time consuming and is not likely to be practical in most situations. Hence, other approaches have been suggested that operate at a more aggregate level.

For example, the actual ULAE costs in past periods, broken down into different categories of fixed and variable expenses, can be collated and then projected forward, allowing for inflationary and other trends, over the expected run-off duration of the relevant claims reserves. Alternatively, the proportion of the total expenses from all sources in past periods (whether connected with claims handling or not) that relate only to claims handling can be assumed to apply to known future expense budgets, which could for example have been specified as part of the insurer's business planning process. Allowance can be made for the reducing proportion of future budgeted expenses that are expected to relate to claims in respect of the current claims reserve.

An example of a projection methodology is described in Newman et al. (1999), in the context of reserving for Lloyd's syndicates, where there is the additional complexity of needing to allocate the ULAE reserve to different "years of account".

Count-based Methods

This type of method involves projecting the future number of claims, categorised by their development status (e.g. reported, open, closed, etc.) and then assigning a per-claim ULAE cost to each of these. Count-based methods seek to overcome two of the potential limitations of ratio-based methods. First, by using only the aggregate amount of claims, ratio-based methods do not take account of the number and average cost of claims. Although larger claims tend to involve a greater absolute level of ULAE, the cost as a proportion of the claim size may be lower than for smaller claims, and ratio-based methods make no allowance for this. Second, ratio-based methods will result in the ULAE reserve changing when the claims reserve changes, whereas in practice there will be an element of fixed ULAE cost that will not vary according to the size of the claims reserve.

Count-based ULAE reserving methods can take various forms, but all will rely on being able to collate suitable data relating to claim numbers. One approach, based on that described in Friedland (2013), involves the following steps. These are applied separately for each reserving category, selected so that the ULAE costs per claim within each category are expected to be broadly similar.

1. **Collate claim number data:** Collate the number of claims for one or more previous periods (e.g. calendar years), split between the number of newly reported claims, the number of claims closed/settled and the number remaining open at the end of the period.

2. **Project claim numbers:** For each future period (e.g. calendar year) over which the current claims reserve will run-off, estimate the future number of newly reported claims, the number of claims closed/settled and the number remaining open at the end of the period. This can be done as for standard Frequency–Severity claims reserving methods, such as by applying the Chain Ladder to claim number reported and settled data, as described in Section 3.5. The diagonals of the projected claim number triangles can then be summed to produce the estimated future calendar period values.

3. **Estimate split of ULAE time across claims lifecycle:** For the different aspects of the claims lifecycle – opening newly reported claims, dealing with already open claims and closing/settling claims – estimate the proportion of ULAE costs that relate to each of these activities. This can be based on time and motion studies and/or judgement of the claims staff. The proportions will depend on the particular context, but purely as an

illustration, Friedland (2013) shows a worked example where the split between these three categories is 70%/20%/10% for newly reported/already open/settling respectively.

4. **Calculate weighted total claim count:** Using the observed claim numbers for one or more previous calendar periods, as collated in Step 1, and the split of ULAE across different activities from Step 3, derive the "weighted total claim count" in each of these previous periods.

5. **Calculate the Average ULAE per weighted total claim count:** This is defined as the actual paid ULAE in each previous calendar period, divided by the Weighted total claim count from Step 4.

6. **Select Average ULAE per weighted count:** After adjusting the Average ULAE per weighted total claim count from the previous step for each previous calendar period so that it represents values in a common time period (e.g. to the first future calendar year), derive a suitable average value for use in estimation of future ULAE costs. This approach assumes that the average does not vary by claim size, which may not always be appropriate.

7. **Calculate the estimated future weighted total count:** This is derived, for each future calendar period, using the projected future claim numbers from Step 2, combined with the assumed split of claim costs from Step 3.

8. **Derive the estimated ULAE reserve:** For each future calendar period, multiply the estimated future weighted count from Step 7 by the assumed average ULAE per weighted count from Step 6, adjusted for future inflation and any other relevant trends (e.g. the average ULAE per weighted count may increase as the reserves run-off, since the more complex claims are likely to dominate). Then sum these to derive the ULAE reserve for the relevant reserving category.

As the above steps are applied separately for each reserving category where the ULAE costs per claim are expected to be broadly similar, the total ULAE reserve will be the sum across all categories. Given the relative immateriality of the total ULAE reserve in many cases, the number of reserving categories used will likely be fewer than used for the core reserving analysis in respect of indemnity and ALAE amounts. In some contexts, a count-based method such as this will either not be feasible due to suitable data not being readily available, or will be deemed too complex given the relative immateriality of the ULAE reserve. In others, for example where there are only a small number of reserving categories and where claim count methods are used for the core reserving analysis, they can be an alternative to some of the other relatively simplistic methods.

A detailed count-based reserving method for ULAE is described in Johnson (1989), although Mango and Allen (2009) argue that the approach has shortcomings. They suggested an alternative method, referred as the "Claim Staffing Method", which involves projecting future claim counts (split according to Opened, Closed and Pending as per the count method described above) together with future claims staffing levels, along with the associated ULAE per staff member cost.

Other ULAE Methods

In addition to the three types of method described above, some other methods have been suggested in various sources. For example, Newman et al. (1999) includes reference to the use of standard reserving methods (such as CL and BF), applied to triangles of Paid ULAE. More recently, Ohlsson (2016) outlines an approach based around an underlying model at an individual claim level that includes both fixed and variable ULAE costs, which leads to a formula for estimating the ULAE reserve, by line of business, that requires only aggregate data. The formula is similar to those used in the ratio methods described earlier in this section, in that it produces an estimate of what Ohlsson (2016) terms the "Expense Reserving Factor" which is multiplied by the observed ULAE from one or more previous calendar periods to produce the ULAE reserve.[3]

6.5 Reinsurance Bad Debt

6.5.1 Introduction

This refers to a type of additional reserve item that is required in some circumstances. It relates to the risk that an element of the credit for outwards reinsurance that has been made within the net of reinsurance reserves will not ultimately be received. It may be referred to as the reserve for Reinsurance Bad Debt, Uncollectible Reinsurance or Reinsurance Counterparty Default risk. This section focuses on the assessment of this reserve, rather than on other related topics such as the consideration by insurers of reinsurance counterparty risk prior to purchasing reinsurance. It is intended to represent a high-level summary of the topic, rather than a detailed analysis of all the matters involved in estimating a reinsurance bad debt reserve. Interested readers requiring this additional detail can refer to the sources cited in this section.

Reinsurance bad debt can arise as a result of a reinsurer's inability to pay, due to suffering financial difficulties, or their unwillingness to pay as a result of some form of contractual dispute. Most insurers will experience delays, to varying degrees, in collection of reinsurance recoveries, although in many cases the balances will be paid within a reasonable period of time. In some cases, the overdue balance may eventually have to be written off in whole or part. Most insurers that have existed for several years will have experienced reinsurance bad debt at some point, although many will only do so very infrequently, if at all.

The amount of overdue balances in respect of paid reinsurance recoveries may be referred to as reinsurance "accruals" in some insurance markets. This is effectively a debtor item on the balance sheet (i.e. an asset). Where the reserving exercise includes an assessment of the reserve required for reinsurance bad debt, the scope may or may not include consideration of any potential bad debt on the accruals element; sometimes it is handled by the

---

[3] For example, in the Simple ratio method – Type 1 in Table 6.1 – the Expense Reserving Factor is the Claims Reserve divided by Claims Paid in the previous calendar period, which is then multiplied by the observed Paid ULAE in that calendar period.

accounting/finance function, rather than the reserving team, since it relates to reinsurance debtors. This will need clarifying if it is not clear at the outset of the exercise.

Regardless of whether reserving practitioners have any direct involvement in the assessment of any potential bad debt in respect of unpaid reinsurance accruals, the fact that this can (at least in theory) arise raises two key questions for the reserving practitioner. First, how have uncollected balances in respect of paid reinsurance recoveries been treated within the reinsurance data used for claims reserving purposes (and by implication the net of reinsurance data) and second, is it necessary to make an allowance for future reinsurance bad debt within the reserve estimates?

In respect of the first point, if any of the uncollected balances (e.g. those that have been written off) have reduced the paid reinsurance recoveries included in the reserving data, then depending on the reserving approach used to allow for outwards reinsurance, this could mean that there will be an implicit allowance within the net reserves for future reinsurance bad debts. A similar issue could arise in relation to treatment of known reinsurance bad debt on outstanding claims, where the outstanding reinsurance recoveries may or may not have been reduced. Hence, this will need further consideration to avoid any double counting with any provision for unpaid balances or additional reinsurance bad debt reserve. To avoid this, some insurers choose to construct data for reserving purposes assuming that all reinsurance balances are fully paid, even where it is known that there have been reinsurance bad debts on accruals or outstanding claims.

In respect of the second point, regardless of whether there are any known or possible bad debts in respect of the paid or outstanding reinsurance recoveries, wherever the net of reinsurance reserves take credit for outwards reinsurance there is always at least a theoretical possibility that some of that credit will never materialise. Whether the reserving practitioner needs to consider this possibility and hence derive a reinsurance bad debt reserve will depend on the jurisdiction and scope of the reserving exercise. In many cases, it will not fall within the scope of the exercise, whereas for others (e.g. under Solvency II, as explained in Section 6.8.2) there will be a requirement for it to be included. Where it does fall within the scope of the exercise, it is commonly allowed for as an explicit additional reserve item.

Where there are known reinsurance disputes, these will usually be treated separately from any reserve for reinsurance bad debt, taking into account the particular issues associated with each dispute. Some disputes may relate to the validity of the whole reinsurance contract itself, whereas others may relate to how the coverage operates, for example regarding how claims arising from particular large events are aggregated for recovery purposes. Any credit that is given for reinsurance in the net reserves will need to take appropriate account of any known disputes, perhaps by taking a probabilistic approach. This will often have already been considered as part of the process of estimating the net of reinsurance reserves, as opposed to being part of the separate estimation of a reserve for reinsurance bad debts. The potential for future disputes may, however, need to be considered as part of the estimation of the reinsurance bad debt reserve.

6.5.2 Methods for Estimating Reinsurance Bad Debt Reserve

The methods that can be used to estimate this reserve can be summarised as follows:

1. Simple proportion method.
2. Reinsurer-specific method.
3. Advanced deterministic and stochastic methods.

These are described further below. The choice will depend on factors such as the complexity of the outwards reinsurance programme, data availability and the potential materiality of the bad debt reserve.

It is assumed for the purposes of the description of these methods that the core reserving exercise has produced an estimate of the reinsurance reserve – that is, the difference between the gross and net reserves. It will comprise amounts relating to reinsurance outstanding claims (i.e. reinsurance case reserves) and to reinsurance IBNR. It may also include an estimate of future reinsurance premiums and commissions that are payable. Any known reinsurance disputes are assumed to have already been allowed for.

The purpose is assumed to be to determine the element, if any, of the reinsurance reserve that needs to be established as a bad debt reserve. It is assumed for the purpose of the methods described below that the bad debt on reinsurance accruals is excluded from the scope of work, as it will generally be easier to assess, since by definition it will relate to specific reinsurers with known amounts of overdue debt.

Simple Proportion Method

This very approximate method involves selecting a proportion of the reinsurance reserves that will not be realised in the form of actual future reinsurance recoveries. This might be done in aggregate across all reserving categories and cohorts, or subdivided in some way. For example, a 1% aggregate proportion will mean that it is assumed that 1% of the reinsurance reserve (i.e. case plus IBNR) will not be recovered – i.e. that the net reserves before bad debt need to be increased by 1% of the estimated reinsurance reserve.

The estimation of the proportion is obviously the key judgement and can be based on a review of any unpaid reinsurance balances, an understanding of the credit worthiness of the relevant reinsurers (e.g. based on their external rating), taking into account the period over which the reinsurance claims will be paid, and a review of the potential for future disputes to affect the amount recovered.

This approach may only be appropriate where the materiality of the reinsurance is relatively low and where the risk of non-recovery is deemed to be low. In some cases, it might be concluded that both the risk of non-recovery and the relative materiality are so low in the context of the overall reserves, that a zero reinsurance bad debt reserve is appropriate. In others, where the reinsurance reserve is very material and there are some known payment issues or disputes with one or more reinsurers, these will need separate consideration and a more detailed approach for the remainder of the bad debt reserve might need to be considered.

Reinsurer-specific Method

This approach is much more detailed than the first method, but the data required ought to be available in many cases. In summary, the approach involves allocating the reinsurance case reserves to reinsurer, determining a bad debt percentage (and hence, amount) for those reinsurers and then inferring the bad debt loading for the IBNR from the implied overall bad debt ratio on the reinsurance case reserves. A summary of the data required as at the relevant valuation date for this method is as follows:

- A high-level overview of the outwards reinsurance programme, for reserving categories and cohorts where it is expected to generate recoveries in relation to the reinsurance outstanding and IBNR reserves.
- The reinsurance unpaid reinsurer balances by reinsurer and age of debt. This might help identify reinsurers with the potential for bad debts or disputes in relation to the reinsurance outstanding or IBNR. Determination of whether any of the relevant reinsurance is collateralised may also be helpful.
- The outstanding reinsurance recoveries by reinsurer – i.e. the reinsurance case reserves by reinsurer. This will require the relevant proportion of the gross outstanding claims to be allocated to the reinsurance contracts, and for the reinsurers on those contracts, along with their shares, to be identified. This should be available in many cases, as there will often be an obligation to advise the reinsurers of these outstanding recoveries.
- Data relating to the security rating of each relevant reinsurer. This will usually be based on security ratings produced by external rating agencies (e.g Standard & Poor's or Moody's). However, in some cases, there will be no external rating available for the particular reinsurance entity, although this might be able to be inferred from another company in the same corporate group where there is a rating, and in others additional data and market intelligence may need to be used to help assess the security of the reinsurer. Data gathered as part of a reinsurance security assessment carried out at the time of placement of the reinsurance can be relevant here.
- Data related to estimated default rates by credit/security rating and term. This might be based, for example, on corporate or governmental default rates published by external rating agencies. These are effectively the cumulative probabilities of default and could be used as a proxy for the creditworthiness of reinsurers with the same rating. Appendix 1 of Bulmer and Working Party (2005) contains tables showing an example of this data obtained from two rating agencies – Standard & Poor's and Moody's. These relate to defaults up to 2004. If these are used for reserving purposes, then up to date versions will need to be obtained from the relevant rating agencies and a decision will need to be made on which source, or a combination thereof, is used for the purpose of estimating the reinsurance bad debt reserve.
- Data related to estimated Loss-Given-Default ("LGD") and/or recovery percentages by credit/security rating and term. The source for this information might be similar to that for the default rates. The LGD represents the loss suffered when a default occurs, and the recovery percentage is the contra of this – i.e. the proportion of the debt recovered

when a default occurs. For reinsurers that are already insolvent, the recovery percentage represents the proportion of their debt they are expected to pay. Where there are reinsurers that have been in financial difficulty and who may have already defaulted on their obligations or gone into liquidation, there are sources available that give indications of the recovery percentage that might be expected. For example, Bulmer and Working Party (2005) gives an example in Appendix 3 of this type of data for a selection of UK companies.

- Details of any set-off balances by reinsurer. It might be possible to reduce any required reinsurance bad debt reserve, by allowing for these in the estimation process.

Once this data has been gathered, this method proceeds as follows:

1. Collate the reinsurance case reserves by reinsurer, net of any premium and claim amounts that can be set-off. This might be grouped by reserving category and cohort, or just done in aggregate. It may also be grouped by term – i.e. the expected duration of the reinsurance recoveries. Any set-off adjustments may need to take into account the relevant provisions of the particular contracts involved, as well as whether it applies across entities in the same corporate group.

2. For each reinsurer in Step 1, determine a security rating. This will be based on a combination of the various data sources mentioned above. This may not necessarily relate solely to external rating agency data, as the reserving practitioner may choose to take into account other data or market intelligence, where relevant. Allowance for the security of any relevant collateralised arrangements may also be made here.

3. For the reinsurance case reserves in respect of each of these reinsurers, using their security rating and an assumed term of the recoveries, determine a bad debt percentage using the default probabilities data, as specified above. This represents the probability of the reinsurer defaulting (in full or in part) on their obligations to the reinsured at some point during the period that the recoveries are due. For reinsurers where there is no credit default probability, make a selection of the bad debt percentage based on the other data collated regarding their financial strength. For reinsurers that are known to be insolvent or in financial difficulties, use other suitable sources to estimate their bad debt percentage. Finally, for reinsurers where there are known or expected disputes in relation to the accruals or outstanding recoveries, take this into account when selecting their bad debt percentage.

4. Unless already taken into account in the previous step, assume an LGD or recovery percentage when a reinsurer defaults. For simplicity it is assumed for the purposes of this description that a single LGD percentage is used, although in practice it could vary by reserving category, rating and term, or be specific to individual reinsurers. For some reinsurers, where LGD and recovery percentage data is deemed unreliable or is difficult to obtain, an assumption could be made, based on judgement, of the appropriate recovery percentage to use.

5. Determine the reinsurance bad debt in respect of case reserves for each reinsurer by multiplying the reinsurance case reserves by the assumed bad debt percentage and the assumed LGD percentage (or 100% − Recovery percentage).
6. Derive an overall reinsurance case reserve bad debt percentage by summing all the individual reinsurer figures from the previous step, and dividing by the total estimated reinsurance case reserves. This might be done by reserving category, type of reinsurance, cohort and term or in aggregate.
7. Select a suitable reinsurance IBNR bad debt percentage. This could be set equal to the case reserves bad debt percentage derived from the previous step, or that figure could be adjusted up or down to allow for the potentially longer term of the reinsurance IBNR recoveries or any known material difference in the security of the reinsurers that are expected to provide the IBNR recoveries compared to those providing the case reserve recoveries. This percentage could be selected by reserving category, type of reinsurance, cohort and term, or in aggregate.
8. Multiply the selected reinsurance IBNR bad debt percentage(s) by the estimated reinsurance IBNR to determine the reinsurance bad debt in respect of the IBNR.
9. Sum the case reserve and IBNR bad debt reserves to derive the estimated reinsurance bad debt reserve.

As mentioned above, the data required for this method may be available in many cases, making it a reasonably detailed but practical approach that can be used where justified by the context and materiality of the reinsurance reserves and the security of the reinsurers.

A number of enhancements are possible to the above approach, two of which are as follows:

- **Bad debt percentage for IBNR:** Instead of inferring a bad debt percentage for the reinsurance IBNR from that derived in respect of the reinsurance case reserves, derive a more specific percentage based on a projection of the reinsurance IBNR by reinsurer. This will require gathering more detailed data in respect of the outwards reinsurance programme, to include all details of contracts that might be expected to contribute towards the reinsurance IBNR. Some of the projection of reinsurance IBNR by reinsurer might be produced by considering the element that arises in relation to IBNER on specific large losses or catastrophes, where it might be possible to allocate the IBNER to reinsurance contracts and hence reinsurer. Other parts of the reinsurance IBNR could relate to specific reserving categories and cohorts where the reinsurers can easily be identified. In practice, a combination of a case reserve-based bad debt percentage for some elements of the reinsurance IBNR and an IBNR/IBNER-specific-based percentage may be necessary.
- **Derive estimated durations for recoveries, including IBNR:** Instead of making approximate judgements regarding how the reinsurance recoveries are split by term, make an estimate of the duration of both reinsurance outstanding and IBNR recoveries, perhaps based on a cash flow projection of gross, net and reinsurance claims as part of the core

reserving exercise. This will allow a more precise selection of term-dependent bad debt factors.

When these two enhancements are used, the method then becomes similar to the method labelled "A Practical Approach" described in Bulmer and Working Party (2005).

Advanced Deterministic and Stochastic Methods

It may be possible in some circumstances to apply a more advanced method than the above reinsurer-specific approach, even after allowing for the enhancements mentioned above. For example, a more granular allocation of the gross IBNR and IBNER to individual reinsurers for each individual reported claim or to an estimated distribution by size of IBNR claims could be made, so as to more accurately identify the reinsurers for the associated recoveries. In addition, a detailed cash flow projection could be developed along with the creation of a model of the outwards reinsurance programme, with the losses being processed through the model to produce the estimated reinsurance recoveries by reinsurer and time period. This could be done on either a deterministic or stochastic basis. Some of these approaches may build on credit risk modelling methods used in financial markets, where there has been considerable research over many years. These methods might involve features such as transition matrices, which allow for the probability of movement between rating categories over time.

Whilst these more advanced approaches are not likely to be justified or practical in many situations, where the assessment of the reinsurance credit risk charge within an internal capital model is based on a stochastic model, it might be possible to use this model to estimate the reinsurance bad debt reserve.

Bulmer and Working Party (2005) outlines an example of a more advanced approach, under the heading of "A Theoretical Approach". That paper also considers various other important aspects of estimating a reinsurance bad debt reserve, including the use of ratings produced by external rating agencies, the assessment of default probabilities and recovery percentages. An example of a simulation approach is given in Pastor (2003).

For companies operating under the Solvency II regulatory system, the technical provisions will need to include an appropriate reserve for reinsurance counterparty default, as discussed in Section 6.8.2.

6.6 Latent Claims

6.6.1 Introduction

The term "latent claims" is generally taken to refer to types of claim where there is a long delay (e.g. greater than, say, five years) between the period when the exposure occurs and the realisation that this exposure has resulted in some form of injury or damage, for example as a result of the delayed onset of symptoms by persons exposed to the relevant source. This usually means that there is a consequential delay between the inception of the policy to which the claim relates and the notification of the claim to the relevant insurer.

In some contexts, latent claims might be deemed to include only types of claim that were not contemplated at the time of the original policy being underwritten by the insurer. However, the delay referred to above is the key feature of latent claims that make reserving for them particularly challenging.

The exposure period itself can be several years long, which results in claims related to a common source potentially affecting multiple periods of insurance coverage simultaneously. This contrasts with individual incident- or accident-related claims which typically affect one period of coverage. Hence, with latent claims it is often necessary to consider how claims are allocated across the exposure period to the relevant insurance policies, which creates additional complexity from a reserving perspective.

Some types of latent claim are referred to as "Mass Torts", particularly in the USA. Another commonly used label is "APH", which is an abbreviation for Asbestos, Pollution and Health Hazard, which are three types of latent claim that often make up the majority if not all the different types of latent claim that impact a particular portfolio.

The main reason why latent claims are considered in a separate section in this chapter is that, as explained later in this section, the mainly triangle-based deterministic and stochastic reserving methods described in Chapters 3 and 4 respectively do not generally work well for this type of claim, and so a different approach is required. For this reason, in a portfolio where both latent and non-latent claims are present, each component is usually considered separately for reserving purposes.

Types of Latent Claim

Examples of specific types of latent claim that have one or more of the features referred to above include Asbestos, Pollution, Noise Induced Hearing Loss and Health Hazards (other than Asbestos). These four types are summarised briefly below. In each case, there is considerable complexity associated both with the nature of the underlying process that gives rise to claims and the way in which they impact on the insurance industry. An understanding of this complexity is helpful before any reserving analyses are conducted. Suggested further reading related to latent claims is included at the end of this section, and details of the reserving methods that can be used for one particular type of latent claim – UK asbestos – are given in Chapter 7.

Asbestos This is probably the most commonly referenced type of latent claim, not least because of the very significant impact of such claims on the insurance industry in a number of developed insurance markets. For example, A.M. Best have estimated (see A.M. Best, 2016b) that exposure to asbestos in the USA will result in an ultimate cost, net of reinsurance, of $100bn to the US insurance industry, and a UK asbestos working party (see UK Asbestos Working Party, 2009) has estimated a total undiscounted future cost of UK asbestos-related claims to the insurance market of around £11bn for the period 2009 to 2050. The majority of asbestos-related claims have resulted from compensation being awarded to persons who have suffered from diseases that are linked to asbestos exposure, such as mesothelioma and asbestosis. The delay between original exposure and the onset of symptoms for these diseases, i.e. the latency period, can be very long, in some cases in excess of 40 years. Further details of how this delay affects the reserving process, along with other asbestos reserving considerations, are given in the UK asbestos section in Chapter 7.

Pollution   The particular type of pollution that can produce latent claims is where there is a gradual underlying process, which takes a long time to exhibit damage to the environment, persons or property. This is in contrast to other types of pollution, which are more "sudden and accidental" in nature, where the damage is typically more immediately apparent. The costs to the insurance industry from this type of gradual pollution arise from a number of sources including clean-up costs, third party property damage and bodily injury. Whilst the scale of the financial impact on the insurance industry of latent pollution does not appear to be as significant as asbestos, it has nevertheless been very material. For example, A.M. Best have estimated (see A.M. Best, 2016b) that in the USA it will result in an ultimate cost, net of reinsurance, of $42bn to the US insurance industry. Taking into account A.M Best's corresponding estimate for US asbestos of $100bn, this implies that asbestos claims will ultimately be less than 2.5 times the volume of pollution claims. However, based on the same source, from a reserving perspective the volume of future claims related to asbestos (i.e. the reserves) as at recent valuation dates is likely to be a greater multiple of those related to pollution, due to the higher proportion of ultimate pollution claims that are estimated to have been paid relative to asbestos. The relativities for individual insurers could be materially different to this, depending on the nature of their exposures.

Noise Induced Hearing Loss   Strictly speaking, this particular latent claim category, abbreviated by NIHL, relates to hearing loss that is caused by exposure to high intensity sound, regardless of the source of that sound. However, in an insurance context, the vast majority of claims arise when that source is occupational, often in an industrial context, in which case it is usually referred to as Industrial Deafness. Hence, references to NIHL in an insurance context often refer specifically to Industrial Deafness, and for the purpose of this subsection, NIHL is taken to mean that. The latent nature of NIHL in an insurance context arises through prolonged exposure to high sound levels, leading to hearing loss that develops gradually over time, and which tends to become apparent in later life, often in retired workers.

Whilst the overall financial impact of NIHL on the insurance industry is materially less than that of asbestos, in the UK the number of claims notified has risen steeply in recent years, leading to an increased focus on this source of claim. Further details are given in UK Deafness Working Party (2013), Association of British Insurers (2015) and Gravelsons and Brankin (2015).

Health Hazards   This refers to a category of latent claim, encompassing several different individual claim types, not otherwise separately identified (and hence in this context excluding Asbestos and NIHL, which are both types of Health Hazard). They are the "H" in the abbreviation "APH". They are often grouped together under the general heading of

Health Hazards, partly as they share a common feature of relating to substances or medical products that are believed to be harmful to health, and partly as they historically have given rise to relatively small volumes of insurance claims, compared with other major latent claims such as asbestos and pollution. Examples of individual Health Hazard claim types include Agent Orange, DES, Chronic Obstructive Pulmonary Disease ("COPD"), blood products and Silicon Breast Implants. Not all of these are currently giving rise to claims, and there is always scope for additional potential types of claim to emerge in future, as medicine and technology continues to advance.

Other Considerations Relevant to Reserving for Latent Claims

The underlying class of business that gives rise to latent claims varies according to territory and type of claim, but will often be some form of liability-related coverage, as opposed to property damage, which usually covers only specified perils. In addition, the coverage provided must be such that claims which are latent in nature can be made against the policy or policies during which the exposure occurred (or for which another relevant claim "trigger" applies, as explained further in Section 6.6.2). This usually means some form of occurrence coverage, rather than claims-made. Latent claims can impact both insurance and reinsurance classes of business, which can affect the nature of the data that is available and hence the reserving methodology that can be used.

Because latent claims often affect more than one accident or underwriting cohort simultaneously, standard development-based reserving methodology that relies on data triangles is not usually suitable as a basis for estimating latent claims reserves. Hence, wherever possible, latent claims will be removed from the data triangles and analysed separately from other claims using the methodology described later in this section. If latent claims are left in the data triangles, they can often have a recognisable impact on the shape of the claims development. "Normal" cumulative incurred claims will typically show an increase in the early development periods, followed by a flattening out to a stable pattern. When latent claims are present, this stable development may be apparent for a number of years, but is often followed by a subsequent increase in cumulative incurred claims, affecting a number of cohorts simultaneously in the same calendar period(s).

If such a pattern exists and is known to be caused by the presence of latent claims, then removal of the relevant claims from the data triangles is preferable, to avoid distorting the estimation process of the reserve for non-latent claims. However, where such removal is not possible, but where other non-triangle data is available for estimation of latent claims, the distorting impact on the triangles can be removed in an approximate way[4] so that the implied triangle-based reserve includes no or minimal implied IBNR for latent claims.

In some situations there may be no prior knowledge of exposure to latent claims, in which case the identification of such a pattern may be an early warning indicator of potential exposure, although there can be other non-latent causes. Whatever the cause, further investigation is usually warranted where such a pattern in claims development is

---

[4] For example, by assuming zero IBNR for cohorts that are believed to have exposure to latent claims, but where the non-latent claims development has stabilised fully.

observed in practice. If the cause is latent, then removal of the latent claims from the core triangle data and use of the methods described below need to be considered.

For many reserving exercises in practice, latent claims do not feature, usually because the portfolio relates to more recent periods of exposure with no known latent claims, or to classes of business where there is not believed to be any such exposure. However, for portfolios of discontinued, or "run-off" business, where the last policy might have been written many years ago, latent claims can in some cases be the major or only reserving category, possibly incorporating a number of different claim types. This will be because all the non-latent claims have been settled several years ago, and so the only remaining types of claim that need to be reserved for are all latent in nature. For such portfolios, an element of the reserving process will involve ensuring that all potential types of latent claim that may affect the portfolio have been identified and an appropriate reserve established. The types of latent claim will include not only those that relate to the reported claims, but also to other types that may affect the portfolio in future.

6.6.2 Reserving Methods for Latent Claims

The reserving methodology used in a particular context will depend on many factors, not least of which will be the materiality and type of latent claim, the jurisdiction in which the claims arise and where the associated coverage is written, and the data that is available. This section summarises in generic terms a number of different types of such methodology. Chapter 7 provides a more detailed description of the specific methodology that can be used for one type of latent claim – UK asbestos.

The general methods that can be used for latent claims reserving include four broad categories – Frequency–Severity (FS), development-based, benchmarking and exposure-based. Each of these is described briefly below. Exposure-based methods are generally regarded as superior to the other methods, but are not always practical, due for example to the lack of suitable data. Whatever method(s) are used, the detail of their application in practice will need to take into account any relevant features of the particular insurance market in which the latent claims are arising, such as how claims are aggregated for the purpose of recovery from insurance policies. For example, in some markets (e.g. UK) asbestos claims tend to be recovered on a per claim basis, whereas in others (e.g. USA) they tend to be aggregated per underlying assured.

Frequency–Severity

This refers to an approach applied at an aggregate level, either for groups of different types of insureds/reinsureds or other relevant portfolio subdivisions, or alternatively across all underlying insureds/reinsureds combined. This is usually done separately for each type of latent claim. The approach is similar to the use of FS methods for non-latent claims, except that the analysis is done across all cohorts combined for the relevant class of business affected by the particular type of latent claim, rather than individually by cohort. The number of future latent claims is first estimated, perhaps based on market-level projections that exist for some latent claim categories in certain countries (e.g. UK asbestos). Then, the average cost per claim is estimated, perhaps split by certain categories (such as head of damage and age), and then combined with the claim number projections to derive the estimated reserve. This type of approach is explained in more detail in the context of UK asbestos in Section 7.7.

Development-based

Although standard triangle-based methods are not usually applicable for latent claims, if the claim amounts are aggregated across cohorts, to form a single row of development data, this can then be projected to ultimate. This projection can be achieved by curve fitting to either the raw gross or net claims paid and/or incurred, or to the development factors. Curve fitting is described in Section 3.2.4, which although this focuses on its use with data triangles with several cohorts, it can also be used for single cohorts. As for non-latent claims data, this approach can be difficult to apply if the claims are still increasing sharply, in which case, benchmarking can be used to validate the results from curve fitting. Development-based methods applied to the claim amount data represent a very simplistic approach for estimating reserves for latent claims, but in some cases low materiality or lack of other suitable data may mean that it is one of the few approaches which can be used. Continuing to monitor the claims development at least enables prior estimates to be validated, and updated if necessary.

Benchmarking

This is a very simplistic approach that can be used as a method in itself, where data is sparse, or as a reasonableness test to supplement other methods. The results derived using only benchmarking are inevitably subject to a material degree of uncertainty, owing to the generic issues associated with using any benchmarking for reserving purposes, which are exacerbated for latent claims due to the often complex underlying claims and exposure allocation process.

A commonly used benchmark factor for latent claims is the survival ratio. This is simply an assumption for the number of future periods (usually years) that an assumed average paid or incurred claim amount per period will apply until the ultimate is reached. Hence, if this average can be calculated from the available development data for a particular latent claim category (e.g. over the past two, three or five years, say) then it can be multiplied by the assumed survival ratio to derive the reserve or IBNR, depending on whether paid or incurred claims respectively are being used. Paid claim survival ratios are generally more commonly used than incurred claim survival ratios, partly as the latter depends on the case reserve adequacy, which can vary significantly between portfolios. Other benchmark factors that can be used for latent claims reserving include ultimate to paid or incurred ratio, and IBNR to outstanding claims ratio. The source of the benchmark factors will depend on the context, but as an example, for US asbestos and pollution, they can be derived from A.M. Best's special reports (e.g. see A.M. Best, 2016b). The specific type of benchmark used will need to take into account the type of portfolio and latent claim for which reserves are being estimated. For example, a reinsurance portfolio might need a quite different benchmark than an insurance portfolio for the same latent claim type. It is not uncommon for the nature of the business underlying the source of the benchmark to be different to the relevant insurance portfolio, and so some adjustment may be necessary to allow for this. Benchmarking can also be used to validate reserves derived using other methods. For example, if the paid survival ratio can be calculated using the derived reserve for a particular portfolio (by dividing that reserve by a calculated average claims paid per annum over recent years), it can then be compared against a benchmark survival ratio calculated from suitable industry or other data.

Exposure-based

This is a form of "bottom-up" approach, which contrasts with the other approaches mentioned above, which are more "top-down". For a particular insurer, their exposure to a particular type of latent claim might arise through several individual assureds. In general terms, an exposure-based approach first involves estimating an ultimate value for the underlying claims that each of these assureds will produce for that latent claim. The approach used to derive that ultimate will depend on the type of latent claim. For example, for US pollution it will typically be applied at the individual pollution site level, with estimated clean-up costs being allocated to insureds (referred to as "Potentially Responsible Parties", or PRPs) who had some involvement with that site. In contrast, for US asbestos it might be based on using a separate projection of claim numbers and average expense and indemnity costs (i.e. a form of Frequency–Severity type approach, as described above, but applied at an individual assured level), allowing for factors such as inflation, dismissed claims and trends in claim propensity. A basis for allocating the resulting ultimate cost for each assured to individual insurers then needs to be determined, taking into account the relevant legal framework. This will depend on several factors, including the way in which the particular latent claims affect insurance coverage (i.e. the definition of the event that activates coverage under the policy – referred to as the "trigger"), how claims are allocated amongst the associated insurers and their shares of the relevant insurance policies. The existence of missing, exhausted or commuted insurance coverage can add complexity to the process, for example. This is repeated for each assured, and then potential additional loadings are added to allow for factors such as new assureds being notified. This general approach will need refining for different latent claim types.

For reinsurance portfolios, an exposure-based approach is more complex than the brief outline given above, as the exposure to a particular latent claim needs to be considered for each reinsured cedant, along with that cedant's own underlying exposure.

Further reference to exposure-based methodology in the context of UK asbestos is given in Section 7.7.

The above approaches are usually applied separately to each different type of latent claim for which there is known exposure in the portfolio. An additional reserve may also be required where there is believed to be potential for further types of latent claim to impact the portfolio in future. These may relate to recognised types of latent claim that are generally known, but have not yet impacted the relevant portfolio, and to currently unknown types of latent claim that might affect the portfolio in future. Other reserving approaches, including stochastic methods, might also be used in some contexts.

6.6.3 Further Reading

Relevant sources for latent claims reserving that are not already referenced in this section include two papers that provide a general introduction to latent claims (Latent Claims Working Party, 1990, 1991), a briefing note on asbestos (Institute and Faculty of Actuaries – General Insurance Practice Executive Committee, 2010) and a detailed paper on UK asbestos (UK Asbestos Working Party, 2004). Two papers that cover exposure-based methodology for US asbestos and environmental pollution respectively are Cross and Doucette (1994) and Bouska and McIntyre (1994). The documentation of the estimation methodology used for some UK Schemes of Arrangement also provides a description of possible reserving approaches for certain types of latent claims. For example see WFUM Pools (2006) and OIC Run-Off Limited (2015).

Further references for UK asbestos are given in Section 7.7.

6.7 Discounting

Discounting of claims reserves for financial or statutory reporting purposes has historically been relatively uncommon in many general insurance markets, with insurers often preferring to hold undiscounted reserves. However, there are a number of situations where it is used in general insurance reserving, including:

- In some regulatory contexts it is a requirement. An obvious example is in the EU under Solvency II, as discussed in Section 6.8.3.
- It is permitted for financial reporting purposes in some markets in certain circumstances. For example, currently in the UK, under UK Regulations (2008), discounting is permitted for general insurance business, subject to certain conditions, one of which is that the expected average interval between the date for settlement of claims being discounted and the accounting date must be at least four years. Hence, in some markets, insurers do discount their reserves for certain types of business for financial reporting purposes, and this can potentially be material in the context of their financial statements (i.e. the difference between the discounted and undiscounted reserves is sometimes significant).
- For transactions involving general insurance liabilities (such as commutations, sale of business or transfer of portfolios) it may be used as part of the process of deriving the value of the liabilities that the parties are willing to agree upon.

Discounting may also be used in other general insurance contexts such as pricing and capital modelling, but the focus of this section is its use in reserving.

When general insurance reserves are not discounted, then it can be argued that they contain an implicit risk margin equal to the difference between the undiscounted and discounted reserves. The uncertainty that often exists in relation to the amount and timing of future cash flows in general insurance may explain why some insurers have preferred not to discount their reserves, but have rather chosen to retain the implied risk margin in response to this uncertainty. When reserves are discounted, in effect they accelerate the emergence of profit compared to when reserves are not discounted[5] (unless a risk margin is added that partially or fully offsets the impact of discounting). However, where the term of the liabilities is short and/or interest rates are low, any margin implicitly included in undiscounted reserves may be relatively small, particularly when considered in the context of the overall uncertainty that may surround estimation of the reserves. On the other hand, where the liabilities are long-tail, the impact of discounting can be very material, so that the selection of the relevant methodology and assumptions is critical.

The remainder of this section first summarises the three key stages in any discounting calculation in a general insurance reserving context after the undiscounted reserves have been estimated – estimating a claims payment pattern, selecting the discount rates and calculating the discounted reserves. It is assumed for this summary that a deterministic reserving method is being used to derive a point estimate for the gross of reinsurance reserves. The section concludes with a brief discussion of some of the other aspects of discounting for general insurance reserving purposes, including allowing for reinsurance and using discounting with stochastic reserving methods. The appropriateness of using discounting in specific contexts is not discussed, as this will depend, for example, on any applicable regulations. The use of discounting is also discussed briefly in a number of the specific contexts covered in Chapter 7 (e.g. in UK Motor and US Workers' Compensation).

6.7.1 Stage One: Selecting a Claims Payment Pattern

A projection of how the undiscounted claims reserve will translate into claims paid in each future period will usually involve selecting a claims payment pattern by reserving category. Depending on the context, the reserving category can be a combination of class of business, currency, claim type and individual large claims or catastrophes. The claims payment pattern can either be expressed in incremental or cumulative form.

The method used to derive the claims payment pattern will vary depending on these categories. A typical (but simplified) approach for a reserving category where the undiscounted claims reserve has been based on applying reserving methods to data triangles is as follows:

1. Consider whether the process for deriving the selected undiscounted claims reserves has involved applying reserving methods to claims paid data triangles.
2. Determine whether any of the results from the previous step provide a suitable basis for the required claims payment pattern. These results do not necessarily need to have been used in the final results selection for the undiscounted reserves.

---

[5] Assuming positive discount rates, as discussed in the discount rate paragraphs in this section.

3. Depending on the conclusions from the previous two steps, apply a suitable reserving method (e.g. Chain Ladder) to the claims paid data triangle from which one or more claims payment patterns can be derived (e.g. different patterns might be needed for different groups of cohorts in some cases). Projection beyond the domain of the data triangle may be required, necessitating selection of a tail factor and/or the use of curve fitting to project the claims payment pattern in the tail.

4. Consider whether any benchmark claims payment patterns are available, perhaps based on projection of market-level claims paid data triangles, which can be used to supplement the results from the previous steps.

5. Based on a suitable combination of results from all of the above, select a claims payment pattern, which may or may not vary by cohort. Although this may start from the first development period, it will only be relevant for the future development periods for each cohort. A review of simple diagnostics such as weighted mean terms can be helpful in deciding upon the final patterns.

6. Compare the selected undiscounted claims reserve for each cohort with the reserve implied by the selected claims payment pattern from the previous step. Reconsider the selected payment pattern if there are material differences.

7. Use the selected claims payment pattern to derive an estimate of the future claims paid for each cohort, by spreading the selected undiscounted claims reserve across the future periods, according to the selected payment pattern. Only the relevant future part of the claims payment pattern for each cohort will be used here, with the incremental future proportions being grossed-up by the estimated proportion paid so far. The concept of "disposal rates" introduced in the context of Frequency–Severity methods (in Section 3.5) can also be helpful here, whereby the claims payment pattern is reformulated so that it represents the claims paid in each period, expressed as a percentage of the outstanding claims at the start of that period.[6]

8. Sum the values for each calendar period across the cohorts (i.e. sum the amounts in the future diagonals of the triangle) to produce the estimated total paid in each calendar period.

9. Review the results for reasonableness, for example by considering numerical and graphical schedules (by cohort and in total) that compare the actual past incremental claims payments with those projected in future by applying the above methodology.

This process will need repeating for each reserving category, and specific components of the reserves due, for example, to individual large claims and catastrophes may need special consideration. The details will depend on the nature of the business and the claim types to which it is exposed. Outwards reinsurance will also need allowing for, as discussed later in this section.

---

[6] For example, if a cohort is at, say, development year 4, and the pattern has 50% paid by then, and 20% paid in year 5, then this will imply 40% of the reserve for that cohort is assumed to be paid in that year, which will be the first future calendar period. In other words, the disposal rate in year 5 is 40%, being 20% divided by the proportion outstanding at the start of year 5 (50%).

Selection of claims payment patterns can be particularly difficult in some circumstances – for example, for new or long-tail reserving categories, where reliance on benchmarks and the use of the practitioner's experience and judgement will be critical. Allowance for the impact on claims payment patterns of external market factors or future legislative changes can also be important. In some cases, this can mean that historical patterns observed from the data need to be modified. The relative materiality of the undiscounted reserves and the likely proportionate impact of discounting for each category will help focus the effort required.

6.7.2 Stage Two: Selecting the Discount Rates

The approaches that can be used to select discount rates in a general insurance reserving context can be categorised in a number of different ways. One such categorisation is to group the approaches as follows, which is as per Actuarial Standards Board (2011):

1. The risk-free approach. Here the rates used are based on selecting a set of risk-free rates. These can be derived in a number of different ways, but in general terms they can be approximated by rates of investment return that are available on fixed income securities that have low default risk and timing characteristics similar to the claims payment pattern derived from the previous stage.

2. The portfolio approach. The rates here are based on the expected investment return from a selected portfolio of assets. The portfolio may either be derived from the actual assets held by the insurer in relation to the relevant reserves or from a benchmark or reference portfolio. If the expected investment return that is assumed is materially different to overall market rates, then this may need to be justified. The rates used may need to be net of investment expenses and taxation, depending on the context.

3. The pre-determined approach. Here, it is assumed that someone other than the reserving practitioner has specified the rates to use (e.g. there may be a client who has requested results to be produced to test the effect of discounting at different rates). Depending on the context, there may be professional obligations affecting the practitioner that may impact upon their ability to use such rates and, if so, the resulting disclosures that will be necessary in any communication of findings.

In a broader context, rather than specifically in general insurance reserving, the Institute and Faculty of Actuaries' Discount Rates Steering Committee ("DRSC"), identified two different approaches to determining discount rates, which are documented in Cowling et al. (2011):

- A "matching" or "market-consistent" type of approach, whereby the rates are selected so that they are consistent with the current market value of assets that, as far as possible, replicate the future economic behaviour of the liabilities to which the discount rates might be applied.

- A "budgeting" type approach, whereby the rates selected are consistent with the expected future returns the party carrying out the valuation or planning exercise believes will accrue from the assets expected to be held to provide for the future cash flows as they fall due.

However, this type of distinction, as noted in Cowling et al. (2011), has not been widely discussed or acknowledged in a general insurance context (at least in the UK), although this may change in future.

A further categorisation of the approaches for calculating the discount rate that is sometimes used is "top-down" and "bottom-up". The former is similar to the portfolio approach described above – i.e. based on actual assets held or on those in a reference or "replicating" portfolio (with possible adjustments for duration mismatches and removal of market risk premium due to credit losses not included in the liabilities). The latter approach uses risk-free rates, possibly adjusted for liquidity differences between the relevant risk-free assets and the liabilities – and hence is similar to the risk-free rate approach referred to above.

Determining which approach is appropriate in practice for a particular general insurance reserving context will depend on a number of factors, including the following (which are not necessarily in priority order, since this will vary according to the circumstances):

- **Nature of liabilities being discounted:** In many general insurance reserving contexts, there will be uncertainty regarding the amount and timing of the cash flows and hence there may be difficulties associated with establishing a portfolio of assets that will match the liability cash flows. This may therefore mean that any matching-type approach that is used is necessarily approximate.

- **Purpose of the exercise:** The approach and the associated discount rates will likely vary according to the purpose of the exercise. For example, a solvency-type exercise might require a matching-type approach, perhaps using risk-free rates, whereas a different approach (and rates) might be used in a commercial or transactional context.

- **Constraints:** The context for the reserving exercise may place certain constraints on the discount rates that should be used. For example, in some circumstances, such as when deriving discounted reserves for financial reporting or solvency purposes, there may be particular requirements regarding the approaches that can be used to derive the rates, or on the specific rates that must be used (e.g. as discussed in Section 6.8.3 in the context of Solvency II). In other circumstances, as envisaged in the pre-determined approach mentioned above, the scope of the reserving exercise may exclude consideration of the appropriate discount rates to use, with a range of rates perhaps being specified by the company or client.

- **Assets held:** The assets held (or planned to be held) by the insurer in relation to the relevant liabilities that are being discounted, and their associated performance may influence the selected discount rates, including consideration of any that do not generate a return as well as the risk of default and any liquidity restrictions. In some financial reporting contexts, the performance of these assets over prior periods may impact on the discount rates that can be selected (for example, in UK Regulations (2008), the rate used must be no greater than that justified by the performance of the relevant assets in both the past year and the past five years).

- **Investment yields and policy:** Information relating to current investment yields for a number of different asset classes (relevant to the term of the liabilities, based on the estimated payment patterns) together with the insurer's investment policy and its potential impact on the investment return that may be generated.
- **Other factors:** These might include the impact of taxation and investment expenses on any assumed rates of return based on a portfolio of assets, and on the inflationary drivers that typically affect a number of different types of general insurance claims. For example, the consistency between the claims inflation and discount rate assumptions may be particularly important in some contexts.

Having reviewed these factors and decided on the appropriate approach to use, the rates selected will either have a term structure, covering the run-off period of the liabilities, or a single rate will be selected.

6.7.3 Stage Three: Calculating the Discounted Reserves

The final stage involves carrying out the discounting calculations based on the selected payment pattern and rates from the previous two stages, separately for each reserving category. Assumptions will need to be made regarding the timepoints to use within each future cash flow interval and any explicit allowance for future claims inflation that has not already been allowed for could also be factored into this stage of the calculations.

If the overall weighted discount rate is positive, then the impact of discounting will be to reduce the reserves compared to the undiscounted reserves. If this is negative, the opposite will be true.

6.7.4 Additional Considerations for Discounting in General Insurance

This section summarises some of the matters that may need further consideration in practice when discounting general insurance reserves.

Allowing for Outwards Reinsurance

In many situations where reserves are being discounted, the results will be needed on a net of reinsurance basis. Although the discounting approaches described above could in theory be applied on a gross or net of reinsurance basis, or to the reinsurance data, the basis used is likely to follow that used to derive the undiscounted net reserves. For example, if these have been derived by deducting the estimated reinsurance reserves from the corresponding gross reserves, then each of the gross and reinsurance reserves could be subject to the application of separate discounting calculations. In this way, any difference in the claims payment patterns on a gross and reinsurance basis can be allowed for, since reinsurance cash flows may lag those of the corresponding gross cash flows. A simplified approach to allowing for this may be used in some circumstances, for example by deriving a mean term for the gross discounted reserves and then adjusting this judgementally when it is used to discount the reinsurance reserves. In general, it may not be appropriate to assume that the overall discount factor that applies to the gross reserves is also appropriate for the net reserves. The extent and nature of the reinsurance programme will affect how important it is to consider these issues, and this might vary by reserving category and cohort.

Impact of Discounting on Analysis of Change

Where changes in reserves between successive valuation dates for a common group of cohorts are analysed on an undiscounted basis, the change in reserves can be compared against the movement in claims paid in order to assess whether reserves have been strengthened or weakened in the relevant period. However, if the reserves have been discounted, allowance also needs to be made for the "unwinding" of the discount in the period (i.e. the expected increase in the reserves allowing for the accrual of investment income at the assumed discount rate). Allowance may also need to made for changes in assumed discount rates between successive valuation dates.

Sensitivity Analysis

In practice, judgement will be involved in selecting both the claims payment patterns and suitable discount rates. In many cases, sensitivity analysis of the impact of alternative patterns and discount rates can be helpful when understanding and explaining the uncertainty attaching to the discounted reserves. For example, where tail factors have had to be used in deriving the claims payment pattern, then their impact on a discounted basis will be less than on an undiscounted basis, particularly for long-tail classes, as the impact of discounting will act to reduce their impact on the level of reserves. Similarly, although a change in the assumed claims payment pattern may not impact the undiscounted reserves, it may do so on a discounted basis. Sensitivity analyses might also involve considering the interaction of the selected discount rates with other variables such as claims inflation.

Discounting and Stochastic Reserving Methods

In a number of contexts, if a full distribution of future cash flows (i.e. reserves) on a discounted basis is required, then a decision will need to be made as to how these can be derived. If a simulation approach, such as bootstrapping (as described in Chapter 4), has been used then it may be possible for discounting to be carried out as part of the simulation algorithm, to derive the discounted reserves for each simulation. By doing this, the possibility, for example, that adverse claims outcomes from the tail of the distribution have longer durations, and hence have greater discount factors (other things being equal), can be allowed for explicitly. This type of analysis will then allow, for example, a percentile value (e.g. 75th) to be derived directly from the discounted simulation results. This may be different to the more approximate approach of discounting a particular percentile value (e.g. 75th) from the undiscounted distribution. Such approximate approaches (or variations thereof) may, however, need to be used where the stochastic methods used do not enable discounting to be embedded as part of the process.

Impact of Professional Standards

Depending on the location of the insurer and the professional designation of the practitioner, there may be specific professional guidance, standards or other relevant material that may need to be considered where discounting has been used. For example, in the USA, the Actuarial Standards Board have issued a discounting standard – ASOP 20 (see Actuarial Standards Board, 2011). Such materials may impact upon the approaches used and the reporting requirements.

6.8 Reserving Under Solvency II

6.8.1 Introduction and Background

The current supervisory, solvency and capital requirements regime for insurers within the EU, known as Solvency II, came into force on 1 January 2016. It has many aspects to it, which can have a significant impact upon insurers across a range of business processes. This section focuses on the reserving aspects – specifically, the components of the so-called "Technical Provisions" (TP) that will appear in an insurer's Solvency II balance sheet. The TP include amounts relating to future cash flows arising from claims, premiums and expenses (related to claims and other sources such as investment management), and in that sense they are broader than just the "claims reserves" (or "loss reserves") that are the focus of much of this book.

The main purpose of this section is to provide a high-level summary of the components of the Technical Provisions ("TP") and of the requirements under Solvency II regarding their calculation. Where relevant, some of the key differences between reserving for financial reporting purposes and reserving in the context of determining the TP under Solvency II are highlighted. The reserving carried out for financial reporting purposes is assumed here as being under the current IFRS/UK GAAP (abbreviated here to IFRS/GAAP). It is worth noting that a new IFRS related to insurance contracts (referred to as IFRS 17 Insurance Contracts) may come into force in future. This is currently expected to be similar (but still different) to Solvency II. Hence, some of the differences referred to here between the derivation of TP under Solvency II and reserving under the current IFRS/GAAP may not apply under the new IFRS for insurance contracts.

The main sources for the requirements in relation to TP under Solvency II are the following three EU regulatory documents:

- European Parliament (2009), which is the relevant European Directive that was originally issued in 2009, and then modified in 2014. This is generally referred to as the "Level 1" Directive or just the level 1 text or document. The content specifically related to TP is contained in Articles 75–86 inclusive.

- European Parliament (2015), which is the regulations issued in 2014 and which is generally referred to as the "Level 2" document or text. The content related specifically to TP is contained in Articles 7 to 61 inclusive, although several other Articles are also relevant (e.g. Article 2 relates to Expert Judgement).

- EIOPA (2014b), which is the guidelines issued by EIOPA in relation to the valuation of TP and which is generally referred to as one of the "Level 3" documents.

There are other associated documents (e.g. Level 3 documents on other subjects such as contract boundaries, which is covered by EIOPA, 2014a) that are also relevant, and local regulators and other market or professional bodies in each country impacted by the Solvency II regulations may also have issued Solvency II-related documents that are relevant for companies and practitioners operating under their jurisdiction. For example, in the UK, the Prudential Regulatory Authority (PRA) Rulebook has a section on TP, and the PRA has also issued Supervisory Statements and other documents that are relevant to TP (e.g. see Prudential Regulatory Authority, 2014, 2015).

Other useful sources referred to in this section include Dreksler and ROC Working Party (2013) and Lloyd's of London (2015e). The latter of these includes references to the relevant sections from the Level 1, 2 and 3 documents referred to above that relate to the various components of the TP.

Before summarising the components of the TP, the relevant entries in an insurer's Balance Sheet on a Solvency II basis are discussed. The TP form the main element of the liabilities in a Solvency II Balance Sheet, which will also include the assets covering the TP. The additional assets in excess of the TP and other liabilities are referred to as the "Own Funds" under Solvency II, and they are divided into "Basic Own Funds" and "Ancillary Own Funds", the latter of which relates to capital that may be called upon in certain circumstances. Own Funds are subject to certain rules that determine what proportion of them are eligible to cover the Solvency Capital Requirement ("SCR"). The SCR is the capital required on a one-year time horizon basis using the Solvency II regulations, which specify use of either a so-called "Standard Formula" or an internal model, or a combination thereof. Although considerable focus is placed by insurers on the determination of the SCR, the determination of the TP is also a critical aspect that influences the Solvency II Balance Sheet and hence the relationship between the eligible capital and the SCR. If the TP are understated, then the insurer may appear more solvent (in a Solvency II context at least) than is actually the case, and vice versa. Hence, accurate reserving in a Solvency II context is just as critical to insurers as it is in financial reporting or other contexts.

The remaining two sections summarise the key components of the TP and the main requirements associated with their calculation, based partly on the summary contained in Dreksler and ROC Working Party (2013) and Lloyd's of London (2015e). The requirements that apply in practice will be those referred to in the relevant regulatory and other documents mentioned above. This summary is not intended as a definitive description; different interpretations of the relevant regulations may apply in practice and there may be important aspects of the Solvency II requirements in respect of the TP that are not covered here. In addition, the regulations could change in future.

6.8.2 *Components of Technical Provisions Under Solvency II*

Under Solvency II, the TP are intended to represent the amount that the insurer would have to pay, net of future costs and benefits from the insurer's outwards reinsurance, in order to transfer the obligations arising under the relevant contracts to a third party. In practice, there is not a liquid market from which such values can be derived, and so the Solvency II regulations specify a basis for derivation of the TP.

This basis requires that the TP are equal to the sum of a best estimate and risk margin in respect of all claim-, premium- and expense-related cash flows, allowing for relevant outwards reinsurance, arising from all business for which the insurer is legally obligated at the relevant valuation date.[7]

The various components of the TP are summarised in Figure 6.1. For ease of explanation, this figure uses widely understood terminology, rather than the precise terminology used in the Solvency II regulations. From the figure it can be seen that each of the Gross and Reinsurance Best Estimate components has two elements – the Claims Provision and Premium Provision, which correspond to the earned and unearned exposures respectively (i.e. the past and future exposures at the valuation date). Each of these has items related to claims (i.e. indemnity amounts), future premiums and expenses. Both allocated and unallocated loss adjustment expenses (i.e. ALAE and ULAE) must be included, as explained further in the expenses subsection below. As also explained further below, an allowance for Events Not in Data ("ENID") may also need to be included.

A risk margin is added using an approach specified by the regulations and discounting is carried out using prescribed term-dependent discount rates.

Although on the face of it this brief description of the TP under Solvency II might suggest that they are fundamentally different to reserves on an IFRS/GAAP basis, the overall reserving methodology, as covered in the other chapters of this book, is still equally applicable for a number of the components of the TP. In fact, part of the process of deriving TP that is currently used by many insurers involves taking the IFRS/GAAP reserves and adjusting them so that they are on a Solvency II basis. In addition, many insurers seek to operate an integrated reserving process that produces IFRS/GAAP reserves and TP (and those required for other purposes/bases), rather than using several independent processes.

The remainder of this section includes paragraphs that summarise the components of the TP, in some cases grouped together for ease of explanation. The final section summarises the main additional requirements regarding the calculation of the TP.

---

[7] Although, as per Article 77(4) of the Level 1 Directive, if the cash flows can be replicated using a financial instrument, the value of the TP can be determined on the basis of the market value of that instrument.

	Claims Provision	**Premium Provision**
Gross Discounted Best Estimate	Earned Claims + Expenses − Future Premiums	Unearned Claims + Expenses − Future Premiums

less   *less*

	Claims Provision	**Premium Provision**
Reinsurance Discounted Best Estimate	Earned Claims + Expenses − Future Premiums	Unearned Claims + Expenses − Future Premiums

plus

Risk Margin

equals

Total Net Technical Provisions

Figure 6.1 Solvency II: Components of Technical Provisions

Gross Best Estimate – Claims Provision – Claims

The Level 1 Directive defines the best estimate component of the TP, at Article 77, paragraph 2, as:

The best estimate shall correspond to the probability-weighted average of future cash-flows, taking account of the time value of money (expected present value of future cash-flows), using the relevant risk-free interest rate term structure.

This definition applies to all cash flows, i.e. those related to claims, expenses and premiums, although the context for this particular section is the claims element.

The definition may at first reading be interpreted as being consistent with the normal meaning of the term "best estimate" used in the context of IFRS/GAAP claims reserves (as referred to in Chapter 1) – i.e. that it represents the average or mean of the range of potential outcomes, except that it refers to all cash flows, rather than just claims, and also that it is specifically on a discounted basis. However, one other important difference arises due to the inclusion of the phrase "probability weighted average of future cash-flows". Various sources (as documented in Section 6 of Dreksler and ROC Working Party, 2013) have interpreted this as implying that some weight needs to be given to losses with low probability, but potentially high severity. Such losses have been described as "binary events" or "Events Not in Data". In some circumstances, it may be necessary for an additional reserve to be added to those derived for IFRS/GAAP purposes, if these do not already make appropriate allowance for these losses. This is considered further in the Events Not in Data (ENID) paragraphs in Section 6.8.3.

If the process used for IFRS/GAAP reserving has produced a gross of reinsurance claims reserve on a best estimate basis by class of business, then, as long as cash flow patterns have also been (or can be) estimated, then the reserve for each class can potentially be discounted and then used as the gross of reinsurance best estimate claims component of the Solvency II Claims Provision, with two possible adjustments. First, an ENID reserve may need to be added, as mentioned above, and second it may be necessary to allow for differences in relation to contract boundaries, as referred to in Section 6.8.3. If allocated loss adjustment expenses are included within this reserve, then it will also include that part of the TP expenses component.

The classes of business used for this component (and the other components of the TP) must comply with the segmentation requirements under Solvency II, as discussed further in the Segmentation paragraphs in Section 6.8.3.

The gross IFRS/GAAP reserve used for financial reporting purposes will (at least currently) usually have been derived using deterministic, rather than stochastic reserving methods. However, the reference to "probability-weighted average" in the Solvency II definition could be interpreted as being more consistent with the output provided by stochastic methods, and specifically those that produce an estimate of the full distribution of future outgo and the associated cash flows. Notwithstanding this, it is believed that many insurers, at least at present, continue to use predominantly deterministic reserving methods to derive the best estimate reserves for both IFRS/GAAP and Solvency II purposes. There is potential for this to change in future, as insurers become more familiar with implementing Solvency II in practice.

If the gross IFRS/GAAP reserves are not on a best estimate basis, then an adjustment will need to be made to place them on a best estimate basis if they are to be used as part of the TP. For example, if they include a management margin in excess of a best estimate, then that margin will need to be identified and then deducted from the IFRS/GAAP reserves.

Note that, since the Claims Provision under Solvency II relates only to the earned exposures, if the data triangles that have been used to derive the gross IFRS/GAAP reserves are not on an accident period basis (e.g. if they use an underwriting period), then the resulting reserves may include an element relating to the unearned exposures, so that element will need to be removed before being used to derive the Claims Provision under Solvency II.

Gross Best Estimate – Claims Provision – Future Premiums

This component relates to gross premiums in respect of earned exposures that are yet to be received. As for the other components, it must be derived on a discounted basis, using a payment pattern that is appropriate for the earned exposures. It is deducted from the sum of the other components of the Claims Provision and hence will reduce the required provision.

Depending on the nature of the business, the future premiums may be based on premium figures collated by the insurer's underwriting and finance teams. In some cases, this may be supplemented by separate projection of premium data, for example using premium data triangles. As for the claims component of the best estimate, the future premium component must comply with the Solvency II segmentation requirements.

Gross Best Estimate – Premium Provision – Claims

Although this part of the TP includes use of the word "premium" in its description, in fact it relates to the gross claim-related cash flows (i.e. indemnity) in respect of the unearned (i.e. future) exposures at the valuation date. If, for example, underwriting year triangles have been used to derive the Gross IFRS/GAAP reserves, then these will include the claims (and possibly ALAE) element of cash flows relating to both earned and unearned exposures. The former relates to the Claims Provision under Solvency II and the latter to the Premium Provision. Hence, if these two elements can be subdivided from the total estimated reserves derived using underwriting year data, then these can be used as part of the calculation of the required claims (and possibly ALAE) element of the Claims and Premium Provision. If accident year triangles have been used, then a separate calculation will be required for the Best Estimate claims component of the Premium Provision, based, for example, on selecting a suitable loss ratio to apply to the unearned element of the premiums.

As for the Claims Provision, the Premium Provision needs to be split according to the segmentation requirements. It must also be discounted and hence cash flow patterns are required. The final point of detail relating to this element of the Premium Provision is that allowance needs to be made for the claims related to "bound but not incepted" business, as explained further in the Recognition of Obligations paragraphs in Section 6.8.3.

Gross Best Estimate – Premium Provision – Future Premiums

This component relates to gross premiums yet to be received in relation to unearned exposures. As for the other components, it must be derived on a discounted basis, using a payment pattern that is appropriate for the unearned exposures, and it must be subdivided according to the Solvency II requirements. It is deducted from the sum of the other components of the Premium Provision and hence will reduce the required overall provision (although since Solvency II requires cash inflows and outflows to be disclosed separately, the future premiums will be shown separately). In some cases, if the business is profitable, this can mean that the Premium Provision is negative.

As for the future premium component of the Claims Provision, depending on the nature of the business, the future premiums may be based on premium figures collated by the insurer's underwriting and finance teams, perhaps supplemented by separate projection of premium data. Additional future premiums may also need to be allowed for in respect of "bound but not incepted" business, as explained further in the Recognition of Obligations paragraphs in Section 6.8.3.

Best Estimate – Expenses

The expenses component of the TP is required (as per Article 78, Item (1) of the Level 1 Directive) to include "all expenses that will be incurred in servicing insurance and reinsurance obligations". This quite broad definition will mean that the provision will need to include expense items that are not typically included in allocated and unallocated loss adjustment expenses (i.e. ALAE and ULAE) under IFRS/GAAP.

The ALAE element will normally be included as part of the claims element of the Claims and Premium Provision. However, the ULAE element will require a separate calculation. The methods that can be used to estimate ULAE are discussed in Section 6.4. As noted there, one of these is a ratio-type method that is referenced in the Solvency II Level 3 document on TP, which can be used subject to certain conditions (as per Guideline 71 of EIOPA, 2014b). All expenses are required to be allocated to homogeneous risk groups and currency, as per the Segmentation of business requirements for TP referred to in Section 6.8.3. Whilst the ALAE and ULAE reserves under IFRS/GAAP could form the basis for the relevant component of the TP, they will need to be discounted, and hence a cash flow pattern will need to be assumed, perhaps based (at least partly) on the claim payment patterns. Consideration will also be needed as to whether all relevant types of expenses required under Solvency II have been included. For example, Article 31 of the Level 2 document provides a list of the different types of expenses that need to be taken into account. These include administrative, investment management, claims management and acquisition expenses, taking into account the overhead expenses incurred in servicing insurance and reinsurance obligations. Depending on the type of business, additional complexity can be introduced through, for example, the need to allow for profit commissions, which may be regarded as a form of acquisition expense. This is discussed further in Dreksler and ROC Working Party (2013).

Determination of the expense items for each relevant component of the TP (i.e. the Claims and Premium Provision components, each by risk group and currency) has to be done in a "realistic and objective" manner and consistently over time. In practice, this will usually involve some level of apportionment, particularly for the overhead expenses. If some of the expense items are specifically related to a particular business segment or category of contract, then these can be allocated directly. For example, a provision for future acquisition expenses may only relate to "bound but not incepted" business (which is explained further in the Recognition of Obligations paragraphs in Section 6.8.3).

The expenses associated with outwards reinsurance and Special Purpose vehicles are included within the gross best estimate of the TP and hence the expenses component of the TP will be the same gross and net of reinsurance.

Risk Margin – Claims and Premium Provision

The risk margin element is included because of the requirement for the TP to represent the amount that insurance and reinsurance undertakings would be expected to require in order to take over and meet the insurance and reinsurance obligations. It is calculated on a prescribed cost of capital basis, intended to represent the cost of providing the necessary funds to support the required SCR over the lifetime of the relevant liabilities. This requires an estimate of the subset of the total SCR that excludes the element relating to new business and part of the market risk. It also requires an estimate of how this subset of the SCR runs-off over time, or for a suitable simplification to be used. This determination and the relevant simplifications are beyond the scope of this book, although the worked example of the extended time-horizon version of the Merz–Wüthrich method in Section 4.6.5 does provide some examples of the possible approaches for estimating how a cost of capital risk margin runs-off over time.

The prescribed basis multiplies the SCR at each future time point by a cost of capital rate that is specified in the regulations (initially set at 6% in the Level 2 document, but subject to review periodically, in accordance with Article 77(5) of the Level 1 Directive) and by a discount factor, using the same term-dependent discount rates as for other elements of the TP.

The risk margin will apply on a net of reinsurance basis in the TP calculations, and so there is no separate calculation of the risk margin on a reinsurance basis.

Further details of the derivation of the risk margin and the associated simplifications that can be used are set out in Guideline 61 of the Level 3 TP document. Additional discussion of this topic is also given in Dreksler and ROC Working Party (2013) and Lloyd's of London (2015e).

Recoverables from Reinsurance Contracts and Special Purpose Vehicles

The discounted cash flows relating to outwards reinsurance and Special Purpose Vehicles (SPVs) need to be allowed for in each of the components of the TP in a similar way as for the corresponding components of the Gross TP.[8] Hence, there will be both Claims and Premiums Provisions on a best estimate basis relating to reinsurance for the earned and unearned exposures respectively and these will each comprise elements relating to claim recoveries (i.e. indemnity), expenses and future reinsurance premiums.

Careful consideration is required when determining the recognition of outwards reinsurance under Solvency II. In the UK, the Prudential Regulatory authority clarified the key principles in Prudential Regulatory Authority (2015). Lloyd's also gave some guidance in Lloyd's of London (2015e), together with detailed worked examples in Lloyd's of London (2015d). In summary:

- For inwards business, all future receipt and payment cash flows within a contract boundary are included in the TP. Contract boundaries are discussed further in the next section.
- For outwards business (i.e. outwards reinsurance), obligations to make future payments that fall within a contract boundary, calculated consistently with the cash flows recognised for the inwards business, are included in the balance sheet. Only future receipts that relate to cash flows of existing inwards business are included in recoverables from reinsurance contracts. Therefore, where an existing reinsurance contract provides recoveries from inwards contracts that are not bound at the valuation date, only reinsurance recoverable cash flows that relate to existing inwards contracts (i.e incepted, or bound but not incepted)

---

[8] For ease of presentation, any further references here to "reinsurance" includes all forms of reinsurance, including SPVs.

should be included, thus ensuring consistency with the inwards contracts valuation. The full amount of any future outwards reinsurance premiums that relate to existing or legally obliged reinsurance contracts should be included on the balance sheet.

• When considering how future reinsurance purchases should be recognised, Guideline 78 in the Level 3 TP document applies. In summary, this states that insurers can recognise future cash flows in relation to these purchases, covering obligations already recognised in the balance sheet, to the extent they are replacing any expiring existing reinsurance arrangements, subject to seven conditions being met, as specified in Guideline 78.

Depending on the nature and complexity of their outwards reinsurance programmes, determining the correct treatment of their reinsurance contracts in the TP calculations will be relatively straightforward for some insurers, but quite complicated for others. Each insurer will need to satisfy themselves that the approach they are using meets all relevant requirements.

Once the reinsurance contracts that are to be included have been established, the overall approach to the calculation of the best estimate Reinsurance TP can either be done directly, by estimating the relevant discounted reinsurance claims, premium and expense cash flows (as per the Gross Best Estimate TP) or indirectly, using, for example some form of simplification (such as reinsurance to gross ratios). The use of simplification approaches is, however, subject to meeting certain conditions and may not be possible in all situations. Further details of these conditions are given in Dreksler and ROC Working Party (2013) and Lloyd's of London (2015e).

Recoverables from reinsurance contracts must be adjusted for losses due to counterparty default risk, taking into account the term of the relevant reinsurance cash flows. The overall approach used to derive this can either use a counterparty-specific methodology, which takes into account both the probability of default of each counterparty and the Loss-Given-Default, or in some circumstances a methodology applied at a more aggregate level. For example, the reinsurer-specific method summarised in Section 6.5 has many features that may be relevant to the calculation required for Solvency II purposes. Section 8 of Lloyd's of London (2015e) includes some relevant guidance in a Lloyd's of London context.

6.8.3 Requirements Related to the Calculation of TP

There are various requirements that impact upon the calculation of the relevant components of the TP. This section provides a brief summary of some of these requirements. Further details are specified in the Level 1, 2 and 3 documents, and, for example, in Dreksler and ROC Working Party (2013) and Lloyd's of London (2015e).

• **Discounting:** As noted above, all elements of the TP are subject to discounting for the time value of money, using the risk-free term-dependent interest rates by currency published monthly by EIOPA for use as at the relevant valuation dates. This means that the cash flows (e.g. Best Estimate Claims and Premium Provisions) also need to be derived in the relevant currencies, which may have an impact on segmentation of business, as referred to below. Matching and Volatility adjustments to the risk-free rates can be made in certain circumstances.

- **Events Not in Data:** As noted in Section 6.8.2 with reference to the Level 1 Directive definition of "best estimate", a possible interpretation of this is that consideration is required of all possible cash flow outcomes, including those that are extreme or rare, or that are not adequately represented in the underlying data, and that this may require an additional allowance to be made beyond the reserves established, for example, for financial reporting purposes under IFRS/GAAP. Although no specific reference is made regarding such additional allowance in either the Level 1 or Level 2 documents, the Level 3 TP document (EIOPA, 2014c) does make reference to *"infrequent, high-severity claims and latent claims"* in relation to Premium Provisions. In addition, various sources (such as Dreksler and ROC Working Party, 2013; Lloyd's of London, 2015e; Prudential Regulatory Authority, 2014) support this interpretation.

  There have been two particular labels assigned to the additional allowance that may be required as a result of this – "binary events" and "Events Not in Data" (ENID) – with the latter now being in more common use (at least in the UK) and hence that is used here. In summary, the determination of an ENID allowance requires the insurer to consider (for both the Claims and Premium Provisions) whether the data that has been used to derive the discounted cash flow estimates makes sufficient allowance (if any) for events that are not adequately captured by that data.

  Whilst such events may often be interpreted as being necessarily extreme or rare, they may equally well simply relate to more commonly occurring events or scenarios that, for whatever reason, are not adequately captured by the data. Data in this context can mean the relevant internal and external data and information that is used to derive the best estimate. Whether this determination results in an additional explicit allowance being required for ENID will depend critically on the context and the process used to derive each relevant component of the TP. Both Dreksler and ROC Working Party (2013) and Lloyd's of London (2015e) give examples of possible types of ENID events as well as methodologies that may be used to derive the required ENID provision.

- **Recognition of obligations and contract boundaries:** In the calculation of the best estimate and risk margin elements of the TP, Solvency II requires insurers to recognise an insurance or reinsurance obligation at a date when the insurer becomes a party to the contract that gives rise to the obligation, or the date when the cover begins, whichever is earlier.

  The obligations that are required to be recognised in the TP in relation to specific contracts are those which arise within the so-called "boundaries" of each contract, which is defined in the Level 2 document, and expanded upon in the Level 3 document on contract boundaries (EIOPA, 2014a). The contract boundaries for a particular contract effectively specify when the calculations underlying the TP need to start allowing for the relevant cash flows and, importantly, when they should stop allowing for them. This is a complex area, and as well as impacting upon the recognition and derecognition of incepted contracts in the TP calculations, one of the potentially significant reserving implications for some insurers, compared to reserving under IFRS/GAAP, is that these requirements effectively mean that the Premium Provision component of the TP may need to allow for claims, premium and expense-related cash flows in respect of contracts that have not yet incepted at the relevant valuation date, but for which the insurer is legally obligated at that date (so-called "bound but not incepted" business). The related obligations that apply to recognition of outwards reinsurance contracts are explained in the Recoverables paragraphs in Section 6.8.2.

Examples of particular types of insurance and reinsurance contracts that may be impacted include delegated underwriting authorities (e.g. as referred to in the London Market sections in Chapter 7), business that is typically renewed prior to the inception date and business that is automatically renewed. Further details are available in the Level 2 document and in the Level 3 document on contract boundaries (EIOPA, 2014a). Other sources such as Lloyd's of London (2015e) and Dreksler and ROC Working Party (2013) discuss the topic in detail.

- **Segmentation of business:** When calculating their TP, non-life insurers must subdivide the obligations into homogeneous risk groups, which should at least correspond to the 28 non-life lines of business specified in Annex 1 of the Level 2 document, together with other lines of business specified there relating to annuities arising from either health insurance or other coverages. The 28 non-life lines cover 12 relating to direct insurance business, 12 corresponding proportional reinsurance lines and four non-proportional reinsurance lines. The best estimate component of the TP must also be calculated separately for cash flows in different currencies. This will have particular implications for insurers writing business in multiple currencies, as is the case, for example, for many organisations in the London Market. Lloyd's of London (2015e) covers the topic of currency groups in a Lloyd's context. Homogeneous risk groups are referred to in the Level 3 document on TP (in Guideline 19) as "encompassing a collection of policies with similar risk characteristics". Some contracts may cover risks across multiple lines of business, which need to be "unbundled", except, as per Guideline 21 of the Level 3 document, where only one of the risks covered by the contract is material.

In addition to the above requirements, there are other aspects of Solvency II that impact upon the overall process for calculation and reporting of TP, including those relating to data quality, comparison against experience, use of expert judgement, validation, future management actions and documentation. All of these and the other requirements in relation to TP contained in the Level 1, 2 and 3 documents and in any other market-specific documents may be relevant to the determination and reporting of TP in practice.

At the time of writing (2017), Solvency II has been in force for a relatively short period of time, and so the approaches used in practice to derive the various components of the TP in accordance with all the relevant requirements are themselves at an early stage of development. It is possible that these approaches may evolve over time, as companies and practitioners gain more practical experience of complying with the TP requirements and receive feedback from their local insurance regulators.

6.9 Miscellaneous Topics

This section covers a number of topics not covered elsewhere, starting with the use of external data sources and concluding with a brief discussion of the determination of Unearned Premium Reserves and the Additional Amount for Unexpired Risk.

6.9.1 Using External and Other Data Sources

There are very few reserving exercises in practice that do not benefit from making some use of data that is either external to the insurer being analysed and/or that is not derived directly from the relevant data for a particular reserving category. Even for well-established entities with large volumes of business and many years of historical claims data, there can still be reserving categories where there is less history or where changes in business mix mean that the historical data is less relevant for the more recent cohorts. In these situations, the use of data derived from other reserving categories for that insurer, or from data that is external to the insurer, can be very useful. For smaller entities, or those with limited development history, supplementing the company's own data with relevant data from external sources will often be an essential part of the reserving process. Even for situations where the entity's own data for one or more reserving categories is deemed sufficient for the exercise, it can still be beneficial to make use of external data sources as a form of benchmark, which can be used to validate the results based on the insurer's own data.

External data can include various sources such as aggregate market data by class of business (for all companies, or groups of companies combined) or data for individual companies by class of business sourced either from published sources or from private data that might, for example, be available to actuarial and other professional services firms who work for a range of different companies. Adherence to relevant confidentiality, contractual or other requirements may impact upon the last of these.

Although the use of such benchmarks can be beneficial in many reserving exercises, they do need to be used with caution. In particular, when selecting which ones to use or when making any inferences drawn from their application, it is important to understand the extent to which the underlying business to which the benchmark data relates is comparable to the particular business that is the subject of the reserving exercise. The nature of this comparability will depend on the context, but can relate to various aspects beyond simply the class of business itself, including insurance or reinsurance coverages, territory, currency, method of acceptance (direct, broker, etc.), "layers" of claim covered, basis of cohort definition (accident, underwriting year, etc.) and where relevant the policy inception date pattern. Where the benchmarks being used are development-based (e.g. ultimate to incurred ratios) then a numerical and/or graphical comparison of the claims development patterns for the benchmark data with that for the insurer that is the subject of the reserving exercise can assist in assessing the suitability of the benchmarks.

Regardless of how comparable the benchmark data appears to be, some uncertainty will often exist in practice regarding its suitability, and this will need to be taken into account in the results selection and communication stages of the reserving process.

Despite this important caveat, there are many circumstances in which the use of external data can be very helpful, and it therefore forms an important part of the toolkit of most reserving practitioners. Various types of benchmarks can be derived from external data; the detail will depend on the context, but some examples are as follows (all usually applied by cohort):

1. Ultimate to incurred or paid ratios (or the inverse – % developed)
2. Ultimate Loss Ratios.
3. IBNR to case reserves.
4. IBNR to premium.
5. Claim frequency.
6. Average cost per claim or claimant.

Examples of external data are given for some of the reserving categories described in Chapter 7.

6.9.2 Allowing for Changes in Characteristics of The business

Within any particular reserving exercise there will be characteristics of the business (i.e. business mix) that vary between the cohorts, so that for example, the insurer might have written more long-tailed business in recent cohorts than was the case historically, or they might have expanded into new territories. In the first instance, this is dealt with by sub-dividing the data into reasonably homogeneous reserving categories, as discussed in Section 2.5.3, with additional points regarding grouping and subdivision of business being given in Section 2.6. This will mean that changes across the cohorts in the proportion of business in each of these reserving categories should not distort the application of the reserving methods to each category.

In some cases, however, within one or more of the chosen individual reserving categories, the underlying characteristics of the business may have changed between the cohorts. For example, unprofitable segments of the portfolio within an individual reserving category may no longer be written in the later cohorts. These changing characteristics may or may not impact upon features that are relevant to the application of the selected reserving methods, such as the expected paid and/or incurred claims development pattern and the profitability of the business (e.g. as measured by the expected ultimate loss ratios). Where they are believed to have such an impact, a decision will need to made as to how the application of the reserving methods to the affected reserving categories should take this into account. This will depend on the particular circumstances, but some examples of possible options are as follows:

1. **Subdivide the reserving category:** Where the changes in the relevant characteristics are such that there are credible subgroupings within the reserving category that can be analysed separately, then the reserving methods can be applied to the individual subgroups. In this context, "credible" will refer mainly to the volume of business within each subgroup across the cohorts, and the associated stability of the development patterns.

2. **Create as-if data triangles:** Where there has been a change, for example, in the mix of business between the recent cohorts and the older cohorts within a reserving category, then this option involves creating so-called "as-if" data triangles, where the data for all the older cohorts is recast so that it is representative of the mix for the recent cohorts. This process will involve, for example, removing claims and premium data from the development history related to business that is not written in the later cohorts. It will create a data triangle that can be used to derive the required reserving assumptions for the more recent cohorts (which will often be most material in terms of the level of reserves), such as development patterns and prior loss ratios. The reserves for the elements of the earlier cohorts that have been removed to create the as-if data triangles will still need to be estimated using suitable reserving techniques (unless the relevant business has fully run-off).

3. **Select weights from history for subdivisions:** In some cases, there may be information available that will enable a weighted average of relevant reserving factors (e.g. development ratios or prior loss ratios) to be derived for each cohort. The weights used can vary according to the mix of business in each cohort. For example, if more recent cohorts have a larger exposure to a subdivision of the reserving category that is expected to be shorter-tailed than the earlier cohorts, then this approach can be used to allow for this feature. The information used to derive the underlying development patterns and/or prior loss ratios can be based on either an analysis of the relevant subdivisions (as per the first option) or on external benchmark sources, for example, from industry data.

Modifications to these approaches are possible in practice, and other bespoke approaches may also be used, depending on the circumstances. The choice of approach will depend on factors such as the data availability and the perceived materiality of the impact of changing mix within the relevant reserving categories.

The next subsection discusses making changes for two specific characteristics – changes in case reserve strength and speed of settlement – both of which can have a direct effect on the claims development patterns.

6.9.3 Allowing for Changes in Case Reserve Strength and Speed of Settlement

There are a wide range of processes that occur within an insurance entity that can impact upon the data used for reserving. Some of these processes are within the control of the insurance entity and others are at least partly influenced by external factors that are difficult for the insurer to control. If any of these processes change over time, they can have a complex impact on the data and hence on the results produced by the application of reserving methods.

Two common examples of such processes are the determination of case reserves on individual claims and the speed of settlement of claims. The former is an example of a process that is usually under the control of the insurer, where for example a change in the strength of the case reserves, as set by the internal claims team, can affect the claims incurred data triangles. In the latter, which is an example of a process that can be influenced by external, as well as internal factors, a change in the speed of settlement can impact upon paid and incurred data triangles, as well as claim number settled triangles.

Identification of changes in these and other processes can be made by various functions within the business, such as claims and finance, and/or from diagnostic analysis of the data used for reserving, before application of the methods. In many cases, there will be a degree of uncertainty about the magnitude, timing and location within the reserving dataset where these changes are occurring. This can in part be dealt with by producing results using a range of alternative interpretations of the impact of the relevant changes. This section discusses some of the ways that these changes can be identified in the reserving data and then outlines possible approaches to allow for the changes when applying reserving methods. Although this section focuses on deterministic methods, most of the points are also relevant to stochastic methods.

Identification of Changes in Case Reserve Strength and Speed of Settlement

This section describes examples of diagnostic analyses that can be applied to data triangles to help identify changes in case reserve strength and speed of settlement. In many cases, unless there is a known and obvious change in these processes, the numerical analyses may not provide clear evidence of a specific change. Hence, such analyses can usefully be supplemented by discussions with relevant staff, particularly the claims team, who may have conducted their own numerical analyses to address such matters, either on an ad-hoc basis or through production of standardised claims management information.

Examples of diagnostics for identifying changes in case reserve strength include the following:

- Creation of triangles of the ratio of Cumulative Paid to Incurred claims, or ratio of Outstanding to Incurred claims. Where this shows that, at the same development points, there is a noticeable change across cohorts in the Paid to Incurred ratio (or Outstanding to Incurred ratio) then this can either represent a change in case reserve strength or a change in speed of payments, or both. Further discussion with the claims staff might help identify the cause, although unless an obvious change has occurred, the results may sometimes be inconclusive.

- Independent audit/review of a sample of case reserves, perhaps by an organisation that is external to the insurer. By reviewing batches of similar types of claims on a regular basis, analysis of the conclusions from successive reviews might help identify any trends over time.

- Analysis at regular intervals of the ratio of final settled claim amount to prior case reserves might help identify whether there are any trends over time in this relationship and hence in the strength of case reserves.
- Within a development triangle, analyse the ratio of the case reserves at the start of each period to the sum of the case reserves at the end of the period plus the claims paid in the period. This will only give an indication of how case reserve adequacy has changed at points of development after there are no new claims being reported (i.e. no IBNR).
- If claim numbers are available and suitable for analysis, then the average case reserve at each development point, after adjustment for claims inflation, can be reviewed. If there is a clear change across the cohorts at one or more development points, then this might suggest a change in case reserve strength. However, there can be many other factors affecting the average case reserves (e.g. change in mix of business, legislative changes, etc.) and so the results can be inconclusive.

Examples of diagnostics for identifying change in speed of settlement include the following:

- Using the triangles of the ratio of Cumulative Paid to Incurred claims, or ratio of Outstanding to Incurred claims, as described above for assessing the change in case reserve strength. Other things being equal, a speed up of settling claims (by number) would usually lead to a faster paid claim amount development pattern, which in turn would cause the ratio of Cumulative Paid to Incurred claims to rise (or the ratio of Outstanding to Incurred claims to fall). As noted above, it may be difficult to infer from this diagnostic alone the cause of changes in these ratios.
- Analysis of the triangle of proportion of claims settled by number at each duration, using an estimated ultimate number of claims derived by projecting, for example, the claim number reported triangle to ultimate. Changes across the cohorts in this settlement pattern might be evident from analysis of this data. The points made in Section 3.5.2 regarding the data considerations for claim number data may be relevant when using this diagnostic.
- Analysis of individual claim data to assess any trends in the average time to settlement (i.e. between date of loss/accident and settlement date). This can be done on various subsets of settled claims – for example, by examining the average delay to settlement for all claims settled for each cohort both within each development period and cumulatively to the end of that development period.

Adjusting the Data and/or Methodology to Allow for Identified Changes

When changes in case reserve strength or settlement speed have been identified by application of numerical analyses and/or by consultation with relevant claims staff then, from a reserving perspective, the key issue is whether an allowance needs to be made in the reserving process for such changes.

This will depend upon the reserving method(s) that are being used and the extent and degree of certainty that there is around the changes in case reserve strength or speed of settlement. At one extreme, if a purely loss-ratio-based reserving method is being used (i.e. an IELR approach) then it may not be necessary to make any adjustments to the process. If, however, a development factor approach is being used and there has been a clearly identified change, then without some adjustment, the development factors derived from one group of cohorts may be inappropriate to apply to the later cohorts. In some cases, perhaps where the change is isolated and reasonably clear, a bespoke ad-hoc adjustment can be made to the reserving methodology to allow for the change. In other cases, it may be sufficient to make alternative assumptions about the impact of the change on the analysis and then to make a selection using judgement from the resulting range of results. However, in some situations, a more formalised analysis may be deemed helpful. One such approach, originally described in Berquist and Sherman (1977), involves adjusting the data triangles in a certain way, before applying standard reserving methods. This is described in the paragraphs that follow.

Berquist–Sherman Adjustments

These adjustments involve changing the original claims paid and incurred data triangles in a certain way, before applying standard development factor CL-type methods, so as to allow for the potentially distorting effect that changes in case reserve strength or speed of settlement can have when such methods are applied to data impacted by such changes. The combined approach of adjusting the data and applying CL-type methods is sometimes referred to as the Berquist–Sherman method, but strictly speaking the reserving method is still a CL-type approach; it is just that the original data is adjusted before the method is applied.

The two types of data adjustment described in Berquist and Sherman (1977) involve changing the case reserves data to allow for changes in case reserve strength or adequacy, and changing the claims paid data to allow for changes in speed of settlement. For the first of these, the approach is designed to adjust the case reserve triangle so that all case reserves at each development point have the same level of adequacy. An outline of one possible procedure for doing this is described in Berquist and Sherman (1977), and is summarised below.

1. Derive the triangle of unadjusted average case reserves by dividing the case reserve triangle by the number of outstanding claims triangle.
2. Derive the triangle of unadjusted average settled claim by dividing the settled claim amount triangle by the number of settled claims triangle.
3. Estimate the historical severity trend by examining the triangles created in the previous step. Berquist and Sherman (1977) show an example where the severity trend is derived based only on the average paid triangle (rather than settled) and is a single value across all calendar periods.
4. Select a "base" set of values for average case reserves from the triangle in Step 1 above. These are assumed to be the case reserves that are set on the current level of strength or adequacy. In the example in Berquist and Sherman (1977), the latest diagonal values are selected.

5. Replace all values in the average case reserve triangle with the selected "base" case reserves, adjusted for severity trend. The example in Berquist and Sherman (1977) does this by using the latest diagonal values and reducing them by the severity trend selected in Step 3 above. Hence, if a particular triangle has an average case reserve value of, say, 11,000 in the latest diagonal of the second development column, and a severity trend of 15% has been assumed, then the average case reserve values in the second development column (except that relating to the latest diagonal) will be $9,565 (= 11,000/1.15), 8,317$ $(= 11,000/1.15^2)$ and so on.

6. This is intended to produce a triangle of average case reserves, where the strength is assumed to be the same, but which also allows for the underlying severity trend. These adjusted values are then multiplied by the number of open case reserves at each corresponding development point in the triangle to produce an adjusted case reserve triangle.

7. The adjusted case reserve triangle is then added to the unadjusted paid triangle to produce an adjusted incurred data triangle, to which a standard CL-type method can be applied.

The results produced by this approach are dependent on a number of assumptions, and the fact that the original case reserve data is replaced with adjusted data can mean that the results are materially different to those derived by applying the same reserving method to unadjusted data. Hence, any results produced by this approach need careful analysis before they are used in practice.

An alternative to changing the whole of the case reserve triangle could be appropriate where, for example, retrospective examination of historical case reserves by the claims team has identified an isolated group of claims that have been set on a different basis to the current basis (in terms of approximate strength), in which case these could be adjusted so that they are on the current basis. Then CL-type methods can be applied to this part-adjusted data.

The second type of Berquist–Sherman adjustment involves changing the data so that all cohorts have the same underlying speed of settlement. One such approach for doing this, described in Berquist and Sherman (1977) is summarised below. If required, this approach can be done at the same time as adjusting the case reserves, to allow for changes in both case reserve adequacy and speed of settlement.

1. Derive the estimated ultimate number of claims by cohort, by projecting the number of reported claims triangle to ultimate using the CL method.

2. Derive the claim number settled (or claim "closed") proportions by dividing the values in each cohort in the claim number settled triangle by the corresponding ultimate number of claims from Step 1. This is sometimes referred to as the disposal rate triangle (e.g. in Friedland, 2010), but this usage is not consistent with the definition of disposal rates used elsewhere in this book (such as that given in the description of Frequency–Severity Method 2 in Section 3.5.3) and hence will not be used further here.

3. Determine an overall claim number settlement pattern by selecting proportions from the previous step at each development point. Berquist and Sherman (1977) give an example where the values in the leading diagonal are used. The aim here is to select a pattern that

will enable the number of claims settled at each development point to be normalised to remove any historical changes in the claim number settlement pattern.

4. Use the selected pattern derive the implied number settled at each development point, which then represents an adjusted claim number settled triangle. Since the example in Berquist and Sherman (1977) uses the leading diagonal for the selected pattern, the leading diagonal in the adjusted triangle will be the same as in the original claim number settled triangle. The values in prior diagonals are derived by multiplying the selected proportion settled at the relevant development point by the ultimate number of claims for the relevant cohort, from Step 1.

5. Define a procedure for estimating the adjusted claims paid (or strictly "settled") amount triangle using the original claims paid triangle and the adjusted claim number settled triangle from the previous step. Berquist and Sherman (1977) give an example where exponential regression curves are fitted to the original claim number settled and claims paid data. The implied curves then define a relationship between claim number settled and claims amount paid. These curves are then used to derive the adjusted claims paid triangle by using the adjusted claim number settled triangle derived in the previous step, to derive the implied adjusted claims paid triangle.

6. This produces an adjusted claims paid triangle, which is intended to have any change in claims settlement pattern removed. One obvious underlying assumption here is that changes in the claim number settlement pattern will be reflected in changes in the claim amount paid pattern, which in some cases may not be the case.

7. The final stage involves either applying a standard CL-type method to this triangle to derived the estimated ultimate claims, or using an alternative linear or exponential regression-type approach as suggested in Berquist and Sherman (1977) and described in Friedland (2010).

As both types of Berquist–Sherman adjustments involve making a number of assumptions that result in changing the original data in some way before applying one or more reserving methods, it should be obvious that such an approach should not be done without an appropriate degree of caution regarding the results that are produced.

Further details of the Berquist–Sherman adjustments are given in the original paper, Berquist and Sherman (1977) itself and in Friedland (2013). A further discussion of the approach is given in Sherman (2007), which also refers to other useful aspects of the original paper such as the suggested list of questions for management that was included in an appendix to that paper, as discussed in Section 5.2.1.

6.9.4 Deriving a Confidence Interval for the Reserves When the Full Distribution Has Not Been Estimated

The approach described here requires a best estimate of the reserves, as well as the prediction error of that estimate. The prediction error could be derived either by using one or more of the stochastic methods outlined in Chapter 4 or by using benchmarking or expert judgement, or a combination thereof. Then, if a suitable statistical distribution is selected, and the parameters derived, it is then straightforward to determine any required percentile or risk measure from that distribution.

If the reserve is denoted by R and the estimated prediction error of that reserve by p.e. (R) then a standard Normal distribution would mean that a 95% confidence interval for the reserves would be

$$(R - 2\text{p.e.}(R)), (R + 2\text{p.e.}(R)). \tag{6.1}$$

However, for many general insurance situations, the Normal distribution may not be appropriate, and a more skewed distribution is likely to be preferable. For example if the coefficient of variation is greater than 50% (i.e. p.e.$(R) > 50\%R$) then the lower bound of the 95% confidence interval for the reserves, as shown above, will be negative, which is usually not appropriate for a reserve value.

A commonly used skewed distribution in a general insurance context is the Lognormal distribution. Here, the parameters are μ and σ with mean and variance as follows:

$$\text{Mean} = \exp\left(\mu + \sigma^2/2\right), \tag{6.2}$$

$$\text{Variance} = \exp\left(2\mu + \sigma^2\right)\left(\exp\left(\sigma^2\right) - 1\right). \tag{6.3}$$

With a mean of R and a Variance of p.e.$(R)^2$, the Lognormal parameters are derived using

$$\sigma^2 = \ln\left(1 + (\text{p.e.}(R)/R)^2\right), \tag{6.4}$$

$$\mu = \ln(R) - \sigma^2/2. \tag{6.5}$$

So, if for example, a best estimate reserve of, say, 100, has been derived, with a prediction error equal to 25% of this, then, using (6.4) and (6.5) will give $\sigma = 0.246$ and $\mu = 4.575$. Then, using standard spreadsheet functions,[9] or other suitable software, values at any chosen confidence level can be derived. For example, the 75th percentile is 114.5 (i.e. 14.5% above the mean) and the 99.5th percentile is 182.9 (i.e. 82.9% above the mean).

There may be other situations where, instead of deriving the prediction error, an estimate for the reserves at one or more confidence levels (as well as the best estimate of the reserves) has been determined. In such situations, there may also be a requirement to derive reserve estimates at other confidence levels. One way of doing this is to select a suitable distribution (such as the Lognormal), solve for the parameters and then derive the reserve estimate at the desired percentile level(s). This can be done easily, for example, using the "solver" functionality in many commercially available software packages. As a simple example, assume the best estimate reserve is again 100 and the value of the reserves at the 75th percentile has been estimated to be 117, then, assuming a Lognormal distribution, it is possible to solve for the coefficient of variation such that the 75th percentile value equates

---

[9] For example, LOGNORMAL.INV in MS-Excel.

to 117. This will be 31%. The reserve value at any other confidence level can then be derived. For example, the 99.5th percentile value is 206.6.[10]

6.9.5 Subdivision of Claims by Size

It is quite common in some reserving contexts to subdivide the claims data for certain classes of business according to claim size. The rationale is usually based on the hypothesis that claims of different sizes have different features that might, for example, affect their paid and/or incurred development pattern.

Two approaches to subdividing data by claim size are:

1. **By claim size "category":** With some London Market classes of business, claims may be divided into three categories – Attritional, Large and Catastrophe. The attritional claims are the low value claims that typically settle reasonably quickly (relative to the other claims); the large claims are the higher value, usually more complex claims that often take longer to settle; finally the catastrophe claims are the claims from large natural or human-made events. In this approach, individual claims will usually only be allocated to one of these three groups, and will not often move between groups. This approach is discussed further in Section 7.4.

2. **By claim size "banding" or "layering":** For some classes of business, individual claims may be divided into layers, according to claim size. In contrast to the first approach, individual claims can fall into more than one layer, and so a decision needs to be made as to whether claim amounts and numbers contribute to all layers that they affect or just perhaps to the highest layer that they affect. This layering type approach may be used in classes such as UK Motor Bodily Injury, and is discussed in more detail in Section 7.2, which relates specifically to UK Personal Motor, although many of the points will be relevant to other classes where this type of subdivision is used.

Whenever a grouping by claim size is deemed helpful, a careful definition is required, along with accompanying rules as to whether claims can affect more than one category and/or move between categories over time. When the data is being used for triangle-based reserving, the rules related to the movement of claims between categories over time becomes critical, as it will impact on the interpretation of the observed claims development.

6.9.6 Unearned Premium Reserve and Additional Amount for Unexpired Risk

As explained in Chapter 3, the most common goal of applying the reserving methods outlined in that chapter is to derive the Claims Reserve (i.e. the total amount relating to

---

[10] These values were derived using the "solver" add-in in MS Excel with the "objective" set such that the LOGNORMAL.INV function at the 75th percentile is equal to 117, and the "Variable" cell to be the coefficient of variation. This then produces a CV of 31% and hence SD of 31, which can be used in (6.4) and (6.5) to derive $\mu = 4.56$ and $\sigma = 0.299$. LOGNORMAL. INV is then used again to derive the 99.5th percentile value of 206.6.

Case Reserves, IBNR and IBNER); the same comment applies to the stochastic methods described in Chapter 4.

This section relates to two other types of reserve – the Unearned Premium Reserve ("UPR") and the Additional Amount for Unexpired Risk ("URR") – that are not typically included as direct outputs from applying these methods to reserving data. Hence, they have not been mentioned in the explanation of the methods or in the discussion of the application of the methods in practice. This section provides a brief reference to some of the considerations involved in the derivation of these two reserve items.

As noted in Chapter 1, UPR and URR (as defined there) are effectively accounting items that may be relevant only in certain countries in particular accounting contexts. This section considers the use of UPR and URR only as they currently apply in a UK context.[11] In other contexts, they may be related to reserve items referred to as "Premium Liabilities" or "Premium Provisions" to distinguish them from "Claims Liabilities" or "Claims Provisions", as they do not relate to claims that have occurred prior to the valuation or reporting date, but rather to the unexpired contractual obligations associated with relevant policies at that date. This section does not consider Premium Provisions when used in this way, since where they do apply, specific definitions and approaches for their calculation will depend on the prevailing accounting and/or solvency regulations. For example, their usage in a Solvency II context is discussed in Section 6.8.

The involvement of the reserving practitioner in the derivation of UPR and/or URR can vary in practice, so that in some cases the insurer's finance/accounting team will be responsible for their calculation, with limited if any involvement from the reserving practitioner, whereas in others they can be included in formal actuarial reserving opinions (e.g. for Statements of Actuarial Opinion at Lloyd's of London, at least up until work done as at 31 December 2015). The remainder of this section provides a brief reference to some of the matters that are relevant to the estimation of UPR and URR.

Unearned Premium Reserves

In summary, this category of reserve represents the proportion of the written premiums at a particular date that relates to future policy periods. Hence, it is not a reserve that represents an estimate of the future cost of claims relating to these policy periods, but simply a proportion of the premiums relating to the unexpired period of risk at a particular date.

The usual approach to determining the UPR is based on time apportionment at an individual policy level. For example, if a 12-month policy has been written with an inception date one quarter before the valuation date, then three quarters of the relevant premium is deemed to be unearned at the valuation date. However, this applies where the incidence of risk over the period of the policy is even; if this is markedly uneven, then a basis that more accurately reflects the incidence of risk may need to be used.

---

[11] Noting that this may change under IFRS 17 Insurance Contracts, as referred to in Section 6.8.

In most situations the proportion of unexpired risk can be determined on a daily basis per policy, using the known inception and expiry dates. Historically, a more approximate method might have been used, and may still be in use in some contexts. For example, the so-called "24ths" method uses data relating to the volume of premiums written in each month during the reporting period (e.g. calendar year). It assumes that all policies are 12 months in duration, with the incidence of risk spread evenly over the policy period. It also assumes that all policies written in each month begin in the middle of the month, which effectively divides the year into 24 half-months, and enables the proportion unexpired at a particular date to be calculated as a fraction, with 24 as the denominator. So, for example, all policies written in May of a particular year will be assumed to begin in the middle of May, which will mean that, at the end of the year (which is the assumed valuation date), they will have 4.5/12ths – i.e. 9/24ths – of their policy period remaining before they expire. Hence, the unearned premium as at the end of the year will be 9/24ths of the premium written in May. Similar calculations are done for each month, with the sum across all months representing the total estimated UPR. Other approximate methods may use different periods, such as quarters (with the so-called "8ths" method). In some cases, different approaches might be used for gross UPR and the corresponding reinsurance UPR.

Where some of the policies have periods other than 12 months, the approximate methods outlined above will need to be modified, or a more exact time apportionment approach will be needed, based on the actual duration of policies.

Even where a more detailed per-policy approach is used, if the incidence of risk is deemed to be sufficiently uneven over the policy period, methods other than those based on time apportionment may need to be used. Examples where the incidence of risk may be uneven include some types of warranty insurance, crop insurance and leasing insurance. The methods used to assess the incidence of risk will vary depending on the nature of the policies concerned.

Additional Amount for Unexpired Risk

This reserve, which is abbreviated here as URR is, like the UPR, an accounting item that applies under certain accounting conventions. One of the other names that may be used for this reserve – the Premium Deficiency Reserve – summarises its purpose. In effect, if the UPR (net of Deferred Acquisition costs) is deemed to be insufficient to cover the claims and expenses in respect of the unearned proportion of risks at the valuation date, the URR represents the additional amount that is required to rectify this. Thus, for profitable business, it would be expected to be zero, whereas if the claims and expenses are expected to exceed the premium, an additional reserve will be required.

Hence, as compared with the UPR, there is a greater degree of estimation involved in calculating the URR, since it depends on an assessment of the claims in respect of the unearned business. This may therefore lead to reserving practitioners having an involvement in the calculation of the URR in some contexts. For example, if an estimate can be made of the ultimate claims in respect of the unearned business, then this can be compared with the UPR (allowing appropriately for expenses as well) to determine if a URR is required. A relatively simplified approach could involve deriving the required ultimate claims estimate by considering the estimated ultimate loss ratios derived for the relevant cohorts using, for example, one or more of the methods described elsewhere in this book. The details will depend on the cohort definition used (e.g. accident or policy year) and an assumption will need to be made about whether the ULRs need adjusting for use in estimating the claims in respect of the unearned business. Other approaches that take into account the unexpired exposure more explicitly are also possible – for example, see Chapter 17 of Hart et al. (2007) for further details.

Whatever approach is used, a decision will also need to be made as to the level of aggregation used for the comparison between UPR and claims on the unearned business – e.g. at a whole of business level or by suitably homogeneous categories of business. This will be governed by any relevant accounting regulations that may apply.

There is limited reference in the reserving literature to the derivation of UPR and URR that can be used to supplement the brief consideration given here. Some examples of reference sources are Vaughan (2014), Canadian Institute of Actuaries (2014), Yan (2005) and Friedland (2013: Chapter 24). Some of these cover the determination of Premium Liabilities under the relevant regulations, which are closely related to UPR and URR, as defined here.

7

Reserving in Specific Contexts

7.1 Introduction

This chapter provides details of selected key features of reserving in a range of specific contexts. For each territory/class considered, matters such as a description of the class, data types and example reserving methods are given. The drafting of this chapter has taken into account input from individuals or firms who have particular knowledge and experience in each of the relevant areas, as noted in the acknowledgements in Section 1.6. The generic matters that are relevant to all reserving exercises, such as understanding the business, checking the data and using homogeneous data groupings, are not repeated here, as they are already covered in the earlier chapters. Similarly, the generic points related to the application of individual reserving methods are also not repeated here, as they are covered in the description of each method in the relevant chapters.

The purpose of collating these details is to give some preliminary information in relation to reserving in the relevant contexts. For each context, it is assumed that the main purpose of the reserving exercise is to determine a point estimate of the reserves for financial reporting purposes. Partly as a consequence of this, the reserving methods referred to are mostly deterministic in nature. In practice, stochastic methods are also likely to be used for some purposes in a number of the different contexts. Where more than one reserving method is mentioned, it is common in most contexts for the final selected results to be based on a combination of different methods, which may vary across the cohorts, as explained in general terms in Section 5.7.

This chapter is not intended to be an instruction manual for the application of the reserving methods described in the earlier chapters, and does not represent any form of recommended approach, guidance or actuarial/professional standard. In practice, for each territory/class there will be many variations and additional detail beyond that described here, which in most cases will be learned through several years of practical experience. Furthermore, in some cases, the application of reserving methods will differ from that described here, perhaps significantly so, due to the specific circumstances.

The various subdivisions of data given for each territory/class are not mutually exclusive and combinations of them may often be used in practice. They are intended as examples only and in many situations only some of the subdivisions might be used, or other subdivisions

410

not given here might be more appropriate. As with all aspects of reserving, the practical constraints that exist and the judgement of the practitioner will be important factors in determining the subdivisions that are used in each reserving exercise.

Despite these important caveats, by setting out some of the reserving features for a range of example territory/class combinations, the intention is to show how the application of reserving methods can vary widely, depending on the context, and in so doing demonstrate that the practice of reserving involves far more than just understanding the mechanics of the various methods that are available.

Notwithstanding these differences in detailed application, it should also be apparent that the same or similar reserving methods are used for many classes, which are often impacted by related issues. Hence, reserving expertise can be utilised across class of business and international boundaries, in conjunction with appropriate specialist and local expertise.

7.2 UK Personal Motor

7.2.1 Category Description

This category relates to insurance coverage purchased in the UK by individuals for vans, cars and motorcycles. It is distinct from UK Commercial Motor, which is insurance purchased by businesses for their motor vehicles. However, many of the points made here are relevant to UK Commercial Motor, and also to reserving for Motor business in other territories. Personal Motor may include usage by individuals for social, domestic and pleasure, and additionally possibly for commuting and business purposes. There are three main types of cover:

1. Third party (TP), which is the minimum cover that is required by law to be purchased by anyone who owns a motor vehicle. It indemnifies the insured driver for their liability arising from injuries that they cause to others (i.e. third parties) and from damage that they cause to other persons' vehicles or property.
2. Third party fire and theft (TPFT) – as TP, plus protection against the insured's vehicle being lost or damaged by fire or theft.
3. Comprehensive – as TPFT, plus cover for the insured's vehicle due to accidental damage. It may also cover the contents of the insured's vehicle, plus medical expenses and possibly personal accident benefits.

Claims can be settled as either a lump sum, and/or, in the case of some types of large injury claims, as Periodical Payment Orders (PPOs). The latter effectively provide the injured party with an annuity in respect of certain aspects of their claim.

7.2.2 Types of Data

Claim amount and claim number triangle data is usually used for the core reserving analysis, with separate tabular data being collated for claims arising under PPOs. If there have been large events affecting several vehicles (e.g. due to flooding or hailstorm) then these might also be identified separately.

Two commonly used exposure measures are premiums and vehicle years. These will be calculated on an earned basis for accident cohort data, and on a written basis for underwriting cohort data.

The subdivisions used for this data are discussed in the following subsection.

7.2.3 Subdivision of Data

Claims data is nearly always split at least between property damage and injury claims, with other possible subdivisions being:

- Type of coverage: TP, TPFT and Comprehensive.
- Source of business: Broker/Direct/Product name/Business grouping, legal entity etc.
- Claim type (High Level): First Party Property Damage ("FP" or "Accidental Damage" or AD), Third Party Property Damage (TPPD) and Third Party Bodily Injury (TPBI).
- Claim size (for TPBI claims): Claims may be split into bands according to size. There are several points of detail relating to this subdivision, which are covered in the next subsection.
- First Party Property Damage: Windscreen (or "Glass") and Non-Glass. Fire and Theft claims might be grouped together and analysed separately from other damage claims.
- Head of Damage for TPPD: Credit Hire and Non-credit hire.
- Head of Damage for TPBI: An example split is between General Damages, Special Damages, third party legal fees and other.

Claim Size Banding – TPBI Claims

For reserving and other purposes, TPBI claims may be subdivided into bands (or "layers") by claim size, for any suitable claim amount type, such as settled, paid or incurred. For example, in some situations, large claims can cause a potentially distorting effect to the data triangles if there is no subdivision by claims size. One relatively straightforward approach to allow for this would be to use two bands – the first for all claims below a selected threshold and the second for the larger claims above that level.

In all cases, after deciding on how many bands will be used, it is also necessary to decide how individual claims will be allocated to each band. For example, claims can contribute to all bands that they affect, or just to the highest band that they impact. These two alternative approaches are explained further below, where it is assumed for this purpose that there are two bands – Band 1, representing claims below £100k, and Band 2, representing claims above £100k.

- **Multiple band contribution:** Claims affect all bands that they impact, so that a claim of size £115k will contribute to both bands (£100k to Band 1 and £15k to Band 2). If claim numbers are being collated by band, then a claim will contribute to the count for each band that it affects.

- **Single band contribution:** Claims only affect the highest band that they contribute a non-zero amount to, with the full amount of the "from ground up" claim being allocated to that band. Hence, in this example, a claim of £115k contributes £115k to Band 2, and nothing to Band 1. For claim numbers, each claim will only contribute to the count for the highest band that it affects.

The same principles apply where there are more than two bands. Some insurers might use the terms "Capped" and "Excess" to relate to claims in lower and higher bands respectively. This terminology is consistent with the multiple band contribution approach, since a large claim will contribute to the Capped category at the banding point (i.e. it is capped at £100k in the above example) and the excess above that will contribute only the excess above the banding point to the Excess category.

Once the number of bands and the approach to allocating claims to bands has been selected, claim amount and number data triangles can then be constructed by subdividing the claims data for each cohort according to the chosen band definitions. When constructing these data triangles, two further points may require consideration:

- **Inflation of layer points:** This relates to whether the layer banding points should vary between the cohorts, to allow for claims inflation. So, for example, the selected banding points might be based on monetary values of the latest accident year, with the banding point for all prior accident years being deflated, with the factors used being designed to act as a proxy for claims inflation over the relevant period. Judgement will be needed to select the inflation rate, taking into account for example any analyses that might have been carried out to estimate historical claims inflation. Some practitioners prefer to adopt an inflation-adjusted approach, so that the banding points for each cohort are in approximate comparable monetary value terms. However, in some contexts, different underlying claims processes are used for different sizes of claim – with the larger claims perhaps being allocated to more senior teams, with a more claim-specific process being used. In these circumstances, fixed thresholds might be used to determine the larger claims that will be subject to a different process and so fixed banding points based on these thresholds could be preferred, which then reduces the risk that the data triangle groups will contain a mixture of different underlying claims processes.

- **Diagonal groups:** Where a multiple band contribution approach is used, then the amount that each claim contributes to the diagonals in the triangle for each band will move up or down, according to how that claim amount changes over time. Similarly the claim number count will also change, according to whether the claim affects each band at each successive diagonal of the triangle. In this case, no adjustment to the historical diagonals is necessary. However, where a single band contribution approach is used, as the claim amount changes over time, claims may affect different bands at successive diagonals. It is then necessary to decide how claims that move between bands should be treated at successive diagonals in the triangle. In theory, various approaches are possible, including: (a) each diagonal in the triangle uses the banding by size in that diagonal; (b) all diagonals use the same grouping based on the size of claim in the latest diagonal only; or (c) all diagonals use the same grouping based on the maximum value of the claim in its development history to date (so that if the value of the claim reduces after this maximum value so that it is now in a lower band, it is not reassigned to that band). In the latter two approaches, the full history is recast at each successive valuation date, so that the historical triangles may change (if claims have moved bands over time according to the banding definition), but there will be a consistent set of claims in all diagonals in the triangle, which might be preferable to using approaches where this is not the case. Variations on these approaches as well as other distinct approaches might be possible in practice, but whichever one is used, a decision will also need to be made as to whether the allocation to bands is carried out using cumulative paid or incurred claims data, and whether the allocation using one of these data types also drives the allocation for the purpose of constructing banded data triangles for the other data type (which would then produce banded paid and incurred data triangles for the same subset of claims).

The banding approach used and the details of various aspects of the approach such as those described above will depend on the individual circumstances and the practitioner's judgement.

Depending on the bands that are chosen, and the specific insurer being analysed, the data triangles for the higher bands can contain relatively limited claims data, which can make it difficult to apply standard reserving methods to estimate reserves for those bands. In this case, one approach is to use some form of extrapolation from the results for the lower band values (e.g. based on estimated claim frequency by size of claim). Benchmarking of large claim propensity can also be an alternative or can be used to supplement the insurer's own large claim analysis. Whatever approach is used for the bands in respect of the large claims, it may also be appropriate to consider whether sufficient allowance has been made for new claims entering the higher bands over development time, and in some cases an additional explicit reserve for such claims may be necessary (which is effectively "pure" IBNR for the relevant bands).

7.2.4 Cohort and Development Frequency for Triangles

Typical examples are:

- accident year cohorts and annual or quarterly development frequency;
- accident quarter or half-year cohorts and quarterly, half-yearly or monthly development frequency; or
- monthly accident cohorts and development frequency.

Claim reporting period cohorts may also be used in some situations.

Quarterly and monthly cohorts can, for example, enable seasonality effects to be allowed for explicitly, which can be noticeable for this class. However, the choice will depend on several factors, including the volume of claims in the relevant subdivision. As an example, for TPBI claims, the UK Third Party Working Party (see Brown et al., 2015) uses quarterly accident periods with monthly development frequency for lower claim size bands and annual accident periods with monthly frequency periods for higher bands.

In the London Market and possibly elsewhere, triangles with underwriting period cohorts might also be collated for UK Motor business (instead of or in addition to accident period cohorts), which then facilitates performance monitoring of blocks of business with similar pricing levels or common capital support.

7.2.5 Reserving Methods

The Chain Ladder method is commonly used for paid and incurred claim amount data, with Frequency–Severity methods being used where claim number data is also collated. The Bornhuetter–Ferguson may also used for more recent cohorts, with the prior assumption being based for example on the results of applying Frequency–Severity methods or on pricing models.

Results for each reserving category or claim type may be blended across a range of methods, with weights varying across cohorts. Graphs might be used to assess the reasonableness of selected ultimate claims. PPO claims are usually considered separately using specialised analysis to estimate reserves for known PPO claims (i.e. related to PPO awards already made) and for claims expected to be settled as PPO claims in the future. This analysis will involve making specific assumptions related to mortality, wage inflation for health care workers, allowance for outwards reinsurance and, where PPO reserves are expressed on a discounted basis, discount rate assumptions.

Standard approaches to allowing for outwards reinsurance can be used for this category, as discussed in Section 6.2. For TPBI claims, making an appropriate allowance for Excess of Loss reinsurance is often a key part of this, with various techniques being used, including those that involve projecting the net of reinsurance data triangles, or the reinsurance triangles themselves, or in some cases stochastic modelling of large claims. The reinsurance credit for PPO claims will usually need specific analysis, particularly where the excess of loss covered is indexed.

7.2.6 Matters Relevant to Reserving

Understanding the case reserving procedures used at the insurer for property and injury claims, and the impact of these procedures on the adequacy of case reserves is usually important in this class, and can vary significantly over time and between insurers. This will include consideration of both the overall philosophy, as well as gaining an understanding of the operational claims practices (such as the use of periodic updates or peer review of case reserves) and the basis for establishing automated initial case reserves before sufficient details are available to determine a more claim-specific estimate.

Changing legal and other external forces can have a material effect on claims frequency, averages and speed of settlement for UK Personal Motor, so that both Paid and Incurred Claim amount and number development can be subject to significant uncertainty and variation between cohorts, particularly for bodily injury claims. An example is the Legal Aid, Sentencing and Punishment of Offenders Act ("LASPO"), which took effect from 1 April 2013, and which is believed to have caused an overall reduction in TPBI claim frequency and average cost (although whether this remains is a matter for debate). Some of these external forces have impacted upon the likelihood of fraudulent claims, particularly those with a TPBI element, and in some cases separate analysis of identified or suspected fraudulent claims may be appropriate to ensure appropriate allowance is made for repudiation of such claims in the reserve estimates.

For some claims, the insurer who has paid part of the original claim to their insured may be able to make recoveries of that part from the insurer of the at-fault insured. The impact of these recoveries can mean that, in some cases, the incremental paid claims development for First Party Property Damage may be negative in the later development periods for one or more cohorts, when the recoveries are received.

Where Frequency–Severity methods are used, the data considerations outlined in Section 3.5.2 are relevant for UK Personal Motor reserving. In addition, if UK Personal Motor claims are subdivided by claim type, then the points made in that section regarding matters such as the treatment of zero (or "nil") claims need to be considered for each claim type, since for example the bodily injury element of a claim may be settled at nil, with other elements being settled at cost.

The other analyses that might be used to assist with the reserving process for UK Personal Motor will vary depending on the context of the exercise, but some examples include analysis of the following:

- The number of claimants per claim for personal injury claims, in triangle format, and the associated impact on average cost per claim.
- The ratio of the number of TPBI to TPBD claims in triangle format, possibly split by claim size.
- Settled and incurred average cost in triangle format, by various claim categories.
- The impact of known and potential changes in the legal environment. A recent example is the Ogden discount rate, which the Lord Chancellor announced in February 2017 would reduce from +2.55% to −0.75%. A consultation was launched in March 2017 covering possibilities as to how, when and by whom the discount rate in relation to personal injury claims should be set. The relevant consultation paper is Ministry of Justice and the Scottish Government (2017).
- Market-wide reports published by the UK Third Party Working Party, as discussed further below.
- Market-wide UK personal injury claims data available in relation to the Claims Portal, which is the technology solution that facilitates the claims process required by the UK Ministry of Justice's Pre-Action Protocols for Low Value Personal Injury Claims. This data is available at Claims Portal (2015).

- Trends in market-wide statistics and reports published by various UK government departments such as the "STATS 19" personal injury road accident data from the Department of Transport (see Department of Transport, 2015 and Brown et al., 2015 for further details).

The UK Third Party Working Party referred to above has for a number of years collated claims data from a large proportion of the UK Motor market and produced reports that summarise their analysis of that data. For example, see Brown et al. (2015) for the January 2015 report, which contains a wide range of different analyses of various aspects of both TPPD and TPBI claims. This report uses TPBI claims divided into 11 layers, ranging from the first layer (in 2010 money) at £0 to £1k through to the top layer at £5m plus, with accident years other than 2010 having all layer points indexed from the 2010 values at 7% to allow for claims inflation. Results are shown for both of the banding approaches described above, with the multiple band contribution approach being referred to as "Type 1" and the single band contribution approach as "Type 2". The working party shows TPBI results split between "Capped" and "Excess", where Capped claims are below £100k and Excess are above that. Results for each of Capped and Excess claims are shown, with both Type 1 and Type 2 banding approaches being used to determine the contribution of each claim to the layers, with the Type 2 results being shown split by individual layer. Where settled claims are analysed by Head of Damage, the individual heads of damage for each claim are allocated to the layer containing the total settled cost of the claim.

The author is not aware of any surveys of reserving practice specific to UK Personal Motor, although the GIROC UK Reserving Survey (see McDonnell and Labaune, 2014) does provide some insights into reserving for UK Personal Lines business, for which it is likely that many of the respondents work for insurers that write motor business. The survey was conducted in 2013/14 and for Personal Lines had 13 respondents, who were all from larger companies. Selected key findings from this survey that appear likely to relate at least partly to the practice used by the respondents in their UK Personal Motor reserving are given below:

- All respondents indicated that they used Paid and Incurred claim amount data for reserving, with many also using claim numbers, claim settled and nil claims data. A smaller number used claim averages and by claimant data.
- The reserving methods used are consistent with those cited earlier in this section – namely widespread use of Chain Ladder, Bornhuetter–Ferguson and to a slightly lesser extent, "Numbers/Averages" (i.e. effectively Frequency–Severity methods).
- Specific features allowed for in the reserving process included seasonality, inflation and trends in matters such as premium rates, frequency/average costs, risk mix and claims process changes.

Further details of issues relevant to UK Third Party Motor claims can be found in Brown et al. (2015). Further information relating to PPOs can be found, for example, in Potter et al. (2015) and IFoA PPO Working Party (2016).

7.3 London Market – Casualty Classes

7.3.1 Category Description

This refers to a broad spectrum of classes of business written in the London Market[1] (including Lloyd's) that are loosely referred to as either Casualty or Liability. Some of these classes are also written in the UK outside of the London Market and in other countries, but the focus in this section is on the reserving features that arise in the London Market in particular.

London Market Casualty business typically provides some form of liability coverage for public or private enterprises, individuals or partnerships. It can include both insurance and reinsurance business and can be written directly or through brokers, on an "Open Market" basis or via a number of forms of "delegated underwriting". The latter refers to an arrangement under which an underwriting entity delegates its authority to a company or partnership to enter into contracts of insurance on its behalf.

As with most London Market business, Casualty classes are typically written on a subscription basis – that is, each risk is placed with one or more insurers/reinsurers, who each take a share in that risk, as documented on the "slip", which is used to summarise the risk. The share taken impacts the premium received and the potential claim sizes for that carrier. Hence, an individual portfolio of Casualty business for a particular risk carrier can be made up of a large number of risks, with varying shares on those risks and hence varying levels of premium and potential claim sizes.

Business can be written on a number of different bases for the allocation of claims to the underlying contracts, with two common approaches being Losses Occurring (which covers claims that have loss dates during the contract period) and Claims Made (which covers claims that are "made" or notified during the contract period, with extensions being possible in some cases). Business can also be written on a Risks Attaching basis, which means that claims which occur during the policy period of the underlying risks that have inception dates in the contract period are covered.

Reserving for Casualty (and other) business written in the Lloyd's market will contribute to the estimation of the Reinsurance To Close ("RITC"), which may involve consideration of factors not discussed here, such as equity between members of the relevant years of account. Further details relating to RITC can be found, for example, in Hindley et al. (2000).

Examples of specific Casualty coverages written in the London Market are:

- **Professional Indemnity:** This class indemnifies a business or individual for the cost of compensating clients for loss or damage resulting from services or advice that they provide. It may be subdivided according to profession, such as lawyers, accountants, architects and engineers. Related coverage for medical professionals and/or medical

---

[1] There is no unique definition of "London Market", but, for example, the London Market Group define it in London Market Group and Boston Consulting Group (2015) as "specialty commercial insurance and reinsurance business backed by London capital, plus business controlled by, but not written by London Market participants". In addition, personal lines business is also written in the London Market, albeit in relatively small volumes.

establishments such as nursing homes and hospitals is usually referred to as Medical Malpractice.

- **Cyber Risk:** This refers to policies providing coverage for first or third party costs, expenses or damages due to a breach (or threatened breach) of cyber security and/or privacy of data, that may or may not include damage to physical property. Other policies that do not specifically provide cover for cyber risk may have incidental cyber-related exposure.

- **Directors' and Officers' Liability:** This indemnifies Directors and Officers of businesses in relation to their liability arising from claims made against them for alleged wrongful acts. When the organisation is a financial institution, this class may be be included in a class referred to as "Financial Institutions", which may also include other types of coverage such as Errors and Omissions.

- **Environmental Impairment Liability:** This provides cover for first party clean-up costs and/or third party liabilities (including associated legal defence costs) resulting from sudden or gradual pollution or environmental damage. Policies can be written on both a claims made or other trigger (e.g. occurrence) basis.

Many other Casualty classes are written by insurers and reinsurers across the market. For example, there are a significant number of separate Risk Codes used at Lloyd's for Casualty classes. This includes business grouped according to criteria such as country of coverage and profession covered, as well as both insurance and treaty reinsurance business. It also includes a number of historical codes no longer in use and some new codes that will be used for the first time for the 2016 Year of Account at Lloyd's. Some of the above coverage example descriptions are based partly on the Lloyd's Risk Code descriptions, which are available at Lloyd's of London (2015c).

7.3.2 Types of Data

Data triangles for paid and incurred claim amounts are the core quantitative data used for reserving this reserving category. The cohort definition is usually "Underwriting Year", which is effectively the same as "Policy Year", in that claims are allocated according to the period in which the policy covering the claim incepted. A "policy" in a London Market Casualty portfolio can itself have policies incepting within it, so that the period of exposure in respect of each cohort can stretch for a number of years, depending on the term of the policy itself, and of the underlying policies that attach to it.

Premium triangle data is also usually collated, as premiums received or "signed" often develop after the end of the cohort. Premiums can either be gross or net of brokerage, which can have a material effect on the derived loss ratios and hence this needs careful consideration, for example to ensure consistency with any benchmark loss ratios that might be used as part of the reserving process. The claims and premium development data for the individual insurer/reinsurer can in some instances be supplemented by additional development data for related specific blocks of business, collated for example from broker renewal presentations or prior history data for particular portfolios.

Other quantitative data that might be collated includes individual large loss data (possibly on a triangle basis), policy and claims databases, relevant benchmark data, plan loss ratios, line size profiles, inception date (and hence earning) profiles and indices for premium rate changes and claims inflation. Typical more qualitative-type data that might be collated includes loss adjuster or claims manager reports for large losses, underwriting or business plan description of classes of business, status of position in underwriting and reserving cycle, commentary on case reserving philosophy and information on key changes in terms and conditions, case reserving strength and delegated vs open market mix etc. between the cohorts being considered.

7.3.3 Subdivision of Data

Triangle data is typically divided by class and often also by currency. The currency split is commonly US Dollars, Sterling, Canadian Dollars and possibly also Euros. Business written in other currencies is often converted into Sterling and added to the UK (i.e. Sterling) currency data.[2] Other subdivisions of the core triangle data that might be used include:

- Separate data for specific accounts/schemes/binders/lineslips.
- Individual large losses (which may affect more than one class and cohort).
- Binder and open market business. The former can sometimes have a longer development tail than open market business, due to the inception date pattern.
- Trigger basis for allocating claims to policies: Occurrence, claims made or other.
- Insurance and reinsurance business, with each of these perhaps split further – e.g. primary and excess for insurance business and facultative, proportional treaty and non-proportional treaty for reinsurance business. Insurance business is often referred to as "direct" business in the London Market and may be grouped together with facultative reinsurance business. In this context, direct business can involve coverage in excess of a particular deductible, and in that sense it may be similar in nature to excess of loss reinsurance business.
- Separate analysis for business which has different key characteristics, such as attachment point, terms and conditions, policy structures etc.

7.3.4 Cohort and Frequency for Triangles

Data triangles are usually collated on an underwriting year basis, with either quarterly or annual development frequency. For some business, accident year triangles may also be collated, although this is relatively uncommon for this category.

---

[2] It should be noted that the underlying policies in respect of the data for a particular currency may originate from one or more different territories to which that currency relates.

7.3.5 Reserving Methods

The Chain Ladder method is very commonly used for this category, applied to one or both of paid and incurred data, with more emphasis often being placed on the use of incurred data, due to the long-tail nature of some of these classes. The IELR and Bornhuetter–Ferguson methods are also very widely used, particularly for the more recent cohorts. The loss ratio/prior assumptions used in these methods are usually based on a combination of business plan/underwriting assumptions, pricing analyses and averages from earlier cohorts (often with premium rate change and inflation adjustments, as described in the worked example in Section 3.4.3). Frequency–Severity methods are only used in isolated cases where claim number data is readily available.

Estimates of ultimate premium may be required for this category, usually for the more recent cohorts. These can be derived using a combination of underwriter estimates (based on knowledge of actual business written and expected to be written for each cohort) and the application of standard reserving methods, such as the Chain Ladder, to premium triangle data.

Benchmarking of development factors and loss ratios is reasonably common for this category, particularly where there is limited development history for the account being analysed. In addition to private benchmarks collated by individual companies or consulting firms, other possible sources of benchmarks include Lloyd's Risk Code triangulations (see Lloyd's Market Association, 2015), the RAA data triangles for the US reinsurance business (see Reinsurance Association of America, 2015) and the UK PRA returns data (available from a number of sources, including A.M. Best, 2015).

Other bespoke reserving analyses and methods might be used for individual contracts, classes or sources of loss. For example, one class that might be deemed to be included in the overall category of London Market Casualty would be UK Motor Excess of Loss, in which case, analysing the PPO claims data separately would usually be appropriate (as discussed in Section 7.2, which covers UK Personal Motor). Other past examples of specific sources of loss which tended to be analysed separately for reserving purposes included claims arising from the collapse of Enron, those related to sub-prime lending in the US and the Madoff case. For some of these sources of loss, identification of all relevant exposures is the first and vital step in seeking to understand the potential for claims to arise, and hence in estimating reserves.

In some cases, the portfolio can be very diverse and consist of several large individual contracts (e.g. reinsurance treaties) and the reserving methodology might then be applied separately to each such contract, with the results being at least partly based on the underlying pricing assumptions used in each case. Some form of higher-level aggregate analysis might also then be done to compare against the contract-specific results.

Although Casualty business, including that written in the London Market, is not normally associated with catastrophe-type losses, there is still potential for these classes to be exposed to large events or common causes of loss that have different features to attritional or individual large losses. Well established casualty catastrophe examples include latent claims such as asbestos and pollution, but there are other possible types such as those arising from product failure/malfunction, corporate failures and more widespread systemic losses arising from common industry practices. For many casualty classes there may be no perceived exposure to casualty catastrophes, but where there are already known losses or identified sources of potential loss, they can be difficult to allow for in the reserving process, particularly when the manifestation of the catastrophe is at an early stage. However, wherever possible, losses from known sources of casualty catastrophes are separately identified and a bespoke reserving methodology is used. Further details of reserving for latent claims can be found in Section 6.6.

Results for each reserving category may be blended across a range of methods, with weights varying across cohorts. Graphs might be used to assess the reasonableness of selected ultimate claims, as long as there is sufficient development history available (which may not always be the case for London Market Casualty business).

7.3.6 Matters Relevant to Reserving

Given that this category relates to a group of classes, there will inevitably be a number of specific matters that apply to individual classes. These are not covered here; rather this subsection focuses on some of the more general matters that apply to this group as a whole, organised under a number of subheadings.

Understanding the Business

As the business within this category can be particularly diverse and can change materially over time within an individual category, gaining sufficient understanding of the business written can be important, including identification of any material changes that have occurred between the cohorts. Meetings with underwriting and other staff can assist with this process, as can a desktop review of relevant documentation such as business plans and class descriptions. In some contexts, examination of detailed policy and claims databases showing individual contract and claims data may be deemed appropriate, in order to gain a deeper understanding of the nature of the business and to understand fully any changes across the historical cohorts. The General Insurance Reserving Taskforce (GRIT) (Jones et al., 2006) identified some example policy and claims database diagnostics that can assist in understanding London Market business for reserving purposes. Although these do not relate specifically to Casualty business, many of them are directly relevant, including:

- Policy: Average and percentiles of policy period by underwriting year, to help identify any lengthening of policy term that can occur in a soft market, for example.
- Policy: Territory mix by underwriting year.
- Policy: Loss attachment mix.
- Claims: Currency mix of incurred claims and loss ratios by underwriting year.
- Claims: Pattern of development of incurred claims for settled large losses – to help identify any obvious pattern of case reserve adequacy for such losses.

Many of these can be examined in graphical as well as numerical form. Further detail and examples are given in Jones et al. (2006), as well as other helpful ideas for assisting in understanding a particular portfolio of London Market business, including example analyses of major open risks, a checklist that can be used when liaising with underwriters to understand the nature of the business that has been written and suggested areas of focus in relation to claims management and outwards reinsurance.

Once sufficient understanding of the business has been obtained, it will influence key aspects of the subsequent reserving process including the subdivisions and reserving methods that are used for the analysis. This may, for example, identify that there have been potentially significant changes in business mix within one or more individual casualty reserving categories that need to be allowed for when applying the reserving methods. The approaches outlined in Section 6.9.2 may be relevant here, and in some cases will involve making use of benchmark data, for example based on the Lloyd's Risk Code triangulations (see Lloyd's Market Association, 2015) mentioned in the previous section.

Case Reserving Process and Philosophy

One other aspect of London Market Casualty business that can be important to understand at an early stage of the reserving process is the approach used for estimation of case reserves, along with any changes in this process across the cohorts being analysed. A range of practices exist that are influenced by the insurer's philosophy for case reserving, and which will impact upon the strength of the case reserves included in the data triangles used for reserving purposes.

There will often be a "market" case reserve figure for each incurred claim, which is usually set by the relevant lead insurer(s). Some insurers may determine that for reserving purposes they wish to use a different value for certain individual claims. They may then choose to allow for this through using specific IBNER values for the relevant claims, or perhaps by making an adjustment to the case reserves when compiling their own internal claims data. In either case, the impact can be positive or negative. Where adjustments are made to the case reserves they will sometimes be separately identified, but in any case their use will impact upon the data triangles used for reserving purposes. For example, where the insurer's reserving philosophy results in additional positive case reserves being added to a number of claims, the incurred claims development will reach ultimate at an earlier stage than would otherwise be the case (i.e. it will be shorter-tailed). With either approach, the impact on the reserving process will need to be considered, particularly where there has been a change in the approach used and where there is limited historical development after the change was introduced.

Data Processing

Data used in reserving for London Market companies can be affected by the source of the underlying business and the way in which data is processed before it reaches the insurer's own data systems. There may be different parties involved such as brokers, agents, service companies and market-wide bureaux. Some data might be sourced from schedules of data, known as bordereau, which summarise the premium and claims data on say, a monthly or quarterly basis. All of this can affect the speed of processing for the premium and claims data, which therefore affects the core triangle data, particularly if the processing speed has changed over time. This will affect the reserving process, since, for example, a change of processing speed will need to be allowed for when selecting development factors if a Chain Ladder type method is being used. It is also important to bear this in mind if benchmarking is being used as part of the selection process for development factors, since, even if the benchmark source appears to relate to the same type of business as the subject of the reserving exercise, differences in the speed of processing may need to be allowed for to avoid using inappropriate benchmark factors.

London Market Casualty business can be quite long-tail for some coverage types, which can present challenges from a reserving perspective, particularly if there is limited development history available. Hence, any issue which affects the speed of premium and claims processing can make this even more challenging for these classes.

Currencies

This category may often contain business written in multiple currencies. Some of the major currencies are identified in the subdivision of data subsection above. The reserving can either be done separately for each currency within each class, or currencies can be combined (or a combination thereof). Where any form of aggregation of currencies is used, the data triangles can be created either by converting all diagonals at the exchange rates at the relevant valuation date, or by converting only the data in the relevant diagonals added since the last valuation. The first approach will mean that the historical converted triangle data may change from one valuation to the next. Although this will not be the case with the second approach, some of the development in the triangle will be due to exchange rate movements, rather than "real" claims development. For this reason, converting all historical data in the triangle using the latest relevant exchange rates is usually the preferred approach, but sometimes data limitations make this difficult.

Exposure Profiles

Whenever the triangle data is organised on an underwriting year basis, some of the claims development after the end of the cohort may be due to claims arising as the exposure continues to be earned.

This will be particularly true for certain types of business written under a delegated authority, such as binding authorities, where some of the underlying policies may not expire until up to three years after the start of the cohort, and longer in certain cases. In addition, some contracts can be written with policy/contract periods of more than one year.

Hence, the exposure profile will impact upon the claims development pattern and so in cases where there is believed to be a significant change in this profile between cohorts, it may be necessary to make explicit allowance within any development-factor based reserving methods. Where detailed inception and expiry date data is available this can be used to assess differences in the exposure profile between cohorts. Where this is not available, analysis of the premium signed development pattern may give a broad indication of the exposure profile. This issue also needs to be allowed for where benchmark development factors are being used if there is believed to be a material difference in exposure profile between the benchmark source data and the underlying business that is subject to the reserving analysis.

Underwriting and Reserving Cycle

Most classes of business are affected, at least partially, by the underwriting cycle, whereby premium rates and terms and conditions change according to loss experience and overall market conditions. This is certainly believed to be the case for London Market Casualty business. Understanding the position of each relevant class of business in the underwriting cycle as at the valuation date can impact upon the judgements made regarding certain reserving assumptions, such as prior loss ratios and development patterns. For example, as noted in Jones et al. (2006), with some classes of business the claims development profile tends to lengthen in the soft part of the cycle. Making a judgement as to whether effects such as this are present when reserving London Market Casualty business can therefore be important.

A number of classes of business, including some London Market Casualty classes, are subject to an additional cyclical phenomenon known as the "reserving cycle". This has been cited as affecting a number of classes of business both in the UK and elsewhere.

In summary, this is the feature that can be observed retrospectively for selected classes/companies whereby successive estimates of ultimate claims for certain cohorts tend to increase over time, whereas for others they tend to decrease. Those that increase tend to be in respect of periods where the market is soft (i.e. premium rates are low and/or terms and conditions are under pressure) and those that decrease tend to be in respect of periods where the market is hard (i.e. the business performs well). A discussion of the reserving cycle is available at Line et al. (2003) and Hilder et al. (2008), which also discuss the impact of the underwriting cycle. Lloyd's have shown the reserving cycle to have been particularly apparent for Casualty business, and to closely follow the underwriting cycle. Further details are available at Lloyd's of London (2015a).

Because the reserving cycle can only be observed retrospectively, as and when the estimates of ultimate claims for particular cohorts are updated, at a specific valuation date there will be uncertainty as to where each cohort is within any perceived reserving cycle. This is complicated by the fact that the cycle itself can change over time, due to changing market conditions and loss experience. Despite these difficulties, consideration of the potential impact of the reserving cycle when reserving for London Market Casualty business may still be appropriate, and in some cases an explicit adjustment to the results produced by applying reserving methods to the relevant data may be necessary (e.g. perhaps by ensuring that the prior loss ratio assumption in a BF method takes into account the estimated position in the reserving cycle of the relevant cohort).

Underwriting Year and Earned Basis

When the triangle data is collated on an underwriting year basis, and that data is projected to ultimate, the estimated ultimate claims will also be on an underwriting year basis and hence relate to both the earned and unearned exposures as at the valuation date. Typically, reserve estimates will also be required for financial reporting purposes in respect of earned exposures only. This will usually only affect the most recent one to three underwriting years, where the exposure is not already fully earned at the valuation date. Different approaches are possible for adjusting the ultimate claims on an underwriting year basis for each relevant cohort to an earned basis. The simplest approach is to use the ULR for the full underwriting year, applied to the earned premium. However, if the claims experience in the earned part of the exposure has been particularly low or high, it might be appropriate to reflect this in the selection of the earned ultimate. For example, if a Bornhuetter–Ferguson approach has been used to derive the full underwriting year ultimate claims, then the prior can be multiplied by the unearned premium to derive the estimated ultimate for the unearned exposures. This ultimate can then be deducted from the full underwriting year ultimate to derive the earned ultimate. Other approaches are also possible, depending on the context and the reserving methods that have been used.

Outwards Reinsurance

As with many classes of business, for some insurers this category can be affected materially by outwards reinsurance, which can take a number of different forms – ranging from proportional treaty quota share and surplus covers, to non-proportional treaty excess of loss and stop loss, as well as facultative covers for selected elements of the portfolio. Separate identification of the impact of the different types of reinsurance may be necessary, to ensure appropriate allowance is made in the estimated reserves net of reinsurance. As explained in Section 6.2, the key judgement for outwards reinsurance involves estimating how much of the Gross IBNR will be recovered from reinsurers – i.e. the Reinsurance or "R/I" IBNR. For proportional reinsurance, this can usually be estimated relatively easily using ratio-type methods. For non-proportional covers, although it is usually straightforward to determine the amount recoverable for paid and reported losses, the credit that should be given for further development of those losses (i.e. the reinsurance IBNER) and for losses not yet reported (i.e. the "pure" reinsurance IBNR) is more difficult to determine. This is particularly true for longer-tailed Casualty classes where losses may take some time to be reported. Typical approaches used to derive the allowance for outwards reinsurance are described in general terms in Section 6.2 and include loss-ratio-based methods, consideration of claims paid, incurred and outstanding reinsurance to gross ratios and of historical loss sizes compared with the reinsurance retention. The approach can vary by cohort, since for example, in respect of those cohorts where it is expected that all losses have been reported, the reinsurance credit will by definition only relate to those losses.

Lloyd's of London Minimum Standards for Reserving

In the Lloyd's market, managing agents are required to meet certain specified Minimum Standards, one of which relates to reserving. Lloyd's have issued associated guidance to provide a more detailed explanation of the general level of performance required. Further details are available in Lloyd's of London (2015b). Although this specifically applies in a Lloyd's context, it may also relevant more generally. The principles include establishment of a robust reserving procedure and appropriately documented process along with the requirement to ensure sufficient reserving information is provided to the Board.

Survey of Market Practice

The 2013/14 survey referred to in the earlier section on UK Personal Motor covered both Personal Lines and London Market reserving practices. Although the London Market element of this did not relate specifically to Casualty business, it does provide some insights into reserving practice that is relevant here, since at least some of the 11 London Market survey participants (who were all from larger companies) are likely to write Casualty business. The survey findings can be found at McDonnell and Labaune (2014), and selected key London Market findings are given below:

- Just over half of the London Market participants said they used some form of "fast close" process, so as to provide finalised results from reserving analyses at an early stage after the valuation date. This process is described in Section 6.3.

- All respondents indicated that they used Paid and Incurred claim amount data for their reserving, with just over a third saying they either already used or were starting to use claim number data for some classes.

- Perhaps not surprisingly, the London Market participants appear to rely quite heavily on qualitative data, such as premium rate change information, exposure/deductible and terms and condition changes and mix of business. However, very few companies seem to have access to good quality exposure-based analytics, such as the policy and claims diagnostics mentioned at the start of this subsection.

- The reserving methods are consistent with those cited earlier in this section – namely, universal use of Chain Ladder and Bornhuetter–Ferguson, with some use of "Numbers/Averages" (i.e. effectively methods based on a Frequency–Severity approach, as per Section 3.5) and Exposure-based methods.

- All London Market participants reserved on an underwriting year basis, with most also converting the results to an accident year basis, often mechanically and often by the finance team.

- Specific features allowed for in the reserving process that are likely to be relevant for Casualty business included inflation and the underwriting cycle. In addition, McDonnell and Labaune (2014) noted that in contrast to Personal Lines participants, over half of the London Market participants specifically allow for the reserving cycle and others say they are aware of it when reserving.

Reserving in the London Market in general terms, rather than specifically to Casualty business, is discussed further in Maher (1995). London Market reserving issues are also discussed in Hindley et al. (2000).

7.4 London Market – Property Classes

7.4.1 Category Description

This category refers to classes of business written in the London Market,[3] which include coverage for property-related risks, some of which are exposed to the risk of damage and other related losses caused by natural and human-made catastrophes.

For the purpose of this subsection, the category is intended to relate to what might be termed "core" property-type classes such as Property Direct and Facultative, Property Risk Excess, Property Catastrophe Excess of Loss, Property Proportional (e.g. Quota Share and/or Surplus). Although a broad interpretation of the description of this category could also include other more specialised classes such as Nuclear, Terrorism, Engineering and Extended Warranty, these are not covered here since they each have specific reserving considerations that would warrant their own separate sections.

This category shares many general market features with the London Market Casualty category described in the previous section. Hence, business is written on a subscription basis, with risk carriers taking a share on each risk they write, and can be written both on an Open-Market basis and through delegated authority arrangements (e.g. binders). Property classes are usually written on a Losses Occurring During or Risks Attaching basis. Some examples of the many and varied property-related classes written in the London Market which are covered in this section are as follows:

- **Property Direct and Facultative:** Contracts written in this class provide coverage for public or private enterprises in relation to damage caused to their physical property and equipment arising from a range of sources such as natural perils, fire and theft. Coverage can include damage to buildings, stock and loss of profit due to business interruption. The "Direct" component refers to insurance business, and the "Facultative" component refers to facultative reinsurance, which typically provides cover for an insurer for a particular part of a direct risk, which for various reasons the insurer is unable to retain, or chooses to reinsure. These two components are often grouped together since they have similar characteristics.
- **Property Treaty Risk Excess of Loss:** This class comprises reinsurance contracts which provide cover for a collection of individual property risks, to protect the insurer against large losses on those risks.
- **Property Treaty Catastrophe Excess of Loss:** This class also comprises reinsurance contracts, but in contrast to Risk Excess of Loss, the protection relates to the accumulation of claims across several risks, arising from individual natural or human-made catastrophic events. The contracts will usually specify the region(s) and peril(s) that are covered.
- **Property Binders:** This is a broad category of property business, whereby the risks are accepted by the insurer via a form of delegated authority known as a binder. Further details of delegated authorities are given in the London Market Casualty category description in Section 7.3. A Property Binders reserving group can include business that relates to

---

[3] See Section 7.3.1 for a definition of the London Market.

one or more regions and may include one or more underlying types of insurance and/or reinsurance business. This can present challenges from a reserving perspective if the mix of regions and type of business has changed significantly over time.

7.4.2 Types of Data

All the types of data identified in the London Market Casualty section are also relevant for Property classes, and are not repeated here. In addition for this category, data relating to known named catastrophes will also be important for reserving purposes, including, for each such event:

- Gross and net of reinsurance claims triangle data by class of business, which will enable them to be removed from the core triangle data. Reinstatement premiums generated by catastrophe losses may also be collated.
- Detailed exposure schedules giving individual contracts that either have had claims paid or reported on them at the valuation date, or which are believed to have some potential to generate claims in future for the specified event. Further details of the content of these schedules is explained in the paragraphs covering individual catastrophe reserving in the reserving methods section for this category (Section 7.4.5).
- Market-level data relating to estimated total losses.
- Where available, a range of different outputs derived using software supplied by catastrophe modelling software vendors. These might for example include estimated claims for the relevant insurer arising from the specified event or from other similar events, associated maps and event footprints, which can be combined with estimated damage ratios and exposure information to derive loss estimates.
- Details of the outwards reinsurance programme which applies to the event.

Some of the above types of data will also be collated for large individual property claims arising from single events.

7.4.3 Subdivision of Data

Triangle data is typically divided by class and territory of coverage, such as Worldwide, US, Caribbean, UK etc. The territory of coverage might suggest a particular currency, which would typically include US Dollars, Sterling, Canadian Dollars and possibly also Euros. Business written in other currencies is often converted into Sterling and added to the UK (i.e. Sterling) currency data.[4] For each class/territory/currency combination, the core triangle data for London Market Property classes may often be separated into Attritional, Large and Catastrophe claim types, with separate methodology possibly being

---

[4] As for Casualty classes, the underlying policies in respect of the data for a particular currency may originate from one or more different territories to which that currency relates.

applied for each of these. Each of the Attritional, Large and Catastrophe triangles (or just the total triangle if this subdivision is not used) may then be subdivided further, including:

• Separate data for specific accounts/schemes/binders/lineslips.
• Individual large losses (which may affect more than one class and cohort).
• Catastrophe events (individually usually plus perhaps grouped for small or older events with low or zero outstanding claims).
• Binder and open market business. The former can sometimes have a longer development tail than open market business, due to the inception date pattern.

7.4.4 Cohort and Frequency for Triangles

Data triangles are usually collated on an underwriting year basis, with either quarterly or annual development frequency. For some business, accident year triangles may also be collated, although this is relatively uncommon for this category.

7.4.5 Reserving Methods

For small Property classes with little or no large loss or catastrophe exposures, standard methods such as Chain Ladder, IELR and Bornhuetter–Ferguson might be used. These are applied to both paid and incurred data, with Attritional/Large/Catastrophe claim types considered together in a single data triangle for each class. For larger classes and those with material large claims and/or catastrophe exposures, the reserving methodology may be applied separately for the Attritional, Large and Catastrophe claim types. However, the methodology may not be significantly different for Attritional and Large, and so these are initially discussed together below, with any potential differences then being highlighted.

Estimates of ultimate premium may be required for this category, usually for the more recent cohorts. These can be derived using a combination of underwriter estimates (based on knowledge of actual business written and expected to be written for each cohort) and application of standard reserving methods, such as the Chain Ladder, to premium triangle data. It is not usually possible to allocate the premiums to one or more of the Attritional, Large and Catastrophe claim types, and so the individual analysis for these will all use the same ultimate premiums. Hence, any loss ratios and prior estimates used in IELR or Bornhuetter–Ferguson methods will need to take this into account. Furthermore, consideration of any material distorting effect on the loss ratios and/or premium development caused by the inclusion of large inwards reinstatement premiums in the premium data for some types of business may be necessary, as may the treatment of any material profit commissions in the data.

As for London Market Casualty business, for some London Market Property portfolios, the analysis might be carried out separately for each of the large individual contracts, taking into account the assumptions made during the pricing of the business.

Methodology for Attritional and Large Property Category

For this category, the Chain Ladder method is often used, applied to Paid and Incurred data. Although the IELR and BF methods are also commonly used for the recent cohorts in particular, the generally short-tail nature of property classes (relative to others with liability exposures) means that more weight can often be applied to Chain Ladder-type approaches than would otherwise be the case. Where the IELR or BF method is used, the loss ratio assumptions are usually based on one or more of business plan/underwriting assumptions, pricing analyses and averages from earlier cohorts (often with premium rate change and inflation adjustments, as described in the worked example in Section 3.4). Frequency–Severity methods are only used in isolated cases where claim number data is readily available.

In addition, if the class is affected by large individual claims, then these might be removed from the core triangle data and analysed separately. The reserves for these large claims will typically be determined by placing reliance on the case estimates set by the claims team, with or without adjustments (i.e. IBNER) where these are justified by the historical case reserve adequacy. An alternative approach is to apply development factors to the latest paid or incurred claims data for the large claims in aggregate, split by cohort, with the factors for example being based on those derived from the core triangle data for Attritional/Large claims. Where significant volumes of large individual claims have been removed from the core triangle data, then consideration will need to be given as to whether an additional loading is required for large claims IBNR.

Although benchmarking of development factors and loss ratios is reasonably common for these classes, in general terms it is perhaps used less than for longer-tailed classes, simply because (other things being equal) less historical data is needed for property classes to derive reserves based on the individual company's own data. Where benchmarking is used for London Market property classes, similar sources as the examples detailed in the section covering London Market Casualty classes can be used.

Other bespoke reserving analyses and methods might be used for individual classes or sources of loss. For example, if the business includes multi-year policies, then this may need to be allowed for in the reserving process, particularly if the volume or average term of such policies has changed materially across the cohorts.

Results for each reserving category may be blended across a range of methods, with weights varying across cohorts. Graphs might be used to assess the reasonableness of selected ultimate claims, as long as there is sufficient development history available.

If the Attritional and Large claim categories are analysed separately, then a similar approach can be used as that described above for the combined category. The advantage of carrying out the analysis separately is that the two categories can sometimes exhibit different development patterns (with Attritional typically being shorter-tailed than Large) which can be important to allow for, particularly if the mix of Attritional and Large claims has changed over time, due for example to a change in contract terms or business mix. The disadvantage of such a subdivision is that the data can be too sparse in some cases.

Methodology for Catastrophe Reserving Category

The approach used for reserving this category will usually involve two elements – the first in relation to known individual catastrophes, and the second for the remaining elements. The proportion of reserves (and hence effort) relating to each element will depend on various factors, most notably the class of business. The subsections below cover each of these two elements.

Reserving Methodology for Individual Catastrophes

The approach used to estimate reserves for individual catastrophe events will depend in particular on the timescale since the event. Immediately after the event, there will be very limited information from which estimates can be derived, but this will gradually improve over time, allowing more insurer-specific analyses to be conducted. Eventually, as the claims advised begin to stabilise, the methodology can also change to reflect this.

The approaches that can be used at the various stages fall into four broad categories – Market share, Catastrophe models, Exposure-based and Development-based. These are described below, together with an indication of the stages at which the approach is typically used. The details of the approach will vary according to the nature of the business, although the focus here applies for catastrophes affecting typical London Market core property classes. In many cases, the methodology used will involve input from a number of different personnel, including claims, underwriting, finance and reserving practitioners. The description below applies to determining an estimate on a gross of reinsurance basis. Outwards reinsurance is dealt with at the end of this subsection.

Market Share   In the period immediately following a catastrophe event, there will be very limited information regarding which of the insured/reinsured risks written by the insurer concerned are exposed to the event, and even less regarding to what extent they are exposed. However, there will often be very preliminary market-wide estimates of the overall economic and/or insured losses expected to arise from the event, produced by various market commentators and, for many natural catastrophes, some of the catastrophe modelling firms. If the insurer/reinsurer can estimate its market share in the relevant period for the classes of business that are expected to be affected by the event, then this can be multiplied by the estimated total insured market loss to provide a very preliminary loss estimate, gross of reinsurance. The selected market share can take into account any emerging information about the potential exposed contracts for the relevant insurer, together with any preliminary insurance market intelligence regarding the event that might be available. Different assumptions for market share and market loss can be used to derive an indicative range for the gross loss. Where the business concerned is inwards reinsurance business, this approach will need to consider how much of the total insured market loss is expected to be retained by the insurance market and how much will be ceded to the reinsurance market, and in particular to the companies and contracts reinsured by the reinsurer concerned. In practice, any estimates produced using this approach are likely to be very approximate. It is not uncommon for the published estimates (based on market share or other approaches) for individual insurers to change over time, as the loss develops and other approaches, as described below, are used to refine the estimates.

Catastrophe Models For many catastrophe events, a number of catastrophe modelling firms produce software which enables the contracts written by a particular insurer/reinsurer to be overlaid with the geographical footprint of a known event to produce an estimate of the insured losses. Like the market share approach, this is useful in the very early period after the event has occurred, but as with that approach, the estimates are often subject to considerable uncertainty. This is partly because the intensity of the event at each specific location and hence the damage caused to individual insured units is very difficult to determine with any accuracy at an early stage after the event. Updated estimates using these models may however be determined at a later stage, as further information develops.

Exposure-based Methods Typically, as more information emerges regarding the insurance/reinsurance contracts that are exposed to the event, the market share and catastrophe model-based estimates can be supplemented by an exposure-based approach. Although the estimates derived from catastrophe models do also allow for the relevant exposures, and so in that sense are also exposure-based, the type of exposure-based approach referred to here takes more account of specific information gathered in relation to the individual exposed contracts. The detail of the approach used and the availability of suitable data will vary according to the materiality of the catastrophe, the context and the nature of the contracts affected (e.g. class of business, insurance, reinsurance, per-risk, per event etc.). As an example, for commercial insurance or reinsurance portfolios, one possible approach involves the following steps:

1. Identify all contracts that may be exposed to the event, and collate all relevant contract details such as sums insured, contract limits, share written and known facultative reinsurance where applicable. Some of these contracts will be known to be exposed to the event (e.g. because claims have been advised on them) and for others there will be varying levels of potential for them to be exposed to the event.
2. Collate all known claims information, based on advices from insureds, loss adjusters and brokers, recording paid and outstanding amounts against each exposed contract.
3. Where possible, determine the maximum amount that can in theory be claimed against each contract for the relevant event, based on information collated in Step 1, taking into account other losses that have impacted the contract. Where relevant, Probable Maximum Loss ("PML") data will also be collated.
4. Where appropriate, begin a process of contacting relevant parties to seek to establish whether individual contracts are exposed to the event and if so, to what extent. The relevant parties will depend on the nature of the portfolio, but typically would include the insured/reinsured themselves and/or brokers, loss adjusters etc. The type of information gathered from this part of the process will range from broad qualitative statements

regarding whether the contracts might be exposed or not, to more quantitative estimates of the amount of loss that might be incurred. For inwards treaty excess of loss reinsurance business, the information might include an estimate of each reinsured company's "From-Ground-Up" ("FGU") loss estimate for the relevant event, net of other relevant reinsurance, that will be ceded to the relevant excess of loss programme. All information gathered at this stage could be documented, so that it can be monitored and updated as more information emerges.

5. Using the information collated in the previous step, determine an estimated ultimate claim (and hence reserve) for each contract, relating to the particular event. Where relevant, this will also include determination of any implied inwards reinstatement premiums that will be due. This estimation process will involve underwriting and claims expertise and can range from a high-level judgemental based approach, to a more systematic approach based on assigning some form of qualitative rating to each contract identified in Step 1, based on perceived claim potential, together with an associated loss level or "burn factor" for each rating. With the latter approach, some contracts may need to be flagged as having unknown exposure in the early stages after the event, with others as being definitely exposed or not exposed, with perhaps two to three intermediate levels of perceived exposure, reflecting the potential claim level.

6. As time develops, and more information emerges and claims begin to be advised, the process of contacting the relevant parties will produce more and more reliable information on which ultimate claim estimates for each contract can be based. In this way, eventually most contracts can be classified as definitely not exposed or definitely exposed, with ultimate estimates for the exposed contracts being based on claims advised and other information available for those contracts.

7. Benchmarking can also be used to assist with the estimation process, where there is knowledge from other sources regarding, for example, the FGU for particular cedants in a reinsurance portfolio.

8. Other analyses can also be carried out to determine, for example, the contracts where there is a large downside risk above the current estimated loss (i.e. the largest difference between maximum potential claim from Step 3 and the estimated reserve against that contract), as well as sensitivity testing of changes in FGU etc.

In many cases, this type of exposure-based analysis for individual catastrophes will be done on a bespoke basis, perhaps with the format and details being collated varying by class of business. However, wherever possible, organising the relevant contract data and associated information into a central database or standardised format can make it easier and faster to derive ultimate loss estimates, apply sensitivity analyses and produce summary reports.

Development-based Methods   As the claims data relating to a particular event begins to be advised to the insurer/reinsurer, development-based methods can be used. For some reinsurance portfolios, an individual event will often affect more than one underwriting year, and sometimes more than one class. It is usual for data to be combined across underwriting years for use in development-based methods, but classes of business with different characteristics would usually be analysed separately. Because there is effectively only one cohort to which the development-based method can be applied, the approach will usually involve fitting some form of curve to the development factors, as explained in the relevant paragraphs of the description of the Chain Ladder method (specifically, Section 3.2.4). This can be applied to both the Paid and Incurred claims data, and will produce an estimate of the factor to be multiplied by the latest Paid/Incurred to derive the estimated ultimate claims. Benchmarking can be used as part of this approach, with factors being derived from previous catastrophe events and from other sources for the same event.

For any individual catastrophe, a blend of the results from one or more of the above methods might be used, depending in part on the information that is available at the relevant time. A graphical review of the paid and incurred claims development against one or more options for the selected ultimate claims can also be helpful in making the final selection.

The selected gross estimates from this process can then be applied to the relevant reinsurance programmes, often by class of business, to determine the estimated net ultimate, IBNR and reserve. Where relevant, allowance is also made for any inwards and/or outwards reinstatement premiums that may be applicable. Sensitivity testing of the impact of changes in the gross catastrophe estimates on the net of reinsurance position can also be carried out, which can be helpful in ensuring appropriate effort is focused on those catastrophes where the net IBNR is most sensitive to changes in gross IBNR. Other matters that might be considered as part of this could include any known reinsurance bad debt or disputes and, where relevant, the impact of alternative scenarios for how claims are aggregated for the purpose of recovery from the reinsurance programme.

Reserving Methodology for Catastrophe Reserving Category – Other Than
Individual Catastrophes

For this category, the data triangles will usually exclude the claims data relating to the individual catastrophes that have been reserved separately using one or more of the approaches described in the previous subsection. The remaining data therefore relates to either older largely settled catastrophes, or more recent smaller catastrophes. For data organised on an underwriting year basis, standard reserving methods (e.g. Chain Ladder and Bornhuetter–Ferguson), can be used for this category. For underwriting years that are not fully earned it is also possible to subdivide the analysis into an earned and an unearned component. The earned element for these years will relate to incurred catastrophes, which will in most situations relate to events already reported to the insurer. For these, a view can be taken in conjunction with the claims team on the reasonableness of the case estimates. In some cases (e.g. retrocession-type portfolios), there may be a relatively small possibility of unreported but incurred catastrophe events, in which case the earned element of the reserves will need to include a loading to allow for this. For the unearned element, one approach is to use a loss-ratio-based method, with the loss ratio assumption being derived from a combination of the values assumed in the relevant business-plan/pricing analysis for the catastrophe element of the claims, and the catastrophe loss ratio observed for more developed under-writing years. An alternative approach is to use a Frequency–Severity type method using historical catastrophe experience or to make use of output from the catastrophe element of the insurer's capital model, where relevant.

7.4.6 Matters Relevant to Reserving

Many of the matters identified in the relevant paragraphs of the Casualty classes section (Section 7.3.6) apply equally well to Property classes, and so only a brief summary is given here, along with any points that apply specifically for Property classes. Hence, understanding the business is critical, particularly if the business mix within individual classes has changed across the underwriting years, for example with different volumes of catastrophe-exposed, or delegated underwriting business being written.

Whilst ensuring that any material changes in case reserving practice are understood, this is arguably less important for Property classes than with Casualty classes, due to the shorter-tailed nature of the business. Although any delays or changes in the speed of data processing will usually become apparent more quickly with Property classes than with Casualty classes, the different sources of data can still have a material effect on the speed of processing and hence on the development data used for reserving. Multiple currencies are equally common for Property classes and hence need consideration.

The impact on claims development patterns of changing exposure profiles can be sig-nificant for Property classes, and so may need to factored into the reserving analysis. The reserving cycle can at least in theory also affect Property classes, but the effect is perhaps less obvious than for Casualty classes.

Where it is necessary to produce results on both an underwriting and earned basis, this is sometimes made easier for Property classes, where at least part of the overall reserve may already be derived separately for the earned and unearned component, as discussed in the reserving methods section (Section 7.4.5). Finally, the approaches described in the Casualty classes section for outwards reinsurance apply equally well to Property classes, although in general terms more of the non-proportional reinsurance IBNR credit may be loss- or event-specific for Property classes.

Survey of Market Practice

Most of the points made in the Casualty classes section in relation to London Market reserving practices identified in the 2013/14 survey apply equally well to Property classes and are not repeated here. In addition, the survey noted that London Market respondents almost universally analysed attritional, large and catastrophe claims separately, which is particularly relevant for Property classes. London Market survey respondents indicated that where they allow for seasonality, it will only be for catastrophe events, which is logical as some of these do have a clear seasonal pattern that can be allowed for in the reserving process. Further details of the survey results can be found in McDonnell and Labaune (2014).

7.5 US Workers' Compensation

7.5.1 Category Description

This category, abbreviated here by "USWC", represents a major class of business in the US for which the policies provide compensation to persons (or their dependents) who are injured at work or acquire an occupational disease. The specific coverage and benefits that are payable vary by US state, which can affect a range of items such as maximum and minimum payments for relevant injuries, rulings for manifestation of claims, statute of limitation requirements and integration of benefits with social security systems.

USWC is provided by a combination of private insurance carriers, state funds, or through full or partial self-insurance. Claims can relate to wage replacement benefits (referred to as "indemnity" or "compensation" payments), medical costs, rehabilitation and other benefits, and can range from small claims with payments of only a few hundred dollars in respect of medical-only accidents to multi-million dollar long-term annuity payments for Permanent Total Disability (PTD) cases. Whilst the delay between accident and reporting to the insurer is usually relatively short, because some claims are payable to injured workers for many years, the claims development pattern for this type of USWC claim can be very long tail – in excess of 30 years in some cases.

Depending on the context for the reserving exercise, this category can relate to USWC on a direct (i.e. insurance) or reinsurance basis, and can further be limited in terms of specific layers of coverage being written, or through self-insured retention levels and/or be protected by outwards reinsurance. It is important to note that employers in the US are required to purchase/provide workers' compensation coverage essentially on an unlimited basis in accordance with the laws of a given state. Employers with operations in multiple states generally are issued a policy endorsement for each state that conforms to the given state's workers' compensation laws.

Actuaries in the US have played a key role in USWC for many decades, in areas such as reserving, rate setting, benefit change assessment and solvency monitoring. They have therefore built up an extensive knowledge of the varied and complex issues that affect USWC. In this brief summary, only a relatively small number of these issues can be included.

7.5.2 Types of Data

Data triangles for paid and incurred claim amounts are the core quantitative data used for reserving this category. Claim number data is also important in order to understand changes in reporting and closing of claims, trends in payments and reported amounts per claim, and where Frequency–Severity methods are used. Exposure data is also usually collated for use with relevant reserving methods. Payroll and/or Full-time-Equivalent (FTE) number of employees are typically used for exposures at self-insured employers and state funds. Direct insurers and reinsurers tend to use premiums for exposures but some direct insurers also consider payroll and/or FTE number of employees.

Other quantitative data that might be collated includes large claims, details of the outwards reinsurance programme and/or self-insured retention levels. Details of individual claims may also be collated to determine reserve estimates using life-assurance type approaches or to understand the cash flows within claims.

7.5.3 Subdivision of Data

This will vary considerably, depending on the context. At one extreme, a single subdivision might be used covering all claim types. This is more common for smaller insurers and self-insured programmes. For other USWC reserving analyses, however, some level of subdivision is often used with the claims data being at least divided into indemnity/compensation and medical claims. For some contexts such as a large state fund or an insurer that writes USWC across several states, there can be many subdivisions. For example, one or more of the following subdivisions could be used:

- Medical Only claims versus Medical on Lost-time claims.
- Compensation type: for example, Temporary Total Disability, Permanent Partial Disability, Permanent Total Disability (PTD), Death, etc.
- By State, as benefits can vary significantly.
- Occupational disease and cumulative injury separated from other injuries.
- Provider type for medical on lost time cases: for example, Hospitals, Physicians, Pharmaceutical, Chiropractors etc.

Some of the above subdivisions will only be used in very specific circumstances, and as for all reserving exercises, the chosen subdivisions will depend on the available data and the context.

7.5.4 Cohort and Frequency for Triangles

The typical USWC reserving exercise in the US uses accident year cohorts, with annual development frequency. Reinsurance portfolios might, however, use underwriting year as the cohort. When monitoring reserve performance by quarter, improved results can be obtained using accident/underwriting year cohorts with quarterly development, at least for younger development ages (e.g. up to two to three years) where interpolation formulas may fail. For large blocks of USWC business, the older accident years might be grouped into a single block. However, when this is done caution must be exercised to ensure the reserving practitioner understands the potential impacts from mixing claims of various ages.

7.5.5 Reserving Methods

The Chain Ladder and Bornhuetter–Ferguson methods, applied to claims paid and/or incurred (or "reported") claim amount triangles, are widely used in USWC reserving.

Frequency–Severity methods are also used in some cases. For the more recent cohorts, the estimated ultimate claims might be based on a "Loss-Rate" method. This involves first deriving the loss rate for each of the older accident years, defined as the estimated ultimate losses derived using, for example, a CL method, divided by exposure. These are then adjusted to make them representative (or "on-level") for the recent cohorts. The factors included in this adjustment will depend on the context, but for example can include benefit changes, exposure (e.g. wages) changes, frequency and severity trend, retention level, and changes in mix of business. Once adjusted, a suitable average of the on-level loss rates can be derived for use as the expected loss rate. This assumption can then either be used to derive the estimated ultimate directly by multiplying it by the exposure for the relevant cohort, or as the prior in a BF method. The latter is very similar to the use of the BF method in conjunction with the CL method described in Section 3.4.

The Loss-Rate method might also be used where the scope of the USWC reserving exercise involves estimating the loss cost for one or more future years, with the estimated future exposure being multiplied by the loss rate derived from earlier cohorts, perhaps adjusted to allow for inflation and for any known benefit changes that will affect these future years.

For some subdivisions of the data, it may be necessary to estimate a tail factor, particularly when development is expected beyond the observable period of the historical data triangles. If so, the curve fitting approaches described in Section 3.2.4 can be used, perhaps supplemented by judgemental examination of the recent trend in incremental claims payments. It is important to recognise that the claims development in the tail for some claim types will be impacted significantly by mortality. Accordingly, in some cases, it may be possible to allow for this explicitly by analysing separately the "pension" or "Lifetime" claims and applying a life-assurance type approach. An example of such an approach, designed to be used for older USWC claims, is explained in Jones et al. (2013). This approach is similar to that used for PPO claims in UK Motor, as referred to in Section 7.2.

As well as using cumulative development triangles, which applies to all the methods described above, incremental data can also sometimes be used for USWC reserving either with or without explicit adjustments for inflation. Specifically, the trend in incremental average payments for older cohorts can be examined to judgmentally remove the observed inflation of the historical development to allow an explicit forward inflation rate that varies from that observed in the data. Such an approach will also need to consider other relevant factors, such as the perceived persistency of claims. As a sensitivity test or when full development data is not available, a paid survival ratio-type method can be used, similar to that used for latent claims, as described in Section 6.6.2.

Where large claims are present, they might be removed from the core analysis, and considered separately, perhaps basing the reserve on the case estimates as set by the claims department.

7.5.6 Matters Relevant to Reserving

As should be clear from the earlier paragraphs in this section, there are a wide range of different contexts for USWC reserving, and the matters that are relevant for a particular exercise will depend on that context. This subsection seeks to identify some of the broader matters that are of general relevance; in any particular exercise there may be several other specific matters that impact significantly upon the reserving process.

Industry Data and Commentary

There are a number of industry sources available to assist the USWC reserving practitioner, including:

- **The National Council on Compensation Insurance (NCCI):** This organisation provides a wide range of information of use to the reserving practitioner, including loss development data, analysis of the impact of benefit changes, research studies and presentations documenting legislative changes and a range of emerging issues on both a country-wide and state basis. Two examples of NCCI published material from 2015 are a research report into the relationship between report lag and claim cost (see Sheppard, 2015) and the 2015 State of the Line, which is an annual review of workers compensation trends, cost drivers, and the significant new developments impacting the industry (see National Council on Compensation Insurance, 2015).
- **Individual state USWC bureau:** Although the insurance departments for the majority of states have designated NCCI as the licensed rating and statistical organisation, there are more than ten that have not, including California, Wyoming, Minnesota and New York.[5] These states have their own USWC bureau, which collate various data and information that is relevant to reserving, including loss development data for example.
- **Other sources:** The Reinsurance Association of America collates data for USWC reinsurance business and provides industry loss development data to subscribers to its service. Reserving practitioners may also access historical development data for the most recent 10 accident years for all US domiciled insurers from Schedule P of US Statutory Annual Statements. Many actuarial consulting practices will also have access to private data to assist them in USWC reserving.

Long-tail and Uncertain Nature of the USWC Development Pattern

USWC claims have varying lengths of claim development dependent on the type and severity of the injury, despite the generally short delay between the date of occurrence and date of reporting to the insuring entity. Specifically, the development of USWC claims range from a short period of much less than a year for minor medical only claims (no lost work days) to a long period for life-time type claims such as Permanent Total Disability and Death claims. For lost time claims (i.e. those involving lost work days), which require indemnity benefits for lost wages, the duration and pattern of development of any individual cohort will be subject to uncertainty due to several factors, such as:

---

[5] See the NCCI State Map on the NCCI website at www.ncci.com for full details.

- The duration of temporary disability cases.

- The potential worsening of injuries increasing the length of temporary disability and/or shifting temporary claims to permanent partial or permanent total claims.

- The impact of mortality on permanent disability cases.

- Changes in the utilisation of the healthcare system as well as innovations in medical devices, treatment and pharmaceuticals.

- The impact of medical inflation and indemnity inflation in US states that have cost of living adjustments.

- The potential for claims to reopen due to changes in medical circumstances.

- Changes in the utilisation of attorneys and/or changes in attorney behaviour.

- Changes over time in an insurer's claim handling practices, including the use of full and final settlements in US states where allowed.

- Reforms to US state workers' compensation laws, which may lead to both intentional changes and unforeseen changes to benefits and claimant/attorney behaviour.

Partly due to these factors, unlike most other casualty or liability classes (except perhaps for those with latent exposures or other annuity-type payments), the development ratios observed for some USWC claim types may not always appear to decrease gradually towards unity. Furthermore, it will often be the case that currently active claims will continue to be payable beyond the observed development period, so that there is no prior history from which to assess the likely future development. These features can make the use of standard development-based methods difficult to apply, and so careful judgement will be needed in practice when these are used, supplemented by use of industry data where appropriate. In some cases, the use of a life-assurance type approach might be appropriate, as referred to in the reserving methods paragraphs earlier in this section.

Discounting

Discounting USWC reserves is typically allowed under US GAAP and allowed in certain circumstances under US Statutory accounting. For consistency, US domiciled insurers and reinsurers often follow the discounting in US Statutory accounting for US GAAP. Some self-insured employers who are not required to maintain Statutory financial statements do elect to record USWC reserves on a discounted basis while others do not. In addition, certain State Funds are required to discount all USWC reserves under state law. For instance, the New York State Insurance Fund is currently required to discount all USWC reserves using a 5% discount rate regardless of the current financial market. On the other hand, most US domiciled insurers and reinsurers, which are required to maintain Statutory financial statements, regardless of whether they also maintain US GAAP or non-US financial statements, either do not discount USWC reserves or only discount "tabular" USWC cases. Tabular cases are defined as those with a relatively fixed and determinable payment stream that is similar to an annuity. Permanent Total Disability cases fall into this category. Other cases that have a payment stream that is not fixed and is easily determinable (known as non-tabular), are generally not allowed to be discounted under US Statutory accounting. However, insurers and reinsurers may discount all USWC reserves for US Statutory accounting if the Insurance Department of their domiciled state approves doing so.

As discussed in Section 6.7, there is uncertainty associated with discounting reserves due to the inherent variability in claims payment patterns and because there is a potential difference between the selected discount rate and the achieved investment yield. For known cases, some insurers may discount the case reserves carried. It is important for the reserving practitioner to be aware of the practice used for the relevant insurer when selecting development factors and comparing to industry statistics.

To conclude, USWC is a dynamic and complex class of business where there can be changes in the legislative and regulatory landscape that impact on the reserving process and judgements therein. The reserving practitioner will need to be mindful of this and the many other developing issues that might affect their analysis both now and in future.

7.6 Asia Motor

7.6.1 Category Description

This category refers to motor insurance written in Asia. These are typically markets which are changing and developing at a much faster pace than in established markets seen in Europe, for example. Awareness of the changes, and consideration of the impact of them on the reserving process, is key to ensuring the approach to reserving remains relevant and current.

The specific details of the coverage and the reserving issues that arise will vary from country to country. This section aims to provide a broad overview of motor insurance reserving topics that apply in the region in general terms, with some examples being given of particular features in individual countries. Any reserving carried out in respect of business in an individual country in Asia may need to take into account additional factors that are not covered here. The four main types of motor insurance generally seen in Asia, provided in respect of cars, vans and motorcycles, are:

1. Act only – this provides cover for Third Party Bodily Injury and Death (TPBI) which is the minimum cover normally required by the regulations.
2. Third Party only – this covers Type 1 as well as Third Party Property Damage (TPPD).
3. Third Party, Fire and Theft – this covers Type 2 coupled with loss or damage to the insured's vehicle due to fire or theft (Fire & Theft).
4. Comprehensive – this covers all the above as well as loss or damage to the insured's vehicle in an accident (Own Damage). Extended benefits such as coverage for windscreen damage, passenger liability and special perils such as flood, windstorm, landslide or subsidence can also be included.

In reserving analysis in the region, motor insurance coverage is commonly subdivided into two categories – Motor Act and Motor Non-Act. Motor Act refers to the TPBI component, which is mandatory under the law, whilst Motor Non-Act includes all other types of losses, namely TPPD, Fire & Theft, Own Damage, as well as losses covered under the extended benefits mentioned above. Motor Act is synonymous with Compulsory Third Party Insurance (CTP) in developed markets such as Australia. Most countries in Asia have introduced CTP, although slightly different terms are used to describe the cover (e.g. in Vietnam and the Philippines it is typically called Compulsory Third Party Liability Insurance, whilst in Singapore it is commonly called Third Party Risk and Compensation Act). In China, motor third party liability insurance was introduced in July 2006, part of which is intended to provide minimum cover. There are two categories: Compulsory Motor Third Party Liability Insurance (CTPL), which covers the statutory indemnity limits, and Commercial[6] Motor Liability Insurance (CMI), which covers liabilities above the statutory limits, and is non-mandatory.

In a number of countries in Asia (e.g. China, Malaysia and Thailand), the premium rates for motor insurance are subject to statutory tariffs, particularly for CTP. Insurers then have to be very selective about identifying profitable segments of the market. This can sometimes mean that availability of cover for some of the more unprofitable segments, such as older vehicles and certain groups of commercial vehicles is limited. As a result, motor insurance pools are often established to provide coverage for vehicle owners who have difficulty accessing insurance. Other pools exist in some markets to fund claims arising from uninsured drivers, as CTP evasion can be quite common.

In some countries in the region, motor insurance policies can have somewhat unusual features that may impact upon the reserving analysis. For example, in Japan and Korea, some policies can have a savings element to them, which can result in premiums being refunded after a period of claim inactivity.

7.6.2 Types of Data

Claim amount and claim number data triangles represent the core data used for reserving analysis. Most regulators require claims data by accident year to be submitted in specific formats for each class of business, from which data triangulations can be collated, if not otherwise available. However, if such data is aggregated across companies for benchmarking purposes, then it is worth noting that in Asia, as in many other markets, this data can be subject to potentially significant limitations – for example, if one or more insurers stop submitting data to the regulator for a particular reason, then unless their claims data is removed from the full historical triangles, the aggregated claims development at subsequent time points could be materially distorted.

Other data that might be collected for reserving purposes includes data on individual large claims if they cause significant distortions to the claims development, and in some cases the full transactional claims database to facilitate more granular analysis.

---

[6] Note that the term "Commercial" in this context does not relate to commercial vehicles, it is synonymous with "voluntary".

Apart from claims data, policy information such as premium and earned exposures are also used for reserving purposes.

7.6.3 Subdivision of Data

Where relevant, data is subdivided by vehicle type – i.e. Private Car, Commercial Vehicle and Motorcycles. Reserving analysis is typically done separately for Act and Non-Act, with Non-Act perhaps also broken down into TPPD, Fire and Theft and Own Damage. In some cases, where a more detailed analysis is justified and credible volumes permit, the analysis might be subdivided further into legal fees, adjusters' costs, workshop repair costs and other costs.

7.6.4 Cohort and Development Frequency for Triangles

Annual accident year cohorts, with annual development frequency are commonly used for reserving purposes. However, insurers with a long business history who might have a large claims database may explore different bases to fulfil different obligations. For example, some insurers may opt for claims triangles based on accident years with quarterly development frequency, in line with their quarterly internal and regulatory reporting requirements. Others may opt for accident quarters or accident months, to give a more granular level of experience monitoring. In some cases, an underwriting period cohort might be used, rather than an accident period. For example, some reinsurers might use this basis, driven by their accounting approach or for other reasons.

7.6.5 Reserving Methods

Three reserving methods are commonly used, applied to claim amount data:

1. The Chain Ladder method based on both claims paid and claims incurred.
2. The Bornhuetter–Ferguson method based on both claims paid and claims incurred.
3. The Initial Expected Loss Ratio (IELR) method (or just Expected Loss Ratio Method).

The Chain Ladder method is invariably deployed as the default projection method especially for businesses that have sufficient claims development. The BF and IELR methods are also often used to supplement the Chain Ladder method, especially for the more recent cohorts. Frequency–Severity methods may also be used in some cases, usually as a check for reasonableness of the results produced by the other methods mentioned above.

7.6.6 Matters Relevant to Reserving

Claims reserving for Asia Motor business is affected by a wide variety of factors including legislation, taxation, capital requirements, market forces, tariff structure and claims management. The impact of these factors on reserving can be qualitative and/or quantitative in nature, and can vary significantly by country and over time. The legal landscape varies significantly across the region, ranging from more litigious common law environments such as Hong Kong to other markets that are less litigious, resulting in shorter-claims development and less exposure to large individual claims. Knowledge of this and of the local legislation that impacts upon motor insurance is an important pre-requisite for any reserving analysis in this region. For example, there is often a time bar (i.e. limitation period) for making TPBI claims (e.g. 7 years in Malaysia, with 3 years in East Malaysia), which can have a significant impact on the claims development. In some cases, the courts can refer to schedules that provide a benchmark for the level of awards made in TPBI claims (perhaps referred to as "compendium schedules"). As a result, any changes made by the relevant authorities to such schedules may have a significant impact on the reserving analysis for TPBI claims. This will need to be taken into account when making judgements as part of the process of applying reserving methods. For example, some insurers may adopt a proactive approach to these changes, by updating all currently outstanding claims to reflect the revised schedules. Communication between the reserving specialist and the insurer's management team is vital to ensure that appropriate allowance is made for any distortions to the development pattern caused by these and other similar changes.

Apart from allowance for known changes in awards schedules, reserving practitioners may also consider the general trend in terms of court award inflation, which needs to be taken into account when making judgements as part of the reserving process.

As in many other regions, in Asia when a TPBI claim is first lodged with an insurer, it is likely that there will be very little information available regarding the nature and extent of the injuries caused in the accident, in order for the claims personnel to base their determination of a case estimate. Unlike for short-tail claims, such as Own Damage where workshop repair costs and adjuster's fees can be often be estimated relatively quickly, it is common for an initial case estimate or "blind reserve" to be established for TPBI claims, given the lack of information. The level of blind reserve is typically set based on the average claims settled in the past year, for example. This will have a considerable impact on the claims development pattern in the early periods. Reserving practitioners will normally take note of the movement in the level of blind reserves over previous years as part of their analysis.

Changes in the taxation landscape in Asia can have a significant impact on reserving analysis. For example, the implementation of a Goods and Service Tax (GST) will affect many aspects of an insurer's business. The immediate impact would be seen for Motor Non-Act, as these claims tend to have higher frequency coupled with moderate severity (compared with its Act counterpart). From a reserving perspective, there is likely to be a distortion in the claims development pattern observed in the financial year when GST was first introduced, which will need to be taken into account when applying development-based reserving methods.

With the introduction of Risk-Based Capital (RBC) frameworks in many countries in this region, the regulators intend the solvency framework to provide a clearer indication of insurers' financial strength. One of the most notable impacts of these frameworks is that claims liabilities need to be booked at 75th confidence level with a risk margin expressed explicitly. Stochastic reserving methods are usually deployed to estimate the required risk margins, provided there is sufficient data. In cases where data for one or more classes of business is sparse, industry data can be used, or even data from neighbouring markets as references. Under the RBC regime in some countries, insurers may need to improve their capital position. They do so in a number of ways, including using Quota Share reinsurance and/or Loss Portfolio Transfers of, for example, long-tail TPBI losses.

Implementation of market initiatives in the region can have an impact on the reserving analysis. One such example involves recovery from other insurance companies of the TPPD components of Motor Non-Act claims. Under a Knock-For-Knock (KFK) agreement, an insurer could recover the costs of a TPPD claim from the insurer of the claimant to whom it paid out the TPPD claims. However, such a process may take a considerable time, as the incentive for such reimbursement is seen to be minimal. Owing to this, the development for TPPD claims invariably will show a recovery in the later periods, when the reimbursements are received. Electronic KFK ("e-KFK") may be introduced in certain markets to enhance the process, through an automated system which will expedite settlement of claims between claimants and insurers, and between insurers in relation to the reimbursement process. The outcome of this can markedly shorten the time to recovery. Reserving practitioners will need to take this into account when projecting the future development pattern of the claims, as applying the historical pattern in respect of future recoveries may no longer be appropriate for the more recent claims.

Similarly, for Own Damage claims, the future patterns of claim development may be significantly affected by the business environment surrounding the process flow of claims from occurrence through to final settlement. Behavioural changes in repair workshops in terms of timing of delivery and costs estimation can have a significant influence on the case estimates used by insurers. Changes to the panel workshops to which insurance companies are aligned can also have a significant impact on the reserving analysis in some cases. Again, this needs to be taken into account in the reserving analysis, particularly when such fluctuations are evident from the claims development pattern. The extent to which this is observable in the data depends on the granularity of the reserving analysis. Communication between the reserving practitioner and the claims team will usually provide valuable insights regarding any significant changes in the overall claims process.

Most insurance companies in the region do not discount their claims reserves for the time value of money. RBC frameworks usually require insurers to value the inflation impact explicitly in the claims projections, if they opt to discount the reserves.

7.7 UK Asbestos

7.7.1 Category Description

This section relates to a specific type of latent claim – asbestos – where the exposure has occurred in the UK. It expands upon the introduction to latent claims and brief reference to asbestos given in Section 6.6. Whilst some of the material provided here is relevant to asbestos reserving in any territory, as noted in Section 6.6.2, some aspects of local insurance markets can have a significant effect on the application of reserving methods for latent claims (e.g. how claims are presented to the relevant insurance policies) and hence the detail provided here relates predominantly to the UK.

There are five main types of medical condition or disease which can be related to asbestos exposure.[7] These are pleural plaques, pleural thickening, asbestosis, asbestos-related lung cancer and mesothelioma. Of these, mesothelioma is the most serious and the majority of sufferers die within two years of diagnosis. It has also had, and continues to have, by far the biggest impact in terms of UK insurance claims, with approximately 90% of the total estimated future costs from 2009 onwards relating to mesothelioma, according to the Institute and Faculty of Actuaries' UK Asbestos Working Party – see UK Asbestos Working Party (2009) for further details. In contrast, pleural plaques is usually asymptomatic and is not compensable in England and Wales. Hence, it does not currently generate associated insurance claims in England and Wales. It is, however, compensable in Scotland and Northern Ireland. For ease of reference, any disease or condition that is asbestos-related is simply referred to as an asbestos-related disease in the remainder of this section, even though strictly speaking some of them (e.g. pleural plaques) might be argued as not being diseases.

Individuals who suffer from asbestos-related diseases have typically been exposed to asbestos for many years. This means that, from an insurance perspective, there is often more than one associated insurance policy that could, at least theoretically, be exposed to claims from any such individual. This is partly why standard reserving methods used for many types of non-latent claims which are applied to data triangles are not usually suitable for latent claims such as asbestos, as a single claim usually affects more than one cohort in a data triangle.

The majority of UK asbestos claims to date have been made under Employers' Liability (EL) policies, with a very small number also being made under Public Liability (PL) policies. Both types of policy indemnify the insured for their liability to relevant third parties arising from injury or disease (including in both cases, death). In the case of asbestos claims under EL policies, the third party will be the employee, who has been awarded compensation against their employer (the insured). For PL, the exposure might arise, for example, where the entity responsible for the exposure was not the claimant's employer, but an occupier of a building (e.g. a local authority) or another party whose work may give rise to the exposure (e.g. a building contractor).

The topic of how an individual UK asbestos claim for each of the five types of asbestos-related disease is shared amongst relevant parties[8] and therefore their EL or PL insurance policies is complex and has been subject to extensive legal argument over many years. It involves consideration first of how these claims are shared amongst employers and other parties, and second how the insurers of these entities respond to such claims. A brief summary of the legal background and other related issues is given in Section 7.7.8. Whilst an understanding of the relevant matters is an important prerequisite for reserving UK

---

[7] These five types of disease are believed to account for virtually all insurance claims, but there are thought to be certain other conditions that can be related to asbestos exposure, such as pleural effusions and laryngeal cancer.

[8] These will be the so-called "tortfeasors" in legal terminology.

asbestos-related claims, for the purposes of this introductory section, it is sufficient to note that any individual asbestos claim can, at least theoretically, be shared amongst more than one insurance policy and hence more than one insurer. This means, for example, that any individual insurer may not be liable for 100% of any one asbestos-related claim.

Insurers that are liable for UK asbestos claims currently expect to continue to pay claims for many further years, not least because deaths from mesothelioma in the UK are expected to continue up to approximately 2050, depending on what projection basis is used. This means that assumptions regarding the future claims notification pattern, average claim costs (including inflation) and, if discounting is used, the discount rate and payment pattern can have a material effect on the level of reserves, as discussed further below

Most reserving for UK asbestos will make at least some use of the considerable volume of research that has been carried out by the Institute and Faculty of Actuaries' UK Asbestos Working Party, which first reported in 2004 as part of the UK Institute and Faculty of Actuaries' annual GIRO conference programme. Several references are made in the remainder of this section to the various papers and presentations produced by the working party, which is abbreviated here by "UKAWP". Before explaining the data and reserving methods that can be used for UK asbestos, it is helpful to summarise some of the key components of the work performed by the UKAWP, where it is directly relevant for reserving at an individual insurer level.

UK Asbestos Working Party

One of the key aspects of the work of the UKAWP is the estimation, at a market level, of the total future cost of UK asbestos claims, from all relevant types of disease. This was first described in their 2004 paper (see UK Asbestos Working Party, 2004) and was subsequently updated in 2009 (see UK Asbestos Working Party, 2009). The latter of these included a post-2009 future claims estimate[9] of £11,316m for all disease types, of which £10,104m (nearly 90%) relates to mesothelioma.

For all asbestos-related disease types, the underlying method used by UKAWP is in effect a Frequency–Severity type approach involving an estimate of the number of insurance claims and an average cost of these claims. Considering mesothelioma first, where the UKAWP's approach is most detailed, the number of claims is estimated by deriving an estimate of the number of future UK mesothelioma deaths, which is combined with an estimate of the proportion of those deaths that will result in an insurance claim, to produce the estimated number of insurance claims. The first of these has been subject to extensive analysis by various entities, including the UK Health and Safety Executive (HSE), certain academics and the UKAWP, with the result that there are numerous models available to the reserving practitioner. Although many of these models provide a good fit to the past data, depending on the model(s) used and the assumptions made, the range of estimated future deaths can be quite wide, but most currently seem to suggest a peak number of mesothelioma deaths of approximately 2,000 per annum occurring between 2016 and 2020, with a tailing off

---

[9] Using what are described in Table 50 of the paper as "Scenario 2" and "Inflation Assumption 2", for each disease type.

which lasts up to around 2050. According to the UKAWP (see UK Asbestos Working Party, 2009) most practitioners probably use a variation of the HSE and UKAWP models, but new models may also need considering from time to time (e.g. see Miranda et al., 2015).

The selected model, along with the assumptions used within it, will produce an estimate of the number of future mesothelioma deaths in each year at the UK population level. This then needs to be converted into an estimate for the number of insurance claims. In the UKAWP's first paper (UK Asbestos Working Party, 2004) reasonable correspondence was observed, at a whole market level, between the trend in notified mesothelioma claims and the trend in mesothelioma deaths between 1972 and 2001. However, this did not continue, since, as highlighted in UK Asbestos Working Party (2007), there was a very sharp increase in the number of notified mesothelioma claims between 2001 and 2007, which was not mirrored in the number of deaths. In UK Asbestos Working Party (2008) it was suggested that the key reason for this was the increase in the number of mesothelioma sufferers that were making insurance claims, rather than an increase in the number of insurers sharing each claim, for example. Subsequently, therefore, in their updated market level estimates given in UK Asbestos Working Party (2009), the UKAWP introduced propensity to claim factors to convert the estimated number of future mesothelioma deaths in each year to an estimated number of future insurance claims. These factors are usually referred to as claimant death ratios (abbreviated here to "CD ratios"). In UK Asbestos Working Party (2009), five different scenarios were selected for future CD ratios by year and age. The UKAWP used these in conjunction with estimates of the number of population mesothelioma deaths to produce a range of alternative estimates for the number of mesothelioma claim notifications in each future year.

The number of claims is combined with the estimated average cost of claims to derive the total cost estimate (before discounting). The cost of any single mesothelioma claim (at a 100% level, before sharing amongst employers and other parties) can be influenced by several factors, which the UKAWP endeavoured to allow for in their average cost model, described in UK Asbestos Working Party (2009). These include:

- The age of the claimant.
- The level of costs under each head of damage. The UKAWP model considered nine different components, including General Damages, Special Damages, Costs of Care, Funeral expenses, Legal expenses etc.
- The level of dependency, if any, between age and each head of damage.
- Whether the claimant is deceased or living at the time of settlement, and if living the selected mortality assumptions.

When deriving average costs for use in the projection of future claim costs for reserving purposes, consideration also needs to be made of how these factors might change in future, allowing for inflation and other claims trends.

The UKAWP model described in UK Asbestos Working Party (2009) allowed for these and other factors and enabled a projection to be made of the average future claim costs by age at settlement and settlement year (assuming a constant mix of living and deceased claimants). These averages are then combined with the estimated number of mesothelioma claims to derive the overall future cost estimate.

For disease types other than mesothelioma, a Frequency–Severity type approach was also used by the UKAWP to derive the market-level insurance cost estimates, with the details varying by disease type, partly driven by the available data. For ease of reference, further information on the approach used by the UKAWP for each disease type is given in the relevant paragraphs in the Reserving Methods section below.

The UKAWP have tended to issue updates to their work periodically, and hence any individual UK asbestos reserving study will need to take into account any such updates that may have been produced after the UKAWP work described here.

7.7.2 *Types of Data*

Insurers with significant exposure to UK asbestos claims will usually maintain detailed individual asbestos claim records. For each claim this data will typically include items such as personal details of the claimant (e.g. name, address, DOB, sex, age at death for deceased claimants), disease type, transactional data relating to claim amounts paid and outstanding and exposure details (industry sector, employers, period of exposure, work location). This data can be analysed for reserving purposes to produce summary information over time, usually by disease type (or at least split between mesothelioma and non-mesothelioma), for key variables such as:

• Number of claims notified by report year.
• Average cost per claim settled and incurred by settlement and report year.
• Number of claims settled at zero cost by settlement year.

  In some cases, the reserving specialist will have access to the detailed claims records, from which the above key variables can be derived, as well as a range of other analyses depending on the context. In other cases, perhaps only the summary information will be available, but this will usually still enable suitable reserving methods to be applied.

Policy data may also be collated, for those policies on which asbestos claims have been made and also in some cases for other policies which are thought to have potential for future asbestos claims to be made against them. For the relevant EL policies, in some cases the underlying exposure data might be collected, with varying levels of granularity ranging from full employee details (dates exposed, age, type of occupation etc.), location and industry market share to summary information showing employee numbers or wage roll over time.

The first UKAWP party paper (UK Asbestos Working Party, 2004) gives further details (in Section 6) of the type of data collated by a number of companies who responded to the working party's original survey.

Where the exposure to UK asbestos is small, higher level data than that described above might be collated. This could include overall summaries of the claims paid and incurred development for UK asbestos for the insurer as whole, as well as perhaps claim numbers where available.

In some situations, where the overall data triangles for the relevant classes of business contain both asbestos and non-latent claims, then, for the purpose of estimating reserves for the non-latent element, it will be necessary to collate paid and incurred data triangles for UK asbestos (and possibly also for other latent claims), so that they can be removed from the data triangles. Assuming that the detailed individual claims records are available for asbestos, these data triangles will, however, not usually be used as part of the reserving process for UK asbestos claims.

The description above relates primarily to insurance exposures. Data collected in relation to claims arising under reinsurance exposures could range from detailed data similar to that described above, for each reinsured company, or could alternatively consist of relatively high-level aggregated data.

7.7.3 Reserving Methods

There are three main types of method used for reserving UK asbestos claims at an individual insurer level:

1. Frequency–Severity (FS) methods.
2. Detailed exposure-based methods.
3. Simple benchmarking methods applied to aggregate data.

According to UK Asbestos Working Party (2004), FS methods were the most common approach used at the time the survey described in the paper was conducted (2004); the author believes they are still in common use today. FS methods will usually, but not always, be applied to data subdivided by type of disease and may also be applied separately for EL and PL coverages (where relevant). The FS approach for EL mesothelioma claims is considered first below, followed by other disease types. Exposure-based and benchmarking methods are then summarised, followed by a brief reference to reserving for PL exposures. All the methods described in this section are assumed to apply on a gross of reinsurance basis. Allowance for outwards reinsurance is considered in Section 7.7.9.

7.7.4 Frequency–Severity Approach for EL Mesothelioma Claims

At a very high level, the approach for an individual insurer can be summarised as follows:

1. Estimate the future number of EL mesothelioma claims notified to the insurer.
2. Estimate the average cost payable by the insurer for each such claim, allowing for inflation and the insurer's estimated share of each claim.
3. Combine the projected number of future notified claims with the estimated future average costs to produce an estimate of the IBNR.

4. Add a suitable reserve for notified but not settled claims (including any allowance needed for IBNER).
5. Estimate the payment pattern for future claim costs, allowing for both currently notified claims and future notified claims.
6. Apply the payment pattern to the total reserve.
7. Select the interest rate assumption to use for discounting for the time value of money, if required.
8. Derive the estimated IBNR and total reserve on both an undiscounted and discounted basis, if required.

Considerable detail is involved in many of these steps, with selected key points being summarised in the paragraphs below.

Step 1: Estimating the Future Number of EL Mesothelioma Claims

This item is perhaps the most critical, but also one of the most uncertain aspects of the overall reserving process for UK asbestos claims. Whilst in theory it might be possible to analyse the historical pattern of an individual insurer's mesothelioma claim numbers and project this forward, the typical approach is to make use of the market-level projections derived by the UKAWP, adjusted where appropriate to allow for the specific exposure profile and claims experience of the individual insurer.

When using the UKAWP analysis to derive reserves for an individual insurer, the practitioner will need to decide on the underlying model and assumptions for population deaths and the claimant death ratio, taking into account appropriate factors based on the insurer's own claims experience and exposure profile. This is a complex area and there are several aspects to consider, including the fact that the projections in UK Asbestos Working Party (2009) explicitly allow for all "future" exposures beyond 2009 and for non-occupational exposures, both of which may need to be adjusted for when used in the context of reserving for a specific insurer. In addition, the UKAWP's population deaths model and CD ratio scenarios apply only to males, and exclude Northern Ireland (NI); hence, if these are used directly, the results may need grossing up to include claims arising from females and NI, depending on the insurer's exposure profile.[10] Periodic updates produced by the UKAWP (e.g. see UK Asbestos Working Party, 2015) provide useful commentary on any observed trends and emerging issues that might also influence these selections.

Once the market-level model and assumptions have been selected, the resulting pattern of future claim numbers can be scaled to produce the estimated number of claimants notifying insurance claims in each future year for the insurer, using a suitable factor, perhaps based on the number of claims notified in the most recent year to the insurer concerned, or averaged over a number of recent years.

The reserving methods survey reported in UK Asbestos Working Party (2004) suggests that some companies use different approaches to that described above for deriving the

---

[10] At a whole market level, in UK Asbestos Working Party (2009), when producing their overall market estimates, the CD ratios were reduced by a factor of 20% to allow for the estimated proportion of claims made against the UK Government, rather than employers. Females and NI exposures were added in by the UKAWP by increasing the results by 5% and 2.3% respectively.

number of mesothelioma claims (at least in 2004 when the survey was conducted). Two such approaches mentioned are exposure-based (either applied to the individual underlying assureds separately, or in aggregate) and using a Generalised Linear Model fitted to historical frequencies. However, it is not believed that these approaches have been used widely in recent years, with most companies tending to use the UKAWP or other market-level projections.

Step 2: Estimating the Average Cost of Mesothelioma Claims

At an individual insurer level, the average cost of mesothelioma claims notified to them will depend on the same market-level factors described in the earlier UK Asbestos Working Party subsection. It will also depend on the number of those claims that settle at nil cost and the insurer's share of the total cost of claims, the age of claimants and the legal jurisdiction (Scotland, Northern Ireland and England & Wales). Analysis of data in relation to all of these factors at both a UKAWP market level and at an insurer level can be used to derive suitable average cost assumptions, for each future year, allowing for inflation and any projected change in age profile of claimants over time. Particular care is needed regarding the inflation assumption, as the estimated reserves can be very sensitive to this, due to the long projection periods involved. One approach is to use insurer-specific data to derive the base level average cost assumptions (as that will implicitly take into account that insurer's share of each claim) and then select a range of future inflation assumptions, perhaps partly based on market-level UKAWP analysis, to derive average costs for future periods.

The analysis might be conducted either for the claim and claim-related expenses in total, or alternatively either partially or fully split by head of damage, with a separate allowance for expenses. An explicit allowance for expenses could involve using a loading to the indemnity cost, or projecting them separately as a monetary amount with a suitable allowance for expense inflation (which may be different to the assumed claims inflation).

When using any market-level 100% average cost data (for example, as produced by the UKAWP), before comparing it with individual insurer averages, an adjustment will need to be made for the fact that EL mesothelioma claims may often be spread across more than one employer and/or insurer (as explained in more detail in Section 7.7.8). The relationship between the market-level 100% average cost value (i.e. the total amount payable to the claimant) and the lower average claim amount payable by an individual insurer will depend on how many other insurers are typically involved in each mesothelioma claim for that insurer. This itself will depend on the nature of that insurer's portfolio, particularly their market share in EL business for the relevant industries over the appropriate time period. Data can be collated by each insurer to calibrate the ratio of the 100% settlement average to the insurer's average. Careful treatment of zero or expenses-only claims is necessary as part of this process, depending on how these have been treated in the claim number projections. For example, if the average costs used are for non-zero claims, then the claim number projections used must similarly relate only to non-zero claims (e.g. by removing the proportion settled at nil cost from the projections derived from claim numbers notified, based for example on analysis of the historical trend in the proportion settled at nil cost).

Step 3: Combine Claim Numbers and Average Costs

Using the results of the previous two stages of the process, the estimated number of claims in each future year is multiplied by the estimated average cost for that year. This is applied by age band where possible. The resulting total will represent the estimated IBNR.

Step 4: Reserve for Notified Claims

To derive the total undiscounted claims reserve, the IBNR from the previous step needs to be added to a suitable reserve for the notified, but not yet settled claims. This can be based on the total case reserves recorded on the insurer's system, perhaps with an adjustment for IBNER, for example based on an analysis of how previous case estimates have compared to the final settled cost. Alternatively, an estimate can be derived using suitable settled number and average cost assumptions. In doing so, appropriate allowance needs to be made for the same points that are relevant for the IBNR claims – namely, the proportion of notified claims that settle at zero cost and the selection of an average cost that allows for the relevant insurer's share of each claim.

Step 5: Estimate the Payment Pattern

To derive the discounted reserves, a payment pattern needs to be estimated. Two possible approaches to this are: (a) deriving a future settlement pattern, based on projecting the past pattern of settled claims in aggregate, or (b) selecting a suitable settlement pattern for notified claims, which can then be combined with the past and future estimated notification numbers and average cost assumptions to derive the settlement pattern.

Steps 6 to 8: Derive the Discounted and Undiscounted Reserves

The final stages of the process involve calculating the undiscounted reserves by adding the reserve for notified claims to the IBNR, and then selecting a discount rate, which is used with the estimated payment pattern to derive the discounted reserves.

In practice, given the considerable uncertainty that exists in relation to the selection of assumptions at each stage of the reserving process, results are often produced using different scenarios or assumptions. For example, the UKAWP produced results using different CD ratio scenarios and inflation assumptions. All of these results were described in UK Asbestos Working Party (2009) as "representing the outcome of reasonable assumption sets that may each be considered central estimates", and which are "not intended to represent optimistic or pessimistic scenarios". This type of approach can also be used for reserving at the insurer level, as can other approaches that might produce results using a wider range of underlying assumptions. The interpretation of any such results will depend critically upon the perceived level of prudence or optimism in the selected scenarios, and this may need careful explanation in any communication of results, if only in qualitative terms. The general points made about communicating uncertainty in Section 5.9 are also relevant here.

7.7.5 Frequency–Severity Approach for other EL Asbestos Diseases

The overall FS approach described here for diseases other than mesothelioma is the same as for mesothelioma, but the volume and granularity of the data available from which the relevant assumptions can be derived, both at a market-level and an individual insurer level is generally lower. Although this creates additional uncertainty in the estimated reserves, the limited materiality of asbestos-related claims for diseases other than mesothelioma makes this less critical for some insurers. Selected key points that impact upon the reserving for each disease type are summarised below. Further details are available, for example, in Section 8 of UK Asbestos Working Party (2009).

Lung Cancer

In UK Asbestos Working Party (2009), the UKAWP derived their market-wide estimates for this disease category using a pragmatic approach based on their mesothelioma model, adjusted to take into account the two key drivers identified by the working party as influencing insurance claims arising from lung cancer – smoking rates and the propensity to claim. Their approach derives a "lung cancer per mesothelioma rate", which is applied to the projection of mesothelioma deaths, at a range of diagnosis ages, to produce an estimate of the future number of lung cancer claims, at a market level. Results were produced using different scenarios for the potential pool of lung cancer claimants compared to mesothelioma claimants and for the propensity to claim. For an individual insurer, these projected claim numbers can be used, scaled to the insurer's own lung cancer claim numbers, together with an examination of the trends in that insurer's lung cancer claim numbers, to derive an estimate of the future number of such claims. The average cost of lung cancer claims can be derived from a combination of market-wide data provided by the UKAWP and the insurer's own experience, allowing for inflation and appropriate sharing amongst insurers and other parties.

Asbestosis

In UK Asbestos Working Party (2004), the UKAWP developed an epidemiological model, referred to by the working party as their "High-Level-Model" or HLM. This was designed to estimate the emergence of asbestos-related diseases other than mesothelioma. The model was used as a basis for the updated projections of asbestosis in UK Asbestos Working Party (2009), which showed three scenarios for the possible future number of claims, two of which continued the apparent decreasing trend in claim numbers between 2003 and 2008. These scenarios can be used for estimating asbestosis claim numbers for an individual insurer, perhaps modified depending on that insurer's own claims experience. In some cases that experience might follow that seen at a market level, where the decreasing trend in claim numbers appears to have reversed in recent years, as shown in UK Asbestos Working Party (2015). Average costs can be estimated in a similar way as for other asbestos-related diseases.

Pleural Thickening

The UKAWP used a judgement-based and less detailed approach than for other asbestos-related diseases to estimate market-level claims costs from this source. Claim numbers were projected by the UKAWP based on three projection scenarios, and then combined with average cost assumptions to derive the overall market-level cost estimates. A similarly less detailed approach is likely to be appropriate for most insurers, assuming that their exposure to this type of claims is relatively small.

Pleural Plaques

As noted earlier in this section, asymptomatic pleural plaques is not compensable in England and Wales, but it is in Scotland and Northern Ireland. Hence, insurers with exposure in these regions will need to consider what reserve is required for this disease category. A pragmatic approach is likely to be appropriate, assuming that claim volumes for the relevant insurer are relatively small.

7.7.6 Other Reserving Methods for EL Asbestos – all Disease Types

Two other reserving methods, mentioned at the start of this section, that can be used instead of, or in addition to FS-type approaches are exposure-based and benchmarking. These are described briefly below. In some cases, it might be appropriate to use different methods for different elements of the portfolio. For example, if an insurer's exposure includes large volumes of mesothelioma claims from a small number of large employers, each with specific exposure characteristics, then one approach would be to use an exposure-based approach for those employers, and an FS-type approach for all other employers combined, separately for each disease type.

Exposure-based Methods

Where it is possible to gather data that describes the underlying exposure of individual insured employers, then this can be used to form the basis of an exposure-based method. In practice, in many situations the relevant exposure periods are often several years prior to the date of the reserving exercise, which means that such data is either unavailable or unreliable. As a result, exposure-based methods are not thought to be used much in practice for UK asbestos reserving, except perhaps for some large individual employers where the relevant data can be obtained. This data can take various forms, but the objective is to gather information that helps establish, for each insured employer, the number and age of their employees who were at risk of exposure to asbestos over time, and the period of such exposure for each employee. Ideally, information relating to the full employment records for all exposed employees would also be gathered, but this may not be available in many cases.

Using this information, the reserving approach is then applied separately for each insured employer. This is usually based on separate estimation of claim numbers and average costs, similar to the aggregate FS-type method described above. The future number of claims is derived by estimating the proportion (by age, and asbestos disease type preferably) of the asbestos-exposed population of insured workers that will generate a claim for that insurer, taking into account the known development of past claim notifications for that employer, as well as wider market-level projections where appropriate. Average costs are then derived based on analysis of historical experience for that employer, perhaps also taking into account the observed average costs across all employers for that insurer, and from UKAWP market-level analysis. Where possible, these averages will take into account any analysis of the exposure data relating to the number of employers and other parties that each claim can be shared amongst.

If only exposure-based methods are being used, then unless all employers with potential asbestos exposure can be identified for the relevant insurer, then an additional reserve may be required to allow for the possible future emergence of asbestos claims from known employers not currently generating asbestos claims and from currently unidentified employers.

Benchmark Methods

These types of method are arguably more accurately described as diagnostic approaches, rather than genuine reserving methods. In the context of UK asbestos reserving, they are typically used either as a broad sense check of the results derived by applying other more sophisticated methods, or where an insurer's exposure to asbestos claims is immaterial and/or where the data available is extremely limited. In some cases, benchmarks are used as a comparative tool to compare the asbestos reserving levels between insurers, although any inferences drawn from such comparisons need to treated with caution due to possible material differences in exposure profile between insurers. Perhaps the most common benchmark used for UK asbestos is the survival ratio; in fact some UK companies publish asbestos survival ratios in their annual report and accounts. This and other benchmarks are described further in the general section on latent claims reserving – Section 6.6.2.

7.7.7 Approach for PL Asbestos – All Disease Types

Any consideration of the reserves required for asbestos claims arising under Public Liability (PL) exposures for a particular insurer will preferably start with gaining an understanding of that insurer's portfolio of potentially exposed PL policies. For most insurers, their exposure will be very limited compared to the potential exposure to asbestos of their Employers' Liability policies. Accordingly, the volume of claims data will also be limited and hence the reserving methodology is necessarily more high-level than for EL, although the overall approach used can be similar. Hence, for FS-type methods, data on the number of claims and average costs can be collated and a suitable estimate made of possible future values in order to derive the reserve estimate.

Some insurers may have more material exposure to PL claims and they will most likely wish to apply more detailed analyses, perhaps at an individual insured level. For mesothelioma claims, such analysis will need to take into account the fact that PL claims are made against those policies that were in force when the malignancy "occurred", rather than when the exposure occurred as for EL.[11] It is also worth noting that the UKAWP market-level estimates for claims arising from asbestos-related diseases only relate to those arising through Employers' Liability policies written by the UK Insurance Market; they do not include any such claims that may be recoverable under Public Liability policies.

A specific point that is made in UK Asbestos Working Party (2008), that may be relevant to reserving for asbestos claims under PL policies, is that the majority of asbestos-related exposure post 1990 will be due to asbestos removal from the workplace or from residential buildings. This could mean that a large proportion of mesothelioma deaths that result from these exposures may be considered "background" or environmental and / or would be more likely to attach to PL insurance policies, rather than EL policies.

7.7.8 Matters Relevant to Reserving – Legal Topics

The matters relevant to reserving U.K asbestos claims is divided into two sections – this first section covers mainly legal topics and the second (Section 7.7.9) covers a number of other relevant general issues associated with the reserving process. The legal topics discussed here are presented only as general information; practitioners who require more detailed or precise legal information related to UK asbestos claims will, for example, need to consult with relevant experts.

Each reserving exercise will almost certainly involve consideration of other additional factors not covered here, related to the specific nature of the relevant insurer's known or potential UK asbestos exposures.

The remainder of this section is divided into five subsections. The first three – Allocation to employers, Allocation to insurance policies and Public Liability claims – relate to mesothelioma, and the next two cover other diseases and concluding remarks (which also refers to other legislative aspects).

Allocation to Employers

Where at least some of an asbestos-exposed person's exposure has been occupational, it will often involve multiple employers (and multiple insurers), so the immediate question arises as to how the "cause" of the disease should be allocated amongst those employers. The problem for the law and, therefore, for the allocation of claims to particular liability policies is that, whilst there is effectively no other known cause for mesothelioma than exposure to asbestos, medical science is unable to identify which particular exposure caused the mesothelioma tumour in any given individual victim. This is known as the "rock of uncertainty", a phrase coined by a judge in one of the leading cases.

---

[11] As explained further in the relevant paragraphs in Section 7.7.8.

Since every individual living in an industrialised society/environment is, at least in theory, exposed to asbestos in everyday life, even where there is only one negligent exposer, as a fact, it is impossible for the claimant to prove that the negligent exposure caused the mesothelioma – it could be exposure in the environment for which no culpable party can be found. To address the injustice which would result, the courts developed a modified test for causation so that if the claimant can show that the negligent exposure made a material contribution to the risk of developing mesothelioma, that was sufficient for the negligent exposer to be found liable – see Fairchild v Glenhaven (2002).

This concept was developed further in Barker v Corus (2006) where the court held that as the gist of the tort was the contribution to risk, that contribution could be divided according to a particular tortfeasor's contribution to that risk, usually on a time-exposed basis.[12] Thus, a part exposer would only be liable for their share of the overall exposure. Within a matter of weeks, the UK Parliament overturned this decision in relation to mesothelioma claims only, with the introduction of section 3 of the Compensation Act 2006.

Thus, despite extensive legal argument, the current position (as defined in section 3 of the Compensation Act 2006) for UK mesothelioma claims is that any of those employers who are responsible for having wrongfully exposed that individual to asbestos are liable for the whole amount of any award for damages, jointly and severally with any other similarly responsible employers. The right of an employer who is deemed liable in this way to seek contributions from other culpable employers is recognised by the Compensation Act 2006.

Allocation to Insurance Policies

The previous subsection deals only with the allocation of mesothelioma claims amongst relevant employers. This subsection covers how the insurance policies purchased by those employers (and other parties in relation to Public Liability claims) respond to those claims. Differing approaches apply according to whether the claim is under an EL policy or PL policy. This is because the triggers for the operation or response of these policies to individual claims are different. Since 1948 most EL policies respond according to when the injury or disease was "caused", whilst a PL policy responds according to when the injury or disease "occurs". In BAI v Durham (2012), also known as The Employers Liability Policy Trigger Litigation or, for shorthand, the "Trigger" litigation, a number of insurers whose EL policies were written on an "injury or disease sustained" basis argued that as the word sustained means "occurred", the approach should be that the policy should respond according to the date of injury, which in a mesothelioma claim meant, at least at that time the date of development of the tumour in the body, taken on the basis of medical research as 10 years plus or minus one year from the date of onset of symptoms. This argument was rejected and regardless of whether the wording is caused or sustained, it was held that the policy or policies in existence at the time the liabilities were generated, i.e. the date of inhalation/exposure, is/are triggered and respond to the claim.

---

[12] In simple terms, in this context the "tort" is the wrongful act of negligent exposure to asbestos and the "tortfeasor" is the entity, such as an employer, who committed that wrongful act.

Thus, the current position (as set out in BAI v Durham [2012]) on this issue is that the causation basis applies, so that the trigger under relevant EL policies for mesothelioma claims in the UK is the date on which the victim was exposed to the asbestos fibres, not the date on which the disease begins to develop or when the victim first shows signs of the disease.

In effect, a victim of mesothelioma can choose to sue any employer that has exposed them. It is said that the claimant can "spike" their claim for damages against any employer that has negligently exposed them to asbestos, and that employer can then similarly "spike" their insurance claim against any of its EL insurers, who can then seek contributions from other relevant parties (e.g. the insured employer and its other EL insurers, and other employers and their EL insurers). This position was clarified in a UK Supreme Court Hearing (Zurich v IEG [2015]) with the overall impact being that, as long as insurers and employers can be located and are not insolvent, each mesothelioma claim will be shared amongst these parties according to their share of the total period of exposure. In practice, this decision created a legal basis for the approach that had been adopted by the industry for many years to share such claims on a time on risk basis.[13]

Where persons with diffuse mesothelioma who were exposed to asbestos either negligently or in breach of statutory duty by their employers are unable to bring a claim for compensation against any employer or associated Employers' Liability insurer, the Diffuse Mesothelioma Payment Scheme, set up by the UK Government, is in place to compensate them.

Public Liability Claims

The basis on which mesothelioma claims are recoverable from Public Liability insurance policies is different to Employers' Liability. This is related to the fact that EL policy wordings typically refer to injury being caused during the policy period, whereas PL policy wordings refer to injury occurring during the policy period. Following Bolton v MMI (2006), the claims are made against those PL policies that were in force when the malignancy "occurred", rather than on the basis of when the exposure to asbestos fibres took place. In Bolton v MMI (2006), the malignancy was taken to have occurred at least 10 years before the claimant displayed symptoms (plus or minus one year). This is known as the "date of injury" and the insurer on risk in the year of the date of injury picks up the claim in full. Since Bolton talked of three possible years (10 years plus or minus one year), in theory the insurers for the policies in force during the relevant period will share the claim on a time on risk basis. However, in practice it is understood that the PL insurer on risk 10 years prior to the symptoms occurring will pick up the claim. Where that is the case for PL claims, time on risk apportionment, as used commonly for EL claims, would not be relevant.

---

[13] However, note that in a claim by an insurer for contribution from an insured which has no insurance for part of the period of negligent exposure, the insurer will remain liable for all of the defence costs under the policy and that element of the claim will not be apportioned.

The ruling in Bolton v MMI (2006) effectively reinforced established market practice and interpretation of the trigger under PL policies.[14]

Other Diseases

In relation to asbestos-related diseases other than mesothelioma, the position regarding how claims are shared amongst employers, other parties and insurers has been subject to less litigation. Where it has been tested in the courts, the legal position does seem to be different to that of mesothelioma, principally because the other diseases tend to be seen as "divisible". This concept can be explained by considering asbestosis, for example, where the disease tends to be regarded as being divisible, that is it is brought about as a result of cumulative exposure to asbestos, causing increasingly severe injury, whereas mesothelioma is regarded as being "indivisible", in that its development is not cumulative, and hence as noted in the Allocation to employers section above, it is not possible to identify the particular period of exposure that caused the disease.

In at least one case in relation to asbestosis (see Holtby v Brigham & Cowan (Hull) [2000]), the position appears to be that the damages payable by an individual employer can take into account the number of years the claimant worked for that employer, so that they are not liable in full for the damages claimed. A similar situation seemed to apply in at least one lung cancer case (see Heneghan v Manchester Dry Docks [2014]) in which due to the particular circumstances that applied there, despite this disease being regarded, like mesothelioma, as an indivisible injury, the court ruled that an apportionment type approach should apply. This may not necessarily apply to other lung cancer cases, where different circumstances might apply, and hence the picture remains uncertain at present.

Concluding Remarks Regarding Legal and Legislative Complexity

It should be evident from the earlier text in this subsection that the issue of allocation of asbestos-related claims amongst relevant parties and their insurers is complex and subject to change. An insurer's claims department will most likely be the best source of up-to-date information on how this legal complexity impacts upon the asbestos-related claims that are payable by that insurer, taking into account the specific nature of their exposure profile. This source can also be used to ensure that any legislative or market-related factors are also taken into account; any developments in relation to the impact of the Employers' Liability Tracing Office (ELTO) on claim volumes and of the Legal Aid, Sentencing and Punishment of Offenders Act ("LASPO") on mesothelioma claims would be examples of possible areas for discussion.

The reserving specialist can therefore benefit from liaison with the relevant personnel in the claims department prior to conducting their reserving analyses, to ensure that their methodology reflects the up-to-date legal and legislative position, where necessary. This can be supplemented, if required, by reviewing summaries of key legal cases and associated or legislative issues, as reported by the UKAWP and other commentators or experts.

---

[14] But note that in the Trigger litigation, Burton J cast some doubt on the 10 year approach, as he suggested 5 years should be used. Hence, at present there remains some uncertainty in the market as to whether 5 or 10 years should apply.

7.7.9 Matters Relevant to Reserving – Other Topics

This section covers a number of other relevant matters (other than legal) – specifically the uncertainty associated with estimating asbestos claim numbers and claim averages, and dealing with inwards and outwards reinsurance.

Uncertainty Associated with Asbestos Claim Numbers

As mentioned earlier in this section, when estimating UK asbestos reserves for a UK insurer, projecting the number of future mesothelioma claims for that insurer is a key uncertainty for the reserving specialist to consider.

One way of approaching this is to use market-level projections produced by the UKAWP, which are then scaled and possibly adjusted further to allow for the insurer's own claims experience and exposure profile. These market-level projections are dependent on several factors, most notably the estimated number of UK population mesothelioma deaths and the propensity for those deaths to give rise to an insurance claim (i.e. the propensity for a mesothelioma sufferer to make a claim for compensation, as measured by the CD ratio referred to earlier). The number of deaths is a measure of the extent of the exposure to mesothelioma in the general population, and the propensity to claim is a measure of what proportion of the affected persons will make a claim for compensation from an insurer.

One way of measuring the propensity to claim, as used in UK Asbestos Working Party (2009), is to analyse the relationship over time between the number of mesothelioma deaths and the number of mesothelioma claimants recorded by the Compensation Recovery Unit (CRU). The CRU is part of the UK Government's Department for Work and Pensions (DWP) and works with insurers, solicitors and the DWP customers to recover (in the context of asbestos-related claims) amounts of social security benefits paid where a compensation payment has also been made. Crucially, the CRU notification requirements mean that they will be informed of all asbestos-related claims giving rise to compensation. Hence, analysis of the CRU data can be helpful in identifying trends in the number of mesothelioma sufferers who are making insurance claims. For example, in their 2015 update (UK Asbestos Working Party, 2015), the UKAWP noted that their initial analysis of the CRU data indicated that the number of CRU mesothelioma claimants had been reasonably stable between 2007 and 2013, whereas in the same period mesothelioma deaths had been increasing. This might imply that the propensity to claim is decreasing. Further analysis by the UKAWP and others in future may either confirm or change this initial view, which may affect the appropriate assumptions to make regarding the propensity to claim.

Reserving specialists who rely on market level analyses for estimating the number of mesothelioma claims for a particular insurer will need to be mindful of developments in the estimation of population mesothelioma deaths and the propensity to claim, as reported by UKAWP and others, so as to ensure their reserve estimates take into account relevant up-to-date information.

One further source of uncertainty that impacts upon estimating claim numbers for both mesothelioma and other asbestos-related diseases includes the fact that the long latency period of these diseases will mean that the claims experience observed up to a particular timepoint will most likely relate to exposures some 30–40 or more years prior to that timepoint, and hence does not enable much inference to be drawn regarding the level of asbestos exposure since those exposure time periods.[15] This creates uncertainty in the estimation of claims arising from exposures in these years, particularly for insurers with significant exposure to employers in the relevant exposure periods. Any changes in working practices in relation to asbestos could also affect the claims experience across different exposure periods. Ongoing monitoring of claims trends will help to identify any need to amend assumptions used in the reserving process.

Other sources of uncertainty are discussed, for example, in UK Asbestos Working Party (2015).

Uncertainty Associated With Average Cost Per Claim

In view of the strong likelihood of new asbestos-related claims continuing to be notified for several years, the impact of inflation on average claim size and hence on reserves can be very material. If reserves are booked on a discounted basis, then the cost of living component of inflation may be partially offset by discounting, but other inflationary drivers such as court award inflation may not. Analysis of trends in average claim cost at an insurer and industry level against a range of possible inflation indices may help select suitable future inflation assumptions, but there will remain significant uncertainty associated with the impact of inflation on future average claim costs.

One other key driver for the future average cost is the assumed age at death, since typically as this increases, the compensation award reduces. Hence, projection of number of claims by age band can help ensure any change in the age profile of claimants is reflected in the average cost assumptions.

Finally, there can be different average mesothelioma costs in different jurisdictions – for example, in Scotland damages in fatal claims are generally higher than in England and Wales, all other things being equal. This adds to the overall uncertainty for an insurer with exposure in more than one jurisdiction.

Reinsurance: Inwards and Outwards

This section briefly considers two related matters – first, the approaches that can be used to estimate the credit for outwards reinsurance and, second, the approaches that can be used for estimating the reserves for UK asbestos affecting portfolios of inwards reinsurance business.

As for all types of business, the approach used for estimating outwards reinsurance, and hence net reserves, will depend to some extent on the nature of the reinsurance programme and the materiality of the reinsurance credit. There are several issues to consider that will affect the reinsurance calculations, including the following:

---

[15] For example, in UK Asbestos Working Party (2015), it is suggested that the past claims experience available for that study enables little to be inferred about the level of exposure from 1970 onwards.

- The nature of the reinsurance – usually either proportional or excess of loss. A ratio-type approach is usually more straightforward to apply for proportional type covers.
- How individual direct asbestos claims, which usually affect more than one year of reinsurance coverage, are treated for reinsurance recovery purposes. Specifically, whether the reinsurance wording allows claims to be "spiked" against a single year of coverage or are the claims spread across the relevant years of exposure. The approach will depend on the wording of the relevant reinsurance contracts, the nature of the underlying insurance contracts and the types of asbestos disease being considered.
- In relation to excess of loss cover, whether the claims are recovered from the reinsurance programme on a per claimant basis, or on some other basis such as per employer, or even for asbestos in aggregate. Again, the approach used will depend on the reinsurance contract wording.
- There may be difficulty in establishing accurate records of reinsurance contracts, due to the long time delays typically involved, and even where the records can be obtained, some of the reinsurers may no longer exist or will have become insolvent.

Further discussion of this topic is given in Section 7 of UK Asbestos Working Party (2008).

In practice, the reserving specialist will need to liaise with the claims and/or outwards reinsurance departments at the relevant company and establish what agreements have been reached with the reinsurers regarding how asbestos claims are recovered from the reinsurance programme. They will then be able to design the approach for estimating the credit for reinsurance accordingly, taking into account the issues referred to above as well as other specific matters that may apply in each case.

With regard to reserving for inwards reinsurance, the approach used will depend on the materiality of the asbestos exposures. For larger accounts, an approach based on analysis at an individual reinsured level may be appropriate. For smaller accounts, higher-level analyses based on extrapolation of historical claims trends may be more appropriate. For analysis done at an individual reinsured level, the approach used will need to consider the same issues referred to above in relation to the credit for outwards reinsurance on direct business. Hence, liaising with the claims team at the reinsurer to determine what approach has been agreed with each reinsured regarding how asbestos reinsurance claims are recoverable from the reinsurance programmes will be an important initial step. The reserving approach for each reinsured can be then be designed to reflect the features of each reinsurance relationship. This will enable a reserve to be estimated for all reinsured companies with known asbestos exposures. In some cases, an additional reserve may also be required if there is thought to be potential for further reinsured companies to advise asbestos claims in future.

Appendix A

Mathematical Details for Mean Squared Error of Prediction

This appendix contains further mathematical and statistical detail in relation to Mean Squared Error of Prediction formulae.

In Section 4.1.3, an approximate formula for the MSEP was given, which breaks it down into components for process and parameter (i.e. estimation) variance:

$$\text{MSEP}(\hat{X}_{i,j}) \approx \underbrace{\text{Var}(X_{i,j})}_{\text{Process variance}} + \underbrace{\text{Var}(\hat{X}_{i,j})}_{\text{Estimation variance}} . \tag{A.1}$$

The conditions under which this formula is valid, and when it becomes an equality is dependent on the assumptions that are inherent within the stochastic claims reserving approach for which it is being used. This Appendix shows how the formula can be developed, according to four different approaches – the first is based on general statistical reasoning and the other three are based on different definitions of stochastic models.

A.1 Derivation of MSEP Formula Based on General Statistical Reasoning

Recall that, as in (4.1), for a single future (and hence unknown) incremental claim, $X_{i,j}$, the MSEP is the expected squared difference between the actual outcome and the predicted value. That is

$$\text{MSEP}(\hat{X}_{i,j}) = \text{E}\left[(X_{i,j} - \hat{X}_{i,j})^2\right]. \tag{A.2}$$

Replacing $(X_{i,j} - \hat{X}_{i,j})$ in (A.2) with

$$(X_{i,j} - \text{E}[X_{i,j}]) - (\hat{X}_{i,j} - \text{E}[X_{i,j}]), \tag{A.3}$$

and expanding, gives

$$\text{MSEP}[\hat{X}_{i,j}] = \text{E}[(X_{i,j} - \text{E}[X_{i,j}])^2] - \text{E}[(\hat{X}_{i,j} - \text{E}[X_{i,j}])^2]$$
$$- 2\text{E}[(X_{i,j} - \text{E}[X_{i,j}])(\hat{X}_{i,j} - \text{E}[X_{i,j}])]. \tag{A.4}$$

Each of the three terms in (A.4) are considered separately. For the first term, note that

$$\text{E}[(X_{i,j} - \text{E}[X_{i,j}])^2] = \text{Var}[X_{i,j}]. \tag{A.5}$$

For the second term, if the stochastic model is constructed such that $\hat{X}_{i,j}$ is an unbiased estimator of $X_{i,j}$, that is if

$$E[\hat{X}_{i,j}] = E[X_{i,j}],$$

then the second term becomes

$$E[(\hat{X}_{i,j} - E[\hat{X}_{i,j}])^2] = \text{Var}[\hat{X}_{i,j}]. \tag{A.6}$$

If the stochastic model is constructed such that future incremental claims are (conditionally) independent of past incremental claims, then the third term will be zero. Combining this with (A.5) and (A.6), allows the approximate relationship in (A.1) to become an equality, that is

$$\text{MSEP}(\hat{X}_{i,j}) = \underbrace{\text{Var}(X_{i,j})}_{\text{Process variance}} + \underbrace{\text{Var}(\hat{X}_{i,j})}_{\text{Estimation variance}} . \tag{A.7}$$

If the model is such that the required conditions are not met, then if the process and estimation variance can be derived, then (A.7) will simply represent an approximation of the "true" MSEP.

A.2 Derivation of MSEP Formula Based on Taylor (2000)

Taylor (2000), Section 6.6 explains how the equality in (A.7) holds for a generic stochastic model, subject to the same conditions. This explanation is helpful in understanding how these conditions manifest themselves in a generic stochastic model, and so is repeated here. Replacing the notation in Taylor (2000) with notation that is consistent with the rest of this book, the explanation begins by defining

$$m_{i,j} = E[X_{i,j}]. \tag{A.8}$$

Assume that the stochastic model for $X_{i,j}$ is formulated, with an error term, $\epsilon_{i,j}$, defined so that $E[\epsilon_{i,j}] = 0$, giving

$$X_{i,j} = m_{i,j} + \epsilon_{i,j}. \tag{A.9}$$

The model is also formulated so that the $\epsilon_{i,j}$ are all stochastically independent, with variance written as

$$\text{Var}(\epsilon_{i,j}) = \sigma_{i,j}^2.$$

Then suppose that $X_{i,j}$ have been observed for the past data triangle – in other words, over the data triangle D_I defined in (2.2), that is

$$D_I = \{C_{i,j} : i + j \le I; 0 \le j \le J\}. \tag{A.10}$$

Then suppose that there exists $\hat{m}_{i,j}$, which is a function only of the past data triangle, D_I, but which is an unbiased estimator of $m_{i,j}$ for all values of i, j regardless of whether they are in the past (i.e in D_I) or the future. The unbiased property will mean that

$$E(\hat{m}_{i,j}) = m_{i,j}. \tag{A.11}$$

Now let

$$\text{Var}(\hat{m}_{i,j}) = \tau_{i,j}^2.$$

For the future component of the dataset, which is being estimated, $\hat{m}_{i,j}$ can be used as a *predictor* of $X_{i,j}$. The *prediction error* for these future elements can then be defined as

$$\zeta_{i,j} = X_{i,j} - \hat{m}_{i,j}.$$

In a similar way as in (A.3), expand this so that

$$\zeta_{i,j} = [X_{i,j} - m_{i,j}] - [\hat{m}_{i,j} - m_{i,j}]$$
$$= \epsilon_{i,j} - [\hat{m}_{i,j} - m_{i,j}]. \tag{A.12}$$

Now, since $E[\epsilon_{i,j}] = 0$ and $E(\hat{m}_{i,j}) = m_{i,j}$, it can be seen that, using (A.12)

$$E(\zeta_{i,j}) = 0,$$

which means that $\hat{m}_{i,j}$ is an unbiased predictor of $X_{i,j}$.

Then, the MSEP is defined as

$$E[(\zeta_{i,j})^2] = E\left[(\epsilon_{i,j} - [\hat{m}_{i,j} - m_{i,j}])^2\right]. \tag{A.13}$$

This can be expanded as follows:

$$E[(\zeta_{i,j})^2] = E\left[(\epsilon_{i,j})^2\right] + E\left[(\hat{m}_{i,j} - m_{i,j})^2\right] - 2E\left[\epsilon_{i,j}(\hat{m}_{i,j} - m_{i,j})\right]$$
$$= \sigma_{i,j}^2 + \tau_{i,j}^2 - 2E\left[(\epsilon_{i,j} - E[\epsilon_{i,j}])(\hat{m}_{i,j} - E[\hat{m}_{i,j}])\right]$$
$$= \sigma_{i,j}^2 + \tau_{i,j}^2 - 2\text{Cov}\left[\epsilon_{i,j}, \hat{m}_{i,j}\right]. \tag{A.14}$$

Now, since the MSEP is being considered for values of i, j in the future part of the triangle, in the third term of (A.14), $\epsilon_{i,j}$ will by definition relate only to these future values, whereas $\hat{m}_{i,j}$ was defined so that it is based only on observations from the past. Hence, $\hat{m}_{i,j}$ will only involve values of $\epsilon_{i,j}$ from the past. Because of the stochastic independence of $\epsilon_{i,j}$, this will therefore mean that the covariance of $\hat{m}_{i,j}$ and $\epsilon_{i,j}$ is zero, and so the third term in (A.14) disappears, which means it can be written as

$$\underbrace{E[(\zeta_{i,j})^2]}_{\text{MSEP}} = \underbrace{\sigma_{i,j}^2}_{\text{Process variance}} + \underbrace{\tau_{i,j}^2}_{\text{Estimation variance}}. \tag{A.15}$$

That is, the same form as (A.7).

A.3 Derivation of MSEP Formula Based on Renshaw (1994b)

Renshaw (1994b) shows how the equality holds for a GLM-type stochastic model and how an approximate version can be derived for certain types of such models by incorporating an approximation for the estimation variance component. In particular, assume that the GLM has the form[1]

$$E[X_{i,j}] = m_{i,j} \text{ and } Var[X_{i,j}] = \phi V(m_{i,j}), \tag{A.16}$$

$$\log m_{i,j} = \eta_{i,j},$$

where $V(.)$ is the variance function and $\eta_{i,j}$ is the so-called *linear predictor*. For a single future incremental value, $X_{i,j}$, the same MSEP formula as (A.12) is used, except that $\epsilon_{i,j}$ is written as $[X_{i,j} - m_{i,j}]$ so that the MSEP becomes

$$E[(\zeta_{i,j})^2] = E\left[(X_{i,j} - m_{i,j})^2\right] + E\left[(\hat{m}_{i,j} - m_{i,j})^2\right]$$
$$- 2E\left[(X_{i,j} - m_{i,j})(\hat{m}_{i,j} - m_{i,j})\right]. \tag{A.17}$$

The first term in this is the process variance and is simply $Var[X_{i,j}]$, which from the model formulation in (A.16) means that

$$E\left[(X_{i,j} - m_{i,j})^2\right] = Var[X_{i,j}] = \phi V(m_{i,j}).$$

For the second term, which is the estimation variance, Renshaw (1994b) uses a Taylor expansion to derive an approximation of $\hat{m}_{i,j}$ to then show that

$$E\left[(\hat{m}_{i,j} - m_{i,j})^2\right] \approx Var(\eta_{i,j})m_{i,j}^2.$$

Further details are available in Renshaw (1994b: 13). $Var(\eta_{i,j})$ is the variance of the linear predictor, which, when this sort of model is fitted using statistical software, is usually available as part of the standard output.

By similar reasoning as in Section A.2 – that is, assuming that future claims are independent of past claims, and $\hat{m}_{i,j}$ is dependent only on past claims – then the third term will be zero. Hence

$$\underbrace{E[(\zeta_{i,j})^2]}_{\text{MSEP}} \approx \underbrace{\phi V(m_{i,j})}_{\text{Process variance}} + \underbrace{Var(\eta_{i,j})m_{i,j}^2}_{\text{Estimation variance}}. \tag{A.18}$$

Further details are available in Renshaw (1994b) regarding corresponding formulae for the MSEP of row totals (i.e. reserves by cohort), diagonals and the total across all future cells (i.e the total reserve).

A.4 Derivation of MSEP Formula Based on Use of Conditional Probabilities

Merz and Wüthrich (2008c), Section 3.1 defines a so-called *conditional* MSEP formula, written as[2]

$$MSEP_{X|D_I} = E\left[(\hat{X} - X)^2 | D_I\right]. \tag{A.19}$$

---

[1] The GLM formulation in Renshaw (1994b) also has an offset term, e_i, which is used for exposure, but the formulae apply equally well without it, as used here.

[2] Merz and Wüthrich (2008c) use $\mathcal{D}$ to denote the set of observations, but D_I is used here to be consistent with the rest of this appendix.

It is logical to condition the MSEP on the known data D_I, since the purpose is to consider how "good" the predictor $\hat{X}$ is, given that known data. The X parameter in this formula represents a generic random variable that is being estimated – it could, for example, be the future incremental claims in a single cell (i.e. $X_{i,j}$ as in the other MSEP formulae in this appendix) or the ultimate claims for a particular cohort (i.e. $C_{i,J}$). Here $\hat{X}$ denotes an estimator for $E[X|D]$ and a D_I-measurable predictor for X. It can similarly relate to a corresponding estimator/predictor of items such as future incremental claims or ultimate claims.

As per Merz and Wüthrich (2008c), if $\hat{X}$ is a D_I-measurable predictor, then (A.19) becomes

$$\text{MSEP}_{X|D_I} = \underbrace{\text{Var}(X|D_I)}_{\text{Process variance}} + \underbrace{\left(\hat{X} - E[X|D_I]\right)^2}_{\text{Estimation variance}}. \tag{A.20}$$

The first term represents the conditional process variance – i.e. the stochastic or random error that will exist regardless of the model being used. The second term, the parameter estimation variance, represents the squared difference between the estimator and the theoretical value of it, $E[X|D_I]$. Since the latter is usually not known (it is estimated by $\hat{X}$) to derive the estimation error it is necessary to use other approaches, such as looking at the possible variations of it around the theoretical value $E[X|D_I]$.

Mack (1993a) defines a very similar formula to (A.20) for the MSEP, expressed in terms of the estimated ultimate claims $\hat{C}_{i,J}$ as follows:

$$\begin{aligned}
\text{MSEP}(\hat{C}_{i,J}) &= E\left[(\hat{C}_{i,J} - C_{i,J})^2|D_I\right] \\
&= \text{Var}(C_{i,J}|D_I) + \left(E[C_{i,J}|D_I] - \hat{C}_{i,J}\right)^2. \tag{A.21}
\end{aligned}$$

At the heart of the Mack model is the approach used to derive these items, as explained in Section 3 of Mack (1993a).

Merz and Wüthrich (2008c) also define the *unconditional* MSEP and explore how this varies according to whether X is independent of D_I or not. The unconditional MSEP is defined by changing (A.20) to

$$\text{MSEP}_X(\hat{X}) = E\left[\text{Var}(X|D_I)\right] + E\left[(\hat{X} - E[X|D_I])^2\right]. \tag{A.22}$$

This formula takes expectations across all possible values for D_I, and so is not conditional with respect to D_I. Then, if X is independent of D_I and additionally $\hat{X}$ is an unbiased estimator of X, so that $E[\hat{X}] = E[X]$, then (A.22) becomes

$$\begin{aligned}
\text{MSEP}_X(\hat{X}) &= \text{Var}(X) + E\left[(\hat{X} - E[X])^2\right] \\
&= \text{Var}(X) + E\left[(\hat{X} - E[\hat{X}])^2\right] \\
&= \underbrace{\text{Var}(X)}_{\text{Process variance}} + \underbrace{\text{Var}(\hat{X})}_{\text{Estimation variance}}. \tag{A.23}
\end{aligned}$$

So, in this case, the parameter estimation variance is estimated by the variance of $\hat{X}$, which can be thought of as the average estimation error. The form of (A.23) is the same as that given at the start of this section of the appendix, i.e. (A.1).

In the second case, when X is not independent of D_I, for the first term of (A.22), $E\left[\text{Var}(X|D_I)\right]$, this can be expanded using a general property of random variables, whereby

$$\text{Var}(X) = \text{Var}(E[X|Y]) + E\left[\text{Var}(X|Y)\right],$$

so that

$$E\left[\text{Var}(X|D_I)\right] = \text{Var}(X) - \text{Var}(E[X|D_I]).$$

Putting this into (A.22) gives

$$\text{MSEP}_X(\hat{X}) = \text{Var}(X) - \text{Var}(E[X|D_I]) + E\left[(\hat{X} - E[X|D_I])^2\right]. \qquad \text{(A.24)}$$

For the second and third terms, if these are written without the outermost expectation operation, for ease of presentation, they become

$$-(E[X|D] - E[E[X|D]])^2 + (\hat{X} - E[X|D])^2.$$

Recognising that $E[E[X|D]] = E[X]$, this be expanded and simplified as

$$-E[X]^2 + 2E[X|D]E[X] + \hat{X}^2 - 2E[X|D]\hat{X},$$

which can be rewritten as

$$(\hat{X} - E[X])^2 - 2\left[(\hat{X} - E[X])(E[X|D] - E[X])\right].$$

Putting the expectation of this back into (A.24), and additionally assuming that $\hat{X}$ is an unbiased estimator of $E[X]$, so that $E[\hat{X}] = E[X]$, (A.24) becomes

$$\text{MSEP}_X(\hat{X}) = \text{Var}(X) + \text{Var}(\hat{X}) - 2\text{Cov}(\hat{X}, E[X|D_I]). \qquad \text{(A.25)}$$

This is the same as the independent case, (A.23) with the additional covariance adjustment which arises due to the lack of independence between X and D_I.

Appendix B

R Code Used for Examples

This appendix provides details of the R code used in the numerical examples. Details of the R software are available at R Core Team (2013). A useful reference source for using R in a range of actuarial applications, including claims reserving, is Charpentier (2014).

This code is included for purely illustrative purposes; it has not been fully tested by the author, and may not be suitable for use in practical reserving exercises. Its use in this book is not intended to endorse the R software or its use in claims reserving.

It should be noted that some changes (e.g. the software library location) may need to be made to the code before it can be used on particular computers.

B.1 How to Load a Data Triangle into R

There are numerous ways to load a data triangle into R, including via the Excel add-in, "RExcel", or by loading to the clipboard and then bringing it into R using read.table function in R. To use in some functions within R, the data must be converted into an object class of the type "triangle". For example, if the data has been loaded into a dataset called, say, "TAdata", then to convert it into a triangle class, use:

```
1 | TAdata<-as.triangle(TAdata)
```

TAdatatotriangle.R

It is necessary to ensure that the data has numerical row and column labels. If it does not, use >fix(TAdata) and then edit manually so that it is in this format, followed by a further use of < −as.triangle to ensure the data is in an object class of "triangle". A similar approach can be used to load any other data triangle into R.

B.2 R Code for Deterministic Methods

B.2.1 Chain Ladder Method Including Curve Fitting

The code shown here demonstrates applying the CL method to the Taylor and Ashe (1983) dataset, including using linear regression to fit the Exponential and Inverse Power curves, and Least Squares for Exponential, to derive a tail factor. It produces the same fitted curves that are shown in the CL method worked example in Section 3.2.8.

```
1  ##  Basic chain ladder example, including linear regression and least squares for
      curve fitting to derive tail factor
2  #Load the Chainladder package
3  library("ChainLadder", lib.loc="C:/Users/David Hindley/Documents/R/win-library/3.0")
4  #View the Taylor and Ashe data that comes with the ChainLadder package
5  GenIns
6  #apply basic chain ladder
7  #delta=1 is col sum avg. If out delta=2 get simple averages
8  chain1<-chainladder(GenIns,weights=1, delta=1)
9  #calc ata just for display purposes
10 chain1_link<-ata(GenIns)
11 #show link ratio estimators - numbers and graph
12 chain1_LR<-coef(chain1)
13 chain1_LR
14 plot(chain1_LR, type="h")
15 #show summary
16 chain1_sum<-summary(chain1_link,digits=3)
17 chain1_sum
18 #show projected triangle
19 chain1_predict<-predict(chain1)
20 chain1_predict
21 #now fit tail factor
22 x<-c(1,2,3,4,5,6,7,8,9)
23 y<-as.data.frame(chain1_LR)$chain1_LR
24 #Exponential curve
25 chain1_model<-lm(log(y-1)~x)
26 chain1_model
27 chain1_fitted<-1+exp(chain1_model$fitted.values)
28 chain1_fitted
29 #plot fitted line
30 lines(chain1_fitted,col="blue")
31 #show fit diagnostics for curve
32 summary(chain1_model)
33 #Inverse Power curve
34 chain2_model<-lm(log(y-1)~log(x))
35 chain2_model
36 chain2_fitted<-1+exp(chain2_model$fitted.values)
37 chain2_fitted
38 lines(chain2_fitted,col="green")
39 summary(chain2_model)
40 #now try least squares fitting method for exponential curve.
41 #starting values for a and b
42 a1=4
43 b1=2
44 nlschain1=nls(y~1+a1/(b1^x), start=list(a1=a1,b1=b1))
45 #summarise fit
46 summary(nlschain1)
47 a2=a1
48 b2=b1
49 weights=c(0,0,0,1,1,1,1,1,1)
50 nlschain2=nls(y~(1+a2/(b2^x)), weights=weights,start=list(a2=a2,b2=b2))
51 summary(nlschain2)
52 fitted.values(nlschain2)
```

CLbasic1.R

B.2.2 Equivalence of Chain Ladder to Linear Regression

The code shown here demonstrates, for selected columns of the Taylor and Ashe (1983) dataset, how linear regression can be used to produce the chain ladder column sum and simple average ratios. This is discussed in Section 3.2.6.

```
1  ##   linear regression example showing equivalence to chain ladder
2  #Load the Chainladder package
3  library("ChainLadder", lib.loc="C:/Users/David
4  Hindley/Documents/R/win-library/3.0")
5  #View the Taylor and Ashe data that comes with the ChainLadder package
6  GenIns
7  #now do linear regression for first two columns.
8  #Use intercept of zero and weight is 1/x as per Section D6 of UK Claims reserving
       manual by Mack
9  x <-GenIns[,1]
10 y <-GenIns[,2]
11 #y~x+0 is a linear regression with intercept zero. weights 1/x gives col sum avg
12 linreg<-lm(y$^~$x+0,weights=1/x)
13 linreg
14 #result should agree with column sum CL of 3.491
15 #Now use weight of 1/x^2, intercept of zero. Then get simple average.
16 linreg2<-lm(y~x+0,weights=1/(x^2))
17 linreg2
18 #above result is 3.566 which is simple average
19 #Now do for columns 4 and 5 which are as used in linear regression graph in relevant
       section
20 x2 <-GenIns[,4]
21 y2 <-GenIns[,5]
22 #y~x+0 is a linear regression with intercept zero. weights 1/x gives col sum avg
23 linreg4<-lm(y2~x2+0,weights=1/x2)
24 linreg4
25 #gives 1.174 as per graph of linear regression shown in CL as linear regression
       section
26 #show plot
27 X<-cbind(x2,y2)
28 plot(X,xlim=c(0,5000000),ylim=c(0,5000000),xlab="Column 4",ylab="Column 5")
29 abline(0,linreg4$coefficients)
```

LinearregressCL.R

B.3 R Code for Stochastic Methods

B.3.1 R Code for Mack's Method

This code produces the results given for the various different alternatives for Mack's method, as given in Section 4.2. The first run of Mack's method ("GenInsMack") will produce the results shown in Table 4.4. The second run ("GenInsMack2") will produce the results shown in Table 4.20, with the tail factor, and with the judgementally selected values for $\widehat{\text{Var}}(\hat{f}_n)$ and $\hat{\sigma}_n^2$. The third run ("GenInsMack3") does the same, except that it allows the MackChainLadder package to fit the tail factor and the values of $\widehat{\text{Var}}(\hat{f}_n)$ and $\hat{\sigma}_n^2$. The fourth run ("GenInsMack4") tests the impact of including the cross-product term, as explained in Section 4.2.8. Finally, the last three runs give examples of using different values of the α parameter, as discussed in Section 4.2.9, which explains the more generalised form of the Mack method given in Mack (1999).

```
1  ##   Mack model example
2  #Load the Chainladder package
3  library("ChainLadder", lib.loc="C:/Users/David Hindley/Documents/R/win-library/3.0")
```

```
 4 #View the Taylor and Ashe data that comes with the ChainLadder package
 5 GenIns
 6 #Apply Mack model,weights=1 just uses same weight for all values in triangle.
 7 #alpha=1 uses actual chain-ladder dev' factors. Other options are possible.
 8 #est.sigma="Mack" means use Mack approximation for last sigma value, as per Mack's
     1999 paper.
 9 #tail=FALSE means no tail factor is used.
10 GenInsMack<-MackChainLadder(GenIns, weights=1, alpha=1,est.sigma="Mack",tail=FALSE)
11 #View results
12 GenInsMack
13 #Now show split of the results between process and estimation (or Parameter error, as
     MackChainLadder refers to it as).
14 GenInsMack$Mack.ProcessRisk
15 GenInsMack$Mack.ParameterRisk
16 GenInsMack$Total.ProcessRisk
17 GenInsMack$Total.ParameterRisk
18 GenInsMack$Total.Mack.S.E
19 #Now show diagnostic charts including residual plots
20 plot(GenInsMack)
21 #Now produce results with a specified tail factor of 1.029 and specified sigma and s.
     e values in tail
22 GenInsMack2<-MackChainLadder(GenIns, weights=1, alpha=1, est.sigma="Mack", tail
     =1.029, tail.sigma=27.5, tail.se=0.01029)
23 #View results
24 GenInsMack2
25 #Now produce results allowing MackChainLadder package to fit tail factor and sigma
     and s.e values in tail
26 GenInsMack3<-MackChainLadder(GenIns, weights=1, alpha=1, est.sigma="Mack", tail=TRUE)
27 #View results
28 GenInsMack3
29 #Now test the impact of including the cross-product term, without a tail factor
30 GenInsMack4<-MackChainLadder(GenIns, weights=1, alpha=1,est.sigma="Mack",tail=FALSE,
     mse.method="Independence")
31 #View results
32 GenInsMack4
33 #Now try with different alpha values.
34 #First try alpha=0 which is equivalent to using a simple average dev factor. Done
     with no tail factor.
35 GenInsMack5<-MackChainLadder(GenIns, weights=1, alpha=0,est.sigma="Mack",tail=FALSE)
36 #View results
37 GenInsMack5
38 #Now with alpha=2 which is equivalent to a regression. Also done with no tail factor.
39 GenInsMack6<-MackChainLadder(GenIns, weights=1, alpha=2,est.sigma="Mack",tail=FALSE)
40 #View results
41 GenInsMack6
42 #Finally with alpha=0 plus tail factor and cross-product term
43 GenInsMack7<-MackChainLadder(GenIns, weights=1, alpha=0,est.sigma="Mack",tail=TRUE,
     mse.method="Independence")
44 #View results
45 GenInsMack7
```

Mackexample.R

B.3.2 R Code for the ODP Model

The code below will produce the results and parameters shown in Section 4.3.8. It also produces the Pearson residuals, which are used in the ODP Bootstrap worked example in Section 4.4.4, and shows residual and Q-Q plots of them, along with a test for Normality, as explained further in Section 4.10.3.

```
1 ##ODP model example
2 #Load the Chainladder package
3 library("ChainLadder", lib.loc="C:/Users/David Hindley/Documents/R/win-library/3.0")
```

```
 4 #View the Taylor and Ashe data that comes with the ChainLadder package
 5 GenIns
 6 #Fit the ODP model to the data. var.power=1 is the ODP; mse.method=formula uses the
      analytical approach with first order Taylor approxmation
 7 #link.power=0 gives the log link function
 8 fit1<-glmReserve(GenIns, var.power=1, link.power=0,cum=TRUE,mse.method=c("formula"))
 9 #extract the fitted model results in glm format from glmReserve
10 fit1_glm<-fit1$model
11 #now view the fitted values
12 fit1_glm$fitted.values
13 #Then view the fitted parameters
14 summary(fit1,type="model")
15 #And now the results
16 fit1$summary
17 #now derive Pearson residuals and view them. Agrees with Pearson residuals table in
      bootstrap numerical example
18 fit1_resid <-residuals(fit1_glm,type=c("pearson"))
19 fit1_resid
20 # plot of Pearsonresiduals
21 plot(fitted.values(fit1_glm),fit1_resid)
22 # qq plot of Pearson residuals
23 qqnorm(fit1_resid)
24 qqline(fit1_resid)
25 #now do shapiro-wilk test for Normality
26 shapiro.test(fit1_resid)
```

ODPexample.R

B.3.3 R Code for the ODP Bootstrap Method

The code below produces the ODP Bootstrap results shown in Table 4.42. It also produces the same residuals, adjusted for degrees of freedom, as shown in Table 4.39 as well as the example results charts given in Section 4.10.4 and an example of the use of maximum likelihood to fit a Lognormal distribution to bootstrap simulation results, as referred to in Section 4.4.5. Finally, the code includes an ODP Bootstrap method using a gamma distribution for the process error, as discussed further in Section 4.4.8.

```
 1 ##Bootstrap model example: includes both ODP and Gamma process errors.
 2 #Load the Chainladder package
 3 library("ChainLadder", lib.loc="C:/Users/David Hindley/Documents/R/win-library/3.0")
 4 #View the Taylor and Ashe data that comes with the ChainLadder package
 5 GenIns
 6 #set seed so can repeat results
 7 set.seed(1328967780)
 8 #Run a chain ladder bootstrap model with 5000 sims and ODP for process error
 9 Boot1<-BootChainLadder(GenIns, 5000,process.dist=c("od.pois"))
10 #View results
11 Boot1summary <-summary(Boot1)
12 Boot1summary
13 #Look at residuals
14 Boot1Resid<-Boot1$ChainLadder.Residuals
15 Boot1Resid
16 #show graphs that summarise results
17 plot(Boot1)
18 #show qunatiles
19 Boot1_quant=quantile(Boot1,probs=c(0.5,0.75,0.9,0.95,0.995))
20 Boot1_quant
21 #now write results to CSV file for importing to Excel
22 write.csv(Boot1summary$ByOrigin,"Boot1_origin.csv")
23 write.csv(Boot1summary$Total,"Boot1_total.csv")
24 write.csv(Boot1_quant,"Boot1_quant.csv")
25 #fit a lognormal distribution to the bootstrap IBNR across all cohorts combined
```

```
26 #first load stats library
27 library(MASS)
28 fit <- fitdistr(Boot1$IBNR.Totals[Boot1$IBNR.Totals>0], "lognormal")
29 #show results of fit
30 fit
31 #now draw actual and fitted cumulative distribution function
32 plot(ecdf(Boot1$IBNR.Totals))
33 curve(plnorm(x,fit$estimate["meanlog"], fit$estimate["sdlog"]), col="red", add=TRUE)
34 #now do with Gamma process distribution
35 set.seed(1328967780)
36 #Run a chain ladder bootstrap model with 5000 sims and now gamma for process error
37 Boot2<-BootChainLadder(GenIns, 5000,process.dist=c("gamma"))
38 #View results
39 Boot2summary <-summary(Boot2)
40 Boot2summary
41 #Look at residuals
42 Boot2Resid<-Boot2$ChainLadder.Residuals
43 Boot2Resid
44 #show graphs that summarise results
45 plot(Boot2)
46 #show qunatiles
47 Boot2_quant=quantile(Boot2,probs=c(0.5,0.75,0.9,0.95,0.995))
48 Boot2_quant
49 #now write results to CSV file for importing to Excel
50 write.csv(Boot2summary$ByOrigin,"Boot2_origin.csv")
51 write.csv(Boot2summary$Total,"Boot2_total.csv")
52 write.csv(Boot2_quant,"Boot2_quant.csv")
```

Bootstrapexample.R

B.3.4 R Code for the Merz–Wüthrich Method

The code below produces both the single year Merz–Wüthrich results shown in Table 4.61 and the multi-year results shown in Table 4.62.

```
1 ##  MW example showing CDR for first year and subsequent years
2 #Load the Chainladder package
3 library("ChainLadder")
4 #View the Taylor and Ashe data that comes with the ChainLadder package
5 GenIns
6 #Apply the Mack chain ladder model to it
7 M <- MackChainLadder(GenIns, est.sigma="Mack")
8 #show results - which agree with Mack worked example
9 M
10 #Then do single CDR results
11 CDR(M)
12 #Results agree with MW example results in multi year worked example.
13 #Now show results for all future years
14 Mall<-CDR(M,dev="all")
15 Mall
16 #Now write results to CSV file for importing to Excel
17 write.csv(Mall,"MWEGCDR_all.csv")
```

MWCDRultexample.R

B.3.5 R Code for Clark LDF and Cape Cod

The code below will produce the results shown in Table 4.64 given in Section 4.7.9.

```
1 ##Clark models example
2 #Load the Chainladder package
3 library("ChainLadder", lib.loc="C:/Users/David Hindley/Documents/R/win-library/3.0")
```

```
 4 #View the Taylor and Ashe data that comes with the ChainLadder package
 5 GenIns
 6 #translate the index so that it is expressed in months
 7 X <- GenIns
 8 colnames(X) <- 12*as.numeric(colnames(X))
 9 #fit a Clark LDF model loglogistic uncapped
10 Clark1<-ClarkLDF(X,G="loglogistic")
11 Clark1
12 plot(Clark1)
13 #fit a Clark LDF model loglogistic capped
14 Clark2<-ClarkLDF(X,G="loglogistic", maxage=240)
15 Clark2
16 plot(Clark2)
17 #fit a Clark LDF model Weibull
18 Clark3<-ClarkLDF(X,G="weibull")
19 Clark3
20 plot(Clark3)
21 #fit a Clark Cape Cod model loglogistic
22 Clark4 <- ClarkCapeCod(X, G="loglogistic",Premium=10000000+400000*0:9, maxage=240)
23 Clark4
24 plot(Clark4)
25 #fit a Clark Cape Cod model weibull
26 Clark5 <- ClarkCapeCod(X, G="weibull",Premium=10000000+400000*0:9)
27 Clark5
28 plot(Clark5)
```

Clarkmodelsexample.R

B.3.6 R Code for Gamma and Other GLM Models

The code below will produce the results shown in Table 4.65 given in Section 4.7.11.

```
 1 ##Gamma model and other glm models example
 2 #Load the Chainladder package
 3 library("ChainLadder", lib.loc="C:/Users/David Hindley/Documents/R/win-library/3.0")
 4 #View the Taylor and Ashe data that comes with the ChainLadder package
 5 GenIns
 6 #Fit the Gamma model to the data. var.power=2 is the Gamma; mse.method=formula uses
      the analytical approach
 7 #with first order Taylor approxmation.link.power=0 is the log-link
 8 fit1<-glmReserve(GenIns, var.power=2, link.power=0, cum=TRUE,mse.method=c("formula"))
 9 #extract the fitted model results in glm format from glmReserve
10 fit1_glm<-fit1$model
11 #now view the fitted values
12 fit1_glm$fitted.values
13 #Then view the fitted parameters
14 summary(fit1,type="model")
15 #And now the results
16 fit1$summary
17 #now derive Pearson residuals and view them
18 fit1_resid <-residuals(fit1_glm,type=c("pearson"))
19 fit1_resid
20 # plot of Pearsonresiduals
21 plot(fitted.values(fit1_glm),fit1_resid)
22 # qq plot of Pearson residuals
23 qqnorm(fit1_resid)
24 qqline(fit1_resid)
25 #write results to csv file for viewing in Excel
26 write.csv(fit1$summary,"Gamma1.csv")
27 ##Now Normal with var.power=0
28 fit2<-glmReserve(GenIns, var.power=0, link.power=0, cum=TRUE,mse.method=c("formula"))
29 #extract the fitted model results in glm format from glmReserve
30 fit2_glm<-fit2$model
31 #now view the fitted values
32 fit2_glm$fitted.values
```

```
33  #Then view the fitted parameters
34  summary(fit2,type="model")
35  #And now the results
36  fit2$summary
37  #write results to csv file for viewing in Excel
38  write.csv(fit2$summary,"Gamma2.csv")
39  ##Now var.power allowed to vary in 1 to 2 interval
40  fit3<-glmReserve(GenIns,var.power=NULL,link.power=0,cum=TRUE,mse.method=c("formula"))
41  #extract the fitted model results in glm format from glmReserve
42  fit3_glm<-fit3$model
43  #now view the fitted values
44  fit3_glm$fitted.values
45  #Then view the fitted parameters
46  summary(fit3,type="model")
47  #And now the results
48  fit3$summary
49  #write results to csv file for viewing in Excel
50  #Note: results not shown in Table in text as they are the same as ODP
51  write.csv(fit3$summary,"Gamma3.csv")
52  #Now Inverse Gaussian with var.power=3
53  fit4<-glmReserve(GenIns, var.power=3, link.power=0, cum=TRUE,mse.method=c("formula"))
54  #extract the fitted model results in glm format from glmReserve
55  fit4_glm<-fit4$model
56  #now view the fitted values
57  fit4_glm$fitted.values
58  #Then view the fitted parameters
59  summary(fit4,type="model")
60  #And now the results
61  fit4$summary
62  #write results to csv file for viewing in Excel
63  write.csv(fit4$summary,"Gamma4.csv")
```

Gammaexample.R

References

Ackman, R.C., Green, P.A.G., and Young, A.G. 1982. *Estimating Claims Outstanding*. General Insurance Monograph. Institute of Actuaries.

Actuarial Standards Board. 2011. *Actuarial Standard of Practice No. 20: Discounting of Property/Casualty Unpaid Claim Estimates*. Available at: www.actuarialstandardsboard.org/wp-content/uploads/2014/02/asop020_163.pdf.

AISAM-ACME. 2007. AISAM-ACME study on non-life long tail liabilities. *Reserve Risk and Risk Margin Assessment Under Solvency II*. 17 October 2007.

Alai, D.H., Merz, M., and Wüthrich, M.V. 2009. Mean-square error of prediction in the Bornhuetter–Ferguson claims reserving method. *Annals of Actuarial Science*, 4(1), 7–31.

Alai, D.H., Merz, M., and Wüthrich, M.V. 2011. Mean-square error of prediction in the Bornhuetter–Ferguson claims reserving method: revisited. *Annals of Actuarial Science*, 5(01), 7–17.

A.M. Best. 2015. *Best's Statement File: United Kingdom*. Available at: www.ambest.com/sales/statementuk/.

A.M. Best. 2016a. *Best's Aggregates and Averages: Property/Casualty, United States and Canada*. Oldwick, NJ: A.M. Best.

A.M. Best. 2016b. *Best's Special Report: Asbestos and Environmental Liabilities*. November 2016.

Antonio, K., and Beirlant, J. 2007. Actuarial statistics with generalized linear mixed models. *Insurance: Mathematics and Economics*, 40(1), 58–76.

Antonio, K., and Beirlant, J. 2008. Issues in claims reserving and credibility: a semiparametric approach with mixed models. *Journal of Risk and Insurance*, 75(3), 643–676.

Antonio, K., and Plat, R. 2014. Micro-level stochastic loss reserving for general insurance. *Scandinavian Actuarial Journal*, 2014(7), 649–669.

Ashe, F. 1986. An essay at measuring the variance of estimates of outstanding claim payments. *ASTIN Bulletin*, 16(1), 99–112.

Association of British Insurers. 2015. Noise Induced Hearing Loss Claims. June 2015. Available at: www.abi.org.uk/News/News-releases/2015/06/Industrial-Deafness-claims.

ASTIN Committee. 2016. *2016 ASTIN Working Party Report on Non Life Reserving Survey*. Available at: www.actuaries.org/index.cfm?DSP=ASTIN&ACT=INDEX &LANG=EN.

BAI v Durham. 2012. BAI (Run Off) Limited (In Scheme of Arrangement) (Appellant) v Durham (Respondent) [2012] UKSC 14.

Bain, D. 2003. A practical implementation of Wright's operational time model. Society of Actuaries in Ireland CPD events, 6 November 2003.

Bardis, E.T., Majidi, A., and Murphy, D.M. 2012. A family of chain-ladder factor models for selected link ratios. *Variance*, 6(2), 143–160.

Barker v Corus. 2006. Barker v Corus (UK) plc [2006] UKHL 20.

Barnett, G., and Zehnwirth, B. (n.d.) *The Need for Diagnostic Assessment of Bootstrap Predictive Models*. InsureWare Pty research paper, Available at: www.insureware.com/Library/Technical/AssessBootstrap.pdf.

Barnett, G., and Zehnwirth, B. 2000. Best estimates for reserves. *Proceedings of the Casualty Actuarial Society*, LXXXVII(167).

Beens, F., Bui, L., Collings, S., and Gill, A. 2010. Stochastic reserving using Bayesian models – can it add value? Presented to the Institute of Actuaries of Australia 17th General Insurance Seminar, 7–10 November 2010.

Benedikt, V. 1969. Estimating incurred claims. *ASTIN Bulletin*, 5(2), 210–212.

Benjamin, S., and Eagles, L. 1997. The curve fitting method. *Claims Reserving Manual*, Vol. 2. London: Institute of Actuaries. Section D3.a.

Benktander, G. 1976. An approach to credibility in calculating IBNR for casualty excess reinsurance. *The Actuarial Review*, 3(2), 7.

Berquist, J.R., and Sherman, R.E. 1977. Loss reserve adequacy testing: a comprehensive, systematic approach. *Proceedings of the Casualty Actuarial Society*, 74, 123–184.

Björkwall, S. 2011. Stochastic claims reserving in non-life insurance: bootstrap and smoothing models. Phd thesis, Department of Mathematics, Stockholm University.

Björkwall, S., Hössjer, O., and Ohlsson, E. 2009. Non-parametric and parametric bootstrap techniques for age-to-age development factor methods in stochastic claims reserving. *Scandinavian Actuarial Journal*, 2009(4), 306–331.

Björkwall, S., Hössjer, O., and Ohlsson, E. 2010. Bootstrapping the separation method in claims reserving. *ASTIN Bulletin*, 40(2), 845–869.

Björkwall, S., Hössjer, O., Ohlsson, E., and Verrall, R.J. 2011. A generalized linear model with smoothing effects for claims reserving. *Insurance: Mathematics and Economics*, 49(1), 27–37.

Bolton v MMI. 2006. Bolton Metropolitan Borough Council v Municipal Mutual Insurance Ltd [2006] EWCA Civ 50.

Boor, J. 2006. Estimating tail development factors: what to do when the triangle runs out. *Casualty Actuarial Society Forum*, Winter.

Bornhuetter, R.L., and Ferguson, R.E. 1972. The actuary and IBNR. *Proceedings of the Casualty Actuarial Society*, 59, 181–195.

Boumezoued, A., Angoua, Y., Devineau, L., and Boisseau, J-P. 2011. One-year reserve risk including a tail factor: closed formula and bootstrap approaches. *Laboratoire De Sciences Actuarielle et Financière*, 2011.7 WP(2144).

Bouska, A. S., and McIntyre, T.S. 1994. Measurement of U.S. pollution liabilities. *Casualty Actuarial Society Forum*, Summer.

Braun, C. 2004. The prediction error of the chain ladder method applied to correlated run-off triangles. *ASTIN Bulletin*, 34(2), 399–423.

Brown, D., Treen, R., and members of the Third Party Working Party. 2015. *Update from the Third Party Working Party – Full Report*. 29 January 2015. Available at: www.actuaries.org.uk/research-and-resources/documents/update-third-party-working-party-full-report.

Bruce, N., and ROC Working Party. 2013. Towards the optimal reserving process: actual vs expected techniques. *GIRO Conference and Exhibition 2013*.

Bruce, N., and ROC Working Party. 2015. Towards the optimal reserving process: the fast close process. *Presented to the Institute and Faculty of Actuaries, Edinburgh 19 January 2015*.

Brydon, D., and Verrall, R. J. 2009. Calendar year effects, claims inflation and the chain-ladder technique. *Annals of Actuarial Science*, 4(9), 287–301.

Buchwalder, M., Bühlmann, H., Merz, M., and Wüthrich, M. V. 2005. Legal valuation in non-life insurance. Conference Paper. *36th ASTIN Colloquium*.

Buchwalder, M., Bühlmann, H., Merz, M., and Wüthrich, M. V. 2006. The mean square error of prediction in the chain ladder reserving method (Mack and Murphy revisited). *ASTIN Bulletin*, 36(2), 521–542.

Budden, M., Hadavas, P., and Hoffman, L. 2008. On the generation of correlation matrices. *Applied Mathematics E-Notes*, 279–282.

Bühlmann, H. 1983. Estimation of IBNR reserves by the methods chain ladder, Cape Cod, and complimentary loss ratio. Unpublished manuscript.

Bühlmann, H., De Felice, M., Gisler, A., Moriconi, F., and Wüthrich, M. V. 2009. Recursive credibility formula for chain ladder factors and the claims development result. *ASTIN Bulletin*, 39(1), 275–306.

Bulmer, R., and Working Party. 2005. *Reinsurance Bad Debt Provisions for General Insurance Companies: With Supplementary Note*. October 2005. Available at: www.actuaries.org.uk/ documents/reinsurance–bad–debt-provisions–general-insurance–companies–supplementary–note–october.

Canadian Institute of Actuaries. 2014. *Educational Note: Premium Liabilities*. Committee on Property and Casualty Insurance Financial Reporting. Document 214114.

Carrato, A., McGuire, G., and Scarth, R. 2015. A practitioner's introduction to stochastic reserving. *GIRO Conference and Exhibition 2015*.

CEIOPS. 2010. CEIOPS' advice for level 2 implementing measures on Solvency II. *SCR Standard Formula – Article 111 j,k Undertaking Specific Parameters*. CEIOPS DOC 71/10.

Chan, K., and Ramyar, M. 2014. How to net down and aggregate gross distribution. *GIRO Conference and Exhibition 2014*.

Charpentier, A. 2012. *R Code for Merz-Wüthrich Method*. Available at: http://perso.univ-rennes1.fr/ arthur.charpentier/merz-wuthrich-triangle.R.

Charpentier, A. (ed.). 2014. *Computational Actuarial Science with R*. Boca Raton, FL: CRC Press.

Chrin, G., and Fell, B. 2010. Are you properly calculating your ceded reinsurance loss reserves. *Casualty Loss Reserve Seminar*.

Christofides, S. 1990. Regression models based on log-incremental payments. *Claims Reserving Manual*, Vol. 2. London: Institute of Actuaries.

Claims Portal. 2015. Executive dashboard. Available at: www.claimsportal.org.uk/en/about/ executive-dashboard/.

Clark, D.R. 2003. LDF curve fitting and stochastic reserving: a maximum likelihood approach. *Casualty Actuarial Society Forum*, Fall, 41–91.

Claughton, A. 2013. Australian reserving methods. *GIRO Conference and Exhibition 2013*.

Conort, X. 2011. Predictive modelling in insurance. SAS Talk,16 June 2011. Available at: www.actuaries.org.sg.

Cowling, C.A., Frankland, R., Hails, R.T.G., Kemp, M.H.D., Loseby, R.L., Orr, J.B., and Smith, A.D. 2011. Developing a framework for the use of discount rates in actuarial work: a discussion paper. Presented to the Institute and Faculty of Actuaries, London, 31 January 2011.

Craighead, D.H. 1979. Some aspects of the London reinsurance market in worldwide short-term business. *Journal of the Institute of Actuaries*, 106(III), 286.

Cross, S. L., and Doucette, J.P. 1994. Measurement of asbestos bodily injury liabilities. *Casualty Actuarial Society Forum*, Summer.

Czapiewski, C., Archer-Lock, P.R., Clark, P., Cresswell, C., Hindley, D., and Shepley, S. 1993. Reserving for outwards reinsurance. *GIRO Conference and Exhibition 1993*.

Dahms, R. 2007. A loss reserving method for incomplete claim data. *Bulletin of the Swiss Association of Actuaries*, 127–148.

Dal Moro, E., and Lo, J. 2014. An Industry question: the ultimate and one-year reserving uncertainty for different non-life reserving methodologies. *ASTIN Bulletin*, 44(3), 495–499.

Davis, H.T. 1941. *Analysis of Economic Time Series*. Bloomington Ind.

Davison, A. C., and Hinkley, D. V. 1997. *Bootstrap Methods and Their Applications*. Cambridge: Cambridge University Press.

De Felice, M., and Moriconi, F. 2006. Process error and estimation error of year-end reserve estimation in the distribution free chain-ladder model. Alef Working Paper Rome November 2006.

Department of Transport. 2015. Road safety data. Available at: data.gov.uk/dataset/road-accidents-safety-data.

Diers, D., and Linde, M. 2013. The multi-year non-life insurance risk in the additive loss reserving model. *Insurance: Mathematics and Economics*, 52(3), 590–598.

Dobson, A.J. 2002. *An Introduction to Generalized Linear Models*. 2nd edn. London: Chapman & Hall.

Dreksler, S., and ROC Working Party. 2013. Solvency II technical provisions for general insurers. Presented to the Institute and Faculty of Actuaries. 23 November 2013 (London).

Efron, B. 1979. Bootstrap methods: another look at the jacknife. *Annals of Statistics*, 7(1), 1–26.

Efron, B., and Tibshirani, R. J. 1993. *An Introduction to the Bootstrap*. Monographs on Statistics and Applied Probability 57. London: Chapman & Hall.

EIOPA. 2011. Calibration of the premium and reserve risk factors in the standard formula of Solvency II. *Report of the Joint Working Group on Non-Life Health NSLT Calibration*.

EIOPA. 2014a. Guidelines on contract boundaries. EIOPA-BoS-14/165.

EIOPA. 2014b. Guidelines on the valuation of technical provisions. EIOPA-BoS-14/166.

EIOPA. 2014c. Guidelines on the valuation of technical provisions. EIOPA-BoS-14/166 EN.

EIOPA. 2014d. Technical specification for the preparatory phase (part I). EIOPA-14/209 30 April 2014.

Embrechts, P., Lindskog, F., and McNeil, A. 2001. Modelling dependence with copulas and applications to risk management. Working paper from ETHZ. Available at: www.math.ethz.ch/embrecht/ftp/-copchapter.pdf.

England, P. D. 2001. Addendum to "Analytic and bootstrap estimates of prediction errors in claims reserving". *Actuarial Research Paper No. 138*, Department of Actuarial Science and Statistics, City University, London.

England, P.D. 2010. Stochastic made "simple". Presented to the General Insurance Reserving Seminar, Institute of Actuaries, 28 June 2010.

England, P.D., and Cairns, M. 2009. Are the upper tails of predictive distributions of outstanding liabilities underestimated when using bootstrapping. *GIRO Conference and Exhibition 2009*.

England, P.D., and Verrall, R.J. 1999. Analytic and bootstrap estimates of prediction errors in claims reserving. *Insurance: Mathematics and Economics*, 25, 281–293.

England, P.D., and Verrall, R.J. 2001. A flexible framework for stochastic claims reserving. *Proceedings of the Casualty Actuarial Society*, 88, 1–38.

England, P.D., and Verrall, R.J. 2002. Stochastic claims reserving in general insurance (with discussion). *British Actuarial Journal*, 8, 443–544.

England, P.D., and Verrall, R.J. 2006. Predictive distributions of outstanding liabilities in general insurance. *Annals of Actuarial Science*, 1, 221–270.

England, P.D., Cairns, M., and Scarth, R. 2012a. The 1-year view of reserving risk: The "actuary in the box" vs emergence patterns. *GIRO Conference and Exhibition 2012*.

England, P.D., Verrall, R.J., and Wüthrich, V. 2012b. Bayesian over-dispersed Poisson model and the Bornhuetter & Ferguson claims reserving method. *Annals of Actuarial Science*, 6(9), 258–283.

European Commission. 2010. QIS5 Technical Specifications. Annex to Call for Advice from CEIOPS on QIS5.

European Parliament. 2009. Directive 2009/138/EC of the European Parliament and of the Council of 25 November 2009 on the taking-up and pursuit of the business of Insurance and Reinsurance (Solvency II) (recast). *Official Journal of the European Union*, 52.

European Parliament. 2015. Commission Delegated Regulation (EU) 2015/35 of 10 October 2014 of the European Parliament and of the Council on the taking-up and pursuit of the business of Insurance and Reinsurance (Solvency II). *Official Journal of the European Union*, 58.

Fairchild v Glenhaven. 2002. Fairchild v Glenhaven Funeral Services Ltd [2002] UKHL 22.

Fisher, W.H., and Lange, J.T. 1973. Loss reserve testing: a report year approach. *Proceedings of the Casualty Actuarial Society*, 60, 189–207.

Forray, S.J. 2011. The Munich chain ladder: overview and example. *Mid-western Actuarial Forum*, Chicago, 28 March 2011.

Francis, L. 2010. An introduction to the Munich chain ladder. *Casualty Actuarial Society Spring Forum*, May 2010.

Frees, E.W., and Valdez, A.E. 1998. Understanding relationships using copulas. *North American Actuarial Journal*, 2, 1–25.

Friedland, J. 2010. *Estimating Unpaid Claims Using Basic Techniques*. Arlington, VA: Casualty Actuarial Society.

Friedland, J. 2013. *Fundamentals of General Insurance Actuarial Analysis*. Schaumburg, IL: Society of Actuaries.

Fulcher, G., and Edwards, M. 2013. Risk: behaviour under control. *The Actuary*. Available at: www.theactuary.com/features/2013/10/risk-behaviour-under-control/.

Gesmann, M. 2013a. Barnett and Zehnwirth PTF using R. Available from line 184 onwards at gist.github.com/mages/4590362.

Gesmann, M. 2013b. Reserving based on log-incremental payments in R. Available at: lamages.blogspot.co.uk/2013/01/reserving-based-on-log-incremental.html.

Gesmann, M., Murphy, D., and Zhang, W. 2014. *ChainLadder: Statistical Methods for the Calculation of Outstanding Claims Reserves in General Insurance*. R package version 0.1.8.

Gibson, E.R., Barlow, C., Bruce, N.A., Felisky, K., Fisher, S., Hilary, N.G.J., Hilder, I.M., Kam, H., Matthews, P.N., and Winter, R. 2009. Reserving and uncertainty: a meta-study of the general insurance reserving issues task force and reserving oversight committee research into this areas between 2004 and 2009. Presented to the Institute and Faculty of Actuaries, 2 November 2009.

Gisler, A., and Wüthrich, M. V. 2008. Credibility for the chain ladder reserving method. *ASTIN Bulletin*, 38(2), 565–600.

Gluck, S. M. 1997. Balancing development and trend in loss reserve analysis. *Proceedings of the Casualty Actuarial Society*, 84, 482–532.

Gravelsons, B., and Brankin, G. 2015. What next for noise induced hearing loss claims? *GIRO Conference and Exhibition 2015*.

Hachemeister, C.A., and Stanard, J.N. 1975. IBNR claims count estimation with static lag functions. *ASTIN Colloquium*, Portimäo, Portugal.

Happ, S., and Wüthrich, M.V. 2013. Paid-incurred chain reserving method with dependence modelling. *ASTIN*, 43(1), 1–20.

Happ, S., Merz, M., and Wüthrich, M.V. 2012. Claims development result in the paid-incurred chain reserving method. *Insurance Mathematics and Economics*, 51(1), 66–72.

Harnek, R.F. 1966. Formula loss reserves. *Insurance Accounting and Statistical Procedures*.

Hart, D.G., Buchanan, R.A., Howe, B.A., and Institute of Actuaries of Australia. 2007. *The Actuarial Practice of General Insurance*. Sydney: Institute of Actuaries of Australia.

Hayne, R., Leise, J., and Working Party. 2005. *The Analysis and Estimation of Loss & ALAE Variability: A Summary Report*. CAS Working Party on Quantifying Variability in Reserve Estimates.

Heneghan v Manchester Dry Docks. 2014. Heneghan (Deceased) v Manchester Dry Docks & Others [2014] EWHC 4190.

Herman, S.C., Shapland, M. R., and CAS Tail Factor Working Party. 2013. The Estimation of Loss Development Tail Factors: A Summary Report. *Casualty Actuarial Society E-Forum*, Fall.

Hiabu, M., Margraf, C., Martínez-Miranda, M.D., and Nielsen, J.P. 2016. The link between classical reserving and granular reserving through double chain ladder and its extensions. *British Actuarial Journal*, 21(1), 97–116.

Hilder, I.M., Paillot, C., Rothwell, M., Sheaf, S., and Toller, J. 2008. The implications of the underwriting and reserving cycles for reserving. *GIRO Conference and Exhibition 2008*.

Hindley, D.J., Allen, M., Czernuszewicz, A.J., Ibeson, D.C.B., McConnell, W.D., and Ross, J.G. 2000. The Lloyd's reinsurance to close process. *British Actuarial Journal*, 6(12), 651–720.

Holtby v Brigham & Cowan (Hull). 2000. Holtby v Brigham & Cowan (Hull) Limited [2000] 3 All ER 421.

Hovinen, E. 1981. Additive and continuous IBNR. *ASTIN Colloquium*, Loen, Norway.

Hürlimann, W. 2009. Credible loss ratio reserves: the Benktander, Neuhaus and Mack methods revisited. *ASTIN Bulletin*, 39(1), 81–99.

IAA. 2010. *Stochastic Modeling: Theory and Reality from an Actuarial Perspective*. Ottawa: Association Actuarielle Internationale/International Actuarial Association.

IFoA PPO Working Party. 2016. A helpful handbook. Draft. June 2016. Available at: www.actuaries. org.uk/documents/information-actuaries-valuing-periodical-payment-orders.

Iman, R.L., and Conover, W.J. 1982. A distribution-free approach to inducing rank correlation among input variables. *Communications in Statistics*, B11, 311–334.

Institute and Faculty of Actuaries – General Insurance Practice Executive Committee. 2010. *Asbestos Briefing Note*. Available at: www.actuaries.org.uk/documents/asbestos-briefing-note.

ISVAP. 2006. Reserve requirements and capital requirements in non-life insurance: an analysis of the Italian MTPL insurance market by stochastic claims reserving models. Report prepared by De Felice M., Moriconi F., Matarazzo L., Cavastracci S. and Pasqualini S., Rome, October 2006.

Jarvis, S., Sharpe, J., and Smith, A.D. 2016. Ersatz Model Tests (June 1, 2016). Available at: http://ssrn.com/abstract=2788478 or dx.doi.org/10.2139/ssrn.2788478.

Johnson, W. 1989. Determination of outstanding liabilities for unallocated loss adjustment expenses. *Proceedings of the Casualty Actuarial Society*, 77, 111–125.

Jones, A.R., Copeman, P.J., Gibson, E.R., Line, N.J.S., Lowe, J.A., Martin, P., Matthews, P. N., and Powell, D.S. 2006. A change agenda for reserving. Report of the General Insurance Reserving Issues Taskforce (GRIT). *British Actuarial Journal*, 12(03), 435–599.

Jones, B.A., Scukas, C.J., Frerman, K.S., Holt, M.S., and Fendley, V.A. 2013. A mortality-based approach to reserving for lifetime workers' compensation claims. *Casualty Actuarial Society Forum*, Fall.

Jones, T., and ROC Working Party. 2007. Quantification and reporting of uncertainty for GI reserving. Paper produced by the Reserving Oversight Committee of the UK Actuarial Profession.

Joubert, P., and Langdell, S. 2013. Modelling: mastering the correlation matrix. *The Actuary*. Available at: www.theactuary.com/features/2013/09/modelling-mastering-the-correlation-matrix/.

Kaas, R., Goovaerts, M., Dhaene, J., and Denuit, M. 2009. *Modern Actuarial Risk Theory Using R*. New York: Springer.

Kirschner, G. S., Kerley, C., and Isaacs, B. 2002. Two approaches to calculating correlated reserve indications across multiple lines of business. *Casualty Actuarial Society Forum*, Fall, 211–245.

Kirschner, G. S., Kerley, C., and Isaacs, B. 2008. Two approaches to calculating correlated reserve indications across multiple lines of business. *Variance*, 2(1), 15–38.

Kittel, J. 1981. *Unallocated Loss Adjustment Expenses in an Inflationary Environment*. Available at: www.casact.org/pubs/dpp/dpp81/81dpp311.pdf.

Kremer, E. 1982. IBNR-claims and the two-way model of ANOVA. *Scandinavian Actuarial Journal*, 1982(1), 47–55.

Kuang, D., Nielsen, B. and Nielsen, J.P. 2008a. Forecasting with the age-period-cohort model and the extended chain-ladder model. *Biometrika*, 95(4), 987–991.

Kuang, D., Nielsen, B. and Nielsen, J.P. 2008b. Identification of the age-period-cohort model and the extended chain-ladder model. *Biometrika*, 95(4), 979–986.

Kuang, D., Nielsen, B. and Nielsen, J.P. 2011. Forecasting in an extended chainladder-type model. *Journal of Risk and Insurance*, 78(2), 345–359.

Latent Claims Working Party. 1990. Report of the Latent Claims Working Party presented to the GISG Convention at Newquay, October 1990. *GIRO Conference and Exhibition 1990*.

Latent Claims Working Party. 1991. Report of the Latent Claims Working Party presented to the GISG Convention at Llandrindod Wells, October 1991. *GIRO Conference and Exhibition 1991*.

Lee, A. 2010. Practical implementation of granular reserving. *GIRO Conference and Exhibition 2010*.

Leong, J., Wang, S., and Chen, H. 2012. Back-testing the ODP bootstrap of the paid chain-ladder model with actual historical claims data. *Casualty Actuarial Society E-Forum*, Summer.

Line, N., Archer-Lock, P.R., Fisher, S., Hilder, I., Shah, S., Wenzel, K., and White, M. 2003. The cycle survival kit: an investigation into the reserving cycle and other issues. *GIRO Conference and Exhibition 2003*.

Liu, H., and Verrall, R.J. 2009a. A bootstrap estimate of the predictive distribution of outstanding claims for the Schnieper model. *ASTIN Bulletin*, 39(2), 677–689.

Liu, H., and Verrall, R.J. 2009b. Predictive distributions for reserves which separate true IBNR and IBNER claims. *ASTIN Bulletin*, 39(1), 214–225.

Liu, H., and Verrall, R.J. 2010. Bootstrap estimation of the predictive distribution of reserves using paid and incurred claims. *Variance*, 4(2), 121–135.

Lloyd's Market Association. 2015. *Loss Ratio Triangulations*. Available at: www.lmalloyds.com/lma/web/Market data and statistics/Loss Ratio Triangulations/Latest Reports.aspx.

Lloyd's of London. 1908. Instructions for the guidance of Auditors. *Addendum to Minutes of the Committee of Lloyd's of London*. 23 December 1908, London Metropolitan Archives, City of London CLC/B/148/A/001/MS31571/068 from Lloyd's of London Collection.

Lloyd's of London. 2012. *Lloyd's Signing Actuaries Forum. 5 December 2012*. Available at: www.lloyds.com/~/media/files/the%20market/operating%20at%20lloyds/valuation%20of%20liabilities/archive/lloyds%20signing%20actuaries%20forum%202012.pdf.

Lloyd's of London. 2014. *Solvency II Model Validation Guidance. April 2014*. Available at: www.lloyds.com/the-market/operating-at-lloyds/solvency-ii/information-for-managing-agents/guidance-and-workshops/model-validation.

Lloyd's of London. 2015a. *Lloyd's Finance Directors Reserving Forum. 3 July 2015*. Available at: www.lloyds.com/~/media/files/the%20market/operating%20at%20lloyds/resources/reserving/fd%20reserving%20forum%203%20july%202015.pdf.

Lloyd's of London. 2015b. *Lloyd's Minimum Standards MS9 Reserving. October 2015*. Available at: www.lloyds.com/the-market/operating-at-lloyds/resources/reserving-guidance.

Lloyd's of London. 2015c. *Risk Code Mappings and Descriptions*. Available at: www.lloyds.com/the-market/operating-at-lloyds/resources/risk codes.

Lloyd's of London. 2015d. *Solvency II. Syndicate SCR for 2016 Year of Account. Worked examples to illustrate the treatment of RI contract boundaries. August 2015*. Available at: www.lloyds.com.

Lloyd's of London. 2015e. *Technical Provisions under Solvency II. Detailed Guidance. July 2015 update*. Available at: www.lloyds.com.

Lo, J. 2011. Extending the Mack bootstrap – hypothesis testing and resampling techniques. *GIRO Conference and Exhibition 2011*.

London Market Group and Boston Consulting Group. 2015. *London Matters – The Competitive Position of the London Insurance Market. Fact Base*. Available at: www.londonmarketgroup.co.uk/.

Lowe, S.P., and Mohrman, D.F. 1984. Discussion of extrapolating, smoothing and interpolating development factors. *Proceedings of the Casualty Actuarial Society*, 71.

Lunn, D.J., Thomas, A., Best, N., and Spiegelhalter, D. 2000. WinBUGS – a Bayesian modelling framework: concepts, structure, and extensibility. *Statistics and Computing*, 10, 325–337.

Lyons, G., Forster, W., Kedney, P., Warren, R., and Wilkinson, H. 2002. Claims Reserving Working Party paper. *GIRO Conference and Exhibition 2002*.

MacDonnell, S., and MUQ Working Party. 2016. Reserve uncertainty framework. Measuring Uncertainty Qualitatively Working Party 2016. *GIRO Conference and Exhibition 2016*.

Mack, T. 1991. A simple parametric model for rating automobile insurance or estimating IBNR claims reserves. *ASTIN Bulletin*, 21(1), 93–109.

Mack, T. 1993a. Distribution-free calculation of the standard error of chain ladder reserve estimates. *ASTIN Bulletin*, 23(2), 213–225.

Mack, T. 1993b. Measuring the variability of chain ladder reserve estimates. *Casualty Actuarial Society Spring Forum, May 1993*.

Mack, T. 1997. Measuring the variability of chain ladder reserve estimates. *Claims Reserving Manual*, Vol. 2. London: Institute of Actuaries. Section D6.

Mack, T. 1999. The standard error of chain ladder reserve estimates: recursive calculation and inclusion of a tail factor. *ASTIN Bulletin*, 29(2), 361–366.

Mack, T. 2000. Credible claims reserves: the Benktander method. *ASTIN Bulletin*, 30(2), 333–347.

Mack, T. 2008. The prediction error of Bornhuetter/Ferguson. *ASTIN Bulletin*, 38(1), 87–103.

Mack, T., and Venter, G.G. 2000. A comparison of stochastic models that reproduce chain ladder reserve estimates. *Insurance: Mathematics and Economics*, 26(1), 101–107.

Mack, T., Quarg, G, and C., Braun. 2006. The mean squared error of prediction in the chain ladder reserving method – a comment. *ASTIN Bulletin*, 36(2), 543–552.

Maher, G.P.M. 1995. Loss reserves in the London market. *British Actuarial Journal*, 1(IV), 689–760.

Mango, D., and Allen, C.A. 2009. Two alternative methods for calculating the unallocated loss adjustment expense reserve. *Casualty Actuarial Society Forum*, Fall.

Marcuson, T.A.G., Turner, J., and Welsh, M. 2014. Peril-based reserving. *GIRO Conference and Exhibition 2014*.

Marshall, K., Collings, S., Hodson, M., and O'Dowd, C. 2008. A framework for assessing risk margins. Presented to the Institute of Actuaries of Australia 16th General Insurance Seminar, 9–12 November 2008.

McCullagh, P., and Nelder, J. A. 1989. *Generalized Linear Models*. London: Chapman & Hall.

McDonnell, S., and Labaune, R. 2014. *GIROC UK Reserving Survey 2014*. Available at: www.actuaries.org.uk/research-and-resources/documents/giroc-uk-reserving-survey-2014.

McGuire, G. 2007. Individual claim modelling of CTP data. *Institute of Actuaries of Australian XIth Accident Compensation Seminar*, Melbourne, Australia.

Merz, M., and Wüthrich, M.V. 2006. A credibility approach to the Munich chain-ladder method. *Blätter DGVFM*, XXVII, 619–628.

Merz, M., and Wüthrich, M.V. 2007a. Prediction error of the chain ladder reserving technique applied to correlated run-off triangles. *Annals of Actuarial Science*, 2, 25–50.

Merz, M., and Wüthrich, M.V. 2007b. Prediction error of the expected claims development result in the chain ladder method. *Bulletin of the Swiss Association of Actuaries*, 1, 117–137.

Merz, M., and Wüthrich, M.V. 2008a. Modelling the claims development result for solvency purposes. *Casualty Actuarial Society E-Forum*, Fall.

Merz, M., and Wüthrich, M. V. 2008b. Prediction error of the multivariate chain ladder reserving method. *North American Actuarial Journal*, 12(2), 175–197.

Merz, M., and Wüthrich, M.V. 2008c. *Stochastic Claims Reserving Methods in Insurance*. London: Chapman and Hall.

Merz, M., and Wüthrich, M.V. 2010. Paid-incurred chain claims reserving method. *Insurance Mathematics and Economics*, 46(3), 568–579.

Merz, M., and Wüthrich, M.V. 2012. Estimation of tail development factors in the paid-incurred chain reserving method. *Variance*, 7(1), 61–73.

Merz, M., and Wüthrich, M.V. 2014. Modified Munich chain-ladder method. *Swiss Finance Institute Research Paper No. 14-65*. Available at: http://ssrn.com/abstract=2489289 or dx.doi.org/10.2139/ssrn.2489289.

Merz, M., and Wüthrich, M.V. 2015a. Claims run-off uncertainty: the full picture (3 July 2015). *Swiss Finance Institute Research Paper No. 14-69*. Available at: http://ssrn.com/abstract=2524352 or dx.doi.org/10.2139/ssrn.2524352.

Merz, M., and Wüthrich, M. V. 2015b. Stochastic claims reserving manual: advances in dynamic modelling (21 August 2015). *Swiss Finance Institute Research Paper No. 15–34*. Available at: http://ssrn.com/abstract=2649057 or dx.doi.org/10.2139/ssrn.2649057.

Meyers, G. 2015. Stochastic loss reserving using Bayesian MCMC models. *Casualty Actuarial Society Monograph Series*, 1.

Meyers, G., and Shi, P. 2011a. The retrospective testing of stochastic loss reserve models. *Casualty Actuarial Society E-Forum*, Summer.

Meyers, G. G., and Shi, P. 2011b. *Loss Reserving Data Pulled from NAIC Schedule P*. Available at: www.casact.org/research/index.cfm?fa=loss reserves data.

Mildenhall, S.J. 2005. *Correlation and Aggregate Loss Distributions with an Emphasis on Iman-Conover Method*. CAS Working Party on Correlation. Available at: www.mynl.com/wp/ic.pdf.

Ministry of Justice and the Scottish Government. 2017. *The Personal Injury Discount Rate: How it Should be Set in the Future*. Available at: https://consult.justice.gov.uk/digital-communications/personal-injury-discount-rate/supporting_documents/discountrateconsultation paper.pdf.

Miranda, M.D.M., Nielsen, B., Nielsen, J.P., and Verrall, R.J. 2011. Cash flow simulation for a model of outstanding liabilities based on claim amounts and claim numbers. *ASTIN Bulletin*, 41(1), 107–129.

Miranda, M.D.M., Nielsen, J.P., and Verrall, R.J. 2012. Double chain ladder. *ASTIN Bulletin*, 42(1), 59–76.

Miranda, M.D.M., Nielsen, J.P., Sperlich, S., and Verrall, R.J. 2013a. Continuous chain ladder: reformulating and generalizing a classical insurance problem. *Expert Systems with Applications*, 40(14), 5588–5603.

Miranda, M D.M., Nielsen, J.P., and Verrall, R.J. 2013b. *DCL Package: Claims Reserving Under the Double Chain Ladder Model*. R package version 0.1.0.

Miranda, M.D.M., Nielsen, J.P., and Verrall, R.J. 2013c. Double chain ladder and Bornhuetter-Ferguson. *North America Actuarial Journal*, 17(2), 101–113.

Miranda, M.D.M., Nielsen, J.P., Verrall, R.J., and Wüthrich, M. V. 2013d. Double chain ladder, claims development inflation and zero-claims. *Scandinavian Actuarial Journal*, 1–23.

Miranda, M.D.M., Nielsen, J.P., and Verrall, R.J. 2013e. *A New R Package for Statistical Modelling and Forecasting in Non-life Insurance*. Cass Business School, 15 July 2013. Available at: www.cassknowledge.com/sites/default/files/article-attachments/DCLpackage1.pdf.

Miranda, M.D.M., Nielsen, B., and Nielsen, J.P. 2015. *A Simple Benchmark for Mesothelioma Projection for Britain*. Available at: www.cassknowledge.com/research/article/asbestos-legacy-simple-benchmark-mesothelioma-projection.

Murphy, D. M. 1994. Unbiased loss development factors. *Proceedings of Casualty Actuarial Society*, LXXXI, 154–222.

Murphy, D. M. 2007. Chain ladder reserve risk estimators. *Casualty Actuarial Society E-Forum, Summer*.

National Council on Compensation Insurance. 2015. State of the line: analysis of workers compensation results. *NCII's 2015 Annual Issues Symposium:* Available at: www.NCCI.com.

Neuhaus, W. 1992. Another pragmatic loss reserving method or Bornhuetter-Ferguson revisited. *Scandinavian Actuarial Journal*, 2, 151–162.

Newman, A., Archer-Lock, P.R., Rix, S., and Wash, C. 1999. *Unallocated Loss Adjustment Expenses*. Available at: www.actuaries.org.uk/documents/unallocated-loss-adjustment-expense-provisions.

Ohlsson, E. 2016. Unallocated loss adjustment expense reserving. *Scandinavian Actuarial Journal*, 2016(2), 167–180.

Ohlsson, E., and Lauzeningks, J. 2008. The one-year non-life insurance risk. *ASTIN Colloquium*, Manchester, 13–16 July 2008.

OIC Run-Off Limited. 2015. Proposal in relation to an amending Scheme of Arrangement – Appendix 2 – Estimation Guidelines. Available at: www.oicrun-offltd.com/html/default.html.

Parodi, P. 2012. Triangle-free reserving. *GIRO Conference and Exhibition 2012*.

Parodi, P. 2013. Triangle-free reserving: a non-traditional framework for estimating reserves and reserve uncertainty. A discussion paper. Presented to the Institute and Faculty of Actuaries. 4 February 2013(London).

Pastor, N.H. 2003. A Statistical Simulation Approach for Estimating the Reserve for Uncollectible Reinsurance. *Casualty Actuarial Society Forum*.

Pezzulli, S., and Margetts, S. 2008. Granular loss modelling. *GIRO Conference and Exhibition 2008*.

Pinheiro, P.J.R., Silva, J.M.A., and Centeno, M.D. 2003. Bootstrap methodology in claim reserving. *Journal of Risk & Insurance*, 70(4), 701–714.

Piwcewicz, B. 2005. Assessment of diversification benefit in insurance portfolios. Presented to the Institute of Actuaries of Australia 15th General Insurance Seminar, 16–19 October 2005.

Plummer, M. 2003. JAGS: a program for analysis of Bayesian graphical models using Gibbs sampling. *Proceedings of the 3rd International Workshop on Distributed Statistical Computing*, 20–22 March, Vienna, Austria.

Plummer, M. 2013. *JAGS Version 3.4.0 User Manual*. Available at: http://sourceforge.net/projects/mcmc-jags/files/Manuals/3.x/.

Potter, E., Brown, K., and members of the PPO Working Party. 2015. *Update from the PPO Working Party – Current Issues in General Insurance. 20 April 2015*. Available at: www.actuaries.org.uk/research-and-resources/documents/a05-ppo-working-party-update.

Pröhl, C., and Schmidt, K.D. 2005. Multivariate chain-ladder. *Dresdner Schriften zur Versicherungsmathematik*.

Prudential Regulatory Authority. 2014. Solvency II: calculation of technical provisions and the use of internal models for general insurers. *Supervisory Statement*, SS5/14.

Prudential Regulatory Authority. 2015. *Solvency II Directors' Update. 14 July 2015*. Available at: www.bankofengland.co.uk/pra/Documents/solvency2/directorsletterjuly2015.pdf.

Quarg, G., and Mack, T. 2004. Munich chain ladder. *Blätter DGVFM*, XXVI, 597–630.

Quarg, G., and Mack, T. 2008. Munich chain ladder: a reserving method that reduces the gap between IBNR projections based on paid losses and IBNR projections based on incurred losses. *Variance*, 2(2), 266–299.

R Core Team. 2013. *R: A Language and Environment for Statistical Computing*. R Foundation for Statistical Computing, Vienna, Austria.

Reid, D.H. 1978. Claims reserves in general insurance. *Journal of the Institute of Actuaries*, 105, 211–296.

Reinsurance Association of America. 2015. *Historical Loss Development Study – 2015 Edition*. Available at: www.reinsurance.org/ProductDetail.aspx?id=147.

Renshaw, A. E., and Verrall, R. J. 1994. A stochastic model underlying the chain-ladder technique. *Proceedings XXV ASTIN Colloquium*, Cannes.

Renshaw, A. E., and Verrall, R. J. 1998. A stochastic model underlying the chain-ladder technique. *British Actuarial Journal*, 4, 903–923.

Renshaw, A.E. 1994a. Claims reserving by joint modelling. *Actuarial Research Paper No. 72* City University, London.

Renshaw, A.E. 1994b. On the second moment properties and the implementation of certain GLIM based stochastic reserving models. *Actuarial Research Paper No. 65*. City University, London.

Rietdorf, N., and Jessen, A.H. 2011. Provisions for loss adjustment expenses. *ASTIN Colloquium*, Madrid, 19–22 June 2011.

Robbin, I. 2012. A practical way to estimate one-year reserve risk. *Casualty Actuarial Society E-Forum*, Summer.

ROC. 2007. Best estimates and reserving uncertainty – Research Oversight Committee Working Party. *GIRO Conference and Exhibition 2007*.

Röhr, A. 2016. Chain ladder and error propagation. *ASTIN Bulletin*, 46, 293–330.

Rothwell, M. 2016. Is your brain playing tricks? *SIAS Presentation*, 6 September 2016.

Saluz, A., and Gisler, A. 2014. Best estimate reserves and the claims development results in consecutive calendar years. *Annals of Actuarial Science*, 8, 351–373.

Schmidt, K.D. 2015. *A Bibliography on Loss Reserving*. Available at: www.math.tu-dresden.de/sto/schmidt/dsvm/reserve.pdf.

Schmidt, K.D. 2006. Optimal and additive loss reserving for dependent lines of business. *Casualty Actuarial Society Forum*, Fall, 319–451.

Schnieper, R. 1991. Separating true IBNR and IBNER claims. *ASTIN Bulletin*, 21(1), 111–127.

Scollnik, D.P.M. 2001. Actuarial modeling with MCMC and BUGS. *North American Actuarial Journal*, 96–124.

Scollnik, D.P.M. 2004. Bayesian Reserving models inspired by chain ladder methods and implemented using WinBUGS. *Actuarial Research Clearing House*, Issue 2.

SCOR. 2014a. SCOR Global P&C reserve triangles – Excel file. Available at: www.scor.com/images/stories/pdf/Inverstors/financial-reporting/presentation/trianglesanalysis2014 finalprotected.xls.

SCOR. 2014b. *SCORs Loss Development Triangles and Reserves as of December 2013*. Available at: www.scor.com/images/stories/pdf/Inverstors/financial-reporting/presentation/2014 trianglesdisclosure.pdf.

Shapland, M.R. 2016. Using the ODP bootstrap model: a practitioner's guide. *Casualty Actuarial Society Monograph Series*, 4.

Shapland, M.R., and Leong, J. 2010. Bootstrap modeling: beyond the basics. *Casualty Actuarial Society E-Forum*, Fall.

Shaw, R.A., Smith, A.D., and Spivak, G.S. 2011. Measurement and modelling of dependencies in economic capital. *British Actuarial Journal*, 16(9), 601–699.

Sheppard, T. 2015. *The Relationship Between Accident Report Lag and Claim Cost in Workers Compensation Insurance*. Available at: www.NCCI.com.

Sherman, R.E. 1984. Extrapolating, smoothing and interpolating development factors. *Proceedings of the Casualty Actuarial Society*, 71, 122–135.

Sherman, R.E. 2007. Updating the Berquist Sherman paper – 30 years later. *Casualty Loss Reserve Seminar*.

Skurnick, D. 1973a. Loss reserve testing: a report year approach. Discussion. *Proceedings of the Casualty Actuarial Society*, 60, 73–83.

Skurnick, D. 1973b. A survey of loss reserving methods. *Proceedings of the Casualty Actuarial Society*, 60, 16–62.

Stanard, J. 1980. Experience rates as estimators: a simulation of their bias and variance. *Pricing Property and Casualty Insurance Products, 1980 Casualty Actuarial Society Discussion Paper Program*, 485–514.

Stanard, J. 1985. A simulation test of prediction errors of loss reserve estimation techniques. *Proceedings of the Casualty Actuarial Society*, 72, 124–153.

Staudt, A. 2012. Two symmetric families of loss reserving methods. *Casualty Actuarial Society E-Forum*, Summer.

Sweeting, P. 2011. *Financial Enterprise Risk Management*. Cambridge: Cambridge University Press.

Tang, A., and Valdez, E.A. 2009. Economic capital and the aggregation of risks using copulas (February 22, 2009). Available at: http://ssrn.com/abstract=1347675.

Tarbell, T.F. 1934. Incurred but not reported claim reserves. *Proceedings of the Casualty Actuarial Society*, 20, 84–89.

Taylor, G.C. 1977. Separation of inflation and other effects from the distribution of non-life insurance claim delays. *ASTIN Bulletin*, 9, 217–230.

Taylor, G.C. 1986. *Claims Reserving in Non-Life Insurance*. Amsterdam: North Holland.

Taylor, G.C. 2000. *Loss Reserving: An Actuarial Perspective*. London: Kluwer Academic Publishers.

Taylor, G.C. 2011. Maximum likelihood and estimation efficiency of the chain ladder. *ASTIN Bulletin*, 41(1), 131–155.

Taylor, G.C. 2013. Chain ladder with random effects. *ASTIN Colloquium, Hague, 21–24 May 2013*.

Taylor, G.C., and Ashe, F.R. 1983. Second moments of estimates of outstanding claims. *Journal of Econometrics*, 23, 37–61.

Taylor, G.C., and McGuire, G. 2007. A synchronous bootstrap to account for dependencies between lines of business in the estimation of loss reserve prediction error. *North American Actuarial Journal*, 11(3), 70–88.

Taylor, G.C., and McGuire, G. 2016. Stochastic loss reserving using generalized linear models. *Casualty Actuarial Society Monograph*.

Taylor, G.C., McGuire, G., and J., Sullivan. 2008. Individual claim loss reserving conditioned by case estimates. *Annals of Actuarial Science*, 3(1–2), 215–256.

UK Asbestos Working Party. 2004. UK asbestos – the definitive guide. *GIRO Conference and Exhibition 2004*.

UK Asbestos Working Party. 2007. UK Asbestos Working Party II 2007. *GIRO Conference and Exhibition 2007*.

UK Asbestos Working Party. 2008. UK Asbestos Working Party update 2008. *GIRO Conference and Exhibition 2008*.

UK Asbestos Working Party. 2009. UK Asbestos Working Party update 2009. *GIRO Conference and Exhibition 2009*.

UK Asbestos Working Party. 2015. UK Asbestos Working Party update 2015. *GIRO Conference and Exhibition 2015*.

UK Deafness Working Party. 2013. UK Deafness Working Party update 2013. *GIRO Conference and Exhibition 2013*.

UK Regulations. 2008. *The Large and Medium-sized Companies and Groups (Accounts and Reports) Regulations 2008 (SI 2008/410)*. Available at: www.legislation.gov.uk/uksi/2008/410/pdfs/uksi 20080410 en.pdf.

Vaughan, R.L. 2014. The unearned premium reserve for warranty insurance. *Casualty Actuarial Society E-Forum*, Fall.

Venter, G.G. 2002. Tails of copulas. *Proceedings of Casualty Actuarial Society*, LXXXIX, 68–113.

Venter, G.G. 2006. Discussion of the mean squared error of prediction in the chain ladder reserving method. *ASTIN Bulletin*, 36(2), 566–571.

Verbeek, H.G. 1972. An approach to the analysis of claims experience in motor liability excess of loss reassurance. *ASTIN Bulletin*, 6, 195–202.

Verrall, R.J. 1996. Claims reserving and generalized additive models. *Insurance: Mathematics and Economics*, 19(1), 31–43.

Verrall, R.J. 2000a. An investigation into stochastic claims reserving models and the chain-ladder technique. *Insurance: Mathematics and Economics*, 26(1), 91–99.

Verrall, R.J. 2004. A Bayesian generalized linear model for the Bornhuetter-Ferguson method of claims reserving. *North American Actuarial Journal*, 8(3), 67–89.

Verrall, R.J. 2007. Obtaining predictive distributions for reserves which incorporate expert opinion. *Variance*, 1: 1, 53–80.

Verrall, R.J. 1991. Chain ladder and maximum likelihood. *Journal of the Institute of Actuaries*, 118(III), 489–499.

Verrall, R.J. 2000b. Comments on "A comparison of stochastic models that reproduce chain ladder reserve estimates". *Insurance: Mathematics and Economics*, 26(1), 109–111.

Verrall, R.J. 2010a. Stochastic claims reserving in general insurance: example spreadsheets. Available at: www.cassknowledge.com/research/article/stochastic-claims-reserving-general-insurance.

Verrall, R.J. 2010b. Stochastic claims reserving in general insurance: online lecture. Available at: http://talk.city.ac.uk/stochasticreserving.

Verrall, R.J., Nielsen, J. P., and Jessen, A. H. 2010. Prediction of RBNS and IBNR claims using claim amounts and claim counts. *ASTIN Bulletin*, 40(2), 871–887.

Verrall, R.J., Hössjer, O., and Björkwall, S. 2012. Modelling claims run-off with reversible jump Markov chain Monte Carlo methods. *ASTIN Bulletin*, 42(1), 35–58.

Wang, S. 1998. Aggregation of correlated risk portfolios: models and algorithms. *Proceedings of the Casualty Actuarial Society*, 85, 848–939.

WFUM Pools. 2006. *The WFUM Pools Scheme – Appendix B – Estimation Methodology for the Valuation of Scheme Claims*. Available at: www.wfumpools.com/SchemeDocumentation.htm.

White, S., and Margetts, S. 2010. A link between the one-year and ultimate perspective on insurance risk. *GIRO Conference and Exhibition 2010*.

Wright, A. 2012. The Munich chain ladder: a personal view. *GIRO Conference and Exhibition 2012*.

Wright, T. S. 1990. A stochastic method for claims reserving in general insurance. *Journal of the Institute of Actuaries*, 117(12), 677–731.

Wright, T.S. 1992. Stochastic claims reserving when past claims numbers are known. *Proceedings of Casualty Actuarial Society*, LXXIX, 225–361.

Wüthrich, M. V., Merz, M., and Lysenko, N. 2009. Uncertainty of the claims development result in the chain ladder method. *Scandinavian Actuarial Journal*, 1, 63–84.

Yan, M. 2005. Premium liabilities. Presented to the Institute of Actuaries of Australia 15th General Insurance Seminar, 16–19 October 2005.

Zehnwirth, B. 1994. Probabilistic development factor models with applications to loss reserve variability, prediction intervals, and risk based capital. *Casualty Actuarial Society Forum*, Spring, 447–605.

Zhang, Y. 2010. A general multivariate chain ladder model. *Insurance: Mathematics and Economics*, 46(3), 588–599.

Zhao, X.B., and Zhou, X. 2010. Applying copula models to individual claim loss reserving methods. *Insurance: Mathematics and Economics*, 46(2), 290–299.

Zhao, X.B., Zhou, X., and Wang, J.L. 2009. Semiparametric model for prediction of individual claim loss reserving. *Insurance: Mathematics and Economics*, 45(1), 1–8.

Zurich v IEG. 2015. Zurich Insurance PLC UK Branch (Appellant) v International Energy Group Limited (Respondent) [2015] UKSC 33.

Index

Printed in the United States
by Baker & Taylor Publisher Services